PHARMACEUTICAL PACKAGING HANDBOOK

PHARMACEUTICAL PACKAGING HANDBOOK

EDWARD J. BAUER
Pittsburgh, Pennsylvania, USA

informa
healthcare

New York London

Informa Healthcare USA, Inc.
52 Vanderbilt Avenue
New York, NY 10017

Printed in the United States of America on acid-free paper
10 9 8 7 6 5 4 3 2 1

International Standard Book Number-10: 1-5871-6151-6 (Hardcover)
International Standard Book Number-13: 978-1-5871-6151-3 (Hardcover)

Library of Congress Cataloging-in-Publication Data

Bauer, Edward J., 1947-
Pharmaceutical packaging handbook / Edward J. Bauer
p. ; cm.
Includes bibliographical references and index.
ISBN-13: 978-1-5871-6151-3 (hardcover : alk. paper)
ISBN-10: 1-5871-6151-6 (hardcover : alk. paper) 1. Drugs—Packaging–Handbooks, manuals, etc. I. Title.
[DNLM: 1. Drug Packaging. QV 825 B344p 2008]
RS159.5.B38 2009
615′.10688–dc22

2008051340

For Corporate Sales and Reprint Permissions call 212-520-2700 or write to: Sales Department, 52 Vanderbilt Avenue, 16th floor, New York, NY 10017.

Visit the Informa Web site at
www.informa.com

and the Informa Healthcare Web site at
www.informahealthcare.com

Preface

Pharmaceutical packaging is a subject that rarely comes to mind when thinking of drugs, medical devices, or other divisions of the health care industry. Packaging done well provides protection, sterility, and safety. Health care professionals and patients hardly give it a thought. Packaging done poorly usually means a package that is hard to open. These perceptions and the almost invisible presence of packaging science in most peoples' understanding of pharmaceuticals was the idea behind this book.

Pharmaceutical products, or more appropriately biopharmaceutical products, and health care in developed countries are wonders of the modern world. Pharmaceutical products and health care in developing countries and remote parts of the world seems like magic. Diseases that were once fatal and chronic conditions that destroyed lives have slowly been conquered by modern medicine. Views of the body, unimaginable for most of the last century with X rays, are now possible with new imaging techniques that let us see the body in exquisite detail. We have come to expect a steady stream of new technology that cures or vaccinates us from ailments and potentially deadly viruses. We take for granted that new and better diagnostic techniques will improve our ability to understand and fix our bodies. We have grown accustomed to transplants, angioplasty, stents to open clogged arteries, joint replacements, and other devices that fix and repair our parts of our body.

The packaging and protection of these modern wonders of pharmaceutical and medical technology are almost as important as the drugs themselves. Without packaging, drugs and medical devices would never leave a factory or a laboratory. Packaging provides containment, protection, and safe delivery of products everywhere health care is needed and makes possible the availability and use of drugs, vaccines and medical devices in hostile environments. It ensures safe delivery of drugs and devices to accident scenes as easily as it does to hospitals. Labels and information contained in packaging communicates and explains to doctors, pharmacists, nurses, health care workers, and patients about how to use a product. It warns you of dangers and communicates how and when to take a drug, what is safe, what precautions to take, and what to avoid when undergoing treatment. This is an amazing set of packaging tasks that few, if any, notice.

Packaging is an emerging science and engineering discipline that touches people everywhere. A combination of natural sciences, engineering, materials science, and other social disciplines contribute to the design, development, and delivery of products, not pharmaceuticals alone. It is a high-technology field that we count on everyday to deliver billions of safe, sterile, and easy-to-open packages that touch every part of our lives.

This book was written as an introduction to pharmaceutical packaging. It has been kept simple and accessible to the average reader with some technical training in chemistry, physics, and engineering. It attempts to introduce you to the many things beyond packaging that are part of the drug, dosage, and regulatory environment. It highlights many of the problems a packaging engineer must face when developing a package for a new product. It uses short explanations of drug composition and interaction with the body to help explain how these issues answer many questions about packaging a drug or medical device. It introduces many issues that are part of the normal compromises and questions surrounding different drugs. It tries to highlight regulatory difficulties by explaining some of the concerns and safeguards various regulations introduce into the package, the product, and the process by which it is made. It provides a short introduction to package-manufacturing processes and the many materials used in pharmaceutical packaging.

Hopefully, the book will help you understand the role packaging technology plays in pharmaceutical and medical device design and development. It tries to introduce you to several basic concepts of packaging.

The book highlights concepts in chemistry, polymer science, packaging and other disciplines to help you understand the product, its composition, what the package must do to protect the product. It provides examples on how the product can change depending on its chemistry and the environment the package must withstand prior to delivery to the patient. It tries to provide you with a practical sense of how the package, the product, and the way it is manufactured all play an important role in producing a safe sterile product.

The book attempts to highlight the whats, whys, and hows that go into pharmaceutical packaging, and attempts to do this while explaining the interactions between the drug or device and the package. It is an introduction to the many diverse skills and needs of pharmaceutical packaging. It highlights the diverse and complimentary skills employed by packaging professionals who are great scientific generalists, that is, people who can combine science, engineering, materials, manufacturing, consumer issues, and societal issues like environmentalism into packaging. All of these factors are part of the input needed to deliver a drug or medical device in a safe and responsible way.

Packaging has become a new stand-alone scientific and engineering discipline within corporations.

Regulation of drugs and medical devices by governments around the world is a big part of packaging. Many reading this book will be surprised to discover the FDA and other regulatory agencies around the world are as critical of the packaging and its performance as they are in examining the efficacy of the drug product.

One liberty taken while writing this book is the use of the words drug and pharmaceutical as synonyms. Technically drug refers to the active pharmaceutical ingredient in a product, and pharmaceutical refers to the finished product. This means that the pharmaceutical is the product being packaged, not the drug.

Hopefully, this book will provide some basic insight into an exciting and challenging science that goes unnoticed by so many.

Edward J. Bauer

Contents

1

Introduction to the Pharmaceutical Industry: An Overview

INTRODUCTION

General Aspects of Drug Packaging

Packaging pharmaceutical products is a broad, encompassing, and multi-faceted task. It differs substantially from food packaging and is equally as challenging. It requires the application of a large amount of scientific and engineering expertise to deliver a product for a world market. Its practice focuses on information and knowledge from a wide range of scientific disciplines, including chemistry, engineering, material science, physical testing, sales, marketing, environmental science, and regulatory affairs to name just a few. This broad general background is needed for the design and development of each and every product produced by the pharmaceutical industry. Packaging is responsible for providing life-saving drugs, medical devices, medical treatments, and new products like medical nutritionals (nutraceuticals) in every imaginable dosage form to deliver every type of supplement, poultice, liquid, solid, powder, suspension, or drop to people the world over. It is transparent to an end user when done well and is open to criticism from all quarters when done poorly. Everyone is a packaging expert, and this is particularly true when one evaluates how something designed to help a person hinders his or her ability to use the product. This book will discuss in detail the many forms of pharmaceutical packaging. It won't describe each and every one, but it will describe the broad families of packaging designed to deliver the many different and unique forms of a product or products to a patient. It will provide an introduction to some of the chemistry of pharmaceutically active molecules and how they must be protected from the environment and from the package itself. It will touch upon the packaging of nutritional products and supplements that are slightly removed from the normal realm of pharmaceutical products but are beginning to play a bigger role in the overall treatment of disease.

Packaging for biologic products can involve a slightly different set of requirements, and some of the unique differences and problems for packaging genetically modified biologically produced products are noted.

Pharmaceuticals use a wide variety of sterilization techniques that vary significantly from those used for foods. An introduction to some of these concepts will touch upon the multiple sterilization processes and the problems they present to the design of drug and device packaging.

Distribution of products is now more global than ever. Mass customization of packaging to permit its use in multiple markets is a topic that needs exposition and discussion.

Environmental issues, including sustainability, will always be a subjective dimension to any packaging design. These topics and many others highlight the breath of knowledge a packaging engineer must master when developing and producing a widely acceptable product.

This is a lot of ground for any book to cover. Hopefully it will provide you with a ready reference replete with examples that provide a starting point for design, development, testing, and execution of a new package for any pharmaceutical product.

The book also provides an introduction to over-the-counter (OTC) packages and products. These are the medicines we keep in our homes and many times carry with us to relieve unpleasant symptoms of things we think of as annoyances to everyday life, like the common cold, or for treatment of common conditions, including rashes, cold sores, dry eyes, and other minor problems.

It will discuss labeling, and how copy and artwork are prepared for all types of packaging. Artwork typically sells a product in the OTC context; artwork creates a feeling about a product, an identity, and in some cases creates in the consumers' mind a reason to choose one product over another.

Amazing, isn't it? So many different requirements, so many facets to packaging, so many scientific, cultural, sociological, and environmental needs. Oh, and by the way, it also has a large regulatory and legal requirement that is outside all of the things mentioned above.

Packaging is an emerging science, an emerging engineering discipline, and a success contributor to corporations. Surprisingly it is something that few corporations have singled out as a stand-alone department or organization. Packaging can reside, or report through research and development (R&D), engineering, operations, purchasing, marketing, or the general administrative department of a company. For the majority of products produced in the food and pharmaceutical industries it is probably the single largest aggregate purchase made by a company of materials critical to the protection, distribution, and sale of the product. Hopefully the contents of this text will provide a new appreciation of how important and complex pharmaceutical packaging is, not just the traditional expectations of product protection.

BRIEF HISTORY

The global pharmaceutical business is one of the most dynamic, research-intensive, and innovative businesses in the world. Today's pharmaceutical industry began to emerge in the mid to late 19th century as a small and unique subset of the chemical industry. For most of the 20th century, the pharmaceutical business paralleled developments in synthesis, catalysis, and manufacturing that were outgrowths of the larger chemical business. Before World War II, advances in chemistry and chemical engineering from Europe, particularly from Germany, drove both the worldwide chemical industry and the smaller pharmaceutical companies. The United States developed its own group of companies that, with only a few notable exceptions, concentrated on the U.S. market, while the Europeans, particularly the Germans, expanded abroad and also set up operations in the United States. The United States and Europe became the two centers of the chemical industry and developed in parallel, as they expanded to meet the growing demand for chemical products in the markets of concentration. Two examples of European influence on U.S. pharmaceutical companies are Pfizer and Merck, both of which have German heritage. Another example of European influence is seen in one of the largest, best-known products in the United States, aspirin, which came from Bayer®, another German company. In fact, aspirin was probably the first of what we would today call a blockbuster drug.

At the end of World War II, the American chemical industry emerged as the dominant force in the world. The U.S. companies exploited the wealth of natural resources available in the United States, and the large volume of knowledge gained from wartime research into rubbers, plastics, and other related chemical and engineering technologies and used that knowledge to expand worldwide.

The pharmaceutical companies followed the same path of expansion, while beginning to develop the unique chemical, chemical engineering, and manufacturing knowledge necessary to produce pharmaceutical ingredients and bring these unique products to market. The world war also produced a burst of knowledge about the production and manufacture of biologic products, notably penicillin. During this period, the United States also began to produce world-class scientific talent necessary to build and sustain the pharmaceutical industry and develop world-class facilities for the development of scientific knowhow in both industry and the universities. The talent was augmented by many émigrés from Europe.

The pharmaceutical industry's strength and reliance on chemistry alone began to change in the 1970s when an entirely new way of developing pharmaceuticals emerged. The new technology was a combination of biology, chemistry, biochemistry, and new cell modification and manufacturing technologies called biotech. This breakthrough technology challenged the traditional way of identifying and developing pharmaceutical products. Genentech was the first company to establish an identity in biotech. It was founded by a geneticist,

Herb Boyer, and began operations in the San Francisco Bay Area during the mid-1970s. Amgen, located in Thousand Oaks California, followed shortly after and was the first of the new biotech companies to introduce a biologically derived pharmaceutical product. These two early leaders not only changed the way pharmaceuticals were developed, they also ushered in a new way of developing pharmaceutical products using genetics, molecular biology, and biochemistry that when combined produced a genetic engineering approach to the treatment of disease. This change in approach has produced a fast and remarkable change in the pharmaceutical companies' core competencies. They have transformed themselves into biopharmaceutical companies.

The primary method the pharmaceutical industry used for drug development during most of the 20th century was to study the chemical reactions of various chemical molecules within the body. The molecules under study came from a variety of sources, both natural and synthetic. Extraction of active chemical ingredients from plants and animals known or identified to exhibit biologic activity is one way the development process progressed. Modifying new or existing chemical entities with human biologic activity was another way the pharmaceutical manufacturers produced compounds and then studied them for their effects in the body to determine efficacy against a disease. Most of the identifications of biologic reactions were carried out in cell culture experiments and in animal studies. This approach relied on chemistry and biochemistry and served the companies well. This approach and method of discovery produced the remarkable array of chemical products we take for granted today.

Going back to the World War II timeframe (1942–1945) another form of product development was taking shape for pharmaceutical products. The type of development was the growth and harvesting of a biologic agent that was then converted to a pharmaceutical product. The best example of this development is penicillin, a product produced from a mold and became the world's first great antibiotic. Penicillin required the development of many of the biologic manufacturing processes needed to grow the mold that produces the active pharmaceutical ingredient. It then required the development of unique chemistries, chemical engineering, and manufacturing skills to extract, purify, and produce the product. These traditional chemical methods and processes, which were the core strengths of the pharmaceutical companies, were applied to turn the raw, dilute material into the injectable and later the oral penicillin product we know today. This traditional path of product development was superseded in the last two decades of the 20th century by biotechnology.

"Biotechnology" is the term applied to developments that come from the combining of biochemistry, molecular biology, genetics, and immunology into pharmaceutical product development and manufacturing. This merging of two distinct sets of scientific disciplines, chemistry and molecular biology, has produced a powerful research engine that creates treatments for disease unknown only a few years ago. This merger of disciplines has changed the way traditional pharmaceutical companies approach drug research. Biotech permits a targeted approach to the development of products, and when combined with computational chemistry, computer assisted synthesis, and a wide range of analytical tools, it gives pharmaceutical companies the ability to study compounds, proteins, enzymes, and other biologically active materials in the minutest detail. All of the major pharmaceutical companies now have their own biotechnology capabilities or have partnered with others to acquire this competency. This combination of scientific methods has opened many exciting opportunities for them and has enhanced the investments made in R&D of their core strengths of chemistry, chemical engineering, and manufacturing. The real name for the industry, and one that it is beginning to adopt in its trade literature, is the biopharmaceutical industry.

It is interesting to note that both Genentech and Amgen see themselves as pharmaceutical companies, not biologic companies. Amgen brought to market the first genetically derived drugs, and Genentech has a major drug pipeline of new products in various stages of development. Eli Lilly and Company led the traditional pharmaceutical companies when they introduced their first biotechnology products in the early 1990s. Lilly introduced the first human health care product using recombinant DNA technology. The modern pharmaceutical industry and parts of the chemical industry now rely on the many advances in the biologic sciences and other key related technologies being led by pharmaceutical company investments in basic science, disease specific research, genetics, computer technology, and other supporting technologies, including packaging, that enable the laboratory discoveries to proceed to commercial products.

The U.S. government also funds R&D at the National Institutes of Health as part of the drug discovery effort [Fig. 1 graph). These funds are in addition to the R&D dollars spent by pharmaceutical companies.

Pharmaceutical companies break their research budgets into two parts: basic research and applied R&D. Research spending by U.S. domestic pharmaceutical companies (Table 1) is indicative of pharmaceutical research worldwide. In 2003, 38% of their budgets went into the basic research and 58% went into applied R&D.

The reach of the pharmaceutical industry is enormous and its impact on a people's lives everywhere is profound. Today diseases that killed millions are routinely treated with antibiotics. Death sentences from diseases like AIDS and leukemia have been put off to the point that these diseases are now treated as long-term chronic problems because modern pharmaceutical products put them into long-term remission. The U.S. death rate from AIDS has fallen by 70% since the mid-1990s with the introduction of protease inhibitors (1). Over the past two decades, the chances of surviving cancer for five years after diagnosis has improved by 10% and stands at 62% today (2). Our understanding of these diseases, resulting from the science that underpins all of the pharmaceutical industry, will continue to lead the way to a brighter, longer-lived, and healthier futures for countless people.

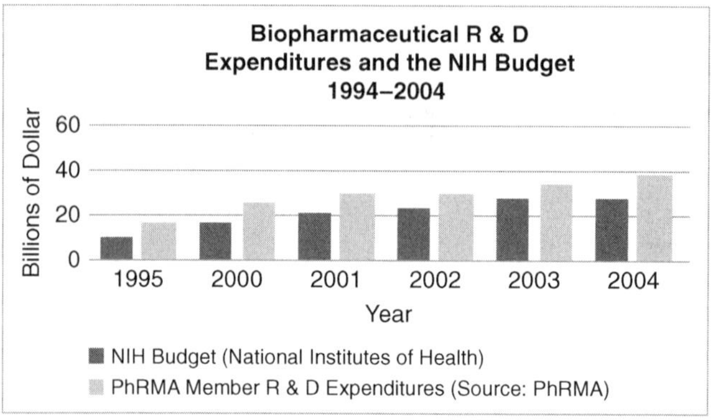

Figure 1 Drug discovery from laboratory to patient.

Table 1 Domestic R&D Spending by Type PhRMA Member Companies: 2003

Type	Expenditure ($)	Share (%)
Basic and applied research	10,382.6.	38.4
Development	15,766.2	58.3
Uncategorized	916.1	3.4
Domestic R&D total	27,064.9	100.0

Source: From Ref. 8.

GENERAL BUSINESS OVERVIEW OF THE PHARMACEUTICAL INDUSTRY

Pharmaceutical products are a very big business. Global sales reached $491.8 billion in 2003 (3). More than 400,000 people go to work each day for a pharmaceutical company in the United States (4). Pharmaceutical companies, or, more correctly, biopharmaceutical companies have a significant impact on our nation's economy. Each job in the pharmaceutical industry produces many others. Economists applying the normal multiplicative effect on the jobs number, estimate the total reach of the pharmaceutical industry was 2.7 million jobs and $172 billion in real output to the U.S. economy in 2003 (4). This output with these related jobs creates a significant addition of 2.1% of total employment in the U.S. economy (4).

Scientists in the United States lead the world in the discovery and development of new medicines. This is due in no small part to the tremendous investments the pharmaceutical companies make in research and development (Table 2). Research spending by the Pharmaceutical Research and Manufacturers Association (PhRMA) companies totaled $39 billion in a 2004 estimate made by this trade organization. It also estimates the total research spending for

Table 2 Research and Development Spending Worldwide 1970–2004 (Pharmaceutical Research and Manufacturers of America [PhRMA] Member Companies' Domestic R&D and R&D Abroad Total Spending)

Year	Domestic R&D ($ in millions)	Annual percentage change	R&D abroad[b]	Annual percentage change	Total R&D ($ in millions)	Annual percentage change
[a]2004	30,643.9	13.2	8,150.3	10.3	38,794.2	12.6
2003	27,407.1	6.8	5,808.3	8.4	33,215.4	7.1
2002	25,655.1	9.2	5,357.2	–13.9	31,012.2	4.2
2001	23,502.0	10.0	6,220.6	33.3	29,772.7	14.4
2000	21,363.7	15.7	4,667.1	10.6	26,030.8	14.7
1999	18,471.1	7.4	4,219.6	9.9	22,690.7	8.2
1998	17,127.9	11.0	3,839.0	9.9	20,996.9	10.8
1997	15,466.0	13.9	3,492.1	6.5	18,958.1	12.4
1996	13,627.1	14.8	3,278.5	–1.6	16,905.6	11.2
1995	11,874.0	7.0	3,333.5	c	15,207.4	c
1994	11,101.6	6.0	2,347.8	3.8	13,449.4	5.6
1993	10,477.1	12.5	2,262.9	5.0	12,740.4	11.1
1992	9,312.1	17.4	2,155.8	21.3	11,467.9	18.2
1991	7,928.6	16.5	1,776.8	9.9	9,705.4	15.3
1990	6,802.9	13.0	1,617.4	23.6	8,420.3	14.9
1989	6,021.4	15.0	1,308.6	0.4	7,330.0	12.1
1988	5,233.9	16.2	1,303.6	30.6	6,537.5	18.8
1987	4,504.1	16.2	998.1	15.4	5,502.2	16.1
1986	3,875.0	14.7	865.1	23.8	4,740.1	16.2
1985	3,378.7	13.3	698.9	17.2	4,077.6	13.9
1984	2,982.4	11.6	596.4	9.2	3,578.8	11.2
1983	2,671.3	17.7	546.3	8.2	3,217.6	16.0
1982	2,268.7	21.3	505.0	7.7	2,773.7	18.6
1981	1,870.4	20.7	469.1	9.7	2,339.5	18.4
1980	1,549.2	16.7	427.5	42.8	1,976.7	21.5
1979	1,327.4	13.8	299.4	25.9	1,626.8	15.9
1978	1,166.1	9.7	237.9	11.6	1,404.0	10.0
1977	1,063.0	8.1	213.1	18.2	1,276.1	9.7
1976	983.4	8.8	180.3	14.1	1,163.7	9.6
1975	903.5	13.9	158.0	7.0	1,061.5	12.8
1974	793.1	12.0	147.7	26.3	940.8	14.0
1973	708.1	8.1	116.9	64.0	825.0	13.6
1972	654.8	4.5	71.3	24.9	726.1	6.2
1971	626.7	10.7	57.1	9.2	683.8	10.6
1970	566.2	—	52.3	—	618.5	—
Average		12.5		16.1		13.0

Note: All figures include company-financed R&D only. Total values may be affected by rounding.

[a]Estimated

[b]R&D abroad includes expenditures outside the United States by U.S.-owned PhRMA member companies and R&D conducted abroad by the U.S. divisions of foreign-owned PhRMA member companies. R&D performed abroad by the foreign divisions of foreign-owned PhRMA member companies is excluded. Domestic R&D, however, includes R&D expenditures within the United States by all PhRMA member companies.

[c]R&D abroad affected by merger and acquisition activity.

Source: From Ref. 8.

pharmaceutical and biotech research to be $49.3 billion when the estimates for non-PhRMA members are added to the member's total. As a percentage of sales, the total for PhRMA members is 18.3% of U.S. domestic sales and 15.9% of sales worldwide (Table 3) (5).

Table 3 R&D as a Percentage of Sales PhRMA Member Companies: 1970–2004

Yr	Domestic R&D as a percentage of domestic sales (%)	Total R&D as a percentage of total sales (%)
2004[a]	18.3	15.9
2003	18.3	15.7
2002	18.4	16.1
2001	18.0	16.7
2000	18.4	16.2
1999	18.2	15.5
1998	21.1	16.8
1997	21.6	17.1
1996	21.0	16.6
1995	20.8	16.7
1994	21.9	17.3
1993	21.6	17.0
1992	19.4	15.5
1991	17.9	14.6
1990	17.7	14.4
1989	18.4	14.8
1988	18.3	14.1
1987	17.4	13.4
1986	16.4	12.9
1985	16.3	12.9
1984	15.7	12.1
1983	15.9	11.8
1982	15.4	10.9
1981	14.8	10.0
1980	13.1	8.9
1979	12.5	8.6
1978	12.2	8.5
1977	12.4	9.0
1976	12.4	8.9
1975	12.7	9.0
1974	11.8	9.1
1973	12.5	9.3
1972	12.6	9.2
1971	12.2	9.0
1970	12.4	9.3

[a]Estimated.
Source: From Ref. 8.

Table 4 Size of Pharmaceutical Markets

Country	Sales to June 2005 ($ billions)	Share of global sales (%)	12-mo change[a](%)
United States	246.4	44.7	7
Japan	60.0	10.9	3
Germany	31.2	5.7	6
France	30.3	5.5	7
United Kingdom	20.3	3.7	3
Italy	19.4	3.5	0
Spain	14.8	2.7	8
Canada	12.7	2.3	10
China	8.6	1.6	30
Mexico	6.9	1.3	11
Top 10 markets	450.6	81.9	6

Note: Sales in the United States are for the 12 months ending in June 2005.
[a]Based on local currencies.
Source: Ref. 10.

Pharmaceuticals are extremely cost efficient when compared with other forms of treatment for disease. Every dollar spent on new medicines reduces the cost of hospitalizations by $4.44 (5) (Table 4). These new medicines also accounted for over 40% of the two-year gain in life expectancy achieved in 52 countries between 1986 and 2000 (6). The next time you wonder about what the pharmaceutical industry is doing for you, ask yourself the following question: What is the value of those two additional years of life?

General Industry Challenges and Trends

The pharmaceutical industry is facing many challenges. The cost of developing new drugs continues to grow. During the 1970s, the cost of bringing a new drug to approval by the U.S. Food and Drug Administration (FDA) was approximately $138 million. Today the cost of bringing a drug to approval by the FDA is more than $800 million (7). A diagram of the cost and timing to bring a drug to market highlights how difficult this process can be (Fig. 2). The potential legal liability, particularly in the United States, when something goes wrong is another enormous problem all of the companies face. Wyeth faced billion-dollar liabilities for its diet drugs Pondamin® and Redux®, and Merck faced large liabilities from its withdrawal of Vioxx® in 2004.

THE EVOLUTION AND STRUCTURE OF THE PHARMACEUTICAL BUSINESS

The structure of the industry and the makeup of companies have undergone a rapid transformation in the last decade, and this process continues at an accelerated pace today. Companies have merged or have been acquired to form significantly larger

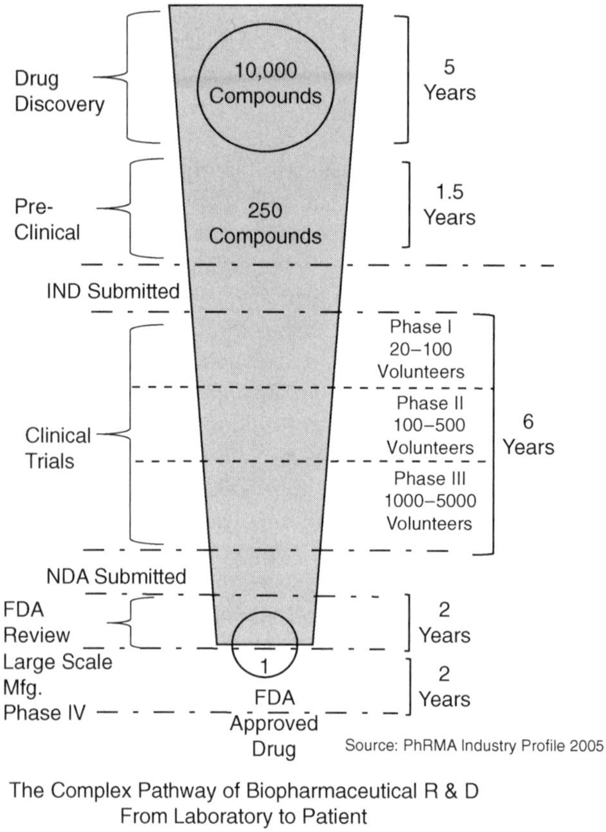

Figure 2 Biopharmaceutical expenditures and the NIH budget.

companies that may be newly named or may retain their traditional name. Remarkably, even after all the merger activity and changes in the shape of the pharmaceutical business landscape the largest company, Pfizer, only controls about 10% of the total U.S. market (based on 2003 sales) (Table 5) (9).

Another market force changing the appearance and development of pharmaceuticals is generic drug products. These products, the same chemical entity or active pharmaceutical ingredient (API) developed by the innovating company, are produced by many companies after the original innovator's patents have expired. Without the high cost of R&D, the competitor can offer the products for sale at significantly reduced prices. The number of new products with profiles that show significant improvement over older drugs is slowing; so generic products in many cases remain the standard of care for the treatment of many ills. A number of the major pharmaceutical companies actively support a dual strategic approach to product offerings and complement their new drug development with a generic drug supply strategy. These products not only produce significant revenues, they

Table 5 Top 10 Pharmaceutical Companies: Five-Year Merger History

Company	Market	Share	
	Based on 2003 sales (%)	Based on 1998 sales (pro forma) (%)	Major component companies
Pfizer	10.1	9.0	Pfizer, Pharmacia, Upjohn, Warner-Lambert, Searle
GlaxoSmithKline	6.6	7.2	Glaxo, Wellcome, SmithKline French, Beecham
Sanofi—Aventis	5.4	5.8	Sanof, Synthelabo, Hoechst, Rhone–Poulenc, Fisons
Merck & Co.	4.8	4.2	
Johnson & Johnson	4.8	3.6	
Novartis	4.3	4.2	Ciba-Geigy, Sandoz
AstraZeneca	4.1	4.3	Astra, Zeneca
Bristol–MyersSquibb	3.4	4.2	Bristol—Myers Squibb, Dupont Pharma
Roche	3.3	3.1	
Abbott	2.8	3.3	Abbott, BASF Pharma (Knoll)
Top 10 corporations	49.6	48.9	

Source: Ref. 11.

also provide the volume manufacturing needed for many raw materials and starting ingredients, used in both new and generic products, to maintain manufacturing costs at a reasonable level. The products also provide significant volume, which is needed in most investment models to pay for the expansion or maintenance of existing manufacturing capacity.

Therapeutic Areas of Concentration

What are the disease-specific areas now receiving the most concentrated investigation? What areas of treatment have produced the most successful products? What ills touch the majority of people in the world, and because of the large patient populations, attract the research and capital needed to understand and treat them? The top 10 therapies based on dollar sales encompass a remarkable set of problems (Table 6). Listed below are the conditions that have major effects on people's health. The top 10 therapeutic treatments based on sales fall into these categories (9):

1. Cholesterol and triglyceride reducers
2. Antiulcerants
3. Antidepressants
4. Antipsychotics
5. Antirheumatic nonsteroidals

Table 6 Top 10 Therapies Based on Global Sales of the Pharmaceutical Class

Therapeutic type	Sales to June 2005 ($ billions)	Share of global sales (%)	12-mo change (%)
Cholesterol and triglyceride reducers	31.6	5.7	10
Antiulcerants	26.3	4.8	3
Antidepressants	20.1	3.6	–3
Antipsychotics	15.5	2.8	11
Antirheumatic nonsteroidals	12.1	2.2	6
Calcium antagonists, plain	11.9	2.2	2
Erythropoietins	11.7	2.1	9
Antiepileptics	11.4	2.1	–15
Oral antidiabetics	10.4	1.9	6
Cephalosporins	9.9	1.8	30
Top 10 therapies	160.9	29.2	5

Note: All therapy classes are World Health Organization code groups. Sales are US dollars for the 12 months ending June 2005.
Source: Ref. 10.

6. Calcium antagonists, plain
7. Erythropoietins
8. Antiepileptics
9. Oral antidiabetics
10. Cephalosporins

As the list clearly shows, a large number of people have debilitating conditions that, when left untreated, significantly reduce the quality of life and life expectancy. Drugs targeted at heart disease are number 1 on the list. The therapies created for these disease-states and those not on the list permit all of us to lead functional productive lives that would not be possible without them.

The top 10 drugs had sales in excess of $55 billion. Seven of the top fifty products were biotechnology products with combined sales of $15.1 billion. (10)

GENERAL WORLDWIDE PHARMACEUTICAL TRENDS

Possibly the biggest challenge facing the pharmaceutical and the health care industries in general is the role the government will play in determining the cost of pharmaceuticals, devices, and treatments for diseases. The United States is the only market in the world that does not have general government price controls. It is also the most heavily regulated market, and the market pharmaceutical companies typically target their drug-approval strategy and development for acceptance and approval by the FDA.

Cost and Pricing Trends

Europe and Japan have led the way in restricting and regulating the cost of pharmaceuticals. This has created problems for pharmaceutical companies in their countries, which if emulated, may also create problems in the United States. Japan has taken a number of steps to restrain drug prices and has been able to reduce total government spending on pharmaceuticals. This restraint has reduced total health care spending in Japan on pharmaceuticals from 30% of the total cost of health care in the early 1990s to approximately 20% today (9). The Japanese impose biennial price cuts on all products to limit annual cost growth of pharmaceuticals in Japan to 3% annually.

Europe has a number of plans and schemes to limit the cost of pharmaceuticals. With the formation of the European Union (EU), the concept of parallel trade has created problems for the pharmaceutical manufacturers. Parallel trade is the practice of purchasing goods in the cheapest countries and selling them in the most expensive. European commission laws permit the movement of goods from one member state to another without restriction. This means that drugs priced in the lowest-cost countries (Eastern Europe) can be moved to the higher-cost countries, i.e., the United Kingdom, Germany, France, and Scandinavia. The EU expanded by 10 countries in May 2004, and many of these countries had strong generic pharmaceutical industries that will take advantage of these higher-cost markets. Germany is the largest market for drugs in the EU. During 2004, they imposed a compulsory discount of 16% on all manufacturers. That discount was up from 6% in 2003. The move applied to all drugs not part of the country's reference price scheme. This is one example of variants being enacted in the other countries. The most notable feature of the controls limits the price of a brand name or patented drug to the same cost as nonpatent medications if the new products are judged equal in treatment or outcome to their generic product equivalents. An example of this method of classifying patented and generic products is a ruling on the overall effectiveness of drugs called statins, which are used for treatment and reduction of cholesterol levels. As one statin loses patent protection, all other products, including all products still on patent, lose it as well in terms of product pricing. All of these schemes have caused most of the companies in Europe to rethink their research strategies, and a number have relocated research to the United States or have canceled expansions in Europe to focus on producing products for markets that pay for the cost of discovery (11).

Generic Products

Another major trend that will affect the pharmaceutical companies and pharmaceutical packaging is the growing use of generic drugs. One of the European pharmaceutical companies, Novartis, has adopted a generic strategy, betting that dollar volume and manufacturing volume will increase more in this area than in the development of new "blockbuster products." Generic drug introductions are at an all-time high, and this trend will continue.

Generics account for approximately 30% of the total volume of drugs dispensed for use but only produce 10% of the global sales revenues (9). Branded products lose sales very rapidly after a generic analogue is introduced. This and the fact that so many blockbuster drugs from the 1990s are losing patent protection in the next few years means generics will be a bigger and bigger part of the prescription drug landscape.

For packaging, the updating and maintaining of materials and labeling for generics will become more and more prominent in the mix of responsibilities. Packaging will work with marketing and sales to begin to develop generic identities for the off-patent products.

OTC Products

The pharmaceutical companies have also pushed for and accepted the fact that many products that require a prescription at introduction will eventually be offered over the counter at pharmacies. In the United States, a wide range of products, from prescription painkillers to female hygiene products, have made the transition from prescription ("Rx Only" on new labeling in the United States) to OTC items. Both generic and branded products will be part of this trend to OTC sales. The trend toward OTC presentations of products is not limited to the United States. Merck and Johnson & Johnson pursued a joint venture in the United Kingdom for OTC sales of the drug Zocor®. This product, a prescription-only drug called a statin and used for lowering cholesterol, was launched as an OTC item in July 2004. Packaging a prescription (Rx) product for OTC sales will be a major responsibility for packaging departments, and as volume grows more emphasis will be placed on designing, developing, and delivering these formerly doctor-directed drugs as consumer products. The responsibility of packaging will be twofold. First, it must convey and provide a simple communication to the user on how to use the product. This may be in the form of how the drug is presented to the consumer on a numbered blister or some more elaborate symbolic presentation for the user that transcends language. Second, packaging will become increasingly involved in the labeling and attendant literature that is needed for OTC products. This labeling and the methods needed to update its content is a major new responsibility that is outside standard structural packaging development.

DEFINITION OF A DRUG

The FDA is very specific about what constitutes a drug. This level of control is evident in the way biologic products, originally regulated by the Center for Biologic Evaluation and Research (CBER), was moved under the jurisdiction of the Center for Drug Evaluation and Research (CDER). These two branches of the FDA work in tandem with each other to evaluate, review, and ultimately approve a drug or biologic product for human use. The fact that the biologic product has therapeutic activity makes it subject to the strict interpretation of the Federal Food Drug and Cosmetic Act of 1938, which has been and continues to be amended as needs and technology change. The Act (12) defines a drug as follows:

A. Articles recognized in the official United States Pharmacopoeia (USP), Official Homeopathic Pharmacopoeia of the United States or the official National Formulary or any supplement to any of them.
B. Articles intended for use in the diagnosis, cure, mitigation, treatment, or prevention of disease in man or other animals: and
C. Articles (other than food) intended to affect the structure or any function of the body of man or other animals; and
D. Articles intended for use as a component or any article specified in clause (A), (B), or (C); but does not include devices or their components, parts, or accessories.

Medical devices, another large portion of the medical industry, are treated in a separate chapter in this book. They have their own packaging and technical challenges that are in many ways similar but distinct in their treatment and evaluation by the agency. Food products with pharmaceutical claims and those that require review and approval by the FDA also must meet the review process albeit at different levels, depending on the claims being made.

Unique food products such as infant formula and medical nutritional foods and food supplements that help manage certain diseases receive high levels of scrutiny. Other products covered in the Act are cosmetics, and these products are making claims that in some cases are becoming more like drugs. Cosmetic products are not discussed in this book.

Throughout this book, the terms "pharmaceutical" and "drug" are treated as synonyms, although they are not. A drug is really an active pharmaceutical ingredient (API) and a pharmaceutical product is an API in combination with other ingredients that are blended or compounded to make the finished product. In some cases, the word "drug" is used in this book as a synonym for "pharmaceutical" and vice versa. The term API is always applied to the active ingredient only.

THE DIFFERENCES BETWEEN PHARMACEUTICAL AND FOOD PACKAGING

Food and pharmaceutical packaging are both equally difficult to do well. Food packaging is far more diverse than pharmaceutical packaging, while pharmaceutical packaging operates in a much more regulated environment. Some understanding of the differences is useful, and for crossover products such as medical nutritional foods, essential. As more and more foods are enhanced with ingredients that can impart a change in the body, or as manufacturers make claims regarding the benefits of food products, which may be considered drug claims by the FDA or the Federal Trade Commission, the amount or regulation and the amount of testing necessary to gain approval of the products increases exponentially. These products will be supplied in familiar food containers appropriate for the type of food product; however, the containers will be required to meet pharmaceutical regulations that were not required when the product was strictly a food. It will become harder and harder to determine if the packaging must follow food or drug regulations.

Food and pharmaceutical packaging follow two different paths in packaging development that not only have many similarities but also have major differences. Food is rarely toxic even when consumed in huge quantities. Nausea and bloating will normally stop someone from overingesting food well before the condition becomes harmful or life threatening. With drugs, overdose is easy and can be fatal. This difference is the reason why labeling is so stringent and the requirements for labeling, discussed in chapters 5 and 11, are so precise. The FDA takes a very dim view of mislabeled pharmaceutical products, so manufacturers are extremely careful about controlling the labeling that goes on any pharmaceutical package.

Drugs, being so toxic, also come under poison-control regulations administered by the U.S. Consumer Product Safety Commission (CPSC). The requirement for child-resistant closures on pharmaceutical products is related to the highly toxic nature of most pharmaceutical products. This requirement creates a great deal of problems for the elderly, who are the major users of drugs. Many elderly patients complain that it is too hard to open or get into a package, and that they must go to great lengths to open the package. Child resistant closures are designed to protect children from poisoning. Unfortunately, many seniors after opening a package with a child-resistant closure only partially replace the closure back on the package and do not engage the closure to the point where it is effective in preventing an inquisitive child from becoming harmed by the container's contents.

Foods don't look alike and certainly don't smell alike. In fact, the appeal to the senses is a primary determinant of which foods we like. We are concerned about the nutritional value of the food we eat, and in recent years, the FDA has promoted new food labeling to detail exactly what the food we eat presents to our bodies in the form of nutrition. This is not the case with drugs. Most pharmaceuticals look alike. Most drugs are packaged in opaque containers that don't permit easy viewing of the contents. There is no sensory component of smell or flavor. This makes labeling of drugs even more crucial. The contents must be accurately described on the labeling. Even with the minor variations in shape and the use of impressed symbols or printing on the outside of a tablet, it can be hard to distinguish between multiple drugs that are part of a patient's regimen. Color helps, but the main way that people can distinguish one tablet from another is through labeling. Accurate labeling is essential for the patient in the use of any prescription product to produce the therapeutic result.

This is especially critical with OTC drugs that have undergone multiple updates to their labeling mandated by new FDA standards to help improve the labels' ability to easily communicate to consumers. OTC packaging of pharmaceutical products has become one of the most difficult forms of packaging. Packaging and labeling of OTC products communicate to the consumer in much the same way they do for food. They have the dual purpose of building brand recognition and communicating the proper use of the product. Many people don't realize how dangerous OTC medicines can be and misuse of these products; or, more properly stated, the improper use or dosage with these products is high. The labeling, with its prominent warnings, alerts even the most casual consumer to the dangers of an OTC product and helps distinguish it from food or candy.

All food is taken or ingested orally. The mouth is a non-sterile orifice, and our digestive systems are structured to kill the majority of harmful organisms that can enter our bodies with food. We do get sick from foodborne pathogens, and in some case severely sick, with *Escherichia coli* or, in extreme cases, botulism. When this happens, it creates headlines and is extensively reported because the occurrence is very rare. Drugs, on the other hand, can not only be taken orally, but can also be administered directly into the circulatory system (parenteral), under the skin (subcutaneous), or across mucous membranes in the nose, throat, and rectum, as well as through the skin with patches or high pressure injections. These methods of ingestion are quite different from any food, and provide the opportunity to introduce harmful or fatal micro-organisms directly into a patient. Drugs and devices must be completely sterile, as opposed to "commercially sterile," the term applied to many retorted (processed) foods like meat, vegetables, soups, and canned products. For food, a complete sterilization, equal to the sterilization required for a drug, would render it tasteless, texture-less, and totally unpalatable.

Drugs are repackaged to a very large degree. This is changing, and in some parts of the world, unit-dose packaging is more common than in the United States. Even so, the pharmacist repackages a large number of products for the patient. This is a requirement that doesn't touch food to a large degree. Packaging must protect the product both in the large containers used for general distribution and in the small containers a pharmacist uses for the repackaged product when it is dispensed from the pharmacy. These same packages must also protect a pharmaceutical product after it gets to the patient's home. Products are held and dispensed from the pharmacy container in which they are supplied far longer than most foods after they are opened.

In many cases, the package not only protects the product but also provides a method for tracking its use or compliance. "Compliance" is a term that is becoming more and more critical; and, for a pharmaceutical, it means the patient follows the dosage regimen specified by the doctor. The term "adherence" is also used in this connotation. Compliance can result in reduced health care costs because the patient gains control over a condition before it becomes much more serious. A good example of a chronic problem that requires constant compliance is hypertension or high blood pressure. Surprisingly, patients typically have a relatively low rate of compliance for these products even though hypertension can cause stroke or heart attack.

DRUG REGULATIONS

Packaging of drugs is highly regulated when compared with food packaging. The USP lists approved packaging for drugs, and this recommendation carries the force of law. There is no single reference for food products. Pharmaceutical products come under a number of specific parts of Title 21 of the Code of Federal Regulations (CFR) that mandate specific procedures for developing, proving, and changing previously approved packaging. The regulation slows and sometimes stops innovation.

Drug packaging is slow to change. The cost of stability studies needed to prove long term packaging safety is extremely high and can take two to five years. Drugs, with packaging defined and approved many years ago, and generic drug packaging is particularly hard to change. Many times the sales dollars and profits generated by a generic drug do not justify the cost of qualifying a new material or a new dosage form. This situation is improving and a good example of the improvement is found when qualifying a new plastic resin. The FDA has always interpreted the "same container closure system" to mean the same plastic resin formulation identified in the original application, or from a suppliers point of view, the same material produced at the same manufacturing facility of the resin manufacturer. The FDA has developed procedures that permit the change of plastic resins if they meet the approval procedures developed by the USP. These protocols permit the establishment of equivalence between two similar types of resin—an example would be high-density polyethylene from two different manufacturers. The procedures permit the qualification and change without prior approval. This approval is always conditional to the material passing real time stability testing with the actual product. More on this topic will be discussed and included in the chapter on regulatory affairs.

Tamper evidence built into the packaging is a much more important issue for drugs than it is for food. This is a direct outgrowth of a tampering problem in Chicago that caused a number of deaths in the mid-1980s. Tylenol® (acetaminophen) packages were tampered with and a poison was introduced. A number of people died. This sparked a major change in how drugs, particularly OTC drugs, were packaged. Tamper evidence is typically costly and requires one or more steps, either in the container-manufacturing process or in the assembly of the container during filling to put the safeguards into place. It also requires education of the public about what to look for and how to identify a package that has been altered or changed. Food products are every bit as vulnerable; however, no regulations now mandate tamper evidence on food products. Food packaging is far more diverse than drug packaging and carries a much smaller profit margin. These two facts make it far more difficult for manufacturers to change food products' packaging by adding tamper-evident features, although when possible, these are included in the package design. A good food example is the use of the breakaway ring on soda and water bottles. Tampering with foods has every bit as much potential to harm the general public as does tampering with OTC products.

The cost of packaging is the last aspect of the differences between drug and food packaging. Food products typically carry a much smaller profit margin, and most food products are produced in much greater volume than pharmaceuticals. The constant pressure on cost, the fact that packaging costs are a much more significant contributor to the cost of a food product compared with pharmaceuticals, and the volume of material used to make a package drives the food packager to be more cost conscious than the pharmaceutical packager. Volume, in this case, is the higher number of units produced for a food product compared with that produced for a pharmaceutical. Pharmaceuticals are costlier than food, and as a result, the percentage that packaging contributes to the total cost of the product is significantly less. The crossover products, such as medical nutritional foods, infant formulas, energy bars, and other similar products, are typically developed and their packaging costs are managed in the same way food products are scrutinized and controlled. Crossover products are the one exception to the general rule regarding the cost of pharmaceutical packaging.

THE FUNCTION OF PACKAGING

All packaging is required to perform two functions, containment and protection. Containment is the first role that any package must play in conjunction with a product. Containment means that the package prevents the product from touching or being exposed to the environment. For a drug package, this means the container completely separates the product from its surrounding physical environment. The package is sealed, preventing the product from entering the environment and the environment from entering the product. It also means the package does not become part of the product or vice versa. The package must remain functionally inert to its contents.

Protection is the second aspect that any package is expected to perform. Protection within the package means the product inside does not sustain physical damage. This could take the form of broken tablets or chemical breakdown caused by light, heat, oxygen, and water vapor.

Along with these two primary functions, a drug package must also provide a number of other features. A short list of some of these protective functions includes:

1. Sterility
2. Reclosure
3. Communication (via the label)
4. Compliance

5. Tamper evidence
6. Temperature control

Each of these items will be addressed in greater detail in various chapters of this book.

Trends in Pharmaceutical Packaging

Pharmaceutical packaging is a demanding and diverse area of package design, development, and engineering. It is undergoing significant change and re-alignment just as the pharmaceutical companies are undergoing change. Emphasis on many of the key aspects of packaging is changing and moving directly into the spotlight of government and consumer scrutiny. This is highlighted by the trends that are affecting packaging directly and how packaging is viewed within the companies and by the users of the products.

Current Trends in Packaging

Packaging is being required to do more and more in many areas within and outside a pharmaceutical company. In the past, its essential roles were containment and protection of the product. In many cases, the company and the consumer paid little attention to the package. Today, packaging's role is being expanded to include branding, communication, distribution control, anti-counterfeiting, poison protection, and much more.

Packaging has emerged as both a science and an engineering discipline that has influence on a product, both within the producing company and with the consumer outside the company. The science portion of this mix is a broad combination of disciplines. It includes polymer science, material science, and analytical chemistry to name a few. These scientific aspects of packaging development are used along with the science of drug discovery as two integral parts of pharmaceutical product development. It has assumed an engineering role by taking laboratory prototypes and in many cases stability sample packaging and converting it into a product and package entity that can be manufactured, filled, sealed, labeled, and distributed safely. It has also assumed the role of a management tool in the manufacturing process. Packaging is the only scientific and engineering discipline within a pharmaceutical company that touches a product from conception to the complete end of a product's life or use, including the recycling or disposal of used packaging.

Packaging is involved in and required to provide guidance and recommendations to researchers and in some cases marketing at the earliest stages of product development regarding materials, packaging options, and sizes of packages when stability studies are begun on an API that shows promise. Many times multiple formulations are part of the study as researchers work to determine the inert materials or excipients needed to dilute the API and allow it to be dispensed safely. This testing of the complete product, including its package sets the course for much of what follows in bringing the product to market. It continues in collaboration with the medical staff and the marketing staff to determine the best method for dispensing the product and the best presentation of the product for the consumer.

Packaging is changing. Its role continues to expand and play a more important part in the delivery of products. This role is being shaped by a number of key trends that affect the way packaging is developed. These trends, which include a shift in the delivery of medical care from the hospital to the doctor's office and the home, place more reliance on the patient, non-M.D. health care professionals, and the products themselves to improve treatment and reduce costs. These existing and emerging trends will significantly change many of the more common functions of packaging and our expectations about what a package is required to do. As health care becomes more expensive and possibly harder to access because of cost, these packaging trends and others that reduce cost will be implemented for cost containment. A good example of one of these trends is the increased approval of OTC products that originally required a prescription. The individual is being permitted to make decisions about the treatment of many diseases and conditions that required a doctor's care just a few years ago. This is both good and bad and its merits are not for review here, however, the opportunity to introduce this choice to the patient has a direct bearing on the reduction of health care costs. It also illustrates the increased burden placed on packaging and labeling needed by the end user.

The issues identified as trends should continue into the foreseeable future. They touch some of the key tenants of packaging, protection, communication, and safety. A short discussion on some of the items in the list below is important to understand how the pharmaceutical business and the packaging surrounding the products are changing to meet new requirements and challenges. More detailed discussions of these issues are part of the specific descriptions given for various packaging options throughout the book.

Influences Impacting Packaging

- Dispensing of product
- Compliance
- Communication of information—labeling
- Tamper evidence
- Radio frequency identification (RFID)
- Anticounterfeiting measures
- Environmental issues
- Unit-dose packaging
- Administration aids
- Growth of the elderly population worldwide
- Generic drugs

- Self-medication
- Product branding
- Graphic development (labeling) changes in pharmaceutical communication
  - Enterprise content management
  - Digital asset management
- Direct to consumer advertising

Dispensing

Dispensing of product can range from a calibrated cup used to take a liquid like cough syrup to a very precise aerosol package that administers a controlled dose of medication for asthma. It can be a polypropylene membrane the patient places on the body to slowly diffuses medication into the body. Patches for smoking cessation are a good example of this type of dispensing. Many products cannot be taken orally and are the ones that require development of a specialized dispensing mechanism to make them work. The trend to build the dispensing mechanism into the package whenever possible is a direction pharmaceutical manufacturers are taking. It is designed to ensure the patient gets the best outcome from the product with minimum effort. By building the dispensing mechanism into the package, the possibility to misuse is reduced. The possibility that the dispenser may be misplaced is eliminated. The patient is presented with the dispensing mechanism, and its use is detailed in the instructions supplied with the product. Common dispensing devices around the home, such as tablespoons and teaspoons, can and do get confused and can result in a problem for the patient. A built-in dispenser eliminates any possibility of patient confusion.

Compliance (Adherence)

Many times the package is required to provide a method to track compliance. Compliance is a measure of how a patient follows directions over multiple days of treatment in the dosage regimen supplied with the product. An example would be directions for taking a product multiple times during the day (e.g., a dosing regimen of 2 tablets 3 times per day) for a number of days to treat a disease. Compliance is one major aspect of drug treatment that is crucial to a successful outcome of the therapy. A good example for the typical patient and consumer is the prescription and dosage regimen of antibiotics for various infections. Not too long ago people would take antibiotics until they began to feel better and then stop taking the drug thinking they were cured. Now doctors emphasize the necessity of taking all of the product prescribed and of completing the multiple days of treatment, not only to prevent recurrence, but also to extend the life of the drug. Bacteria are very adaptable entities that can and do develop resistance to antibiotics in a number of ways, one of which is surviving a partial treatment of an antibiotic regimen. Tuberculosis, a disease that was conquered by antibiotics, has begun to make a comeback and drug-resistant strains have been identified. Compliance packaging and the monitoring of the patient has the ability to reduce our overall health care bill by providing proof of intervention into multiple conditions that are far more costly to treat when they are left partially treated and require additional treatment or hospitalization.

Another example is hypertensive products for reducing high blood pressure. These products, taken every day, can prevent a stroke, a debilitating and sometimes fatal outcome of the controllable chronic condition. Compliance packaging may in the future be a requirement to monitor a patient's continued treatment of a condition.

Communication of Information—Labeling

Another trend in pharmaceutical packaging is the increased emphasis on communication required for any product. This communication is especially true for OTC products, but it is also required of prescription products and medical devices. The labeled packaging, defined as the label, carton, insert, or electronic media such as a CD or other digital forms of information that is part of the package, provides the health care professional and the patient with the most complete set of instructions, warnings, cautions, and side effects for the product. To the surprise of most people, all drugs have some side effects, sometimes mild, sometimes severe, and all side effects are required to be detailed in the labeling on the product. The labeling communicates these facts to the physician and patient. The warnings range from a statement about mild discomforts such as dry mouth when compared with a placebo to a "black box" warning, the strongest emphasis the FDA can place in labeling to highlight potential problems concerning the use of a product. The use of pictograms and other nonliterate forms of communication is another part of this growing trend. In the E.U., labeling in all languages of the Union must be part of the package communication on both pharmaceuticals and medical devices. Today the number of languages required on products marketed throughout the E.U. is 13. This number is increasing to at least 20, and as new countries are admitted and become part of the E.U., even more may be required. The union is expanding to the east and a number of countries in Eastern Europe have already joined the E.U. and begun the task of harmonizing regulations. Part of the joining together is consistent regulations regarding the number of languages required on pharmaceutical packages. The resulting increase in required languages will result in an increase in the amount of packaging or printed material used in a complete package to increase the "billboard," the amount of area that can be printed for communication in and on the package. Multiple strategies to include all languages on packages are already used, and these will increase with the new regulations.

Tamper Evidence

Tamper evidence in packaging has become a major area of concern and emphasis for all pharmaceutical companies and consumers. The Tylenol® scare in Chicago during the 1980s was a wake-up call for the public, the government, and the companies. Laws and regulations were passed and codified to ensure that packages provide visible evidence to the consumer that they have been opened. The public has been educated to look for these features in packaging. They have become accustomed to seals under bottle caps, bands on bottle caps, and other safety measures that communicate whether the product has been opened when it reaches their possession. Tamper evidence is the last line of defense to inform the consumer something may be amiss with a package.

Tamper evidence is only required on drug packaging. This is surprising, because if you think about it, the potential harm to someone from contaminated food is just as great. The problem with extending this to all foods is one of scope and cost. Tamper evidence is present on many products; however, it is an increased cost that the manufacturer typically accepts as a way to provide a better product or to match the expectations of consumers regarding product safety.

Radio Frequency Identification (RFID)

Radio frequency identification (RFID) involves using a computer chip encoded with information that can produce a radio signal and broadcast the information to sensors at multiple points in the supply chain. The chip can be active (powered by a battery and intermittently or continuously broadcasting) or passive. Passive RFID labels or tags respond to the broadcast radio energy at a specific frequency and emit a radio pulse or signal back to a receiver that detects and processes the large amount of information encoded on the chip about the product and its packaging. RFID is being explored for two different pharmaceutical packaging applications: the first is to thwart counterfeiting and the second to track the product through the supply chain at the case and pallet level.

For anticounterfeiting, the tag is programmed with a unique number that is encoded and encrypted. The idea behind the anticounterfeiting information encrypted on the chip is that the pharmacist or dispenser of the product can read the code and then pass it through an agency or run it against a national database and verify the encrypted information.

The second application of RFID identification is as a substitute for barcodes. The information on the chip is used to identify, control, and verify the amount of product from the manufacturer through the supply chain to the retailers' shelves. In some cases, it is also being used to identify product or packaging for disposal. Wal-Mart and Target are two large US retailers along with Metro Stores in Europe who have wide ranging and active RFID initiatives that mandate the use of RFID identification labels on cases and pallets of products. The number of products with this information interface is rising each year because of the corporate mandate and initiative of Wal-Mart and Target. RFID offers the promise of automatic identification and tracking of a product when it moves in and out of warehouses and storage areas in retail stores, and as a way to track the bulk packaging waste to the recycling center within the retail store. Consumer concerns about privacy and the potential that retailers could track individual consumer-buying habits, restrict the use of labels at the individual product level. The other limitation of the technology, specifically how well a passive tag can be activated and then read, limits the usefulness of the technology at the individual-unit level in retail. For health care products, particularly those of high value, the usefulness of tracking items is under study at Harvard Medical School and at other health care institutions.

Anticounterfeiting

Anticounterfeiting is an extremely serious and perplexing topic most pharmaceutical companies face. Counterfeit pharmaceuticals and medical devices are very prevalent in all parts of the world. It is a looming problem that affects not only drugs, but software, aircraft parts, auto parts, and a host of other products. The opportunity for people to represent worthless or dangerous look-alikes as prescription products or as OTC products in all parts of the world continues to grow. The U.S. Commerce Department estimates that counterfeit products worth over $5 billion are sold in the United States each year. This amount is growing rapidly. RFID, mentioned previously, is the latest technology under study to fight this problem. Many packaging schemes, from holograms to reactive inks, have been tried and are in use; unfortunately, even the latest anticounterfeiting measures can be defeated or copied, and educating the public to look for a unique feature that identifies the product as genuine is very difficult. Few products offer the profit potential that pharmaceuticals offer, so the incentive to capitalize on worthless copies of a product will continue. Even hard-to-make parenteral products, products designed for injection, have been counterfeited and have reached the market.

Environmental Issues

Packaging in the United States and Europe strives to impart a minimal environmental impact by minimizing materials and placing a focus on using materials that are easily recycled. Packaging is always first and foremost about using a minimal amount of material to provide optimum protection and safety while delivering the product and environmental awareness only reinforces this dynamic. Both areas of the world along with many others are now concerned about what to do with packaging after it completes its functional life. This trend is also growing in other parts of the world. Major retailers are beginning to discuss and in some cases enunciate corporate environmental and sustainable responsibilities. They are setting goals for their products that take into account the environmental impact of making and disposing of a product, and are requiring manufacturers to help them meet their environmental sustainability goals.

This environmental trend for packaging has been around for the last 20 years and is expected to grow. As consumer awareness increases and information that highlights raw material sources, resource use, sustainability, recycling, and ultimately disposal are made part of the general public consciousness, packaging will have to change. Until packages can be consumed with a product, they will be a major contributor to waste. Individual delivery of packaged product, brought about by purchase over the Internet or by phone from catalogs and delivered to your door will add to the amount of packaging used by consumers. Instead of a case of product being opened at a retailer, 24 individual cases, albeit smaller, will be required to deliver the product to one's door. This change is one potential area for major growth in the volume of packaging used and its increased contribution to the waste stream. Products purchased online and delivered to one's door require more packaging than those obtained from the pharmacist or purchased OTC. These products greatly increase the amount of secondary and tertiary packaging required to survive a small package distribution system. The percentage of products purchased online is increasing, and the total impact this may have on packaging and the environment will grow.

Branding Prescription and Generic Products to Establish Consumer Identity

Branding and brand name have become leading ways pharmaceutical companies set their products apart from their competition. The first and still most effective way to communicate with doctors and patients is through the sales representative of the company. The representative relies on literature and samples to communicate the company's message about the benefits and attributes of the pharmaceutical product. The literature used and the samples provided to the doctor for the patients are designed to be coordinated in working together for improved doctor and patient communication. The brand look and design allow the patient to identify the product being sold as the same as the sample received in the doctor's office. This is important for the elderly or for patients who take a large number of medications. A simple method that helps people to identify prescription products from OTCs medicines makes life better for them. Branding, which includes all the visual identifications used on a product are extremely important to people. It also permits the product to establish a presence and recognition in the marketplace that is important after patent expiration, when the product becomes a generic drug.

Initially new products are protected by patent and have no competition, but today it is not unusual for multiple drug companies to conduct parallel R&D on a class of drugs. Statins, the number 1 prescribed products in the world, are a good example of this. Multiple companies developed multiple variations of the statins, and, as a result, multiple companies supply statin analogues for treatment of cholesterol. Each is different, and each has patent protection of the unique API. Providing an easy method for patients to recognize what they are using becomes important because not everyone reacts in the same way to similar but not identical drugs in a class of product.

As a drug completes its patent life, generic manufacturers copy the original product and provide a low cost generic drug alternative. All pharmaceutical companies work hard in marketing and communication to establish their particular brand of product as the gold standard for the therapeutic treatment of the condition or disease. They want to deliver this message to consumers. If they are successful, consumers will continue to ask for a product by name after patent expiration and won't accept generic substitutes. This softens the blow of patent expiration and permits pharmaceutical manufacturers to maintain manufacturing volume crucial for low-cost supply.

Many new prescription products have a potential for over the counter sales. Here it is crucial to establish the brand name with consumers. Examples of products that required a prescription during patent life and then moved to OTC sales after the patent expired would be the nondrowsy antihistamines (Claritin®) and proton pump inhibitors for heartburn (Prilosec®). These were prescription products when introduced. At the end of their patent life, with a documented history of safe usage by millions of people for many years, they were considered safe by the FDA to be approved as OTC products. People still look for these products by brand name.

Direct-to-consumer Advertising

Advertising of prescription products directly to the consumer has become a standard marketing tool used by all the pharmaceutical companies. For packaging, the consistency of product labeling and supporting literature are an area of emphasis. Packaging of prescription products was a dull, bland, and somewhat repetitive exercise not long ago. Product specified by the doctor and needed by the patient did not require anything special in terms of graphics and packaging for promotion. Advertising to consumers has changed all of that. Now companies use television, print, and Internet advertising to make people aware of their product and the benefits it provides in the treatment of any number of conditions. The consumer asks for the product by name and relates to the images and information provided in the communication media to be sure it's the right product. Direct-to-consumer advertising does not automatically generate sales of a product. Its value and the major benefit this form of communication provides are creating consumer awareness of medical conditions and one of the new treatments available. Patients will discuss the condition with their doctors and inquire if the symptoms they exhibit are similar to those described in the advertising. The doctors may advise or determine that they don't have the condition, or if they do, it doesn't need treatment, or that other products available for treatment of the condition are a better choice for the patient. Only 35% to 40% of people inquiring or requesting a drug by name actually end up with the product.

Direct advertising also is playing a role in helping health care professionals remain abreast of new treatments and products in the pharmaceutical pipeline. Today we are overwhelmed with information and advertising, if nothing else, is a start for many toward awareness of a product they may need. Doctors are swamped with information just like all of us in today's digital information age. Simple direct communication to them, which prompts their interest and encourages them to learn more about new treatments is a good outcome for direct advertising.

Information Technology and Graphics Trends

Enterprise content management and digital asset management. This is the digital age. Information is designed, stored, and disseminated digitally. For a pharmaceutical company the amount of information needed to communicate to a patient or doctor all information required is daunting. It is also expensive and difficult to manage, when variations in claims and marketing approach proliferate for a product depending on the part of the world in which it is marketed and sold. Two new information technology tools, digital asset management and enterprise content management, are at the forefront of technology and systems needed by companies to manage all the information and changes occurring in pharmaceutical products, particularly pharmaceutical product labeling. Complementary technology provided by the Internet and XML, or eXtensible Markup Language, are components of these two broader and more significant technology changes.

Enterprise content management. "Enterprise content management" (ECM) is the broad term applied to all forms of communication within a company and how these forms will be managed now and in the future. Global companies, including global pharmaceutical companies, produce an enormous and rapidly growing volume of content. Knowledge workers spend an inordinate amount of their time just looking for the information they need to do their jobs. In pharmaceuticals this information problem leads to a lack of consistency in branding and messaging of products. It is also slowing project teams entrusted with product development as they struggle to collaborate and use information as quickly as it is developed. It also slows down Web sites in changing and delivering the appropriate information about a product in multiple countries or regions. Everything from e-mail to visual images and all the data developed or collected within a company is digital or being converted to digital information. This mountain of information requires a strategic vision to avoid redundant technical development and to archive and save the information in a form that is easily accessible when crucial to business functions. Packaging creates a large amount of product-related information that can be used multiple times in many different ways to minimize the cost of its creation and maintain the consistency of graphics, the color, and the message it conveys in multiple languages around the world. Specifications, validations of packaging materials and systems, bills of material, incoming inspection reports and many other items are all part of the packaging contribution to manufacturing and maintenance of a product. All of the documentation mentioned and all of the labeling and graphics used in marketing and packaging a product reside in multiple file formats with multiple generic and industry-specific standards. All of this information requires a strategic vision and a simple method to organize it for all to use. This is the promise of a true ECM system within a company.

Packaging will play a major role in any ECM system development. The breadth and depth of information considered part of packaging and the related marketing collateral used for a product cannot be managed in any other way. ECM and its subset digital asset management (DAM) are becoming crucial to eliminating waste in the development of packaging specifications and labeling.

As companies manufacture more and more products in multiple plants or in multiple regions of the world, the ability of a packaging group to access and share information becomes more and more important. Quick access to available information, communication of the information to multiple end users within and outside the company, and the ability to track and change items broadly without making changes to each and every item are the goals driving the use of ECM. These systems permit the company to communicate all the information needed by suppliers, multiple offices, and Web sites simultaneously instead of requiring the creation and maintenance of the information for each individual requirement or need. An example of this electronic asset capability is the communication of a product insert to the Web site for the product, to the printer supplying the insert to manufacturing, and to every doctor and salesperson needing that information at any time. Today's world demands this type of information availability, and ECM systems will begin to manage that job.

DAM. DAM is a subset of ECM. This is a component that organizes all the digital assets of a company as a library within the ECM system. Packaging for these assets includes all labeling, graphics, sales aids, promotional literature, product inserts, and images associated with each product. It is a simple repository is needed that permits sales, marketing, regulatory affairs, quality assurance, and packaging to quickly find and use these assets effectively. It also reduces the risk that a location or department may use a version or revision of a critical piece of information that is out of date or incorrect. By structuring information and then marking it with XML, a company can quickly change and update information everywhere, and feed that revision immediately to any department, supplier, or customer authorized to see and use the information. This is in stark contrast to hard copy files and multiple digital file formats stored on a server. It is a complete catalog of assets that permits easy access for all parties everywhere.

SUMMARY

This is an amazing collection of things to cover in one volume. This book is an introduction to and overview of these topics, and it should remove many of the questions and unknowns about pharmaceutical packaging.

REFERENCES

1. CASCADE Collaboration, Determinants of survival following HIV-I seroconversion after the introduction of HAER. The Lancet 2003; 362:1267–1274.
2. Stein R. From Killer to Chronic Disease: Drugs Redefine Cancer for Many. Washington Post. January 29, 2003; Section A:A1.
3. Class S. Health care in focus: the pharmaceutical industry is seeking a new prescription for success as it faces pricing pressures, challenges from generics, and consumer disenchantment. Chemical and Engineering News. December 6, 2004; 82(49):18–29.

4. DeVol R, Wong P, Bedroussian A, et al. (Eds). Biopharmaceutical Industry Contributions to State and US Economies. Santa Monica, CA: Milken Institute Publications, 2004:1. Available at: www.milkeninstitute.org/pdf/biopharma_report.pdf
5. Pharmaceutical Research and Manufacturers of America. Pharmaceutical Industry Profile. Washington, DC: PhRMA, 2005.
6. Pharmaceutical Research and Manufacturers of America. Pharmaceutical Industry Profile 2004. Washington, DC: PhRMA, 2004.
7. Grabowski H, Vernon J, and DiMasi J. Returns to research and development for the 1990s new drug introductions. Pharmacoeconomics 2002; 20(suppl 3):11–29.
8. Pharmaceutical Research and Manufacturers of America. PhRMA Annual Membership Survey. Washington, DC: PhRMA, 2005.
9. Class S. Health care in focus: the pharmaceutical industry is seeking a new prescription for success as it faces pricing pressures, challenges from generics, and consumer disenchantment. Chemical and Engineering News. December 6, 2004; 82(48): 20.
10. Class S. Pharma 2005. Chemical and Engineering News. December 5, 2005; 83(49): 20.
11. Class S. Health Care in Focus. Chemical and Engineering News. December 6, 2004; 82(48): 22.
12. Federal Food, Drug, and Cosmetic Act §201. Definitions 21 USC §321 chapter 2. June 25, 1938.

2

Pharmaceutical Dosage Forms and Their Packaging Requirements

INTRODUCTION

Before one begins to understand pharmaceutical packaging, some basic information is required regarding physiology, chemistry, drug delivery pathways, drug characteristics, pharmaceutical Current Good Manufacturing Practice (CGMP), the Food and Drug administration (FDA), and other topics. This varied and complex information plays a role in the design and required performance of pharmaceutical packaging.

Pharmaceutical drug products, the compounded products administered to the patient, are made up of active pharmaceutical ingredient(s) (API) and diluents or excipients that dilute the API to a safe and efficacious point, increase the volume of the product to make it easy to administer, or modify the solubility of the product to make it available for the patient's body to metabolize, while protecting the active ingredient from decomposition or change. The physiologically inert excipients also act as a second layer of protection for the product after the packaging by providing coatings, buffers, or free radical scavengers that protect the API from change. They also act as solvents, lubricants, and binding agents that aid in the manufacture and dosage administration of the product. All pharmaceutical products and specifically their APIs must deliver the certified therapeutic result throughout the stated shelf life of the product. This expectation is built into the FDA regulations surrounding drugs and medical devices and the consumer's expectation that the product will perform as specified and described in the information contained on the product's packaging.

The problems surrounding these products for both manufacturing and packaging come from the types of chemical compounds being delivered to the patient. The API molecules, both traditional chemical compounds and larger biopharmaceutical molecules, have inherent problems relating to their stability, purity, and sterility. The formulation of the product and the development of its packaging must meet the unique needs of each new product. This background combined with an understanding of drug physiology and drug dosage forms is required to put packaging into an understandable context.

STABILITY

Drug products are designed to provide effective and safe treatment to a medical condition. Safe treatment includes maintaining the effectiveness of a product over its stated shelf life. This means the product and package must maintain both the integrity and effectiveness of a drug from the time of manufacture and packaging to the point where the product is consumed. The first and most important requirement for packaging in this context is protection of the drug from chemical change. A bulk package used during manufacture must not interact with a drug during manufacture, and final or individual dose packaging must be proven to not have interacted with the product in subsequent testing of the finished product before New Drug Application (NDA) approval and its release for sale. The package also must not interact in a significant way with the product during distribution and storage throughout its stated shelf life. The package must protect the drug from any harmful effects during distribution and ensure that a patient will receive the expected benefit that the product is proven to provide. When a product completes normal quality control testing at the point of manufacture, confirming it meets all required specifications for approval and release, the job of packaging really begins. The package must protect and maintain the product in this final manufactured form throughout its stated useful life. Shelf life is typically the period of time the drug product retains 90% of its activity or potency. At the end of this time period, which is determined by chemical and biological testing during drug development, the drug should be discarded. This period of time or the final date of drug effectiveness or shelf life is clearly marked on every package.

CHEMICAL CHANGE

Most drugs are organic molecules that are subject to change through a number of chemical pathways that can be triggered by light, heat, moisture, or oxidation (1). Many active pharmaceutical ingredients are extremely reactive in very small amounts and must be diluted to present the body with the correct therapeutic concentration needed. The amount of active ingredient contained in a tablet, solution, or suspension is typically diluted with other ingredients called excipients that may provide protection from the harmful effects of moisture and oxygen for short periods of time, and also provide the physical bulk needed to produce a dose volume that can be easily handled and consumed. In the case of liquid products, water, organic solvents, oils, and alcohols are the typical excipients used to produce a volume of liquid large enough to be measured and consumed. Topical products are similarly diluted with a variety of materials that facilitate application to the skin or affected area.

Biopharmaceutical products are different from the complex chemical molecules that are the active ingredients in traditional drug products. These are typically very large molecules that include proteins and enzymes. The variations in packaging for these products are similar to many of the classical pharmaceutical compounds and will be treated in the same way in this chapter, with references to them in the various dosage forms.

Small molecule APIs, the molecules that produce the therapeutic effect, are very unique chemicals. Their biological activity is typically dependent on two different characteristics of the molecule, its chemical composition, and the exact configuration of the chemical constituents within the molecule. In biological products, because the molecule is very large, its shape or folded stereochemistry is important as active sites are only presented with chemical binding or reaction in the body when the molecule is folded in one specific configuration. Proteins can and do have multiple ways to form and fold, and maintaining the proper structure is important.

Chemical APIs typically come in the form of optical isomers (1). Optical isomers are forms of the same molecule that are mirror images of each other. Just as you have a right and a left hand that are mirror images of each other, these molecules have left and right orientations of key active groups around a central carbon atom. Surprisingly, one form of the molecule, either the levo (l) form or the dextro (d) form of the same molecule, is pharmaceutically more active than the other. In almost all biological systems, stereochemical specificity is the rule rather than the exception, since the all important body catalysts, a group of chemicals called enzymes, are optically active. Hence, the compounds they interact and work with must be optically active as well.

Whenever a compound contains a carbon atom bonded to four other different chemical groups, it can be considered a derivative of methane (CH_4), and what we know about the methane molecule can be applied to that molecule, regardless of how simple or complex it may be. This carbon atom may also be called a "chiral carbon," a term that indicates it is surrounded by four different entities (Fig. 1). By definition, a chiral carbon cannot be converted from one stereochemical configuration to another by rotation of bonds. A chiral carbon has four different chemical substituents attached to it, and because of this fact, it can exist as two mirror image forms (Fig. 2).

Analytical examination of a carbon atom bonded to four other atoms indicates that the bonds form the shape of a tetrahedron. In fact, this idea was proposed during the 19th century before the direct determination of molecular structure was possible. The evidence for this structure was derived from optical activity of molecules with a formula, CWXYZ, which creates physically similar molecules that rotate light in different directions. The letter symbols WXYZ each represent different chemical groups distributed around the central carbon

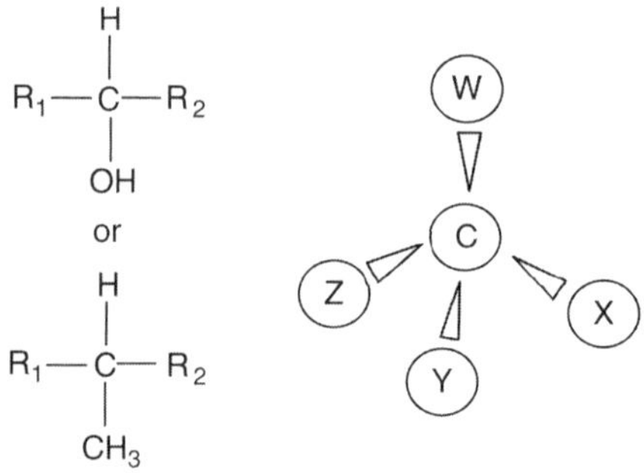

Figure 1 Representation of a chiral carbon with different groups around the central carbon.

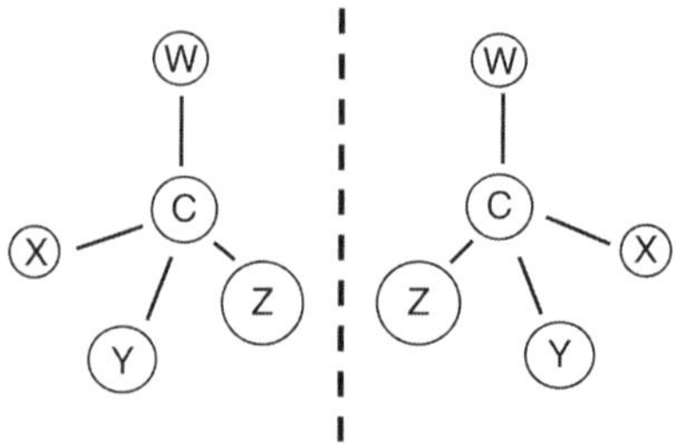

Figure 2 A representation of enantiomers as mirror images of each other.

atom in a tetrahedral structure. These molecules are called enantiomers (Greek: enantio-, opposite).

Light possesses properties that are best understood when it is represented as a wave function. The vibrations of the wave occur at right angles to the direction the light travels. There are an infinite number of planes passing through a line of propagation of light, and ordinary light passes through all of these planes. When light is plane polarized, typically called polarized light, it means it has been transformed into light that is vibrating in only one of the planes. An optically active substance is one that rotates polarized light, so that when the light emerges after passing through the substance or a solution of the substance, it is rotated, which means it is vibrating in a different plane. The instrument used to measure the rotation is called a polarimeter. The polarimeter uses two lenses; the first lens is used to polarize the light, then a tube is used to hold the substance to be measured, and finally a second lens that can be rotated. Light enters the first lens and is polarized into one plane of vibration; it then passes through the substance and is rotated either left or right in its plane of vibration if the substance is optically active, and then passes through a second polarizing lens. By turning the second lens to the left or right until the maximum amount of light emerges, the amount of rotation created by the substance between the two lenses can be measured. If the rotation of the plane of light measured by the rotation of the second lens is to the right (clockwise), the substance is "dextrorotary," and if the rotation is to the left (counterclockwise), the substance is "levorotatory." The two terms are derived from Latin, in which *dexter* means right and *laevus* means left. Lactic acid extracted from a muscle will rotate light to the right and is known as dextrorotary lactic acid. When a starch is fermented to ethyl alcohol, one of the by-products, 2-methyl-1-butanol, rotates light to the left, and is known as levorotary 2-methyl-1-butanol. The two forms of 2-methyl-1-butanol, for example, have identical boiling points, melting points, densities, refractive indices, and all other physical constants that can be measured except this rotation of plane-polarized light.

A large amount of the research effort is often expended in developing a synthesis pathway of a complex drug molecule to produce a chemical reaction that favors the formation of the most pharmaceutically active molecule and minimizes the formation of its optical (mirror image) analogue. A common product that exhibits this difference in activity is epinephrine, where its levo (l) form is 15 to 20 times more active than the dextro (d) form of the molecule (Fig. 3) (1).

Proteins and biopharmaceutical molecules display some of these same properties in the way they fold or twist into a complex shape. The folding or twisting presents or makes available different surface groups that determine the molecules' pharmaceutical activity.

The active pharmaceutical ingredient of a drug can be sensitive to a number of physical and chemical conditions that a packaging engineer must be cognizant of when designing a package. Many of the chemical compounds that act as drugs are inherently unstable. Protection from oxygen, heat, moisture, and light are the most common environmental factors that a pharmaceutical product requires. These molecules can and will undergo change after exposure to some or all of these factors. When a molecule is biologically reactive, it follows that it must also be chemically reactive as well. Heat and light can change the drug

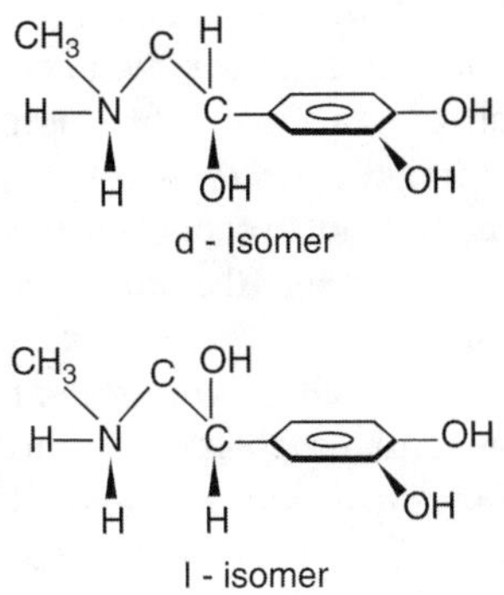

Figure 3 Dextro and levo forms of epinephrine.

product from one optical isomer to the other by breaking chemical bonds, permitting rearrangement, or it can begin the breakdown of the molecule into smaller molecular components. Rarely are these components or degradation products harmful, but they may reduce or eliminate the effectiveness of the drug. The body reacts to a drug compound and then metabolizes or breaks the material down by chemical reactions in the body; these degradation products are usually different from those produced by environmental factors and time during the shelf life of the drug. Many molecules are only active in the dextro (d) or levo (l) forms mentioned earlier, and the packaging problem is to prevent energy needed to break bonds from entering the package and permitting the molecule to change to the wrong form, which after conversions may be slightly more stable than the active enantiomer. The interconversion of enantiomers (configurational enantiomers) is slow because bonds must be broken. The pharmacist and the pharmaceutical manufacturer are responsible for understanding the inherent instability of the molecule, and using this knowledge are required to formulate a product that overcomes or minimizes the instability long enough for the product to reach the patient. During packaging at the point of manufacture and packaging used in a compounding pharmacy, it is essential to provide product protection. Packaging is a major part of the layers of protection built for a drug to minimize chemical change. Typical conditions encountered by the product such as heat and light can provide the necessary energy to begin a change in a product. Packaging can insulate the product from these factors and contribute significantly to the total shelf life of a particular molecular form.

THERMAL PROTECTION

Short-term protection of a product from heat is all that the packaging can provide unless designed to maintain a temperature range for slightly longer times, perhaps three or four days when a product is shipped to a remote part of the globe. This is common for vaccines, and the thermal protection required is one of the unique forms of packaging used by pharmaceutical producers. Labeling is also a key packaging feature that highlights the required storage conditions of a product to prevent exposure to environmental factors that degrade the drug.

Protection from heat (or freezing) is required throughout distribution and storage of the product by the health care professional or the patient before use. Warehouse conditions are monitored and documented (required by the FDA) by pharmaceutical manufacturers and distributors to protect the products from temperature extremes. This monitoring is also carried out on the mode of transportation used to ship the product such as a truck or cargo container to ensure that the product is maintained in an environment certified during drug development as safe. It is not unusual for packaging to be color coded to highlight the proper storage conditions. Unfortunately, there are no standard conventions to this color code system signifying the conditions each color represents, and each of the manufacturers that use color-coded labels or containers has their own unique system.

CHEMICAL REACTIONS

The chemistry behind the reactions that cause change in drugs is in most cases relatively straightforward. Hydrolysis, oxidation by environmental oxygen or other substances that remove electrons, and light, which increases the energy state of a molecule, permitting it to undergo change, are the most common culprits that packaging must stop or retard to deliver a safe and effective product to the consumer. In many cases, the packaging must protect the product from more than one of these risk factors, and understanding the materials used to produce the product and the dosage form of the product are essential for development of a package that is safe and protects the product for long periods of time. Most pharmaceuticals are stable for a minimum of two years from the date of manufacture.

Packaging must be involved early in the exploratory research of a biologically active molecule to provide expertise and direction on how to protect the product even before the product reaches the Investigational New Drug (IND) stage of review. Drug discovery exploratory group is charged with the task of screening numerous compounds for biological activity and beginning the process of testing a compound for possible product applications. This group will often automatically place the API into a stability program immediately after it shows promise as biologically active. This is done for a number of reasons, including decreasing the time necessary for development of potential efficacy, understanding of the chemical kinetics behind stability of the molecule, and providing long-term stability data for an IND application or, later, an NDA to the FDA. The same data are used when completing the dossier of information required by regulatory agencies outside the United States. Many times these data represent the best and most complete picture of the molecule's behavior over extended periods of time, particularly if kinetics or chemical structure is complex and multiple degradation pathways exist. Small chemical changes can cause marked differences in the performance of a drug. Many times these changes are very hard to detect. Small changes in color, solubility, separation, or other observable physical changes are used to measure the effectiveness or potency of a product. Along with the pure molecule, the API is combined with multiple formulas (combinations) of inert ingredients called excipients to produce stability samples in tablet, liquid, gel, or ointment form, depending on the condition being treated and the most effective administration route for the patient. These additional ingredients dilute or modify the bulk volume of the product and provide physical characteristics needed for manufacture or protection from harmful environmental effects during manufacture. They become an integral part of the product and may carry over into the storage life of product characteristics that provide the first line of defense against chemical change. They accomplish this by providing or maintaining a chemical condition that stops or hampers product degradation. An example of this is a pH buffer, which is used to maintain an ingredient in the acid (H^+), base (OH^-), or pH neutral environment that creates the best stability conditions. The excipients may also be a second line of defense for the product after package protection from the environment has been slowly compromised. An example of this would be oxygen scavengers formulated in a product's excipients and then packaged in plastics that prevent the formation of free radicals that are typically the precursors of degradation. The scavengers eliminate any atmospheric oxygen that slowly seeps into the container through the oxygen barrier plastic (2) and thus extend the product shelf life or long-term efficacy. They also provide short-term product protection after the packaging is opened by the patient.

A standard group of excipients, diluents, and other inert materials are added to the active molecule to reduce its API strength to the level believed to be biologically effective and least toxic (3). Toxicity is determined by in vivo testing and later in animal testing to develop a profile of the most effective range of doses that produce the therapeutic effect. This extensive toxicity profile must be completed before an IND application is submitted to the FDA for testing in humans. Biologically active substances that complete toxicology screening and manufacturing suitability are submitted to the FDA in the form of an IND application, and with the agency's approval, are then used in clinical trials on human subjects. The various steps and levels of clinical testing and the requirements for each phase of testing from the IND application to NDA are covered in chapter 5 on "regulatory affairs."

Early packaging development and the introduction of multiple forms of packaging for a product typically follow a discussion and review of the product with both the discovery group and the marketing group to produce a best estimation about how the drug will be presented or administered to a patient. Manufacturing is also part of this process; they typically input how and where a product is planned for manufacture and how that manufacturing operation may influence the packaging considered or available for the product. For example, tablets undergo stability testing in bottles, blisters, and, if required, bulk containers. Bulk containers are necessary when preparation and tableting take place in one location and tablets are packaged finally for the patient in multiple locations, which may mean in multiple countries.

Moisture Protection—Protecting the API from Hydrolysis

Hydrolysis is the chemical reaction or process in which a molecule is split into two different species on reaction with water. One of the cleaved species acquires the hydrogen ion (H^+) and the other the hydroxyl ion (OH^-) or a positive hydrogen ion and a negative hydroxyl ion. This is distinct from hydration of a molecule in which water is added or absorbed from the atmosphere by a substance, with no chemical change. Hydrolysis in organic chemistry is the opposite of condensation.

Physiological irreversibility of hydrolysis is used by the body in metabolic pathways, since many biological processes are driven or derive their energy through the cleavage of pyrophosphate bonds; the metabolism of adenosine triphosphate is a prime example of this hydrolysis. Under most physiological conditions, which involve dilute aqueous solutions, the metabolic reaction is essentially thermodynamically irreversible if the concentration of metabolic starting material is low.

Many pharmaceuticals contain ester or amide functional groups that can react with water, leading to the cleavage of the chemical bond and the formation of two new species (Fig. 4). The chemical process that cleaves a compound into two or more simpler compounds through the uptake of the hydrogen ion (H^+) or hydroxyl ion (OH^-) parts of the water molecule on either side of the chemical bond cleaved is a major pathway for the degradation of many drugs. Common products that fall into this chemical degradation process include aspirin, anesthetics such as procaine, tetracaine, and cocaine, and heart stimulants such as atropine, a drug used to increase heart rate. Products like these contain chemical linkages called amides or esters, the stable salts of the base material. The ester or amide group is particularly susceptible to hydrolysis. Reaction of the functional group with water causes cleavage of the bond in the molecule and the creation of two new compounds. Amide groups after hydrolysis form acids and amines. Esters after hydrolysis produce an acid and an alcohol.

The first line of defense for a drug containing a functional amide or ester group is the use of a buffering agent introduced as part of the excipients. The buffering agent can be a solid in a tablet or a liquid in solutions. It can also be an oil used in the formulation of an ointment, lotion, or cream. These chemicals minimize the formation of an acid or base condition needed to facilitate hydrolysis reactions.

Hydrolysis can occur slowly in pure water, but the addition of a small amount of acid or base can act as a catalyst to increase the rate of the decomposition reaction. The addition of a buffering agent to a drug product in solution minimizes the potential change.

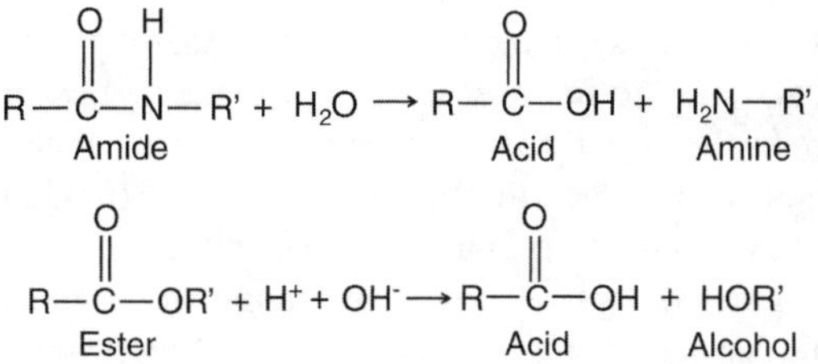

Figure 4 Hydrolysis of esters and amides into acids and amines or alcohol.

Another method of protection from hydrolysis is the introduction of a surfactant or an organic solvent such as propylene glycol. The surfactant or organic solvent can bind to the active groups and prevent them from coming in contact with the hydrolyzing agent. Barbiturates are a class of drugs that use blended diluents containing propylene glycol to make them more stable.

In products that are somewhere between a drug and a food, similar protection is required. Thiamine and niacin (niacinamide) are two common nutritional supplements that must be protected from hydrolysis. Protection of these ingredients in food products with specific label claims is common, and the same methods of protection used for pharmaceuticals are also used in a supplement or food product.

Oxidation—Reactions with Oxygen

Oxygen is one of the most common environmental factors to cause drug degradation. Oxidation is the removal of electrons from a molecule and does not always require the presence of oxygen. When oxygen is involved, the molecule can produce free radicals that speedup the degradation process. This is one place where a little oxygen exposure (oxygen-induced free radical) can go a long way in speeding degradation. Pharmaceutical packaging can use many different approaches to block or eliminate oxygen or the free radicals it produces and provide the product protection needed. All or some of the potential options eliminate or minimize environmental oxygen from degrading the API. Examples of a few of the approaches include using a metal or glass container that is impervious to oxygen, using plastics with a material barrier to oxygen, or using plastics, paper, or composite materials that incorporate an oxygen scavenger in combination with the oxygen barrier. It should be noted that any plastic package has some permeability, so you are describing a very slow controlled leak when you discuss oxygen protection and plastic containers. One of the newest methods of increasing oxygen barrier in plastics is the use of an oxygen scavenger in the packaging material. The scavengers are found in the walls of the container.

The closure or cap on a bottle must be engineered to prevent oxygen entry. When a screw-type closure is used, the gasket would be made of a material that prevents the slow introduction of oxygen to the headspace of the package, or it may contain a scavenger similar to the material contained in the plastic bottle itself. Other techniques include a heat induction seal with a foil containing multilayer liner in a closure and the introduction of an inert gas on the inside of the container during filling. When an inert gas is used many times, the product itself is manufactured under a laminar hood to blanket the manufacturing process with an inert gas to further reduce residual oxygen that would be carried into the package by the product.

Oxygen scavengers and other free radical scavengers can inhibit the effects of oxygen coming in contact with the product. These materials include sulfites, thiosulfites, and ascorbic acid in water-based systems. In oil-based systems, palmitates such as ascorbyl palmitate or hydroquinone are used to absorb the small amount of oxygen that find its way into a product.

Heavy metals are classical catalysts of oxidation reactions when hydroxyl or hydrogen ions (hydronium ion) are present. The use of reactive agents to remove these ions when a heavy metal is present is another effective strategy to prevent oxidation. Obviously the removal of the metal catalyst from the drug mixture is required if the concentration is high, but even the very small amounts that are carried over from a manufacturing step may require removal until they are below required levels. Unless stability or toxicology data indicate otherwise, removing extremely small amounts of metal with extensive purification steps, usually multiple recrystallizations, is expensive and difficult to manage on a large scale to remove minute traces of these materials. For these materials, usually present in the part per million or part per billion ranges, the alternatives for protection from moisture or oxygen are more effective.

LIGHT PROTECTION

Light is an energetic waveform that can provide the energy necessary for a substance to react or change configuration. The higher energy level can rupture a chemical bond and is another pathway to the formation of free radicals. The smaller or shorter the wavelength of light, the more energetic it is. Ultraviolet light is the most energetic form of light and is present in sunlight. Visible light is less energetic but can still provide the energy necessary for a chemical reaction to take place. Racemization, the change of a compound from one optical isomer to another, is a change that can take place when light energy is provided to an active molecule. Light can also generate heat, or can be absorbed selectively by materials in the formulated product. These materials, although not the API, can transfer energy to the drug product by simple collisions, and this pathway can also lead to stability degradation and product change. Steroids such as hydrocortisone, prednisolone, and methyprednisolone are examples of drugs that are susceptible to changes from light. Their degradations take place through free radical reactions, and compounding these products with antioxidants is an effective way to increase stability and shelf life. Packaging these products in opaque containers, or in containers that block the harmful wavelengths of light associated with degradation, is a typical solution to this problem. If the product is a liquid, or if the product must be measured from the dispensing container, labels that have small graduated openings or windows may be used to protect the product from most of the environmental light to which it may be exposed.

MATHEMATICAL METHODS AND ACCELERATED METHODS FOR ASSESSING SHELF LIFE

A number of methods are available to predict or develop a good approximation of the shelf life of a product. The methods include storage of a product at an elevated temperature to increase the chemical reactivity and develop an idea of how a product degrades. The Arrhenius equation is used as part of the process to develop a correlation between time and temperature to predict the potential shelf life of a product. This can predict the effect of temperature on chemical stability and can also be used to determine the effect of a catalyst on decomposition.

The Arrhenius equation:

$$k\text{-}A \times \exp^{(-E/RT)}_{a}$$

Note: k is the rate coefficient, A the frequency factor constant, E_a is the activation energy, R is the universal gas constant, and T is the temperature (in degrees Kelvin).

Computers and their ability to manage data have greatly simplified this predictive process. It may not be necessary for a researcher to determine the mechanism of degradation initially. Testing of the API and testing of the formulated drug in multiple accelerated aging conditions produce an understanding of the drug concentration effect on stability and possible effects of concentration and excipients on the degradation products. After the data are obtained, they can be analyzed and extrapolated to provide a good approximation of shelf life. These techniques are particularly useful when multiple diluents and excipients are part of the mix and their combined effect is difficult to calculate. This mathematical technique cannot be used for a final shelf life determination used for final product labeling but can be used to develop shorter-term shelf lives acceptable to the FDA and used during the introduction of the drug, while long-term testing at standard temperatures takes place. This accelerated technique is also valuable for validation of the first production lots of a product being brought to market. Generally, final shelf life listed on a product indicates that the product still exhibits 90% of its potency at the expiration of end date on the product label or package. For products that are unstable at room temperature, the same testing is carried out at both refrigerated and frozen conditions to produce the understanding of decomposition products and develop the calculation of potential shelf life.

PURITY AND STERILITY

Purity and sterility are standard expectations for a pharmaceutical consumer. They are attributes that are required by the FDA and outlined in regulations and guidelines on CGMP for the manufacture, processing, packaging, or holding of human or veterinary drugs. Purity refers to the idea that the drug product in the package is pure and free from contaminants. You are getting exactly what you expect in its purest formulated form for combating the disease or condition. This was not always the case, and one of the main reasons for the passage of the Food, Drug, and Cosmetic Act that created the modern FDA was to regulate the purity of drugs.

Sterility is another expectation of the consumer and all regulatory bodies around the world. Drug products, depending on how they are prepared, can still contain harmful organisms. There are numerous methods employed to eliminate harmful organisms, but for them to be effective, scrupulous attention must be paid to all parts of the manufacturing process, including the procurement of raw materials and packaging to ensure that the final product does not contain something harmful. Processes used to sterilize a drug begin with an assumption of a known bioload or bioburden, the amount of harmful organisms occurring naturally in the ingredient(s) or introduced in the manufacture or packaging of the product. If a naturally derived material is heavily contaminated with an organism, the sterilization process may be overwhelmed and not able to fully eliminate all the harmful organisms in the sterilization cycle.

Drug Purity

Drug purity or the lack thereof was one of the main reasons Congress passed the initial regulations governing medicinal products. "Snake oil," as many drugs were referred to at the turn of the 20th century, was a derogatory term that still persists. It meant that medicinal products magically achieved their performance with unique ingredients that were company's or individual producer's secret. Unfortunately, many of the original products touted as medicine contained harmful and dangerous ingredients, including opiates (narcotics) that relieved many symptoms but did not cure disease. Over the past hundred years, this possibility has been eliminated. Pharmaceuticals must stand multiple layers of review and testing and pass them all to achieve any claim for efficaciousness and safety. All the materials used in the product are examined during the review.

Purity is the first major concern of a manufacturing operation. Following the rigorous review and approval process all drugs undergo, the next most important requirement for producing a drug is the diligent focus within any manufacturing operation to follow and maintain all the performance and testing standards that were defined, proven, validated, and submitted to the FDA during the product's developmental phase. This scrupulous focus on detail starts with the raw materials used to manufacture the API, the excipient materials that are part of the compounding products, and the packaging materials that contain and protect the product. Raw materials rarely are elemental; they are usually partially compounded intermediate products that are subject to some degree of variability. The manufacturing organization and a separate department, the quality assurance organization, must be constantly vigilant to maintain the specifications and standards that define all the component materials or packaging in all phases of product production and testing.

Quality Assurance

The quality assurance organization, although located within the manufacturing operation and part of its day-to-day function, is required by law to remain outside the manufacturing organizational structure. This is to ensure that their decisions on product quality and purity are not influenced by the people responsible for making and supplying the product. Quality assurance is responsible for testing and/or certifying incoming materials, intermediate components, and other materials needed for manufacturing and to meet the defined and validated specifications for each of the product components that are part of the manufacturing process. This includes all packaging components and the finished product. Quality assurance is also responsible for making sure that the label applied to any product is correct right down to the latest revision. They are responsible for any variable information that is required in manufacturing, such as the lot or batch number of the product, the time and date of manufacture, and the expiration date expressed on the labeling. Record keeping for manufacturing a drug is extensive and under the purview of the quality assurance organization as well. Each time a product is produced, the batch record and the documents used by manufacturing throughout the production process are reviewed by the quality organization to ensure that they are correct. The batch record not only provides the quantity and type of raw material used for manufacturing the product, it also contains the detailed instructions of how to bring the ingredients together in each step of the process. The batch record receives a physical sign-off (initials or actual signature) by technicians and operating personnel charged with making the product and ensures that each step in the process was carried out correctly.

Quality assurance certifies that the technicians or operators, manufacturing employees actually involved in the production of products, have received the necessary education and training indicating they understand and are capable of producing the product correctly. Quality assurance or the manufacturing operation must maintain detailed training records of each employee involved in the manufacture of pharmaceuticals. The technicians (operators) in every step of the manufacturing process must sign or initial each step specified in the batch record, stating that the operation was done properly in the proper sequence with the correct ingredients. This affirmation involves two operators or technicians, each of whom is required to check that the raw materials used are correct, the correct manufacturing equipment is being used, the cleanliness of the equipment meets the required standards developed during validation, the weight or amount of material added at each step is correct, and the components are brought together in exactly the way as specified in the manufacturing instructions and records.

Throughout this process, quality assurance monitors and reviews adherence to the specified instructions. It may be in spot checks of the records for products in process or it may be sample collection and testing of intermediates created at various stages in the process. Each of these checks assures that the process is proceeding correctly. At the conclusion of the manufacturing process, quality assurance tests the finished product and signs the product release documents to certify that the product is suitable for sale and use. This process is required each time a product is produced and includes a review and approval of the records for each manufacturing cycle or lot of a product. Following completion of the manufacture of a product, the records used during manufacturing are reviewed again, step by step, to ensure that nothing was overlooked or skipped. This means that each notation by the operators in manufacturing is checked to ensure that the step was completed and done at the proper time and that no problems were noted or observed during each step in the process. The operators in manufacturing are responsible to note any variation or unusual condition observed as the product is produced. This could be as simple as a slight variation in the color of an ingredient, a longer than expected time for a product to mix or react, or a processing disruption like the loss of agitation or heat at some point in the process. These deviations are noted and examined by quality assurance to determine if the variation compromises the product's quality in any way. Any deviation noted during manufacture must be reviewed and cleared before a product is released. This may involve additional testing or other investigatory steps that prove the product has not changed in any material way from the product that was approved by the FDA. The deviation records are an important part of any manufacturing facility's record-keeping responsibility and must be maintained and archived. They are subject to review and evaluation by the FDA in periodic inspections and audits of manufacturing facilities.

Drug Sterility

Drug sterility refers to microbial contamination of products. Beyond the assurance of purity in a drug product, this second and sometimes more difficult aspect of manufacturing is the ability to control the maintenance of sterility throughout manufacturing and packaging plant. Medical products are assumed to be sterile in the package; however, this may not always be possible, or it may be a question of degree. In 2004, a large portion of the U.S. supply of influenza vaccine was declared unfit for use because of bacterial contamination. Recalls of other products with high levels of bacterial contamination have included such common over-the-counter products as milk of magnesia, baby lotion, and alcohol-free mouthwash.

The ingredients for any product, including the API, the excipients, and the packaging, are all potential sources of microbial contamination. The facility that produces a product is another potential source of microbial contamination. The people who work in the facility are also possible sources of contamination.

Drug sterility in a manufacturing operation requires diligent attention to multiple sources of contamination. The manufacturing equipment, particularly areas within the equipment that are hard to clean, can be ideal areas for bacteria to flourish. The raw materials used in the manufacture of a product, particularly natural products, or materials derived from natural products, are another source of potential contamination. The atmosphere is another potential contributor of airborne spores and microorganisms. Liquid and topical products are most susceptible to contamination, but all forms of pharmaceutical products can be contaminated if rigorous attention to sterility is not maintained. Manufacturing operations are required to maintain constant monitoring of the biolevel, the amount of microbial contamination of raw materials, and the process water used in their manufacture. The air within a manufacturing suite or manufacturing area must conform to a set limit regarding contamination. Class 10,000 or class 100 rooms are areas in the facility that are positively pressurized with highly filtered [HEPA (high efficiency particulate air) filter] air, where the amount and size of airborne particles are limited to a standard specification. Class 100 areas are defined as areas where each cubic foot of air must contain less than 100 particles, 0.5 μm in size or larger. Air filtration, cleanliness, and constant monitoring achieve this level of environmental control.

Products, particularly liquid products and topical solutions, may contain preservatives to ensure that sterility is maintained. Some of the common preservatives are alcohols, phenol, hexachlorophene, benzoic acids, and benzoic esters. These materials provide both antimicrobial and antifungal properties to the liquid. The Code of Federal Regulations (CFR) requires the use of preservatives in multidose vaccines [21 CFR 610.15(a)]. Preservatives containing mercury, a common example being thimerosal (an organomercurial), have been used since the 1930s, but are now being removed from vaccines. A theoretical potential for neurotoxicity heightened concern about mercury compounds, and this potential combined with the requirement for an increasing number of immunizations of children younger than six years has resulted in a joint effort by the FDA and the manufacturing companies to eliminate or reduce to trace amounts thimerosal used in vaccines for children.

The problem of bacterial contamination of packaged liquids has very real and swift consequences. A good example is the bacillus *Pseudomonas aeruginosa*; this microbe is extremely virulent and grows in ophthalmic solutions (4). *P. aeruginosa* can cause blindness in 24 to 48 hours after introduction into the eye. This organism typically is introduced into an ophthalmic solution from droppers or containers that have had contact with infected materials. Multiple dose packages used by doctors can and do come in contact with these dangerous organisms, and without preservative an infection can be passed to multiple patients from one infected package.

In all solutions, the most common bacillus of the *Staphylococcus* group is a real threat for contamination of the product during manufacturing. Testing of products for bacterial contamination is recommended by the FDA and other regulatory organizations and often is standard practice for oral, ophthalmic, and topical solutions and suspensions. Testing in oral solutions most often looks for *Escherichia Coli*; in topical preparations, testing is for *P. aeruginosa* and *Staphylococcus aureus* along with some measure of the total bacterial count. This testing includes periodic environmental testing of the areas used for preparation and packaging of a product and is an important way manufacturers identify, disinfect, and control potential contamination before it starts.

Sterilization is a step all products undergo to minimize or eliminate potential contamination. Terminal sterilization, the use of heat, is the most common method used by drug manufacturers to eliminate bacterial contamination, microbes, and spores in products. This process, which heats the product to a temperature above the survival point of any potential bacterial contamination, kills or deactivates any microbes and renders them harmless. Terminal sterilization cannot be used for all products. The heat used to kill the microbes in many cases is greater than the amount of heat energy needed to cause the product to break down or change chemical composition. When this happens, a number of other forms of energy are used to inactivate the organisms. These include radiation (both high and low energy, e.g., electron beam and ultraviolet light) and gas treatments of the product with ethylene oxide that also kill potentially dangerous organisms.

Another method of sterilization has slowly been gaining favor for liquids. This method or approach to sterilization is called aseptic filling and packaging or aseptic packaging. When this method is employed, the drug product and its package are sterilized separately, often using different methods of sterilization for the product and the package. The two sterile components are then brought together in a sterile or aseptic environment. These operations require extremely clean environments and the use of "clean rooms," that is, class 100 areas mentioned earlier. Training of personnel about how to properly gown (dress) and prepare for entry into an aseptic packaging area along with control of the clothing used in the area and positive pressure with filtered air inside the room are a few of the requirements for employing this technique and for this type of sterilization to be successful. Aseptic packaging relies on process control not only of equipment but also of people to achieve a repeatable result.

Another sterility concern is contamination of product by pyrogens. There are many potential pyrogens, but the most important one of concern in drug production and packaging is endotoxin, a residue from gram-negative bacteria. Pyrogens can be present in the raw drug product and the water, or they can come from contamination during processing of the product. A number of different methods and schemes are employed to eliminate pyrogens in the raw materials or during processing.

Glass and other components used for packaging parenteral products undergo depyrogenation before filling. Glass may pick up endotoxins from organisms in water used to clean the glass at the end of manufacture. Depyrogenation of glass and other components is normally done using a heating cycle of 250°C for 45 minutes or by heating and washing the materials in a strong alkali or oxidizing solution. The process of washing surfaces with detergent has been used in a limited number of cases. Following this preparation procedure for packaging materials, the sterile unfilled packages are carefully handled and protected from possible recontamination before filling.

Producing sterile or extremely clean packaging materials and handling of packaging components to maintain sterility is a key element in achieving sterility in products. Production of packaging components in conditions similar to those used to fill or package the drug product is a standard procedure for container manufacturers. This includes production of packaging in clean rooms or laminar flow enclosures that provide adequate air filtration and positive pressure around the manufacturing equipment and the people assembling packaging components. Constant monitoring of aseptic area environmental conditions and culture testing of product is also employed to identify, minimize, or eliminate package contamination. Testing of packaging components during the development phase of a product for bioburden, which is the amount of potential microbial contamination the package may present to the final sterilization process, is standard procedure. Sterilization challenges of product and package with known amounts of biological contamination is another technique used to prove that the sterilization cycle is capable of eliminating biological contaminants over a wide range of conditions. Periodic retesting of finished unfilled packages to verify that the bioburden remains within established limits is also required.

DRUG PHYSIOLOGY

Pharmaceutical physiology is the route the therapeutic product employs to gain access to the body. Pharmaceutical physiology can be very simple and straightforward when a product is applied topically to the skin, or it can be much more involved when the pharmaceutical product is administered parenterally or across mucous membranes. The primary package for many drugs is in many cases the vehicle used for drug administration. Prefilled syringes are used to quickly administer a drug by injection. Prefilled and premeasured packages of drugs, designed for use with large volume parenterals and pumps, can permit the introduction of multiple drugs in intravenous drips on well-regulated dispensing schedules. Aerosols, atomizers, and other packaging components may be the primary method of administering a drug through the nasal mucosa or the lungs. Patches or topical creams and ointments may be used to administer a product through the skin or vaginal mucosa. Each method of administration has its advantages and disadvantages and each needs to be understood in at least an elemental way to understand the many packaging requirements for pharmaceuticals.

Oral Administration of Drug Products—Gastrointestinal Methods

The first method everyone associates with for taking medicine is swallowing a tablet or liquid. This is drug administration through the gastrointestinal tract. A tablet, capsule, caplet, or liquid is swallowed and then absorbed through the stomach and intestines. All drugs must be dispersed to the molecular level for them to be absorbed in the gastrointestinal tract. True solutions of a drug product are absorbed the fastest, while suspensions, tablets, and capsules take somewhat longer for the body to break them down to the molecular level. Solid forms of the pharmaceutical product must first break up and dissolve before bioavailability, the systemic absorption of the API(s), can begin. A tablet generally uses a coating of some type to protect the product or to mask a very bad taste. The coatings, usually sugar based, eliminate the bad taste in a person's mouth and generate a slippery surface to make swallowing easier. They also provide a colored glossy surface that improves the appearance of the tablet. Tablets are made in a wide variety of shapes (Fig. 5).

The surface of the tablet may be printed after coating to improve identification of the particular drug product. The coating on the tablet must first dissolve or disintegrate before the tablet itself can begin to break down into a powder and finally dissolve in the fluids of the stomach. The rate of this breakdown can be fast or slow. A tablet that breaks down quickly will permit the buildup of the API in the bloodstream more quickly than a tablet that is slow to dissolve.

Another method of administering a solid product to the gastrointestinal tract is through the use of a soft gelatin capsule. The capsule contains powder or granules and dissolves quickly in the stomach, permitting fast action of the

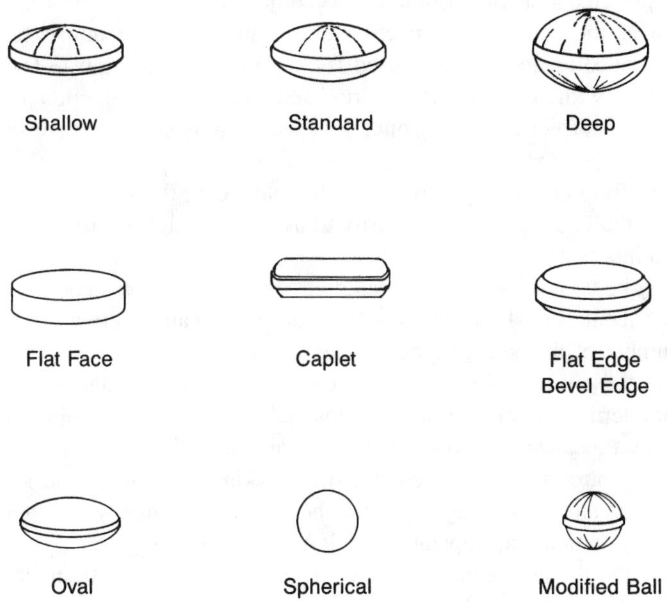

Figure 5 Chart of tablet shapes.

product. Most gel capsules break down quickly and behave like a liquid form of the product.

Some products require slow dissolution and delayed bioavailability to extend the time the product remains effective. It may also be required in products that are designed to be absorbed in the intestines, or if the medication causes irritation to the stomach. In cases where extended slow release of the drug product is needed, a layered or specially coated tablet or granules within a capsule may be part of the formulation to match bioavailability to the patient need.

Liquids, taken orally, display many of the same problems their solid counterparts exhibit. They may smell or taste bad; syrups or flavorings are used to mask these unpleasant characteristics. Color is usually added to a liquid product to improve its appearance and appeal, particularly in products for children.

Direct Injection of Drug Products

Injection is the fastest and most direct way to administer a drug. This makes the effects and benefits of the drug quickly apparent to the patient. Directly injecting a drug also eliminates any problems with absorption by the stomach or intestines. Injection eliminates the need for the patient to swallow, and can be administered even when the patient is unconscious or unwilling to voluntarily take a drug. This method also has a number of limitations; for example, suspensions cannot be injected directly into the bloodstream because they could block capillaries before dissolution. This method carries with it the greatest degree of risk from infection and the possible risk of introduction of other contaminants into the body. Disinfection at the site of the injection and maintenance of antiseptic conditions with the needles and syringes used for the injection is a must. The most serious problem with this method of drug delivery is the possibility of administration of an incorrect dose, and once given, it is almost impossible to correct.

Injections into fluid-containing portions of the body, e.g., the spinal cavity and the eye, require the highest degree of purity to avoid potential sensitivity of nerve tissue to irritants or toxic materials.

Injections may be used to produce a local effect. Injection of anesthetic by a dentist or a doctor subcutaneously into a muscle produces a localized effect for a limited time. A number of methods to localize the release of a drug in a specific tissue are used, and these methods are reviewed during drug and package development when a long term level of drug must be maintained systemically either in the entire body or in a localized area or organ. An example of this would be the implantation of birth control drugs by injection under the skin for sustained release. These methods keep the amount of drug applied to the local area relatively constant and are called electroporation and implantation administrations.

Electroporation is the use of electric shock at a specific location to create pores on the tissue surface (this can be skin or organ tissue as well as muscle and connective tissue) that permits direct injection of drug into affected area, as opposed to a systemic buildup of the drug through larger doses that slowly increase the concentration and the effect of the drug throughout the body.

The second administration may be implantation of a device that contains a measured amount of drug that is released slowly over time in the localized area. A good example of this method is an implant for the eye, which treats some forms of macular edema. Circulation to the retina and the eye in general is very limited, so the amount of drug required to systemically treat a condition is so high that it cannot be introduced.

Topical Administration of Drugs, Transdermal Methods

An increasingly popular method of introducing drugs, hormones, and chemicals to a patient is through the use of transdermal patches. This method of administration is found in nicotine patches for smoking cessation, birth control hormones, painkillers such as fentanyl, nitroglycerin for heart problems, and drugs to treat common ailments like seasickness or motion sickness.

The transdermal method of application uses dosage of the drug product contained in a patch. The patch is applied to the skin in an area that is consistent with the labeling of the product, typically in an area hidden by clothing, and the drug slowly diffuses through the outer layers of skin and reaches the bloodstream through the capillaries under the skin's surface (Fig. 6). This is a systemic administration of drug through the skin into the circulatory system. By controlling the rate of absorption and diffusion of the drug and its carrier through the outer layers of skin called the stratum corneum, the drug moves into the body and bloodstream as it passes through a convoluted path of intercellular channels. The drug then passes through the epidermis and into the blood circulating in the skin. The skin plays a very large role in the relative rate of diffusion, permeation, and absorption. Factors such as skin hydration, skin elasticity, oiliness, age, and temperature determine the diffusion rate.

Diffusion rates of transdermal drugs can be modified through the use of solvents like dimethyl sulfoxide and a number of its analogs as well as laurocapram. The use of surfactants is another way to increase permeability of the drug through the skin. Surfactants make the drug "wetter than water," and this characteristic helps carry it through the skin layers quickly and effectively.

Problems with transdermal patches include the patient's tolerance and sensitivity to the adhesive used on the patch, particularly if the adhesive contains any natural latex rubber. The extreme sensitivity of people allergic to these substances has resulted in all adhesives being made from synthetic materials, which are not as prone to reaction with the body's immune system.

Topical Administration

The topical administration of drugs directly to an outbreak of disease is the simplest and easiest method. Application of ointments to the skin for various

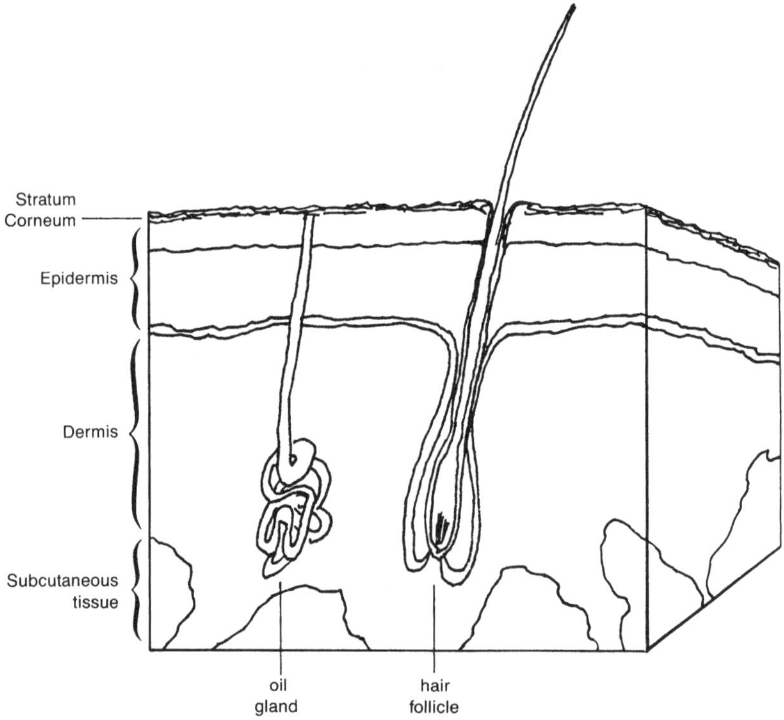

Figure 6 Cross section of the skin.

rashes, administration of drops to the eyes and nose to treat a number of conditions, application of hormones to the genitalia, and application of materials to the inside of mouth or on the lips to treat cold sores are examples of topical administration of drugs. The API is applied directly to the problem area and interacts with the disease or condition at the location of trouble. Packaging for products of this type may involve applicators and other mechanical delivery devices, such as droppers for the eyes or applicators to brush or wipe product onto the affected area.

Administration of Drugs through Mucus Membranes, Inhalation, and Nasal Administration

One of the most intriguing methods of drug administration is through the use of vapors or small particles created by aerosol action. The nose and respiratory system present a large area for potential introduction of a drug with minimal tissue thickness intervening between the tissue wall and the circulatory system. This allows the absorption and permeation of the drug into the body systemically at rates that approach those found with direct injection for bioavailability. Inhalable insulin has been approved to treat both type 1 and type 2 diabetes. For diabetics, this method is a tremendous improvement over direct injection of insulin. The new method eliminates the constant pain and irritation experienced during insulin injection. Drugs to treat asthma are the most visible forms of common products using the inhalation method for dosing. Other drugs such as antibiotics, heart medications, and anesthetics are also administered this way. Following the injection of anesthesia, a controlled mixture of gas and oxygen is administered to patients to maintain unconsciousness during a surgical procedure or operation.

The most critical aspect of drug administration to the lungs by inhalation is control of the particle size of the mist. The particles must be between 1 and 5 μm in order to be carried by the gas flow (breathing in) to sites in the lungs that permit interaction and administration through the respiratory system. The importance of particle size cannot be understated. If the particles are larger than 5 μm, they will not move to administration sites in the lungs and will probably be caught and moved to the digestive tract by ciliary action. Some percentage of the particles are going to agglomerate during administration, and this characteristic must be considered and measured to ensure that the proper dose of product will be delivered to the active sites within the lungs. The small particle size and the drug itself are formulated to mitigate ciliotoxicity, which would impair ciliary activity in the respiratory mucous membrane cilia. The drug must also have solubility or be treated to have solubility in the fluids, such as the natural surfactant in the lungs that permits the movement of the tissue without pain, for rapid incorporation into the bloodstream. A good example of differences in solubility is found in epinephrine, where the bitartrate form of the drug is more quickly absorbed than the hydrochloride or sulfate salts, the bitartrate form being the most soluble form of the compound.

Packaging examples for this type of drug administration would be inhalers for asthma patients. The liquid contained in the package is atomized or broken up into extremely small particles when dispensed. The patient dispenses the product and breathes simultaneously to move the drug into the lungs.

The other common method of administering drugs through inhalation is the use of nasal sprays. These can be sprayed or placed in the nose by dropper. The nose is lined with mucous membranes and this presents a path for the drug to enter the body. Drugs administered this way may be systemic or topical for treatment of a condition in the nose and sinus cavities. Antibiotics, antihistamines, steroids for allergy, and asthma drugs are examples of various classes of products that can be administered this way. Flu vaccine, a biologic, designed for administration through the nose, has also been approved for use. Solutions prepared for administration through the nose are formulated typically with a pH of 5.5 to 5.6 to match the natural pH of the nasal environment and have osmotic properties that maintain normal ciliary activity.

Rectal Administration of Drugs

The rectum, which refers to a relatively large portion of the lower intestine, is an excellent area for administering drugs. Veins in the lower portion of the intestine bypass the liver and permit a drug to be delivered directly into the vena cava. Drugs administered this way are absorbed in much the same way as drugs administered through other body cavities. Formulation of suppositories for this type of administration is quite precise and somewhat tricky. The body temperature of 98.6°F limits the range of melting or dissolution points of a suppository to 1° or 2° below the body temperature to ensure that the product melts or dissolves inside the lower intestine and is available for absorption thought the mucosa. The drug partitions itself between the inside of the rectum and rectal fluids and across the intestine lining. Bioavailability of the API is criticalalong with the time taken by the API to move across the membrane and into the bloodstream. Permeation across the mucosa can be modified or enhanced using bile salts that increase or decrease the rates of the products that move across the mucosa. Surfactants and other wetting agents are also effective in improving this permeation. Anti-inflammatory materials also aid in drug administration and minimize any reaction of the tissue to the drug. The most common drugs administered via this route are sedatives, tranquilizers, and analgesics.

Packaging for rectal products may require some unique features. Since these drugs are designed for dissolution in the body in a very narrow temperature range, the products must be protected from heat and must not be subjected to temperatures above their melting points. Temperature indicators on cases or individual packages can confirm that the product has been protected from heat. Clear blisters, which permit instant examination of the suppository, are another method for examination of the product before administration. The physical changes in appearance of a product that has melted and then resolidified are sometimes obvious and apparent on visual inspection.

DOSAGE FORMS OF DRUGS

Drug dosage forms can be solid, liquid, or gaseous. Each of these physical states may encompass a wide variety of manufacturing processes and require a wide variety of packaging. A general overview is provided to introduce the reader to the majority of forms used. A few examples of each physical state include true solutions and suspensions for liquids, powders or granules for solids, and ointments that may be emulsions, gels, or pastes. Although some anesthetics are true gases, the majority of gaseous dosage forms are really the dispersion of a solid or a liquid by an aerosol, where the gas carries the aerosol droplet to the intended patient's mucosal interface.

Dosage forms for each physical state of a drug product also vary on the basis of method or site of administration. Here again the physiology of the administration site must be matched to the physical characteristics of the drug product for repeatable control of the dosage of the drug's API.

Dosage forms and preparations do not always fit into nice, neat, and easily divided categories. Powders, a solid form, can be tablets, capsules, suspensions, and emulsions to name a few possible physical states. They can also be found in solid, liquid, and aerosol dosage forms. The wide range of possibilities provides the patient with multiple ways of ingesting a drug (systemic administration) or applying a drug to a local problem (topical application) with a large number of variants between these two extremes. As each dosage form is considered, the form and the product being delivered present unique packaging problems. The difficulty in packaging many drugs lies in the formulation of the product to interact with various parts of the body. The modifications to diffusion rates, absorption, adsorption, and permeation present problems in choosing primary package materials, manufacturing and packaging methods, along with protection of the product through the distribution chain.

As stated in the beginning of this introduction, products can be solid, liquid, or gas. This basic classification will be carried through the separation of the dosage forms but will surprise most with the rich variation and crossovers of different forms from one physical state to another.

The most common form of administration and the most common dosage form is a tablet or its variation such as a capsule, caplet, or powder. Liquids may be in the form of true solutions (Newtonian), suspensions, or emulsions. Gases will be identified, but the majority of the discussion for this physical form will focus on aerosols. Each of these forms will be discussed under the broad physical categories of solid, liquid, or gas. Packaging naturally flows from these distinctions and will be discussed as part of the overall discussion of the dosage form.

Solids

Solids comprise a wide variety of materials. They can be powders, granules, microencapsulated particles, and agglomerated powders (this refers to how a particle is increased in size). An even longer list of other solid and semisolid forms includes tablets, caplets, pellets, ointments, lozenges, creams, or some form of capsule containing the solid, emulsified, gelled, or suspended forms of the drug to name just a few. Suppositories and chewable tablets round off this list.

Powders

Powders are mixtures of very fine (small particle size), dry, chemicals and drugs (APIs) intended for oral or topical use. They are the starting materials for the manufacture of tablets and other dosage forms. One major advantage of a powder over a compressed dosage form is its extremely large surface area. This physical property permits the powder to dissolve, permeate, or disperse much more quickly than a compacted tablet. If a drug cannot be compressed into a form small enough to be easily swallowed, the powder form may be mixed with a beverage or food to make drug administration easy for the patient. Older patients and children may experience difficulty in trying to swallow tablets or capsules; the use of powder form is a viable alternative to eliminate the problem and to encourage use and compliance.

Powders were one of the first forms of drug preparation and date to very early humans. Roots, herbs, naturally occurring salts (e.g., sulfur compounds), and other naturally occurring substances were ground together to produce medicinal powders by a shaman or witchdoctor. By the middle ages, image of the person preparing these mixtures had changed to that of an alchemist with his or her mortar and pestle–grinding materials. This slow change from antiquity to the modern doctor is an example of how medicinal powders always have been recognized as effective treatments of problems and how they have been accepted and used. Oral powders or granules are still used today, and in some cultures in the Far East, are considered the standard preparations for dispensing a remedy. If you ever have the chance to visit a Chinese pharmacy, take the time and make it a point to stop and look around. Traditional pharmacies in Hong Kong, Taiwan, and Mainland China contain a wide variety of very unusual items that are ground together to produce a remedy, many times while you watch.

Powders that contain relatively nonpotent drugs may be provided in bulk containers. These products purchased over the counter include talcum, baby powders, tooth powders, laxatives, douche powders, dietary supplements, and dusting powders. Those that require a measured dose use the cap as the measuring device or a common household item like a teaspoon. Many products are not measured and are applied directly to the skin.

Powder manufacture and preparation. Producing a solid API normally starts with producing a powder form of the chemical. It includes the production in powder form of all the excipient materials used to dilute and adjust the properties of the product. The finished blend of powdered materials may be used in powder form but most often is compressed into tablets. Powders are prepared in a controlled manufacturing environment that bears little resemblance to the mental picture of an alchemist or pharmacist grinding a concoction with a mortar and pestle.

Powders may be the result of a precipitation reaction or other chemical reaction that produces a solid after extraction from a solvent, or they may be derived from naturally occurring products that are already solids. The natural products are ground to very fine consistency and then separated and purified to extract the API. Both methods produce coarse or caked solid product of aggregated ingredients that must be separated and reduced to particle size. Separation of materials and contaminants relies primarily on physical differences of the materials to effect the separation. Solubility, precipitation under controlled conditions, solvent extraction, and distillation are few examples of the physical processes used for material separation and as part of the isolation process to obtain the key active ingredient.

Separation of a powder into its constituent components is accomplished in a number of ways. The raw powder may be solubilized and a chemical reaction to precipitate either the wanted or unwanted ingredients then employed. The wanted ingredient can then be collected as the precipitate or extracted from the solvent. This works best when the contaminants remain in solution, while the valuable drug product precipitates at a specific temperature or is precipitated from the solution by changing it (chemical reaction) into an insoluble salt. A variation of this method can be simple melting of the raw product and selective separation at different temperatures to extract the active drug. This method is limited because the API may change because of heat into an unwanted substance or an unwanted form of the molecule.

Another variation may be the use of solvent extraction. Here, some or all of the wanted or unwanted materials may be soluble in an organic solvent, while the other materials in the mixture may be soluble in water. By dissolving the material in water/solvent combinations, the wanted and unwanted materials can be separated.

Another form of separation used with solubilized powders is spray drying. The solid is already free from unwanted contaminants or other ingredients but is in solution, much like salt or sugar dissolved in water. Spray drying entails taking a saturated solution of the product and atomizing or spraying the product into a vessel with controlled temperature and humidity to remove the solvent. The wanted product falls out of the solution much like salt or sugar would emerge when a saturated solution of either dries. If the process and the material have the right physical properties, a very uniform powder may be produced with this method, eliminating the mechanical grinding of coarse material and the problems associated with separation or classification of different-sized particles.

Once the coarse product is purified and isolated from contaminants, it must go through a process that reduces its particle size and renders the final product uniform in reactivity or bioavailability. Common methods used to reduce a drug's particle size include roller mills, ball mills, and hammer mills (Fig. 7). All these mechanical methods of physical breakdown create heat, and this heat may limit mechanical breakdown for some materials that are reactive or have low-melting points.

The resulting powder from one of these mechanical methods of crushing or pulverizing the raw product are then introduced into some type of cyclone or fluid bed that permits separation and classification of the powder by particle size.

These methods work best with powders that are somewhat crystalline and can be reduced to very small particles. Powders that are spray dried for separation are already in a fine form, which permits introduction into a cyclone or fluidized bed for separation and classification by particle size.

Fibrous powder substances do not yield to this type of processing and require other methods of separation and preparation. These substances contain connective material that does not break down easily in the normal methods used to prepare a powder.

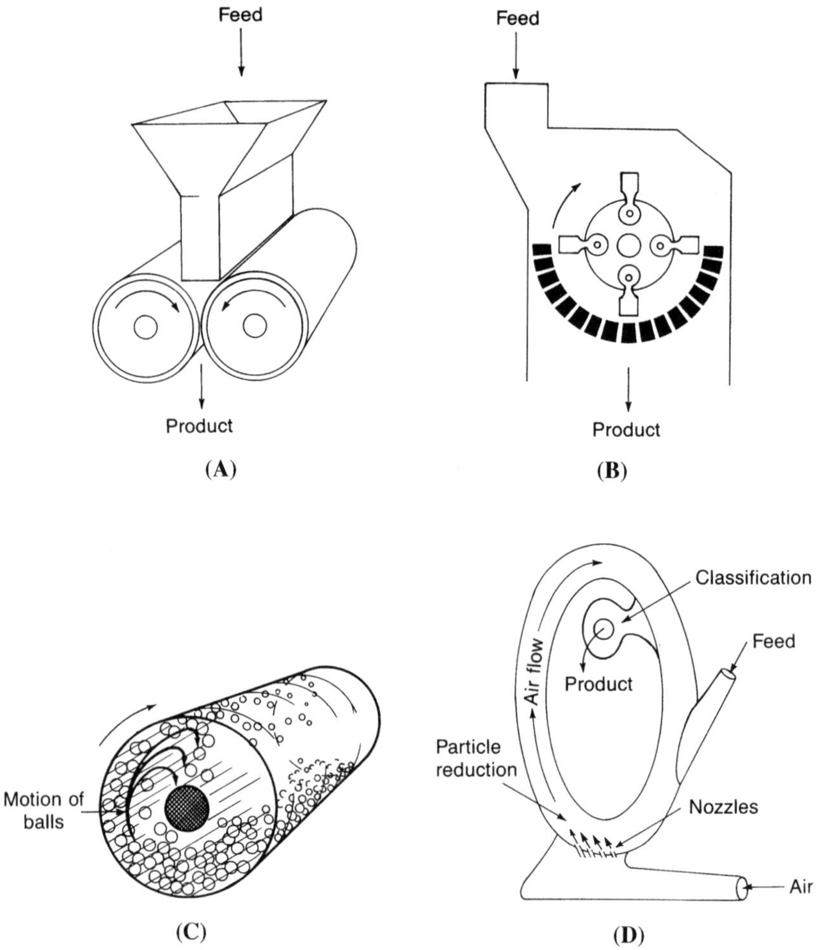

Figure 7 Particle size reduction methods for granulation.

Extremely hard powder substances present a different problem. They may be too hard and cause abrasion or mechanical breakdown of the grinding equipment.

Following the separation and classification of the various particle sizes of both the API and the excipient materials used to prepare a drug; the dry substances are mixed in exact proportions to begin the process of producing a finished drug. The API along with its powdered excipients is dry blended using a number of mechanical methods. Rotating drums with screws and baffles on the inside surface is one way to blend the powders. V blenders, mixers with two cones joined together to form a V, are other common pieces of equipments used to dry blend powders. The arms of this mixer are rotated across the axis of the V, permitting material to fall to either side of the V on each rotation. The random falling on each side of the V affects mixing. One surprise with this type of mixing is the possibility that the materials can reach a point where they begin to fall back into the unmixed constituents. Another method of blending powders is a fluidized bed with some type of mixing apparatus. Here the powder is pushed into a cloud through the use of air, and the materials mix in the dry dust cloud much the same way materials mix in a liquid solution. The blended powder is then ready for dispensing as is or ready to be tableted.

Lyophilized powders—sterile powders designed for reconstitution and injection. Some powders are supplied for parenteral administration (direct injection). These products are made of materials that are not stable in solution over extended periods of time. The extended period of time may be minutes or hours. Many drugs and biologics will not tolerate the preparation methods discussed above for powder preparation. These powder materials are converted from liquids to powders using a process called lyophilization, or freeze drying, as it is commonly known. This process removes the solvents used in drug preparation by evaporation at ambient or lower temperatures while under vacuum. This dries the product and prepares the resultant powder for packaging in the same step. Lyophilization in almost every case takes place in a vial, or the final packaging container. The vial, containing the dry powder, is sealed with an elastomer and ring in the same way a solution would be sealed in a vial. The doctor or health care professional reconstitutes the product with solvent, sterile water, or an isotonic solution immediately before injection. Products of this type have limited time of useful activity after reconstitution, and the labeling on the container is very specific in this regard. Some products may be reconstituted and used in multiple dose applications and some may even be stable for short periods (1 or 2 days) under refrigerated conditions. Other products may only retain useful activity for minutes or a few hours after being solubilized for injection. Heparin, an anticoagulant, is an example of a widely used lyophilized product.

Tablets

Tablets are probably the most common form of drug delivery. A tablet contains an API with or without diluents and other materials to make a solid dosage form that can be swallowed (Fig. 5). Tablets are also prepared for administration in the buccal (cheek) pouch (subcutaneously) and under the tongue (sublingually). Nitroglycerin tablets are a good example of a tablet placed under the tongue for administration. Effervescent tablets are another variation of tableted drugs.

There are two types of tablets, compressed and molded, which identify the manufacturing process used for their preparation.

Compressed tablets are made using three different methods of preparation: wet granulation, dry granulation, and direct compression. Each of these methods first prepares the materials used in the tablet into a form that is uniform and

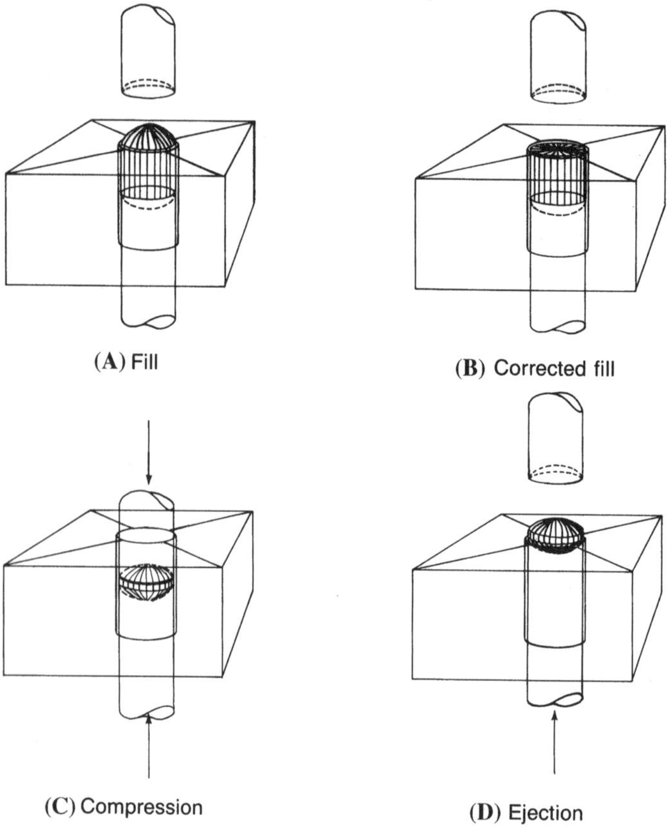

Figure 8 Diagram of tablet making by compression.

flows freely in its dry state. Both wet and dry granulations improve powder flow and compressibility into tablets. The majority of tablets are made by compression. A hardened steel punch and die compress the granulated powder along with excipients, diluents, lubricants, and other materials into a hard solid dosage form (Fig. 8). Tablets can be in the shape of capsules and carry the name caplets. Extremely large tablets called boluses are used in veterinary applications.

Tablets come in a wide variety of shapes and sizes. A chart of various tablet shapes displaying the wide variety and different types of tablets commonly found is shown in Figure 5.

Tablets also incorporate color materials or derive color from the API or other ingredients used in their manufacture. They can come in a wide variety of colors and may be printed on one or both sides. The colors and the markings are used for identification. Tablets may also be impressed with a name, a number, or another designation that permits easy identification.

Compressed tablets. Majority of tablets made use of a compression process (Fig. 8). The API is mixed with a diluent, lubricant, binder, and, possibly, disintegrating agent and compressed under extremely high pressure to form the tablet. A diluent is added when the amount of active ingredient is very small or if the active ingredient is difficult to compress into tablet form. Diluents used in compressed tablets include well-known materials such as lactose, starch, microcrystalline cellulose, and dibasic calcium phosphate along with a variety of other lesser-known materials. In tablets containing a small amount of API, the tableting process and the performance of the tablet to administer the drug are determined by a diluent, sometimes called filler. Hydrophobic APIs present a bioavailability problem in tablet form. The low water solubility creates a problem that is overcome through the use of water-soluble solid diluents. These break down in the stomach or the intestines, permitting the API to react with acid or base conditions in the lining of either organ and then be absorbed.

Binders, the materials that hold tablets together, are other ingredients in their formulation. These materials include acacia, gelatin, sucrose, methylcellulose, carboxymethylcellulose, hydrolyzed starch pastes, and povidone. These materials add to the cohesiveness found in the diluent. Binders may be added dry, but they are much more effective when added in solution before granulation. When tablets are produced by direct compression of the ingredients without a granulation step, microcrystalline cellulose is the binder most often used.

Lubricants are another extremely important class of materials used in the formulation of a tablet. Lubricants make it possible for a tablet to be ejected from a tableting press and reduce the friction of the dry powder as it flows into the dies and undergoes compression. The material also prevents the adherence of the tablet to the dies and punches. Typical materials used as lubricants include talc, stearic acid, vegetable oils, and metallic stearates. Most lubricants are hydrophobic and may slow the rate of a tablet's dissolution and disintegration. Because of this unwanted property, lubricants are used sparingly, and at times, the minimum amount used may create manufacturing problems. The minimum lubricant level creates problems of excessive wear of die and punch sets, or of broken tablets during tablet ejection from the die and punch. Two liquid materials, lauryl sulfate and polyethylene glycol, which are more miscible, have been tried to overcome this problem. Neither of these materials possesses the lubricating properties found in more hydrophobic ingredients.

Glidants are other types of materials used in tablet formulation. These materials improve powder flow or fluidity and make the movement of powder from blending through the tableting process much easier. Silicas, primarily colloidal pyrogenic silicas, are most often used to provide or improve this physical characteristic of a formulated or blended powder mixture.

Disintegrating agents are also used in tablet formulation. These agents speed the breakup of a tablet, making the API available quickly after ingestion. Modified starches and cellulose are the two most common disintegrating agents used in tablets. Other materials used include microcrystalline cellulose (this is used for other tablet properties noted above), alginic acid, and cross-linked povidone. The effectiveness of any of these materials relies on a number of properties, including its concentration, how it is added to the tablet powder, and the degree of compression or compaction of the tablet.

A lesser-used method for tablet disintegration is through the use of effervescent materials. The effect of placing these tablets in a liquid activates a chemical reaction and speeds the breakup of the tablet.

Color may be added to a tablet formulation for identification and for visual appeal or aesthetic value. The FDA has approved a number of FD&C and D&C dyes for this use. These dyes are typically absorbed into insoluble aluminum hydroxide and are called lakes when in this form.

Finally, sweeteners may be added to a formulation to counteract an unpleasant taste. This problem is normally overcome by coating the tablet.

The manufacturing process for tablets follows three paths: wet granulation, dry granulation, and direct compression. There are number of steps in each of these operations; these steps comprise primarily of mixing of the API and other ingredients followed by compression into a tablet. Direct compression is only used when the powder form of the drug, obtained by spray drying or other means, has good physical attributes that permit direct compression into a tablet (Fig. 8).

The first manufacturing operation is mixing the ingredients in a step generally referred to as wet granulation or dry granulation. These steps are required to improve the flow of the powder mixture and to improve the compressibility of the powder for making tablets.

The first method, wet granulation, starts with the mixing of the drug with all its additives in large blenders, various types of stirred mixers, and fluidized beds. A water solution of the tablet binder is added to the completed mixture and produces a wet agglomerate of the materials suitable for processing. This wet mass is then sieved or screened to improve consistency and dried with warm air in a fluidized bed or an oven before breaking the mixture into granules for tablet processing. The dry mixture is again screened before the tableting operation. For large volume products, a screw extruder or continuous mixer can be used before the drying step of the granulation.

Dry granulation is used when the API and the other ingredients in the mixture have inherent binding properties. This process is used primarily with drugs that are sensitive to moisture. The mixture of powders is compressed into large and usually poorly formed solids. The resulting solid mixture is then milled or broken down into granules and screened to the desired particle size for tableting (Fig. 7). Dry granulation eliminates both heat and moisture during processing. Dry granulations are sometimes produced by passing the ingredients through high-pressure rollers to produce thin cakes of product that can then be milled and screened to the correct granule size. In both operations, the milled and screened powder will be under tight moisture control to minimize the buildup of static electricity and to maximize the tableting properties.

Excipients are available that eliminate the need for granulation. Directly compressible excipients include sucrose, dextrose, and cellulose specially prepared to enhance their properties of fluidity and compressibility. The use of these excipients permits direct compression of materials, eliminating all the problems of wet and dry granulation. The major problem of using this method is the sensitivity of the excipients and the API to minor physical changes such as humidity, age, heat, and other factors that can alter their fluidity and ability to be compressed. This sensitivity may cause major problems in the manufacture of tablets.

Regardless of preparation method, the granulation then goes through the same basic process of producing a finished tablet (Fig. 8). A cavity usually cylindrical in shape and open to accept a punch at both ends provides the forming chamber for a tablet. The lower punch is inserted into the cylinder and the cylinder is filled with a measured amount of the granulation. Any excess granulation (powder) is scraped off in this filling step of the die. The second punch is then driven into the cylinder, and the two punches compact the granulation under high pressure to achieve the desired compression into a physically robust tablet. This is important because the tablets exit the machine and are collected in bulk containers for additional processing or packaging. The punches used to produce the tablet may have raised areas that produce scoring lines or identifying marks on the compressed tablet. Following compression, one of the punches is withdrawn, and the other punch moves through the cylinder to eject the tablet from the die. Multilayered tablets can be produced by adding multiple granulations to the die cavity with multiple compaction steps between each addition followed by a final compression or compaction. This can also be done by placing a partially compressed tablet into a second tableting machine, adding the additional ingredients and putting the new material and tablet through a second compression step.

Molded tablets. Molded tablets are made in an entirely different way from compressed tablets (Fig. 9). In this manufacturing method, the active ingredients and diluents are mixed with powders or solutions of lactose and/or powdered sucrose. The powders are usually moistened with a water/alcohol mixture where, in most cases, the amount of alcohol is quite high. The amount of alcohol is determined by the solubility of the ingredients and the desired hardness of the finished tablet. This mixture is then placed into a mold and allowed to dry or is force dried. Drying may take place in the mold or the solution may be more gel-like and retain the basic molded shape; in this case, the tablet may be allowed to dry outside the mold. Molded tablets are quite friable, requiring care in packaging and dispensing to prevent their breakup.

Tablet coating. Tablets are coated for a number of reasons. The coatings can protect the ingredients from moisture, oxygen, or light. They mask undesirable odors and tastes, improve the ability of a person to swallow the tablet, and may improve the appearance of the tablet.

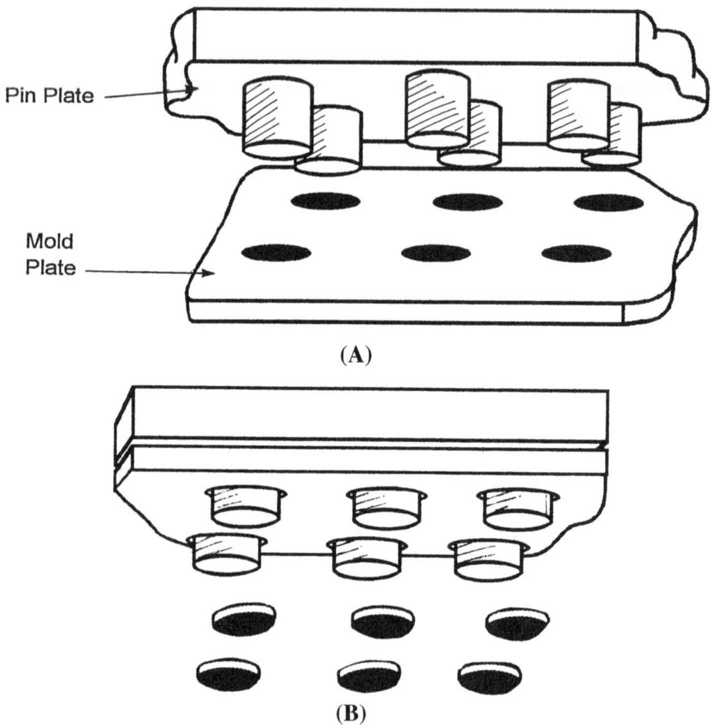

Figure 9 Molding tablets.

Coatings for tablets are made from sugar solutions that typically contain starch, calcium carbonate, talc, and titanium dioxide suspended in a gelatin or acacia. The coating will contain any colorant used in the process. In some cases, water-protective coatings made from shellac or cellulose acetate phthalate are applied using nonaqueous solvents before the sugarcoating.

Tablets are placed in a revolving vessel. The vessel may be called drum or pan. Pan coaters are vessels shaped like a large round ball with a wide opening tilted at a high angle (Fig. 10). The vessel is rotated on a very high vertical axis.

Drum coaters use the same tumbling motion found in pan coaters but are more horizontal in setup (Fig. 11) Drum coaters or pan coaters equipped with spray nozzles are used for film coating of tablets. This is when the coating is sprayed directly on the tablets exposed by tumbling, and multiple spray applications slowly build up the tablet coating.

Coating materials such as shellac, which will not solubilize the tablets, are ladled or poured directly onto the tablets (typically in a pan coater) and then dispersed by the mixing and tumbling action of the pan or drum. This is the first coating step.

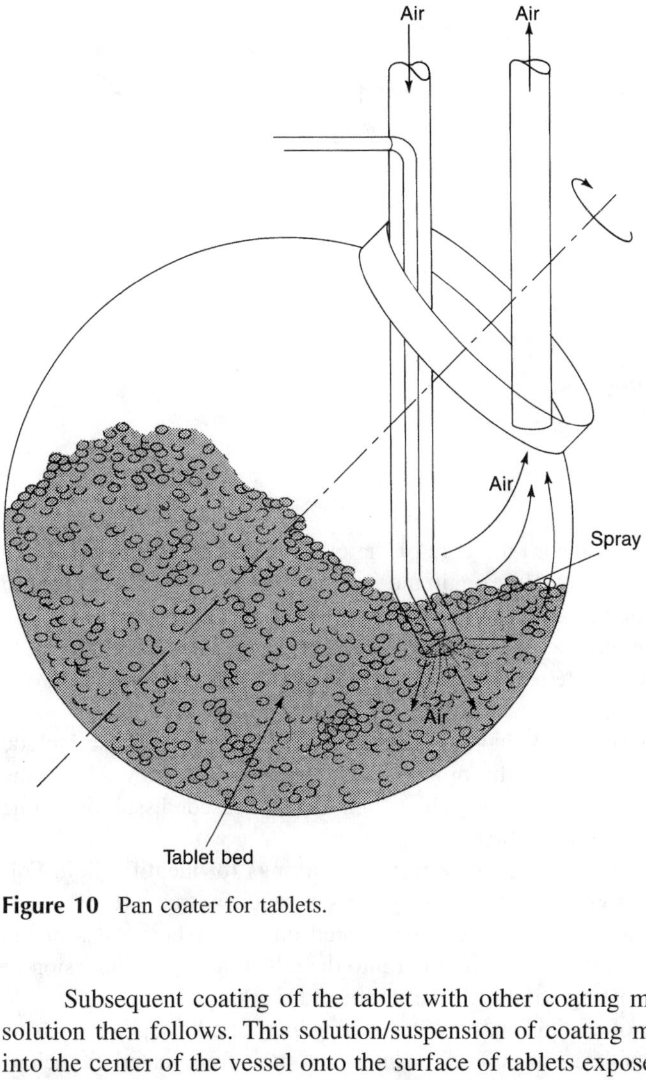

Figure 10 Pan coater for tablets.

Subsequent coating of the tablet with other coating materials or a sugar solution then follows. This solution/suspension of coating materials is sprayed into the center of the vessel onto the surface of tablets exposed by tumbling. As the tablets tumble, they are slowly coated with multiple injections (sprays) of the coating solution that adheres directly to or transfers from coated tablets to uncoated tablets. Drying air is constantly moving across the tablets through either the side openings in the drum or the pan. Air may also be introduced through an opening in the turning vessel and exhausted through perforations in the pan or drum. The multiple repetitions of solution in the spraying and drying process permit the coating to slowly build up on the outer surface of all tablets in the drum or pan.

In a fluidized bed-coating operation, the air movement is designed to agitate the tablets with drying air; keeping them somewhat suspended keeps the

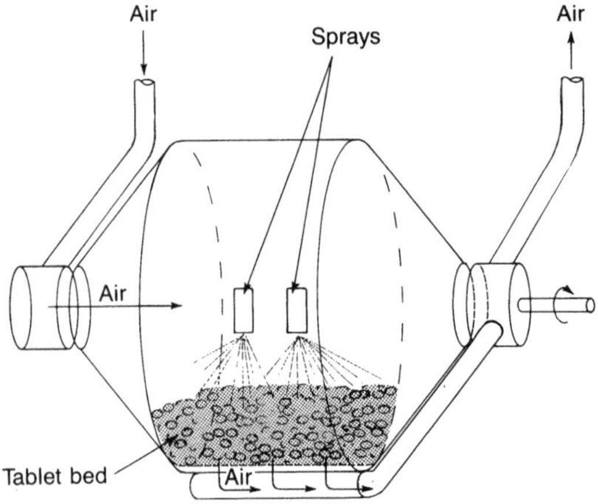

Figure 11 Drum coater for tablets.

tablets moving upward toward the core of the cylinder and then outward to the walls of the vessel (Fig. 12). Here again the coating solution is slowly sprayed in multiple cycles to slowly grow into a uniform coating.

Coated tablets may be polished and further coated with wax or shellac to improve appearance. Dilute solutions of wax in solvent or shellac in solvent are used and do not disturb the sugarcoating.

Sugarcoatings on tablets have a number of disadvantages. These include the length of time needed for the process, the increased bulk they add to the tablet, the need to waterproof the tablet, and the increased dissolution time caused by waterproofing the tablet.

Sugarcoated tablets may receive printed markings for identification. Following the coating steps the tablets may pass through a true offset printer designed to mark the tablet. They can be printed on one or both sides of the tablet. This is done if markings are not put into the tablet during compression or if the thickness of coating would fill the impressed markings.

A second type of coating, called a film coating, may also be applied to tablets as an alternative to sugarcoatings. Film coatings are made from materials such as hydroxypropyl methylcellulose, methylcellulose, or hydroxypropylcellulose mixed with propylene glycols and cellulose acetate phthalate in both aqueous and nonaqueous solvents. This material is sprayed on the surface of the tablet and forms a thin protective film in a much shorter time than the pan and drum methods described for sugarcoatings. The thin film means the tablet maintains its original shape and any markings or grooves pressed into the tablet in compression remain visible.

Delayed release tablets. Tablets may be coated with materials usually called enteric coatings to stop tablet dissolution or release in the stomach. Drugs that

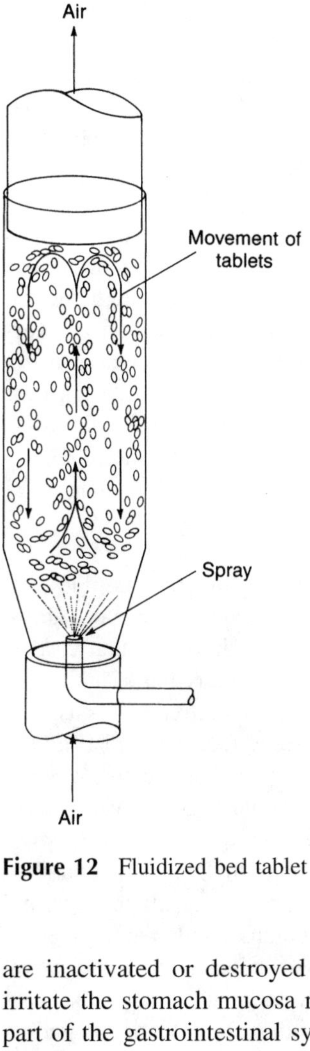

Figure 12 Fluidized bed tablet coating.

are inactivated or destroyed by stomach acids or drugs that may damage or irritate the stomach mucosa receive this treatment to protect them in the upper part of the gastrointestinal system and permit them to disintegrate in the intestines. A delayed release of this type solves the stomach administration problem but also delays the time from ingestion to activation. Tablets like this may take an hour or more before bioavailability of the API to the body begins.

Extended release tablets. Extended release tablets are designed to dissolve or make the API available to the patient in a controlled way over an extended period of time. Many terms are used to describe this type of product, including "prolonged action" or "sustained release." Here the granules contained in a tablet have been coated in the powder preparation step to dissolve at different intervals and with exposure to different gastrointestinal chemicals. Initially, a measured dose of the drug is made available to the body by treating some of the API to dissolve in the stomach and then this level or concentration of drug is maintained over an extended period of time by additional API prepared to dissolve in the intestines. As the various coatings slowly dissolve, they replenish the API that is being metabolized. This type of action is very beneficial in products for colds, especially when one wants relief from symptoms over an extended period of time, for example, to sleep overnight.

Chewable tablets. Chewable tablets are produced by compression. The major characteristic of these tablets is that they are designed to be chewed and tasted. Antacids, children's vitamins, and some antibiotics are manufactured for this type of administration. Chewable vitamins have been promoted and are very popular with children and may come in shapes resembling cartoon characters or other easily recognized toys. The tablet is designed to provide a pleasant taste and odor in the mouth, with minimal residue that can be easily swallowed. Manitol, sorbitol, and sucrose are the standard binders or fillers used in formulating these tablets along with various colors and flavors that enhance the appearance and taste of the tablet.

Lozenges. Lozenges are another type of tablet designed for slow dissolution or slow disintegration in the oral cavity (mouth). They contain one or more active ingredients formulated in a sweetened and colored base similar to that used in chewable tablets. They are also prepared using a gelatin base that is molded in the shape of the lozenge. Molded lozenges are sometimes referred to as pastilles, and compressed lozenges may be called troches. Normally, a lozenge treats a local problem such as an infection in the mouth or throat, but in some cases, contains active ingredients that are absorbed and provide a systemic effect as well.

Capsules. Capsules are a solid dosage form similar to tablets in that they present the patient with single or multiple units that contain all the pharmaceutical products necessary for therapy. They differ significantly from tablets in their manufacture and assembly. Capsules begin with powders or granules that are loaded into a hard or soft shell package made from gelatin that is soluble in the body. The capsule protects the product from a number of potential degrading exposures and protects the patient from bad taste, odor, or possible tissue irritation in some parts of the gastric system. Capsules come in a wide range of sizes (Fig. 13), starting with the smallest size listed or called size or number 4 (four)

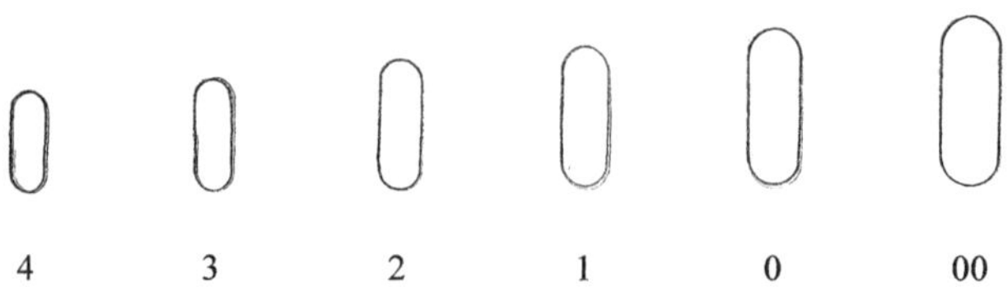

Figure 13 Standard capsule size comparison.

and increasing to the largest size listed as size 000 (triple zero). A double zero (size 00) is the largest size acceptable to human patients.

Hard capsules are made as two separate halves in a unique manufacturing process. The empty capsules are transported to the final filling location and are filled using a number of different methods to separate and fill and to reunite two separate halves of the capsule after filling. Soft capsules are made, filled, and sealed on the same equipment and usually contain a liquid or semiliquid interior surrounded by a hardened gelatin. Soft capsules or "soft gel" (the name has a number of phonetic variations) have become very popular because they deliver drug products to the body faster than sugarcoated tablets. Many consumers report that they are easier to swallow than hard capsules or tablets.

Hard capsules. Hard shell capsules are made from gelatins with very high gel strength (Fig. 14). The gel, after drying, is hard to the touch while still being somewhat pliable if squeezed. The gel is not brittle and can resist some physical abuse. The hard gelatin used to make these capsules is derived from pork skin, bone, or, in some cases, starch. The most common gelatins are manufactured

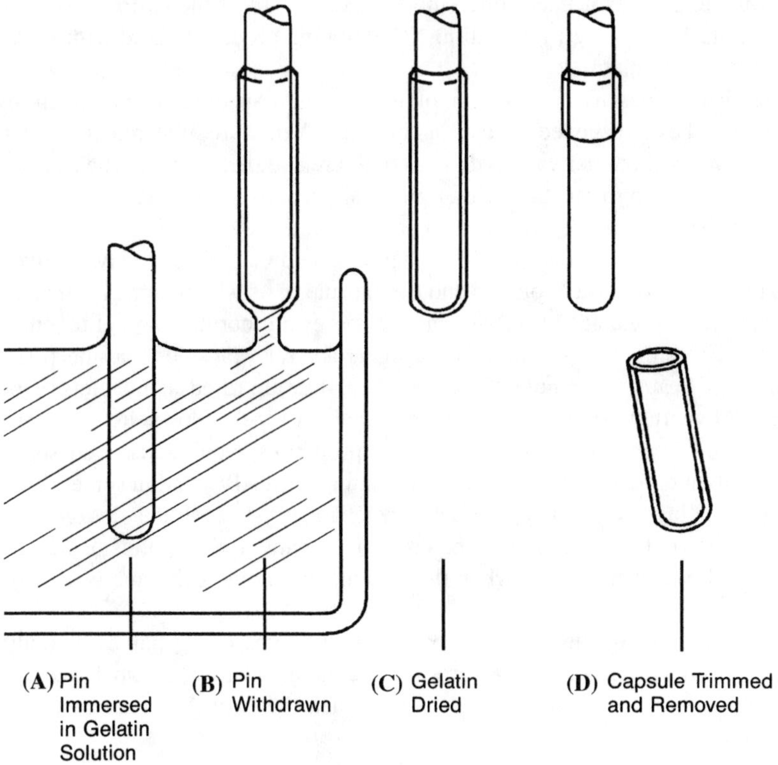

Figure 14 Making of hard gelatin capsules.

from acid processing of pork skins. Bone gelatins are derived from alkaline processing. Both bone- and pork-derived gelatins are blended to attain the desired clarity and toughness, properties needed for manufacturing the capsule shell, and for withstanding the mechanical forces generated in filling and sealing the capsules. Starch materials may also be used to produce a suitable hard capsule.

Hard shell capsules may contain colorants (approved FDA dyes and lakes), titanium dioxide, iron oxide, or some opaquing agents. Sucrose may be added as a hardening agent, and preservatives may be added to stabilize the capsule or to protect the intended product to be filled in the capsule. Gelatin capsules in the finished state contain 10% to 15% water. This level of moisture is obtained by precise control of the environmental conditions and the temperatures used in the gelatin bath and the temperature of the pin inserted in the gelatin. Variations in temperature, concentration of the gel, and humidity result in varying thicknesses of the gel capsule. The gel capsule body or cap is stripped or removed from the dipping pin and trimmed to size. The two haves of a capsule are produced in two separate operations. After trimming, the two halves are mated for storage and shipment. Control of manufacturing conditions is crucial for maintaining the dimensional tolerances needed for a smooth and tight fit of the two halves after initial manufacture through final filling. The dipping process is used with pork- and bone-derived gelatins.

Starch capsules use injection molding for their method of manufacture. Two separate dies are needed for caps and bodies. A mixture of starch and water is forced into a mold under extremely high pressure and partially set. The capsule then continues drying until the correct physical properties are obtained.

All capsules must be protected in storage until they are filled. Too much moisture can make the gel capsule soft or pliable and may cause the two halves to stick together. Too little moisture and the capsule parts will become brittle and be susceptible to breakage. The two capsule halves are normally fitted together for storage and shipment by the capsule manufacturer. Capsules may be filled with powders, beads, or granules. In some cases, they are filled with tablets (e.g., diltiazem a hypertensive and heart calcium channel blocker) or pellets that are coated to achieve enteric or extended release properties. Nonpareils, inert sugar beads, are often used as a starting point for coating with APIs, and in some cases, additional coatings to modify the delivery characteristics of the dosage and achieve the desired presentation of the drug to the patient. Semisolids or liquids may be filled in a capsule, and when this is done, a sealing technique is used to prevent leakage.

Hard gelatin capsules usually consist of two telescoping halves or body pieces. Generally, indentations or grooves are molded into the two halves to provide a positive sealing or locking feature when the two halves are mechanically mated after filling.

The two halves may be joined and fused using a variety of thermal techniques to seal the capsule, or they may be sealed ultrasonically (Fig. 15).

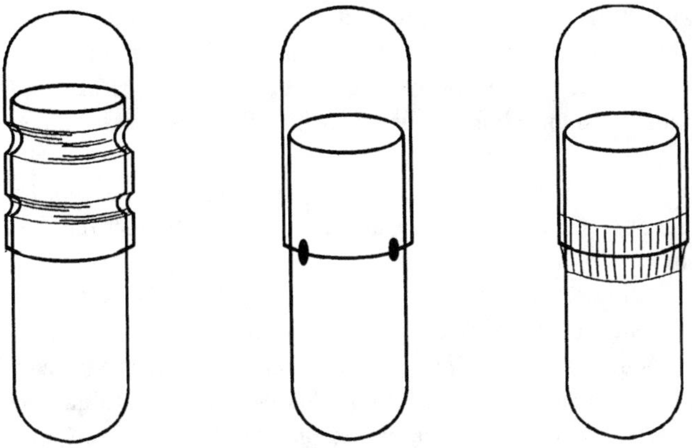

Figure 15 Examples of sealed capsules with different methods of sealing.

Banding is another method used to seal capsules. A layer of gelatin (it can be more than one) is applied over the seam between the cap and body. Liquid fusion is another way capsules may be sealed. In this method, filled capsules are wetted with a water/alcohol mixture that penetrates the area of the seam; when the capsule is dried, the two halves are fused together. Starch capsules are most often fusion sealed with this method. Sealing of capsules prevents tampering with the contents and prevents separation during shipping and handling of the capsules.

Filling of hard gelatin capsules is accomplished by first separating the body and cap of the capsule. The two halves of the capsule body are supplied in an assembled orientation and are separately filled and reassembled in the process. The drug with its excipients or diluents may be filled as a powder but more modern high-speed capsule fillers form a small plug by compression of the material being filled and insert it into the capsule halves. The powders used in filling capsules are generally formulated with the same diluents, glidants, lubricants, and other materials that modify the powder or granules producing the same attributes needed for wet and dry granulations for tablets. The powders or granules must exhibit many of the same physical properties needed for control through transport, filling, and sealing in the manufacturing equipment. In some cases, when the density of the formulation is very low, an additional granulation step may be required to increase the density of the granules or powder.

Hard starch capsule shells are supplied as separate halves. The two separate halves are fed into separate sections of the filling machine, oriented during filling, and then joined after being filled with the powder or granule.

Compaction or compression of the drug material inside the capsule is critical to maintaining proper dosing and dispensing requirements. The amount of compression in forming the plug may cause problems with drug delivery, and if the ingredients are hydrophobic, a wetting agent or some other ingredient that enhances solubility or enhances the breakup of the granule or powder and promotes the dissolution of the hydrophobic ingredient into the body may be necessary.

Soft shell capsules. (Soft Gels® or Gel Caps® are examples of multiple trademarked designations for the capsules.)

Soft shell capsules have emerged as a preferred form of administration for many products. This dosage form has the advantage of presenting the drug in liquid form for faster uptake and effect. The liquid centers are easier to produce and make uniform when compared with the tumbling action required to mix dry powders. Liquids are easier to measure or meter into the capsule than powders or granules. The liquids present a drug that is already in solution or suspension and thus much more available for uptake by the body. The uptake is enhanced because the drug is already dissolved or suspended in a hydrophilic liquid. Soft shell capsules have become a very popular form of dosage and have supplanted caplets and tablets in popularity for a number of over-the-counter products. Part of this preference is derived from improved speed and absorption of the product contained in a soft shell capsule; liquids are typically absorbed faster than solids. Although most soft shell capsules contain a liquid, they can also be filled with a paste, powder, or even a tablet.

Soft shell capsules are made from gelatin, the same material used for hard shell capsules, but with additional polyol plasticizers such as sorbitol or glycerin (Fig. 16). The shell of the capsule is much thicker than the hard shell capsule, and the amount of softness or hardness of the shell is determined by the ratio of plasticizers to gelatin. The shell of the capsule may contain dyes, titanium dioxide, opacifiers, pigments, and preservatives. The soft shell capsule can be printed or impressed with identification for the product, manufacturer, or product strength. The soft gel capsule shell normally contains between 6% and 13% water in its composition. Flavors are sometimes added to a soft shell capsule, especially if the dosage is designed to be chewed and swallowed.

Soft shell capsules are filled with liquids that do not attack the gelatin shell and prevent interaction of the drug with the capsule shell. The original choice of liquid for dissolution or suspension of active ingredients in a soft shell capsule was a vegetable oil, usually in a partially polymerized form of the oil called oligomer, a low level of molecular-weight polymer, permitting it to remain a liquid. These oils have been slowly replaced by low-molecular-weight polyethylene glycols that do not exhibit as many bioavailability problems as the natural oligomers. The materials are miscible in water but are nonaqueous, so they do not interact with the gelatin shell.

Soft shell capsules are produced on equipment (1) that combines the gelatin shell and the liquid interior in a continuous process (Fig. 16). A rotary die process is most often used to produce this type of capsule. The soft gelatin,

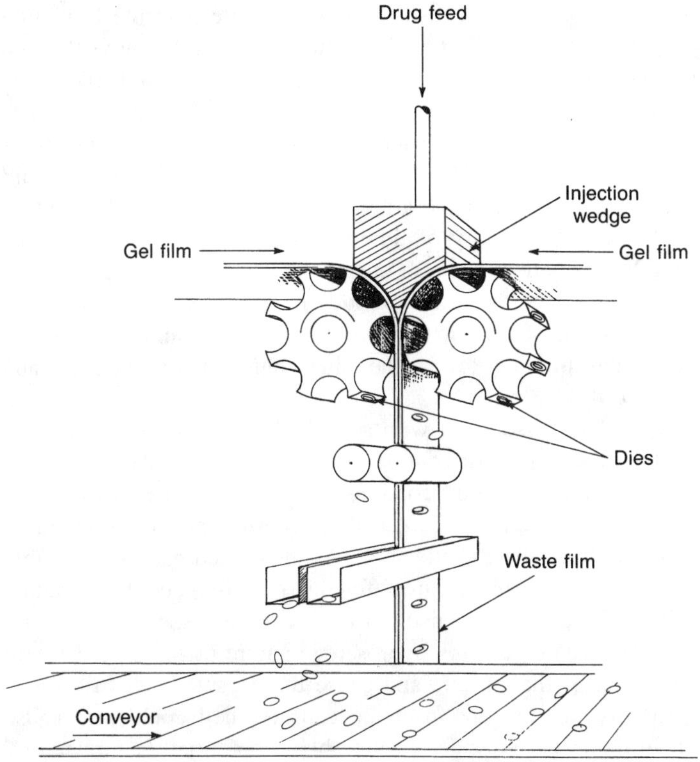

Figure 16 Diagram of a rotary die process for forming soft shell capsules.

usually in the form of semisolid sheets, is brought together as two separate pieces. These pieces form the two halves of the capsule shell. As the two sheets of material come together, liquid is dispensed between them, and the sandwich immediately goes through two separate dies that form the capsules and seal the halves of the capsule together. The gelatin material that is not part of the capsule is recycled in the process after the completed capsules are cut or pushed from the gelatin sheet.

Other processes that may be used for forming soft shell capsules include reciprocating dies or plates. The gelatin film may be produced in a separate process and stored; however, normal procedure is to produce the gelatin and cast a gelatin film in the process immediately before the capsule-forming step. The gelatin formation into a sheet is much like a plastic extrusion process. The semisolid liquid flows through a metering die set to a specific thickness and then into a drum that cools and solidifies the gelatin material. Control of temperature, moisture, and the types of plasticizers used all contribute to the strength and properties of the gelatin sheets fed into the capsule process. The gelatin sheet may be lubricated with mineral oil to move easily over the various machine surfaces. When this is done, the oil must be removed after capsule formation with a wash by organic solvent. Following formation, the capsules move through a drying process that sets the final amount of moisture in the gelatin between the 6% and 13% levels mentioned earlier.

Gelatin is an excellent oxygen barrier and when combined with pigments or dyes can also provide protection from light. Soft shell capsules make liquid medications easily portable and provide a more accurate controlled dosage that may not be possible with a quantity of liquid measured by a separate device when administered by the patient. Gelatin breaks down quickly in the stomach, and the liquid presentation of the drug provides fast therapeutic action when compared with the amount of time it takes to dissolve a sugarcoating from a tablet and the additional time it takes for the tablet ingredients to break up and dissolve in the gastric fluids.

The typical problems associated with soft shell capsules include embrittlement if the shell of the capsule dries out or actual dissolution of the shell if it is stored in high heat and high-humidity conditions for extended periods of time. The drug product carried in the liquid center of the capsule must be soluble in a material that does not attack the capsule shell. As mentioned earlier, vegetable oils and now polyethylene glycols are used for this purpose. The drug product itself may be hydroscopic and draw moisture from the capsule shell even when it is contained in a nonpolar liquid. This hydroscopic nature of the drug can pull moisture from the shell of the capsule and cause the capsule shell to become brittle. The availability of moisture from the capsule shell could also cause changes to occur in the drug molecule. This problem is normally addressed by microencapsulation or by using a form of the drug that is not soluble in water.

Non-oral soft shell capsules. Soft shell capsules are a unique dosage form when used in nonoral application. Pediatric and geriatric patients that have trouble swallowing or cannot swallow a capsule may receive a drug rectally using the soft shell form. The gelatin shell dissolves and the drug is absorbed in the same way as a suppository. This method of administration is useful in patients with gastrointestinal problems that would be compounded by the introduction of the drug by that delivery route.

Implants or pellets. Another form of solid drug product is an implant. These pellets or small cylinder-shaped rods of the drug are designed for subcutaneous implantation surgically under the skin. Their advantage is that a measured release of the drug is provided over an extended period of time. The pellets or cylinders are sterile forms of the drug that may or may not contain excipients and typically are highly purified forms of the drug. The implant dissolves slowly over time, providing a constant systemic administration of the drug to the patient. Drugs prepared in this form are usually supplied as a kit that includes a medical device that is designed to implant the drug under the skin. A common drug administered in this form is female hormone therapy.

Eye implants for administration of drugs within the eyeball are another form of this type of drug delivery. They combine the compressed drug (tableted) with a holder and a suture to permit attachment of the implant within the eye. The advantage of this system for ophthalmic drugs is the topical administration of a drug to a part of the body with low levels of circulation where systemic introduction of the substance to the body would create other complications or problems.

Systems like this capitalize on the ability of a medical device and dosage form to work in combination to overcome physiological problems and present a drug product to the affected area over an extended period of time.

Suppositories. Suppositories are a solid drug dosage form that dissolves, melts, or softens inside the body to release a drug product. Suppositories may be designed for the vaginal, urethral, or rectal orifices of the body. Suppositories can deliver a drug for both topical and systemic action in the body. Most suppositories are made with cocoa butter; however, gelatin, hydrogenated vegetable oils, and fatty acid esters of polyethylene glycol may also be used. The choice of the base ingredient for the suppository affects the delivery of the drug. All base materials are designed to melt or dissolve quickly in the body, but fat-soluble drugs may be inhibited in their action when blended with a high-fat material like cocoa butter. The site of administration also dictates the type of base material used. Cocoa butter produces an unwanted residue and is not suitable for vaginal suppositories. Normally, a water-soluble base is used for administration in this part of the body. Conversely, water-soluble bases are not suitable for rectal administration because the rate of dissolution and drug release is too slow.

Suppositories are prepared by mixing the API with the base material in either a solid or liquid form. Melting the base at a low temperature and then dispersing the drug in the liquid achieves the finished compounding and is an alternative to dry mixing of the ingredients. The liquid suppository material is then placed in a mold and allowed to cool and solidify.

Suppositories must be stored at controlled room temperature (25°C) and preferably never higher than 30°C. Suppositories made with water-soluble bases such as gelatin and polyethylene glycol must be protected from both moisture and elevated temperature.

Liquids

Solutions

Solutions are just what the word says they are: one or more drug substances dissolved in a solvent or solvents. The drug material is dispersed to the molecular level by the solvent or solvents. Solutions are more uniform than powder mixtures, so this form of drug dosage is considered more uniform than others when given to the patient. The problem with solutions is that many drug substances are prone to breakdown or display some form of chemical instability while in the liquid form. Depending on the solubility of the molecule or materials in question, liquid forms of a drug usually are more bulky than the solid form of the same compound. This is one of the benefits tablets and capsules have over liquid solutions. Packaging the liquid form of a product typically produces a package of greater bulk than the same product in solid form, and the package has more problems to overcome as part of the filling and sealing process. Liquid products containing molecules that are light sensitive are more susceptible to photolytic breakdown in the liquid form, and packaging must shield the product from those wavelengths that would attack the molecule or possibly all light.

Products that use a solvent or mixture of solvents where one is volatile require protection from heat. This is a major concern along with the performance capability of the container and closure system to withstand increased internal pressure at elevated temperatures. Leakage and contamination at the seal is another concern with liquids, particularly if the diluent or carrier used in making the product can support bacterial growth.

Solutions are designated or labeled for their specific method of product administration. Oral solutions would be administered through the mouth, while a topical solution would be applied only to the specified local area of the body.

Solutions are required for the injection of drugs. This form of solution used in parenteral applications is considered an injection. Many products, particularly vaccines, are stable in a solution suitable for injection while others have limited stability after being solubilized. This second set of materials that are unstable in solution are diluted and solubilized immediately before administration, normally in the vial or container that contains the stable powder or granule form of the product. Most products of this type are prepared from a solution in a process called lyophilization or freeze drying, where the water or solvent used to manufacture the product is removed, leaving a stable and sterile form of the drug in an uncapped vial. The vial is normally sealed in the same way that a liquid-filled vial is, with an elastomer stopper and an aluminum ring. Prior to use, water for injection (WFI), or an electrolyte solution is added to reconstitute the powder to a liquid form that can be administered parenterally. Products administered this way fall into a category of injections less than 100 cc in volume. A different set or requirements are used for products greater than 100 cc in volume.

Oral solutions. Oral solutions are designed for administration through the mouth. Oral solutions may or may not contain a sweetener, coloring, or flavoring. Oral solutions are usually aqueous solutions with added diluent materials to improve or ease administration of the drug or multiple drugs dissolved in the solution. Many oral solutions are first prepared in a concentrated form that must be diluted before administration. Some solutions contain cosolvents other than water, and when these solutions are diluted, the possibility of precipitation of the dissolved ingredients may occur if the dilution is done improperly.

Solid materials or mixtures of soluble solids may be dispensed and labeled for oral administration and are usually prepared by dilution with water; such products (e.g., potassium chloride) would be plainly labeled "for Oral Solution."

The term "syrup" is applied to oral solutions that contain a high concentration of sugar or other sucrose sweeteners. Originally, this term was only applied to solutions that were very near saturation with sugar, but over the years, it has been applied to any sweet and slightly viscous liquid and has been extended to oral suspensions as well.

Oral solutions may contain, in addition to sweeteners, glycerin or polyols like sorbitol to modify mouth feel, taste, solubility, and crystallization of the solution constituents. Solutions may also contain ingredients that inhibit the growth of bacteria, mold, or yeasts. Sweeteners are not limited to sucrose or sugar, aspartame and other sugar substitutes are used along with thickening agents (e.g., hydroxyethylcellulose) to modify viscosity and mouth feel and are used to treat diabetic patients.

Oral solutions that contain alcohol are usually referred to as elixirs; in fact, the proper use of the term "elixir" requires that the solution contain alcohol. Some products may require a large amount of alcohol to achieve solution, and these solutions may create a pharmacological effect in a patient because of the amount of alcohol administered with the drug. This problem is usually overcome by using propylene glycol or glycerin along with alcohol and water to minimize alcohol complications.

Topical solutions. Topical solutions are designed for application directly on a problem. Topical solutions are referred to as "lotions." Most topical solutions are designed to be applied directly to the skin, although a smaller number may be applied to the mucosa in the nose or to other parts of the body as specified on the labeling. The majority of topical solutions are aqueous based, although other solvents such as alcohol or a polyol such as propylene glycol may be present.

Tinctures. Tinctures are unique solutions prepared from vegetable sources or from synthesized chemical substances that are solubilized in water or water/alcohol solutions. Tinctures typically have established standards that do not correspond directly to the solubility of the material; a tincture solution is adjusted to the established standard of concentration or proportion of the drug required by the USP standard (3) for the product. The most common tinctures represent 10 g of the drug solubilized in 100 mL of solution or tincture, with this concentration adjusted following a chemical assay of the potency of the initial solution.

Suspensions. Suspensions are very similar to solutions, and many times people use the terms interchangeably even though they represent two totally different types of products. A suspension is a liquid product that contains solid particles of another material suspended or dispersed throughout the liquid phase. The particles are not soluble in the liquid. Suspensions may be labeled with terms or titles similar to solutions that more accurately designate the type of product administered, such as oral suspension or topical suspension. The term lotion has also been applied to topical suspensions (e.g., calamine lotion).

Good examples of oral suspensions include milk of magnesia and many liquid antacid products. These products are stable suspensions and are supplied in ready-to-use form, although some agitation or shaking may be required to uniformly redisperse the package contents if it has been sitting for an extended period of time.

Oral suspensions and other types of suspensions may contain antibacterial additives to protect against mold, yeast, or bacteria contamination. Oral antacids are susceptible to contamination when the user drinks the product directly from the bottle instead of using a cup or other administration device that prevents direct introduction of bacteria into the product.

Suspensions may contain sweeteners, viscosity adjusting materials, wetting agents, clays, surfactants, polymers, and other ingredients that prevent hard settling of the insoluble particles and improve the ability of the suspension to be administered.

Some suspensions are prepared for sterile injection, including ophthalmic and otic suspensions. These materials are diluted just prior to injection with WFI or some other suitable diluent. As a general rule, suspensions should not be injected intravenously or intrathecally. The suspended particles can clog blood vessels before they dissolve. Packaging for suspensions is the same as for true solutions.

Transdermal drug delivery. "Patche" is the common term used to describe transdermal drug delivery as a dosage form (Fig. 17). The patch is applied directly to the skin for the purpose of delivering a drug through the skin to the circulatory system for systemic treatment of a condition or disease. Nicotine

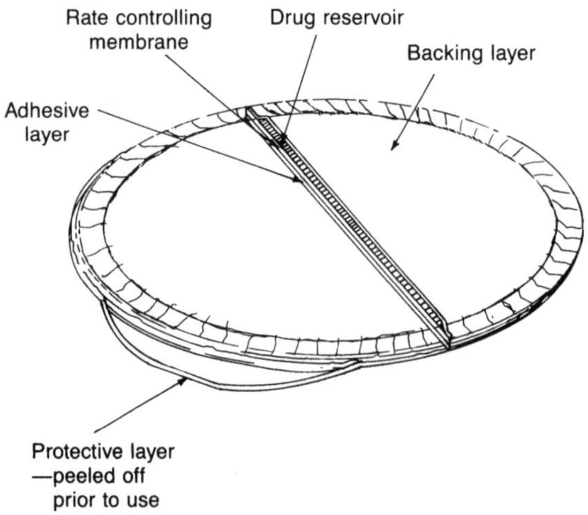

Figure 17 Diagram of a transdermal patch.

patches are probably the most common form of these devices recognized by the general public along with hormone products.

Transdermal patches are uniquely designed and constructed with multiple layers of material and drug product to achieve the desired delivery of an API to the body. The system looks very much like a Band-Aid®. The transdermal patch has an outer layer to protect the contents after application, a membrane or layer of material to control the rate of drug diffusion or administration through the skin, and some type of adhesive to hold the patch on the skin. The drug is typically contained between the outer layer of the patch and the rate-controlling membrane or layer; some people refer to this area of the package as the drug reservoir. The strength or advantage of a transdermal system of drug delivery is its ability to achieve a steady-state concentration in the patient as long as the patch is applied. This could also be described as a steady state or constant rate of delivery to the body. Most transdermal systems are described in terms of their release rate, and this is dependent on the membrane and drug formulation contained in the reservoir as well as the size or area of the patch. These factors determine the amount of drug delivered in a steady-state manner to the circulatory system.

Ophthalmic preparations. A separate class of dosage forms is defined for the eye. Ophthalmic materials may be solutions or suspensions and may be dispersed in aqueous, nonaqueous, or petroleum bases (4). Ophthalmic preparations must contain antibacterial agents to prevent the growth of microorganisms that may be introduced to the product through contact with the eye during dosage or administration of the product. Ophthalmic preparations are normally supplied as ointments, suspensions, or solutions. Large-volume solutions called collyrium, which are used to wash the eye, do not contain active ingredients but must take into consideration the isotonicity of the lacrimal fluid in the eye. The concern regarding isotonicity and pH is very important when a drug must be reconstituted from a solid for instillation in the eye. The material used for dilution is chosen to match the pH and isotonicity of tears so as to minimize any discomfort to the patient. A number of lyophilized drugs, which are unstable in solution form, must be reconstituted with an appropriate diluent for instillation into the eye. Solutions to reconstitute ophthalmic preparations are prepared with the necessary additives to match the needs of the eye. Packaging of these diluents materials is just as rigorous and difficult as packaging a drug product for use in the eye.

Ophthalmic ointments. Ophthalmic ointments are preparations for the eye that contain a drug or drugs dispersed primarily in a petrolatum base. These materials are sometimes dispersed in water-soluble bases that are more appropriate for water-soluble drugs.

All ingredients for an ophthalmic ointment are presterilized and then compounded under aseptic conditions to produce the ointment. The USP (3) is very specific about the ingredients meeting sterility requirements along with the test methods required to prove lot-to-lot sterility. Most manufacturers routinely test ophthalmic products for sterility prior to release to ensure that a microorganism contaminant has not found its way into the product as an improperly sterilized raw material or in the aseptic manufacturing process. Ophthalmic ointments contain a preservative that will prevent growth of microorganisms in the product and will destroy microorganisms introduced to the product after opening and use. A few ophthalmic drugs are bacteriostatic and do not need preservatives.

The active ingredient in an ointment is added to the base as a micronized powder or as a solution. The ointment must contain no large particles and must be certified to contain no metal particles. Metal particles could be introduced to the product during the manufacturing process.

Any ointment must be nonirritating to the eye, and the choice of diluent or carrier agent must exhibit this property. In addition, the final ointment must be compatible with the secretions bathing the eye and must permit diffusion of the active ingredient through the interaction of the product with tears. The ointment base is adjusted to maximize product stability and product compatibility with the eye over the storage life of the product.

Ophthalmic solutions. Ophthalmic solutions are formulated and manufactured in much the same way as ophthalmic ointments. They must be free of particles or foreign matter contaminants and are usually packaged in a container that is designed for instillation of the product into the eye. Some understanding of lacrimal fluid (tears) and the pH of the eye is needed to know how this type of dosage is formulated.

Lacrimal fluid is isotonic with the blood. This means that the fluid we call tears is not plain water but really a solution of salts and proteins that correspond to an isotonicity value equal to that of 0.9 % sodium chloride solution. The eye can tolerate lower and higher values (0.6–2.0%) without exhibiting discomfort noticeable to the patient. A product may be formulated to be hypertonic, which is outside of this comfort range, in order to speed uptake and interaction with the eye. Hypertonic products are administered in small doses that are quickly diluted by the lacrimal fluid (tears) to minimize the time of patient discomfort.

Solutions may be buffered to enhance the effectiveness of the drug ingredient and its suitability for use in the eye. Drugs that perform best as undissociated free bases, that is, products that are normally salts (e.g., alkaloid salts), are most efficacious at pH levels that maintain the undissociated state. Adding a buffer to a product attains a compromise pH level that balances stability and effectiveness. The use of a buffering agent requires many considerations. Normal tears have a pH of 7.4 and possess some buffering capacity in their makeup. Thus, when a drug is added to the eye as one or two drops of product, the normal buffering action of the tears in combination with the buffer in the solution is sufficient to neutralize the hydrogen or hydroxyl ions in the product. Alkaloid salts are an example of a weakly acidic material and typically have a weak buffer capacity. Here, the tears can dilute a small dose of product added to the eye quickly enough to avoid discomfort. Many drugs are not stable at pH 7.4, so buffering solutions are chosen that permit the final product to be as close as possible to this value while preventing precipitation or rapid deterioration. Another reason to add a buffer to an ophthalmic solution is to minimize the increase in pH of the solution from the release of hydroxyl ions from a glass package.

Ophthalmic solutions may be thickened using methylcellulose, polyvinyl alcohol, or other thickening agent. This is done to prolong contact of the drug with the eye.

Drugs that lose effectiveness when buffered or would not be stable in the normal range for an ophthalmic solution are supplied in many cases as a dry lyophilized powder. Adding an isotonic diluent to the product and immediately administering the liquid to the eye overcome the short-term stability of the product in solution.

Sterility of ophthalmic solutions is of great importance. The typical method of producing these products is with sterile filtration under aseptic conditions. The filter retains and removes any bacteria. The container into which the product is filled has been presterilized by radiation or autoclaving and is maintained in the sterile aseptic state until it is opened, filled, and sealed. Autoclaving or heat processing is always a favored method for sterilization; however, many drugs are not stable in high-heat conditions, or cannot be buffered or adjusted to maintain stability at sterilization temperatures. Normally, sensitive materials like this are prepared in a single-use container or in containers designed for use by one patient only.

Ophthalmic suspensions. Ophthalmic suspensions are similar in makeup and properties to the description supplied for solutions (4). The difference is that solid particles are suspended or dispersed in the liquid used as the carrier media for instillation into the eye. The particles or powder used in the suspension must be micronized to prevent irritation or possible scratching of the cornea. These products must be treated with extreme care to ensure that the particles have not caked or agglomerated into a mass that would cause harm to the eye. Packaging of a suspension is normally in a clear bottle that permits the doctor or patient to ascertain that the condition of the product is suitable for administration.

Gases

Drug products may also be gases. It is rare that we think of the true gas as a drug, but some examples would be oxygen to aid patients with emphysema or gases that are used in combination with other drugs for anesthesia. Halothane, isoflurane, sevoflurane, and nitrous oxide are gases used for anesthesia. Packaging of gas products for this use is not typical packaging dispensed by the doctor or patient. Instead of focusing on gases in this section of drug dosage forms, a more appropriate topic is aerosols and how they deliver a product to the body in a form that is similar to a gas infusion.

Aerosols

Aerosols used to dispense pharmaceutically active ingredients are products that are packaged under pressure. When released by a valve contained in the packaging, the therapeutic agent is released as a mist or very fine spray. Aerosols come in many forms and are used for topical applications as well as nasal, lingual (mouth), or inhalation applications. The term "aerosol" has been applied to a wide variety of products that are supplied under pressure, including foams, ointments, and semisolid fluids.

Aerosols are considered a dosage form for a number of reasons. The mixture of product and propellant and their possible interactions, the potential change of a molecule under pressure, and the multiple specialized components employed to make an aerosol-dispensing container all are scrutinized as a drug delivery system. Further complications are introduced when a precise metered dose of product is required from the aerosol container.

The first and most common thought about using an aerosol is in the administration of the product to the lungs as an inhalation aerosol (Table 1). This is one of the most demanding applications of aerosol packaging because the product must be released and broken up into extremely fine particles (<5 μm in size) to penetrate the lungs. Aerosols for other administration areas such as the nose and throat may produce much larger particles, some as much as several hundred micrometers in size.

The makeup of an aerosol product consists of a container, some form of propellant, some form of solution or emulsion that contains the active ingredients and diluents, a valve that controls the release of the product, and some form of actuator that releases the valve, directs the spray of product, and in some cases, provides the mechanical mechanism to break the liquid into the required particle size. These components have the capability of producing a wide variety of effects on the products dispensed, including metered or uniform doses, distribution of particle size, spray pattern, temperature of the spray, wetness of the

Table 1 Aerosol Therapy Overview: Relationship Between Particle Size, Target Area, and Mode of Transport

Particle size	Target area	Mode of transport
<1 μm	Particles are exhaled	Remain in gaseous state
1–5 μm	Peripheral bronchial passages	Particles form a sediment
5–10 μm	Upper respiratory passages and central bronchial passages	Particles rebound
>10 μm	Upper respiratory passages	

Source: From Ref. 5.

spray, shape of the spray emerging from the actuator, and in the case of aerosols not dispensing a fine spray, fluid viscosity or foam density of the applied product (Table 1).

Aerosol containers. Aerosol containers present some unique advantages for dispensing products. They are convenient and dispense product with the "touch of a button," and can easily maintain sterility of the product throughout the life of the dosage because they do not come in contact with the affected area of the patient. These characteristics along with the complexities and requirements needed to design the proper drug, excipient, and propellant mix, plus all the associated mechanical hardware needed to make the package work move them into their own dosage form as defined in the USP (3). These packages are highly specialized containers that dispense product in a unique way and open additional avenues for drug dosage that are not available by any other means. They also provide the capability for a patient to use inhalation products in a convenient portable form. Asthma sufferers in particular benefit from the ability of these containers to dispense epinephrine or other drugs directly into the lungs to counteract asthma attack.

Aerosol containers may be made from glass, plastic, or metal. Many of these materials are used in combination with others to achieve benefits that a single material cannot provide. An example would be the coating of a glass container with plastic to minimize breakage of the container. Thermosetting materials in the form of coatings are used on the inside and outside of metal containers as a polymer barrier to insulate the metal from interacting with the product or the environment. Coatings, inks, and varnishes on the outside of metal containers provide a method for labeling the product.

Glass aerosols are the most difficult to develop. They require a balance between maintaining the pressure inside the container and maintaining maximum impact resistance of glass under internal pressure. A glass aerosol when broken has the potential to hurl shards of glass over a large area.

Plastic aerosols do not suffer from the breakage problems encountered with glass. They do suffer from slow creep of the plastic material, which, in an extreme case, could misshape the container and make it unsuitable for use. All plastics have some diffusion characteristic and are not hermetic. This means that the propellant or the liquid drug product they contain can slowly diffuse through the walls of the container. Plastics are more commonly found as containers that are not under pressure and use a pump mechanism to deliver the product to a specially designed actuator for breakup into a mist for aerosol administration.

Metal containers (cans) are the standard containers when discussing an aerosol container. Metals used to make aerosol container include steel, aluminum, stainless steel, and tin-plated steel. Metal containers used for aerosols almost always have the metal substrate coated with some type of interior coating (epoxy, polyamide, polyester, vinyl, etc.) to insulate the metal from the liquid and the propellant it contains.

Types of aerosols. An aerosol may be a two-phase or a three-phase system consisting of gas and liquid or gas, liquid, and solid, respectively. The two-phase system consists of a solution of the active ingredients and diluents mixed together with both a liquefied propellant and a vaporized propellant.

Two-phase aerosols. Two-phase aerosols are the most popular dosage forms now in use for drug solutions (Fig. 18). The solvent that holds the ingredients is typically a mixture of cosolvents such as water, alcohol, propylene glycol, or polyethylene glycol that are soluble in a propellant that is liquid under pressure. The vaporized phase of the propellant provides the initial push of the product from the container. Probably the best way to understand an aerosol is to recognize that there are only two phases formed or contained in the aerosol, liquid, and gas. The advantage of a two-phase system lies in the ability to greatly

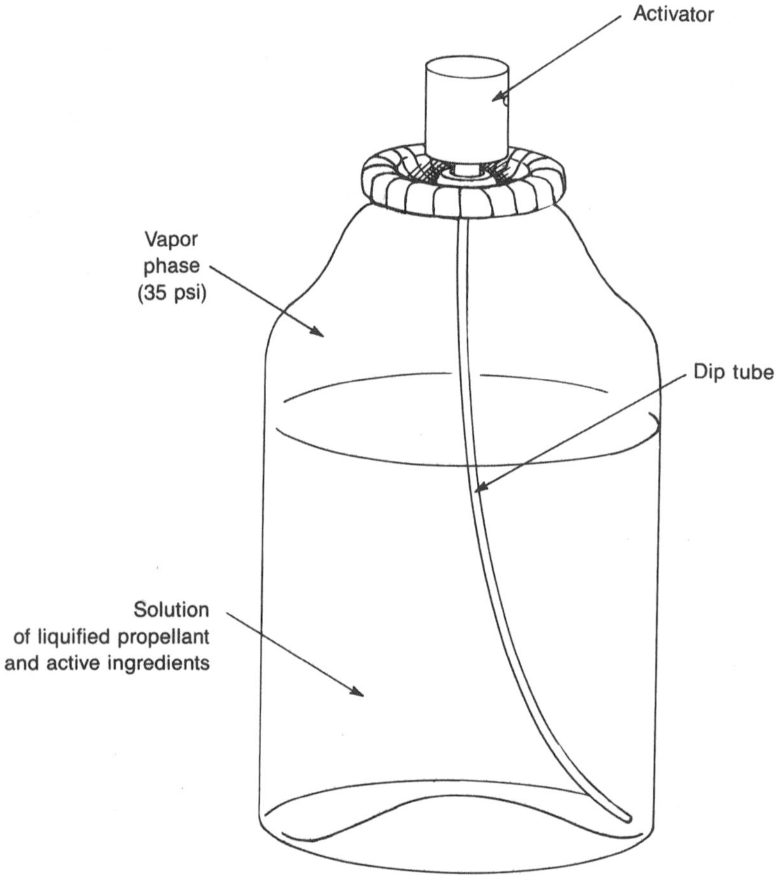

Figure 18 Two-phase aerosol container.

enhance the mist formation by discharging both the liquid and the propellant together. The maximum effect that can be achieved in the formation of a mist comes when the drug is soluble in the pure liquefied propellant. This and the fact that many drug substances are soluble in the cosolvents mentioned earlier make this method of aerosol formation the most popular form for drug solutions. As the combination of solution and propellant emerge from the actuator, the rapid evaporation of the liquid propellant creates a very fine mist of the drug.

A modified type of two-phase aerosol system is one that produces a dry spray or a powder spray. Here, the micronized particles of the drug and any excipients used to enhance properties are dispersed directly in the propellant. When the aerosol is sprayed, the propellant immediately evaporates, and only the powder reaches the patient.

Three-phase aerosols. A three-phase aerosol container consists of a similar arrangement of components with the addition of suspended or emulsified solid along with the vaporized or liquefied propellants, or if it is a true solution, the propellant is immiscible in the liquid (Fig. 19). In a three-phase system, the propellant is not miscible in water, or the suspending agent for the solid and three distinct phases are formed: the drug solution, the liquid propellant, and the propellant gas. For this type of aerosol container, the propellant only pushes the liquid from the container. The design of the actuator breaks up the liquid and forms the type of mist or spray that emerges from the nozzle.

Another method to improve the characteristics of the mist is to shake the container to disperse the propellant in the liquid. If the liquid propellant floats on top of the liquid, the discharge of both propellant and liquid and the propellants' subsequent evaporation can enhance the mist formation beyond that achieved at the actuator. When this characteristic of dispensing both solution and propellant is desired, the solution in the aerosol is formulated with both water and alcohol to modify its density and create the slight density differences necessary for the propellant to float on top of the liquid. The choice of propellant is critical to a system for it to float on the liquid. Most three-phase aerosol systems have the propellant sitting below the liquid on the bottom of the container.

Aerosols may use one other method to achieve the small particle mist-dispensing effect. By feeding the vaporized propellant to the actuator in a separate stream, the vaporized propellant is mixed with the drug in the actuator, while the propellant inside the container pushes the stream of liquid. The stream of propellant moving to the actuator is much larger than the stream of the solution, and the interaction results in a very fine dry spray of particles coming from the actuator. This method of particle size dispensing uses the minimum amount of propellant to achieve the effect. A spray from an aerosol produced this way does not feel cold to the patient when applied.

Aerosol foams. When an aerosol produces a foam presentation of the product, the solution phase inside the container becomes more complicated. Foams are

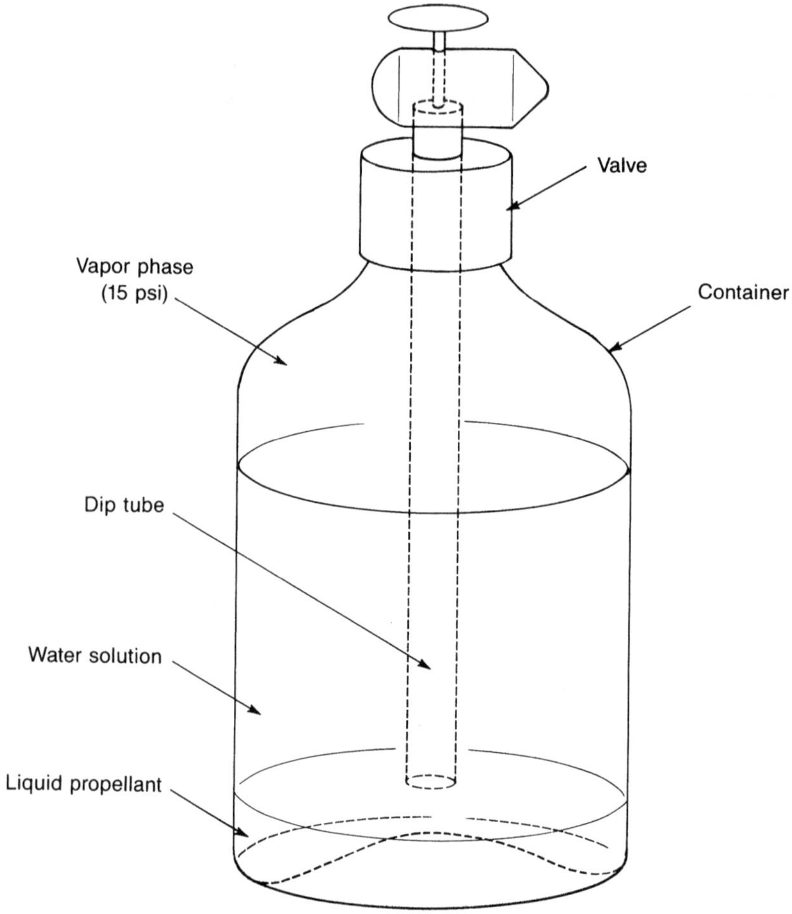

Figure 19 Diagram of a three-phase aerosol system.

produced from emulsions. A continuous phase in the solution consisting of the API, surfactants, and other liquids is one part of this mixture, with the propellant and emulsion being the other phases. The product dispersion and its subsequent foam production by the aerosol result when the product is shaken before use. Foams can also be produced by using soluble gas dissolved in the emulsion. Typically nitrous oxide or carbon dioxide is used to achieve this type of emulsion.

Isolated aerosol systems. Aerosols can be designed to isolate the propellant from the solution or suspension. Aerosols with the propellant separated from the product contain a bladder or bag that holds the propellant under pressure and can expand as the actuator is engaged. The bladder expands, forcing the liquid from the container. Systems using this aerosol-producing method are useful for dispensing ointments.

Another more unusual approach maintains the propellant and the liquid in separate compartments within the aerosol container. The propellant gas is released as a separate stream and forms a venturi with the dip tube in the container drawing the solution out of the container. The propellant is mixed with the liquid in the actuator just before the portion of the actuator that creates the spray characteristics.

Propellants. Propellants are liquefied gases that have a vapor pressure exceeding that of atmospheric pressure. Propellants used for aerosols have undergone a major shift in the last 10 years. Originally, fluorocarbon gases that are derivatives of methane, ethane, and propane were used as propellants. These liquefied gases were very inert with most drug substances and very safe in the manufacturing environment. With the concern over the interaction of these materials with ozone in the atmosphere at high altitudes, other compressed or liquefied gases have replaced most of these propellants. Now, typical low-molecular-weight liquefied hydrocarbon gases are used as propellants in aerosols, for example, are butane and pentane.

Compressed gases are also used to power aerosols. The typical compressed gases chosen as propellants include carbon dioxide, nitrogen, and nitrous oxide. As mentioned earlier, a mixture of gases either compressed or liquid or a combination of liquid and compressed gas is chosen to achieve the optimum pressure for dispensing and to maintain that optimum pressure throughout the administration life of the drug product from the container.

Mechanical parts of an aerosol container. The mechanical components of an aerosol container work in combination with the propellant to produce mist, spray, foam, or ointment delivery. These parts include the valve, actuator, and mixing cup for inhaled aerosols and for nonpressurized aerosols, a pump.

The valve is one of the more unique parts of any aerosol container (Fig. 20). It must be inert to the drug and the diluents and excipients used to dissolve or suspend the active ingredient. It must regulate the flow of the product and possibly the propellant from the container. A valve may be designed for continuous delivery of a product or designed to deliver a metered dose of the product. Continuous delivery is most often used for topical products, and metered doses are most often found for oral and nasal inhalation. Surprisingly, the ability of these valves to deliver an accurate dose is quite good. They compare favorably in uniformity to tablets and capsules.

Pumps used to produce metered doses of an aerosol product have become very popular over the last 10 years as an alternative to gas propellants. Most consumers associate them with environmental-friendly packaging and recognize that no fluorocarbons or other potentially harmful gases are used to dispense the product. They are quite accurate in the metered delivery of dosage and strong

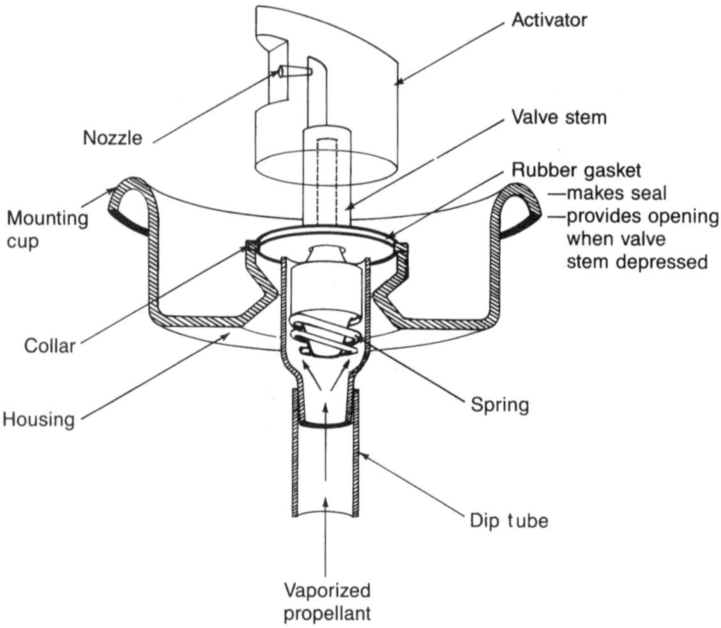

Figure 20 Diagram of the valve and mechanical parts of an aerosol container.

enough to push the solution or suspension through an orifice to produce the desired particle size for delivery. One drawback to pumps is the need to prime them prior to their first use. This involves filling the metering chamber contained in the valve to assure a uniform dose. Pumps must also be primed if they have not been used for an extended period of time.

Most valves and interior pump components are made from plastics, elastomers, and stainless steel. The stainless steel is most often found in the spring of the valve. Glass balls may be used for valve opening and closing mechanisms.

Actuator is an extension of the valve. It is attached to the valve stem and most often is used to designate the direction of the spray along with providing the pressure point to actuate or open the valve. Actuators are designed with an orifice that produces the proper presentation of the drug product. They may also contain an expansion chamber prior to the orifice to enhance the formation of the aerosol mist.

Manufacture of aerosols follows one of two general processes. The first process fills the container with a metered volume of the drug solution or suspension; it then chills the drug product and meters the propellant into the container as an extremely cold liquid. As the propellant begins to evaporate (boil), the container closure containing the valve assembly is crimped or double seamed into place. As the container warms, the remaining propellant partially evaporates, while some remain in solution, and the container becomes pressurized.

This method of manufacture has the added advantage of evacuating the head-space of the container of unwanted atmospheric gases. These gases are displaced as the propellant evaporates from the surface of the liquid in the container.

The second common method of manufacture is called pressure-fill. When this method is used, the container is filled with the solution and closed with the valve assembly. The propellant is then forced into the container through the valve or, in some cases, under the valve cap with immediate sealing. When this method is employed for filling, the headspace of the container (the unfilled area above the liquid contents) must be evacuated of air if the presence of oxygen or other atmospheric gases are deleterious to the contents.

Oral and nasal aerosols intended for inhalation dosage. Not all drugs are supplied in an aerosol container. Pharmaceutical inhalants are drugs that are supplied to be dissolved or atomized into a mist or vapor for therapy. Drugs may be dispersed into vapors, mists, or aerosols using an atomizer or vaporizer. In a device of this type, the medication is usually dissolved in water and the water heated and pushed through a system that creates a fine mist or evaporates the liquid and presents a saturated atmosphere to hood or other device for inhalation. Devices like these are most often used to relieve bronchial or nasal congestion and possibly some congestion in the upper respiratory area. Another method of using an aerosol uses this technique. The aerosol is sprayed into a hood or enclosed space and the patient then breathes the mist.

Nebulizers and oral aerosols are another aerosol dosage form designed to push a drug product deep into the lungs or bronchial tubes. Products supplied for use with this form of dosage therapy provide symptomatic relief from asthma and bronchial ailments. This method of administration is another way to provide systemic delivery of a product to the body when the product degrades in the gastrointestinal tract or is not persistent in the correct state long enough for absorption.

Oral aerosols are designed to go deeper into the lungs than standard sprays. These products, usually for asthma relief, combine a metering valve with some type of cup or delivery device that also mixes air into the dosage being delivered to the patient. This means that the patients not only must manipulate the device but they must also perform a breathing inhalation at the same time for the drug to reach the affected area. There are large number of variations in the administration devices supplied with these specialized aerosols, and the variations between them coupled with different directions about how to use the different devices are complicated for patients to follow.

Both solutions and suspensions are administered in this manner. Both types of drug products require development of a very small particle size to avoid coalescing into a form that cannot reach the targeted pulmonary area. When a suspension is involved, the product must not only break into extremely small particles but must also be designed not to agglomerate. The use of surfactants and lubricants along with minimizing the amount of water present are the most common methods used to minimize or stop this problem.

Nasal aerosols are slightly different than the bronchial or pulmonary aerosols just discussed. These products can reach the recesses of the sinus cavities without moving into the lungs. They are typically an alternative to drops or sprays and have the advantage of not placing a concentrated dose of product in a very small area of the nose or sinus cavities, thus avoiding mucosal irritation. Products designed for nasal aerosol administration do not come in contact with affected areas, which improve maintenance of sterility.

Many nasal aerosols are produced by the mechanical action of a metered pump and a small orifice tip that interacts with the solution under pressure to break it into a fine mist. Hay fever and allergy medications are dispensed in this manner with budesonide being one of the more popular allergy medications delivered with this system.

Topical aerosols. A wide variety of products are delivered as topical aerosols or sprays directly on the affected area of the body. This method of application is very soothing when treating a burn or other skin problem where the application of a liquid or ointment to the affected area would create additional irritation.

For burns, the cooling effect of the gaseous propellant along with the fine mist of the drug produces a soothing effect. Because of these advantages in application, a wide variety or products are supplied as topical aerosols, including local anesthetics (e.g., benzocaine), antiseptics that may also contain or be an antibiotic, burn preparations that combine both antiseptics and anesthetic ingredients, and possibly a protective film that acts as cover or dressing for the area. Additionally, steroids such as prednisone, calamine lotion, and sunburn ointments, preventatives such as zinc oxide, and medications for athlete's foot all use this dosage form.

Because of the wide range of conditions treated, a wide variety of aerosols are employed for topical use. Powders, mentioned earlier as a micronized drug product suspended directly in the propellant, are used for a variety of topical treatments of the skin and particularly for athlete's foot.

Foams and ointment administration from an aerosol are also part of the range of dosage presentations in topical aerosols. Foam aerosols for rectal or vaginal administration of a variety of products are available.

Topical aerosols most often use hydrocarbon propellants to push the large volume or viscous formulations of foams and ointments from the container. All hydrocarbon propellants are extremely flammable, and warnings are prominently displayed on the containers.

SUMMARY

When one reviews the needs of stability, sterility, purity, and dosage forms, one is presented with a wide variety of possible packaging needs. The primary needs of product protection and containment are required for all forms and types of packaging, while methods of dispensing become another packaging requirement with different types of product and different product forms [i.e., solid, liquid, or gas (aerosol)]. Mechanical aspects of package design for convenience, simplicity of use, controlled dosage, reconstitution, and administration all add to and increase the complexity of the packaging problem.

The physiological aspects of the various dosage forms also place a wide variety of needs on the design and delivery of any potential package. Parenterals, ophthalmic preparations, intravenous fluids, and injectable products require high standards for sterility and purity. In many cases, the packaging may contain the mechanical means for administration. Tablets and capsules typically require moisture protection and possibly protection from mechanical or physical damage during transport. The solution to these problems may be as simple as cotton under the cap of a bottle to prevent the tablets from moving or as complex as a barrier container with modified atmosphere packaging to prevent moisture or oxygen ingress. Both solids and liquids may require protection from light.

This list of problems requires a broad and comprehensive understanding of the strengths and weaknesses of various packaging forms. It also requires an understanding of the regulatory and USP (3) monograph requirements for approved drugs.

FURTHER READING

The National Formulary, NF 23. Prepared by the Council of Experts and Published by the Board of Trustees. Rockville, MD: United States Pharmacopeial Convention Inc., 2005.

Berins ML, ed. SPI Plastics Engineering Handbook of the Society of the Plastic Industry, Inc., 5th ed. New York: Van Nostrand Reinhold–Chapman & Hall, 1991.

Bakker M, Eckroth D, eds. The Wiley Encyclopedia of Packaging Technology. New York, Chichester, Brisbane, Toronto, Singapore: John Wiley and Sons, 1986.

Hanlon JF. Handbook of Package Engineering. 2nd ed. Lancaster Basel, PA: Technomic Publishing Inc., 1992.

REFERENCES

1. Jenkins WA, Osborn KR. Packaging Drugs and Pharmaceuticals. Lancaster Basel, PA: Technomic Publishing Inc., 1993.
2. Hernandez RJ, Selke SEM, Culter JD. Plastics Packaging—Properties, Processing, Applications, and Regulations. Cincinnati, Munich: Hanser Gardner Publications, Inc., 2000.
3. The United States Pharmacopeia, USP 28. Prepared by the Council of Experts and Published by the Board of Trustees. Rockville, MD: United States Pharmacopeial Convention Inc., 2005.
4. Bartlett JD, Jaanus SD. Clinical Ocular Pharmacology. 3rd ed. Boston, MA: Butterworth-Heinemann, 1996.
5. Stedman's Medical Dictionary. 27th ed. Philadelphia, Baltimore, New York, London, Buenos Aires, Hong Kong, Sydney, Tokyo: Lippincott Williams & Wilkins, Wolters Kluwer Company, 2000.

3

Vaccines and Biologically Produced Pharmaceuticals

INTRODUCTION

Biotechnology is an exciting area of pharmaceutical research and development. It comprises drugs, proteins, and vaccines. The promise that it holds has captured the imagination and interest of the press and general public. The potential benefit from new products developed by this technology is staggering. It offers the possibility for cures of genetic conditions and for repair and healing of severely damaged cells that cannot repair themselves. It is the brave new world of multidisciplinary science, medicine, and from a packaging perspective, a new range of products to protect. As the technology and the products it manufactures move to commercialization, packaging must develop new methods, materials, and delivery systems. The pharmaceutical industry considers these technologies and the products they manufacture to provide an opportunity to dramatically improve health care on a broad spectrum of problems little understood as few as 10 years ago.

Vaccines are in an equally exciting position. The introduction of new vaccines that treat a wide range of diseases, including bacterial meningitis, a severe disease of the brain and spinal canal, is equally as exciting as the biologic drugs. The manufacture of vaccines is undergoing a major change as it shifts from antibody production in eggs to products produced by recombinant RNA and DNA techniques. The research into diseases like HIV has led to the possibility of developing a vaccine to immunize against this deadly killer. Our understanding of the immune system in the post-HIV world has improved dramatically. Targeted vaccine development into areas that permit the body, properly immunized with a vaccine, to fight off many deadly killers is again at the cutting edge of medical technology.

How are biologic products made and how are they packaged? The whole idea of biologically produced materials is foreign to the general public. There is a large amount of fear when one discusses genetically modified crops, genetically modified animals, or cloned animals outside the scientific community. Genetically altered crops are resistant to disease, permit prudent use or elimination of chemical herbicides, have extended food crop plants' ability to cope with drought, and improve yields dramatically. They are criticized and banned outright in some places for fear of potential unknown harm associated with them. The European Union has prohibitions and strict labeling requirements on genetically altered crops in the human feed chain, while the United States accepts and regulates these products closely and permits their use. Consumer attitudes are much the same; the American consumer has not made this a major environmental issue, while European consumers are extremely sensitive to the subject.

Pharmaceuticals fall somewhere in between the two extremes. People are excited about their promise and are less critical of the possible long-term questions that surround the technology. Biologics are not new materials; we use a number of biologically derived products everyday without fear or questions about safety. Alcohol, antibiotics (penicillin), and a large number of dairy products are the result of bioengineering. Our ability to harness biology in much the same way we have harnessed chemistry and physics is the next great frontier of pharmaceutical science. The human genome project and the information it provides to researchers in the medical and health care field are unprecedented. This information begins to establish understanding of genetic conditions and diseases and begins to unlock secrets regarding why people react differently to drugs.

BIOLOGIC PRODUCTS

The scope of biotechnology is very large; it is usually defined as the use of living cells (organisms), including mammalian cells, in the manufacture of products. This definition places alcohol, antibiotics, dairy products, beer, and vaccines under the umbrella of biotechnology.

Regulation of pharmaceutical biologic producers and products was enacted in 1944 under the Federal Public Health Service Act (58 Stat. 682) and later amended. The groups of products licensed under the Act are detailed in Table III, part F, and are generally known as biologic products. They are regulated by the Food and Drug Administration (FDA) as defined in the Code of Federal Regulations (CFR 21 parts 600–680) that are referred to or called biologics (Table 1). These regulations are the basis of federal control of biologic products with the exception of a few diagnostic aids. The regulations are administered by the Center for Biologics Evaluation and Research (CBER) to evaluate and control biologic products. The diagnostic aids regulations are administered by the Center for Devices and Radiological Health of the FDA. This control by CBER is not to

Table 1 Biologic Products Regulated by CBER and CDER

CBER	CDER
Vaccines	Cytokines
Plasma expanders	Monoclonal antibodies
Blood products (blood derivatives)	Interferon
Whole blood	Enzymes
Gene and cellular therapies (somatic cells)	Growth factors
Includes viral inserted (vectored)	
Antitoxins and allergenics	Proteins for therapeutic use Produced and extracted from animal or microorganisms (does not include clotting factors)
Manufacturing reagents Antibodies, cytokines, proteins	
In vitro diagnostics	

Abbreviations: CBER, Center for Biologics Evaluation and Research; CDER, Center for Drug Evaluation and Research.

be confused with that of the Center for Drug Evaluation and Research (CDER) that regulates finished drug product derived from biologic processing (1).

The current use of the term comes from two extremely significant advances in the technology. These advances were the development of recombinant DNA (rDNA) technology and new methods that permitted the production of large quantities of monoclonal antibodies. Additional technology has also been developed that has led to cloning and transgenic animals, plants, and gene therapy (2). These technologies have huge potential applications in pharmaceuticals along with antisense DNA, but they remain creatures of the laboratory for the most part and are not the normal production mechanisms used for biologic products now in the market.

BIOLOGIC DRUGS

Biologic drugs are produced using mammalian cells and gene-modified bacteria. A number of drugs like insulin have been used for years. In the last 25 to 35 years biologics have become an area of intense interest. Biologic drugs target specific receptors or targets to block or change the way the body responds to a disease or genetic condition (Table 2). Older drugs, mainly small molecule compounds, require relatively high-dose levels to reach a potency level that acts systemically in treating the disease but may cause harm to other parts of the body or in the case of a number of genetic conditions like arthritis or psoriasis suppress the immune system (3). The more specific the biologic target, the less chance the drug has to interfere with other biologic functions, and this presents the theoretical possibility of it being safer to use.

Table 2 Biologic Classes of Products and What They Treat

Biologic product type	Condition treated
Insulin and insulin analogs	Diabetes
Human growth hormone	Natural growth hormone deficiency
TNF blockers	Blocks cytokines in treatment of rheumatoid arthritis and psoriasis
Erythropoietins	Treatment of anemia
Interferon-α	Treatment of hepatitis B and C
Interferon-β	Treatment of multiple sclerosis
Cancer antibodies	Treatment of metastatic cancers (e.g., breast cancer)
Enzyme replacement	Treatment of mucopolycharidosis, Fabry disease and Gaucher's disease
GCSF	Treatment of low levels of white blood cells that fight infection (neutropenia)
Antiviral antibodies	Treatment of respiratory viral infections (syncytial) in premature infants
Follicle stimulation hormones	Treatment of infertility
Recombinant coagulation factors	Treatment of bleeding caused by hemophilia
Teriparatide (rDNA origin)	Osteoporosis
Other biologics	Treatment of genetic conditions and immune system responses like asthma, cystic fibrosis, ischemic stroke, acute coronary syndrome, wet macular edema, sepsis

Abbreviations: TNF, tumor necrosis factor; rDNA, recombinant DNA; GCSF, granulocyte colony-stimulating factors.

Biologic drugs interact with a number of sites within the body to achieve their effect. The generic name of the drug provides an indication of what it does. The suffix used in the name denotes the origin or type of a biologic drug.

Examples are as follows:

- *mab* is a monoclonal antibody
- *ximab* is a chimeric (mouse-human) monoclonal antibody
- *zumab* is a humanized monoclonal antibody that reduces the amount of mouse contribution to less than 10%
- *umab* is a human monoclonal antibody
- *cept* is short for receptof and means an antibody fusion protein that mimics an immunoglobulin

Examples from actual generic names include etanercept (Enbrel) for rheumatoid arthritis or efalizumab (Raptiva) for psoriasis.

Biologic drugs are difficult to identify and to produce. Their long development cycle and the exacting conditions needed to manufacture the product contribute to their high cost. In general, the biologic agent acts to trigger some of the following reactions:

- Activating T-cells
- Inhibiting memory or active T-cells
- Blocking migration of T-cells to a specific organ or area of the body; for example, blocking migration of T-cells to the skin in the case of psoriasis
- Inhibiting the production of certain chemicals in the body
- Inhibiting cytokines, for example, etanercept used to treat rheumatoid arthritis blocks tumor necrosis factor (TNF)
- Blocking conversion of one cytokine into another
- Stimulating blood-producing cells
- Increasing the level of a hormone
- Adding a missing enzyme (insulin)

This list is not complete and does not include vaccines that will be discussed later in the chapter.

Biologic drugs tend to be large molecule products that are almost impossible to reproduce chemically as is the case of small-molecule pharmaceutical products. The drugs are produced in living tissue or cell cultures that must be maintained without change. The unique cell culture that produces the large molecule has the potential to mutate, particularly if the culture is some form of bacteria. Another problem is the life span of any unique culture. All things die, and modifying and maintaining a pure strain of a particular cell culture to produce the drug or biologic over extended periods of time present other unique challenges.

As mentioned, biologic products are proteins and other very large molecules. This presents a different set of problems for administration of these drugs. Our bodies are very good at digesting proteins and breaking down other complex molecules, making most of these products unsuitable for oral ingestion. This means that the majority of biologic drugs are injected or infused into a patient. When first introduced as nontraditional therapies, a patient had to go to a doctor's office or a health care facility for administration of the product, thereby increasing treatment costs. This fact and the resulting questions of cost and accessibility of drugs that are very expensive have created a number of packaging challenges and changes in how to protect and dispense the drugs and administer them to patients.

Packaging of biologics presents a number of unique challenges. These include refrigeration, interaction with the primary packaging material, and customization of the package to a delivery device suitable for use by the patient. Most of the products are susceptible to changes when exposed to light and temperature. The changes are most often in structure (folds) and bond breaking. Good examples are two of the leading TNF biologics etanercept (Enbrel) and adalimumab (Humira). One of the drugs etanercept is lyophilized (freeze-dried) and then immediately refrigerated between 2°C and 8°C (36–46°F), the other adalimumab remains in liquid form but also is chilled immediately. The drugs must be protected from light, and when traveling, the drugs must be maintained chilled, not frozen, with some types of cold packs. Maintaining cold chain integrity of any drug is difficult and is a combination of packaging designed to protect and maintain the required temperature range and a well-documented cold chain. The cold chain is a series of distributors and transportation carriers with records proving that the product never experiences any excursion outside its required temperature range. Cold chain distribution is also used for vaccines.

Protection from light is another component of packaging these products. The large molecule is sensitive to change, and energy from light may permit the molecule to refold to a lower-energy state.

The primary packaging materials are another issue with this type of product. The large molecules can bind or adhere to the inside surface of a package, reducing the delivered dose. The majority of biologics are packaged in glass to eliminate or to minimize this problem. Because biologics are still relatively new, packaging development using plastics and other materials is a work in process. The drugs are expensive, and experimentation with new materials is costly.

Biologics are mixtures of a number of different molecules very similar but not exact copies of each other. The large molecules may also have unique structures or organizational patterns (folds) that affect the way they interact with the body. This is analogous to the different mirror images of small-molecule pharmaceuticals, which react in different ways with the body and produce differing degrees of efficacy. Large-molecule biologics have structural arrangements (folds in the backbone of the molecule) that express different three-dimensional patterns (reactive sites) to the body. These variations effectively rule out generic biologic drugs, called biosimilars, because it is almost impossible to produce an exact copy of an existing product.

Biosimilars (4) pose a difficult question for regulatory agencies and governments in Europe and the United States. Just as generic products are less expensive than their branded counterparts, biosimilars hold the promise to be less expensive than the original product.

Packaging development for biologics has emphasized solutions for two problems. The first problem is the need to adjust a biologic drug dose to match the recipient's body weight. These drugs are very specific in their effect and are generally prescribed on a per-kilogram-of-body-weight basis. Packaging development to minimize the amount of product supplied to the physician or patient while providing a sufficient quantity to cover a wide range of body weights has resulted in multiple solutions to this dilemma.

Biologics, as mentioned, are almost always injected into the patient. One major cost reduction has permitted administration of many of these drugs to take place at home and not in the doctor's office. Packaging has enabled the development of automatic "pens," which are simple handheld injection devices that make it possible to administer a biologic drug for treatment of TNF safely and easily at home.

This is an area of packaging development that will expand to meet patient and product needs while making biologic drugs easy to use, safe, and simple to transport (5).

VACCINES

Vaccines are one of the greatest achievements in medical science. A vaccine teaches your body's immune system how to resist infection and disease. It is based on the idea that by presenting the immune system of the body with a weakened or modified version of a disease, it will not only destroy the vaccine components that mimic the disease but will also remember the danger and produce memory cells that protect a person, sometimes for a lifetime, from any chance of being infected by the disease.

Measles, mumps, pertussis (whooping cough), diphtheria, and rubella are a few of the diseases that killed tens of thousands in the 19th and early 20th centuries but are rarely mentioned as dangerous any more. Parents lived with the fear that their child would contract polio and be paralyzed for life or could possibly die because of the infection. Today polio is another disease that has been conquered in the Western Hemisphere.

Influenza or flu remains a dangerous infection for the very young and the elderly, but yearly flu shots (vaccines) reduce the severity of infection in those that receive them and have markedly reduced deaths in the most vulnerable populations in our society.

Everyone has a built-in protection against disease called the immune system. This complex biologic adaptation allows our body to recognize and combat disease. When your body fights off a viral or bacterial invader, you get naturally acquired immunity. This comes at a price because you are sick for a period of time, while your body fights the infection, but as it eventually wins the fight, you recover. Your body goes through a progression of identifying the invader and producing antibodies to fight it. Normally it takes the body about a week to learn how to identify and fight off a dangerous microbe. If the microbe is extremely virulent, this may be too long and the body may be overwhelmed by the invader. This means you are sick for a number of weeks, while your body works to fight off the infection. In severe cases, the infection causes death before the body can win the fight.

Vaccines are biologic products that prepare your body for the fight against the disease, and the preparation is a fight your immune system is guaranteed to win (6) (Table 3). Traditional vaccines contain weakened or killed microbes or parts of microbes that rally your immune system into action. It is an unfair fight because the microbes have been treated in a way that makes them harmless to the body. Your body quickly overcomes these disease analogs and in the process learns how to recognize, fight, and defeat the disease if it ever reenters your

Table 3 Overview of Vaccines

Vaccine type	Disease	Advantages	Disadvantages
Live attenuated vaccines	Measles, mumps, rubella, polio (Sabin vaccine), yellow fever	Produce a strong immune response Often give lifelong immunity with one or two doses	Remote possibility that the live microbe could mutate back to a virulent form Must be refrigerated to stay potent
Inactivated "killed" vaccines	Cholera, flu, hepatitis A, Japanese encephalitis, plague, polio (Salk vaccine), rabies	Safer and more stable than "live vaccines" Do not require refrigeration; more easily stored and transported	Produce weaker immune response than live vaccines Usually require additional doses or booster shots
Toxoid vaccines	Diphtheria, tetanus	Teach the immune system to fight off bacterial toxins	
Subunit vaccines	Hepatitis B, pertussis, pneumonia, caused by *Streptococcus pneumoniae*	Targeted to very specific parts of the microbe Fewer antigens so lower chance of adverse reactions	Difficult and time consuming to identify the best antigens for an immune response
Conjugate vaccines	*Haemophilus influenza* type B, pneumonia caused by *S. pneumoniae*	Allow infant systems to recognize certain harmful bacteria	
DNA vaccines	Currently in clinical trials	Produce a strong antibody and cellular immune response Relatively easy and inexpensive to produce	Still in experimental stages
Recombinant vector vaccines	Currently in clinical trials Veterinary approved use: vaccine for West Nile virus in horses	Closely mimic a natural infection to stimulate a strong immune response	Still in experimental stages
Combination vaccines	Diphtheria Tetanus Pertussis Measles Mumps Rubella	Reduce a doctor's visits Improve protection for the child and community Reduce injections (needlesticks) Reduce cost	

body. This process is called artificially acquired immunity. Vaccines do not fight a disease; they prepare you to never get the disease.

When a large group of people in your community has been vaccinated, the chances of the disease occurring in the community or area are greatly decreased. This is called herd immunity, and it means that if a large number of people in a group are protected, the group as a whole is much less likely to get the disease. This type of immunity is not a permanent thing. If all the members of a group do not get vaccinated or if continued vaccination in an area stops, the disease can come back with terrible results. An example is found in the United States when because of safety concerns many parents did not get their children vaccinated against measles. The disease returned in 1989 with an epidemic of 55,000 cases and 155 deaths attributed to measles.

The most famous vaccination occurred when Edward Jenner, a rural English physician, noticed that milkmaids infected with cowpox did not become infected with smallpox, an extremely virulent and deadly disease. Cowpox is a mild virus that infects both cows and people. Jenner injected a boy with the serum he obtained from cowpox blisters. Six weeks later, he injected the boy with serum from a smallpox blister, and the boy showed no signs of the disease. From this insight and the first clinical trial, modern vaccines were created. The name vaccine comes from this first treatment or first vaccination. The word vaccine is derived from the Latin word *vaccinus*, which means pertaining to cows.

Combining the vaccine with an aggressive World Health Organization (WHO) campaign to vaccinate people whenever and wherever smallpox broke out in the 1950s, 1960s, and 1970s resulted in the eradication of the disease. The last known case of smallpox occurred in Somalia in 1977.

Every bit as dramatic is the reduction in polio cases in the United States. During 1954, the year before the polio vaccine was available, doctors reported 18,000 cases of paralyzing polio, and three years later, a widespread program of vaccination had reduced that number to 2500. The WHO is working to rid the globe of polio just as it did with smallpox in the 1970s. Following the same types of programs used against smallpox, the WHO was able to reduce the total number of cases of polio worldwide to 537 in 2001. The disease has been completely eliminated from the Western Hemisphere.

Infectious diseases preventable with the use of vaccine are as follows:

- Anthrax
- Cervical Cancer
- Diphtheria
- Hepatitis A
- Hepatitis B
- *Haemophilus influenzae type b (hib)*
- Human Papillomavirus (HPV)
- Influenza
- Japanese Encephalitis (JE)

- Lyme Disease
- Measles
- Meningococcal
- Monkey Pox
- Mumps
- Pertussis (Whooping Cough)
- Pneumococcal
- Poliomyelitis (Polio)
- Rabies
- Rotavirus
- Rubella (German Measles)
- Shingles (Herpes Zoster)
- Smallpox
- Tetanus (Lock Jaw)
- Tuberculosis
- Typhoid Fever
- Varicella (Chicken Pox)
- Yellow Fever

Vaccines are extremely cost effective. The small price for a vaccination eliminates the need for potential hospitalization or major home care. It is not unusual to see cost benefits for vaccines quoted at five times to hundreds of times the actual cost of the vaccine. Vaccines are preventative medicine at its best. The minor inconvenience of obtaining a vaccination in one of its many forms eliminates weeks of illness and possible hospitalization.

Types of Vaccines

There are a number of different types of vaccines—all of which are difficult to package. Each type of vaccine is produced in a different way and has different requirements for proper protection and distribution of the material.

Types of vaccines

- Live attenuated
- Inactivated
- Toxoid
- Subunit
- Conjugate
- DNA
- Recombinant vector

Live Attenuated Vaccines

Live attenuated vaccines are some of the most successful against tropical diseases. Because the vaccine contains live but weakened microbes or virus, it produces one of the strongest responses by the body's immune system to the disease. This type of vaccine usually confers immunity with one dose and rarely requires a booster or subsequent revaccination. This is important in remote areas where access to any health care is almost nonexistent. Typically a health care worker will touch these areas infrequently. The chance of getting people in remote regions back for multiple booster shots of a vaccine is small.

A live attenuated vaccine offers the best chance for success in remote areas and populations. One group vaccination of a village or remote group not only confers artificial immunity on each individual but also provides a measure of herd immunity or community immunity to the village and geographic area. New members of the community and members of the community not present for the vaccination program benefit from an area free from the disease and free from individuals becoming infected and passing the disease on to other members of the community. The new members of the community require vaccination, but unless they come in contact with the disease from someone outside the group, they are much less likely to contract the disease.

Packaging of a live attenuated vaccine is challenging. It requires not only the standard development of a vial or ampule for injection but also the development, testing, and certification of bulk containers needed to maintain a uniform product temperature during shipment.

The products require refrigeration to maintain product temperatures between 2°C and 8°C. Live attenuated vaccines cannot be frozen, and they rapidly lose potency when subjected to heat. Because these vaccines are used in some of the most remote parts of the world, pallet-load quantities of product must be maintained at these temperatures for as much as 96 hours without any external refrigeration. A majority of shipments are to the tropics, which makes this requirement difficult to achieve. Specially lined pallet-size containers, using polystyrene foam as the insulation material, is the outer protection most often employed. Along with insulating the shipping container, provisions within the container are made for gel packs or cold packs that maintain the temperature range inside the package.

In some cases, the bulk pallet packaging is subdivided to permit a breakdown of the shipment at a distribution point and further transport of smaller packages to remote areas. The packaging may also require a complete kit for administration. The kit includes all the syringes, chemicals, and coverings needed for administering the vaccine. The patients may range from infants to adults. Design and development of cost-effective administration kits that are easy to use form part of the packaging design for mass immunization.

Primary packaging of vaccines is for the most part the same as that used for parenteral drugs. Glass vials and stoppers are the primary packaging components for the vaccine solution. The packages must be inert to interaction with the vaccine and any adjuvant, a compound added to a vaccine to improve its effectiveness. Along with the vaccine and its adjuvant, the packaging must be suitable for any diluents used in its manufacture.

Because the products are shipped to areas beyond normal cold chain distribution, many of the packages include some types of temperature indicators. These indicators change color when subjected to heat beyond the acceptable range for different periods of time. This hopefully alerts the health care worker to a problem and prevents the use of products that are no longer effective.

Inactivated Vaccines

Inactivated vaccines use the same microbes as live attenuated vaccines. Instead of using a microbe or virus that is weakened, this type of vaccine uses a microbe or virus that has been "killed." The disease-causing microbe is killed with heat, chemicals, or radiation and cannot harm the body. This makes the vaccine easy to dry and transport and very stable over a wide temperature range. These vaccines are usually supplied in freeze-dried form making them very accessible to people in remote areas or developing countries. They do not require refrigeration.

Packaging of inactivated vaccines requires the addition of sterile water or saline solution to the "kit" used in the field. Clean water is a major problem in the developing world, and the vaccine must be reconstituted with a diluent that will not transmit another disease. Chemicals for sterilizing needles and other administration items are sometimes packaged with the vaccine, but most often these are supplied separately.

A major problem with inactivated vaccines is the milder response of the immune system to the product. These products do not produce a response strong enough to create lifelong immunity. The products almost always require booster shots, with the same packaging as the initial dose.

Toxoid Vaccines

Toxoid vaccines teach the body how to fight a harmful chemical called a toxin. Toxins are first produced in a bioreactor and then inactivated by treatment with formalin, a combination of formaldehyde and sterile water. This produces a harmless substance called a toxoid that can be used to train the body to handle the more dangerous toxin from the infection. The immune system learns how to lock on to the toxin and make it inactive. This is the body's way of producing a chemical block.

Packaging for toxoid vaccines is similar to that used for most inactivated vaccine products. They may be liquids or can be freeze-dried and are supplied in glass vials with stoppers, the typical parenteral packaging unit.

Subunit Vaccines

Subunit vaccines take the idea of an inactivated vaccine one step further. These vaccines use only one component of the offending microbe or virus to train the immune system. The part of the microbe used for subunit vaccines is the antigens. These are the chemical fingerprints found on the surface of a bacterium or virus that alerts the body to an invader and initiates the internal mechanism that identifies and fights the offending microbe. Some of the newer forms of subunit vaccines take this even further by using only the epitopes, the very specific part of the antigen that disease-fighting T-cells identify and bind to. The major advantage of subunit vaccines is the fact that they only use a part of the microbe, and thus have a lower risk of adverse reactions than other types of vaccines.

Packaging may come into play during multiple parts of the manufacture of subunit vaccines. After the antigens are identified, a bacteria strain is cultured and grown on an industrial scale to produce large quantities of the antigen. The bacteria are treated with various chemicals and enzymes that break down the microbe and permit the harvesting and separation of the antigens. A subunit vaccine normally contains more than one antigen and may contain 20 or more antigens.

Reusable packaging is needed if a time lag is incurred between the different steps in the process, and then more standard packaging that addresses protection and transportation needs is applied to the completed product. Parenteral vials, ampules, and prefilled syringes have all been used to package these products.

Hepatitis B vaccine is a subunit vaccine that uses rDNA technology to produce the antigens. Genes from hepatitis B that encode the antigen are placed into baker's yeast. This is then grown and harvested for the antigens.

Conjugate Vaccines

Conjugate vaccines are a way to develop vaccines that are effective for infants and young children whose immune system cannot recognize or do not react to other forms of a vaccine. Bacteria that have a polysaccharide coating are able to hide their identifying antigens under this shell. Infants and young children's immune systems do not recognize this type of invader and do not respond to the disease.

Conjugate vaccines are a form of subunit vaccines. They link the polysaccharide to antigens or toxoids that the infant or young child's immune system can identify. This permits the immune system to react to the polysaccharide and identify and defend against the invading bacterium.

Packaging for these vaccines is similar to that used for subunit vaccines.

DNA Vaccines

DNA manipulation is still an experimental technique undergoing development to produce the next generation of vaccines. DNA vaccines are in development and testing for malaria, HIV, herpes, and influenza.

Scientists using this technique can now identify the genome of an offending bacterium or virus and use only those parts of the genome that produce the microbe's antibodies. These are the parts of the gene that can be engineered into an effective vaccine. By using only the genes within the offending microbe that produce the antigens and placing them in a form that can be taken up by some cells within the body, they stimulate these cells to make and display on their surface or release into the bloodstream the same antigens that the dangerous organism displays. DNA vaccine systems make the body's own cells replicate and display on their surface the dangerous microbe's antigens. These cells with foreign antigens on the surface or the free antigens they release then stimulate the body's immune system to recognize and defeat the invading organism. In effect, our body's own cells become a bioreactor to create the vaccine. Think of it as internal vaccine-producing factories built into our cells.

These vaccines are safe because they contain only a fraction of the offending bacterium's gene. They are relatively easy to make and are much less expensive than vaccines made with current techniques. These vaccines also promise to reduce the current manufacturing cycle for a new vaccine. This is important for vaccines used to prevent influenza, which now require the Center for Disease Control (CDC) and others to predict what strains of flu are included in the vaccine each year. Vaccines can contain some but not all the possible influenza components, and a determination of which five to include is made by the CDC each year. Shortening the period between identification and delivery reduces the possibility that the virus will mutate, and the vaccine originally designed for flu season will be ineffective.

The vaccine is administered by needle and syringe, or it can be shot into the patient using high-pressure gas. The high-pressure direct injection method forces DNA-coated gold particles through the skin and into the cells. This type of vaccine is sometimes called a "naked DNA vaccine." Another technique mixes the DNA vaccine with molecules the body absorbs quickly into cells.

Recombinant Vector Vaccines

Recombinant vector vaccines use attenuated or noninfectious microbial DNA cells that produce the desired antigens and introduce them into the body using a host organism (7). The term "vector" refers to the host virus used to carry the microbial DNA into the subject's body and inject it into cells. In this technique, the attenuated or benign virus receives genetic material from the disease-causing virus. It then carries that DNA into other cells. The advantage of these vaccines is that they closely mimic the real disease-causing microbe and create a strong immune response in the body.

This technique is used in a little different way for bacterial response. A benign or attenuated bacterium is given the antigen-producing DNA from a dangerous microbe. The bacterium uses the genes to produce and display the antigens of the dangerous microbe on its surface. The benign bacterium with the new genes is then treated and destroyed by the body in the same way as the dangerous pathogen would be treated while conferring immunity on the subject. An equine recombinant vector virus was approved in 2005 for veterinary use. This technique remains experimental in humans. Both the FDA (1) and the European Medicines Agency (EMEA) (7) are following developments in this technology closely.

Combination Vaccines

Combination vaccines are used to deliver multiple products to the subject in one dose. Children are required to receive a DTP (diphtheria/tetanus/pertussis) injection and an MMR (measles/mumps/rubella) vaccination before attending school in almost every state.

Combination vaccines provide a number of benefits. They reduce the number of a doctor's visits a child requires to gain immunity to many dangerous diseases and thus reduce the overall cost of health care. They are of benefit to parents because they reduce the number of times a child must see a doctor or health care professional for the initial vaccination and booster shots, improving the likelihood of complete immunity in the child.

The amazing thing about the child's body is the number of T-cells and B-cells the immune system maintains to fight an infection. There are billions of these cells circulating constantly in our bodies. There is more than enough on hand to fight the multiple components of a combination vaccine.

Other Vaccine Components

Vaccines contain ingredients designed to enhance and improve the immune response of the body. These ingredients are called adjuvants. The only types of adjuvants approved for use in the United States are aluminum salts. The compounds perform a number of different actions. They bind the antigens in the vaccine; they deliver antigens to the lymph nodes, one of the key components of the immune system; they help retain the vaccine at the site of the injection; and they are taken up by the macrophages permitting these cells to better present the antigen to the lymphocytes. The slow release of the vaccine caused by the aluminum salts increases the body's response to the antigen, as does the help it provides in presenting the antigen to the lymph nodes.

Other components in vaccines are antibacterial compounds, thermal and chemical stabilizers and buffers to help the vaccine maintain potency when exposed to high temperatures, and diluents to standardize the vaccine dose.

By law vaccines in multi-dose vials must contain a preservative (Table 4). The Code of Federal Regulations [21CFR 610.15(a)] requires that a preservative be present in levels that are nontoxic and persistent enough to maintain sterility of the product throughout its shelf life (8). The antibacterial and antifungal compounds are designed to keep the vaccine sterile. Infections have been transmitted from patient to patient by contaminated vaccines. The vaccines become contaminated through multiple punctures (needlesticks) through the seal of multi-dose vials. The addition of antibacterial or antifungal preservatives to keep a vial of product sterile while in use is part of multiple dose–packaging design. The FDA (8) has approved a number of these compounds, but one material, thimerosal, has been controversial in recent years.

Thimerosal is an organomercurial that has been used in vaccines since the 1930s as a preservative. The body can metabolize this substance into ethylmercury and thiosalicylate. It has been extremely effective in preventing microbial growth in vaccines. During the late 1980s and early 1990s questions regarding its safety in infants were raised. Particularly, questions about a link between vaccines containing thimerosal and autism, attention deficit hyperactivity disorder (ADHD), and speech and language delay were examined. The Institute of Medicine, specifically the Immunization Safety Review Committee,

Table 4 Preservatives used in U.S.-Licensed Vaccines

Preservative	Vaccine examples (trade name)
Thimerosal	DT
	Td (several manufactures)
	TT (several)Influenza (several)
2-phenoxyethanol and formaldehyde	IPV (IPOL; Sanofi Pasteur, SA, Lyon, France)
	DTaP (Daptacel; Sanofi Pasteur Ltd., Lyon, France)
Phenol	Typhoid Vi polysaccharide (Typhim Vi, Sanofi Pasteur, SA, Lyon, France)
	Pneumococcal polysaccharide (Pneumovax 23, Merck & Co. Inc.,
	West Point, Pennsylvania, U.S.)
Benzethonium chloride (phemerol)	Anthrax (Biothrax; BioPort Corporation, Michigan, U.S.)
2-phenoxyethanol	DTaP (Infanrix; GlaxoSmithKline Biologicals, Rixensart, Belgium)
	Hepatitis A/hepatitis B (Twinrix; GlaxoSmithKline Biologicals, Rixensart, Belgium)

Abbreviations: DT, Pediatric Diphtheria and Tetanus vaccine; Td, Adult/Adolescent Tetanus and Diphtheria vaccine; TT, Tetanus Toxoid; IPV, Inactivated Polio Vaccine; IPOL®, Poliovirus Vaccine Inactivated, produced by Aventis Pasteur SA; DTaP, Pediatric Diphtheria, Tetanus and Acellular Pertussis vaccine.
Source: U.S. FDA, CBER.

examined this issue and on October 1, 2001, issued a report stating that the evidence gathered was insufficient to accept or reject a causal relationship between this chemical and the disorders. In 2004, the Committee issued a final report using epidemiologic evidence from the United States, Denmark, Sweden, and the United Kingdom stating that the data favor a rejection of any link between these disorders and thimerosal.

The FDA has worked with manufacturers to eliminate thimerosal (8) from vaccines and thus limit infants' exposure to mercury. Because of this effort, the majority of vaccines that once contained thimerosal are now supplied in thimerosal-free versions, and new vaccines entering the market are thimerosal free or only contain minute traces of the product.

Next Generation Vaccines

Vaccines are going to undergo significant change in the near future, and the packaging for vaccines is going to change as well. Conventional shots for immunity will be replaced by skin patches and sprays up the nose. Genetic engineering of the various viruses also hold out the possibility that common foods can be modified to express the antigens or other markers necessary to trigger an immune response within our bodies. Transgenic animals may express milk or other consumables that act as vaccines (6).

The packaging changes are not far off. Already insulin and flu vaccines are available in inhalable form. Sophisticated inhalers, developed with new packaging, provide a simple and easy way to dose insulin without the needlesticks. A flu vaccine for adults that is administered as a nasal spray has proven effective and is available in limited quantities. These two examples are only the beginning of how new biologic drugs and vaccines will be administered in the future.

SUMMARY

Biologics and vaccines, particularly the newer vaccines, are an area that will undergo a great deal of package design and development. Our ideas about injecting a vaccine into the body are giving way to a number of new methods for introducing the vaccine into the body which are based on the many new methods used for producing the vaccine products.

Biologic drugs will need significant package development. Protection from light and heat will be the obvious changes, but the most significant changes will come from new methods of delivery that make their use simple and foolproof for the patient.

FURTHER READING

Skin Therapy Letter.com 2005—Skin Care Guide.com. Available at: http://www.skintherapyletter.com/treat/psoriasis/biologic_drugs.html.

Diaz-Borjon A, Weyand CM, Goronzy JJ. Treatment of chronic inflammatory diseases with biologic agents: opportunities and risks for the elderly. Exp Gerontol 2006; 41 (12):1250–1255.

Felcone LH. The Long and Winding Road to Biologic Follow ons. Biotechnology Healthcare. 2004:20–29. Available at: http://www.biotechnologyhealthcare.com/journal/fulltext/1/2/BH0102020.pdf

REFERENCES

1. Kraniznor M. Overview of the Drug and Biologic Review Process. U.S. FDA, 2000. Available at: http://www.fda.gov/oia/embslides/drugreview/sld001.htm.
2. U.S. FDA. Drugs@FDA, Glossary of Terms. CDER, Updated January 2007.
3. About.Com: Arthritis, Biologic DMARDs—Dosage—Side Effects—Interactions—Warnings, The New York Times Company, About, Inc. 2008. Available at: http://arthritis.about.com/od/brms/Biologic_DMARDs_Dosage_Side_Effects_Interactions_Warnings.htm.
4. Gabriel SE, Coyle D, Moreland LW. A clinical and economic review of disease modifying antirheumatic drugs. Pharmacoeconomics 2001; 19(7):7715–7728.
5. Rucker, L, Purvis L. AARP Overview on Biologic Drugs. AARP Public Policy Institute, June 2007.

6. U.S. Department of Health and Human Services, National Institutes of Health, National Institute of Allergy and Infectious Diseases. Understanding Vaccines, What they are and How They Work, 2003. (NIH Publication No. 03-4219). Available at: www.niaid.nih.gov.
7. European Medicines Agency (EMEA), Committee for the Medicinal product for Human Use. Concept Paper on the development of a guideline on live recombinant vector vaccines. London, 2007. (Reference Document EMEA/CHMP/308139/2007). Available at: http://www.emea.europa.eu/pdfs/human/vwp/30813907en.pdf.
8. U.S. Federal Food and Drug Administration, Center for Biologics Evaluation and Research. Thimerosal in Vaccines. Available at: http://www.fda.gov/cber/vaccine/thimerosal.htm#pres.

4

Medical Foods

INTRODUCTION

Medical foods is a topic few people associate with pharmaceuticals. Many would consider them a new direction for pharmaceutical products. Medical foods in reality are products prescribed by a physician for a patient with a medical problem. The medical food may be used to manage a disease or health condition, or in the case of an infant, it may be the sole source of nutrition for the child during its first year of life. Medical food sustains the patient under the care of a physician over an extended period of time (1). Infant formula provides nutrition to babies when the breast milk is unavailable or inadequate to meet the infant's nutritional needs. Both types of products have a variety of unique characteristics and regulations required for manufacture and packaging. These requirements go beyond those normally associated with foods and highlight some of the problems of multiple agencies and multiple regulations affecting production.

Medical foods have high–unit volume packaging, meaning that the number of packages produced is much higher in unit volume than the number of units produced for most prescription drugs. They are regulated differently from standard foods and fall into an intermediate regulatory-defined space between consumer food products and true pharmaceuticals. As a product group, they are used in hospitals, nursing homes, and individual patients' homes. They are administered by a wide variety of caregivers who possess a wide range of training and skills. The packaging of both the food products and the medical devices required for administration, if tube feeding or parenteral infusion is used, requires understanding of the product composition; the types of diseases they treat; the manufacturing, processing, and packaging operations that create them as well as of the varying conditions under which they are administered.

Packaging of these products must be easily understood by a wide variety of caregivers with varying levels of training. The packages and the devices used to administer the products must clearly differentiate between enteral and parenteral feeding. The packages must protect the product without any adverse interaction and provide a method to measure how much nutrition is being supplied to the patient in a 24-hour period. Packaging is an integral part of the design of a product and the method used to manufacture it. In aseptic manufacturing, it is integrated into a complete product delivery system and receives as much scrutiny as the product and the equipment. Packaging of parenteral nutritional products is similar to that used for parenteral drugs, and the differences between the two parenteral products are discussed in this chapter.

Medical foods come in two different forms, enteral and parenteral (1). Enteral nutrition is taken through the mouth and digested through the gastrointestinal (GI) tract; it can also be supplied to a specific section of the GI tract. Parenteral nutrition is supplied directly into the circulatory system (1). Enteral products are covered by the Federal Food, Drug, and Cosmetic Act (FFDCA) of 1938 and are classified as foods. This definition has been modified over the years, but enteral products remain a specialized class of food. Infant formula is covered by specific sections of the federal code. Parenteral products are fed directly into the patient's circulatory system and are considered drugs by the Food and Drug Administration (FDA). Parenteral products require preapproval by the agency even though they resemble enteral food products in chemical composition.

HISTORY OF MEDICAL FOODS

Medical foods have developed rapidly over the last 50 years. During the 1930s and 1940s elucidation and understanding of how metabolic pathways worked, including their role in sustaining life, were the initial steps in understanding the role played by food and how it was used by the body. Part of this understanding lay in the unraveling of metabolic disorders and the role different food components played in causing or mitigating the disorder. Further understanding of the critical roles played by amino acids, fatty acids, vitamins, and minerals was needed before a medical food could be formulated. Additional understanding of the complete cycle of metabolism and the contribution of all types of nutritional components such as proteins, carbohydrates, vitamins, minerals, fats, and fiber also improved the formulation of medical foods.

Part of this increase in knowledge was the explanation defining a number of metabolic diseases and the role specific foods or specific ingredients in food played in their manifestation. Medical nutritional products were developed for two reasons: to support hospitalized patients who exhibit metabolism disorders and the recognition that hospitalized patients have specialized physiological and nutritional needs and requirements. Some of the key requirements for medical nutritional support of hospitalized patients include the following:

- An increased need for protein by patients suffering from severe trauma or sepsis.
- Malfunctions of the organs that process food or food metabolites (e.g., the kidneys, stomach, pancreas, liver, and intestine).
- Obstruction or resection of the bowl or GI tract that restricts the patient's ability to absorb nutrients.
- Diseases of the GI tract.
- Treatment procedures for a disease or injury such as cancer, AIDS, and burns that physically impair the GI tract and the patient's ability to ingest nutrition. This includes dysphagia and the complete loss of appetite during radiation and chemotherapies.
- Surgical procedures, including those that result in mechanical or anatomic changes in the GI tract.
- Neurological disorders such as stroke, Alzheimer's disease, cerebral palsy, Parkinson's disease, and multiple sclerosis.
- Patients, particularly infants, with inborn errors in metabolism, which include homocystinuria, galactosemia, fructosuria, hyperlipoproteinemia, phenylketonuria (PKU), and maple syrup urine disease (MSUD).

In the last point of the list above, the patients must avoid specific nutrients or food components to prevent illness, mental retardation, or death, and they must ingest specific and increased amounts of metabolites needed to stimulate critical metabolic pathways.

REGULATORY REQUIREMENTS OF MEDICAL FOODS

From the very beginning of the FDA in 1906, there have been problems with defining the interface between drugs and foods. This has been a particularly difficult area to regulate because foods can and do make claims that change physiology and thus fall into the definition of a drug. The FDA, throughout its history, has pursued regulatory and legal action against products and companies producing medical foods and infant formula that make unique or special dietary claims that are similar to drugs without scientific substantiation. It is very difficult to make this distinction and is even more perplexing when one considers normal (consumer) foods come with labeling that makes health claims (disease prevention claims) such as lowering cholesterol or claims made by dietary supplements that are permitted as "structure function claims" as defined by the U.S. Congress in 1994. Composition, manufacturing, and distribution of these pseudo-foods also fall into the jurisdiction of a number of other agencies within the federal government structure, including the United States Department of Agriculture (USDA), while much of state law supports or mimics parts or all the federal regulations (2). This is one reason medical foods and infant formulas are restricted by the unique regulatory demands and testing needs, including clinical testing to a select group of manufacturers. Most of the regulatory aspects are detailed in the following sections of this chapter and are separated from the regulatory overview of chapter 5.

MEDICAL FOODS

Medical foods are the marriage of traditional food products and pharmaceutically proven nutritional needs formulated into food products with specially designed nutrient attributes. These two industries come together in unique products that shape a stabilizing effect or actual improvement in a patient's condition through nutrition. They may be used exclusively or with other medications and food supplements to treat a disease or chronic condition. They are used in hospitals to supplement or sustain patients in need of nutrition. Sick people do not always have the appetite or the ability to consume enough calories and other nutrients needed to combat disease, sustain and enhance the immune system, recover from surgery, or maintain body weight. The use of both enteral and parenteral medical foods for hospital patients, nursing home patients, and people in home care have greatly contributed to their overall improvement and well-being during a recovery from surgery or disease.

Medical nutritional products are used to sustain patients in a coma or in situations in which they cannot feed themselves. They also provide unique combinations of nutrients that address and help manage conditions like diabetes or provide a unique blend of digestible ingredients to improve the well-being of patients suffering from a number of different medical conditions.

Medical foods are an emerging new product area for pharmaceutical companies and food companies. Both manufacturers see the opportunity to extend their traditional strengths and product offerings into a new area.

Medical food products must not be confused with or promoted by companies to satisfy a consumer demand for health and wellness alternatives to existing food products. *Medical foods are distinct from products marketed to the public at large and are used under the supervision of a physician.*

Pharmaceutical companies see these products as another method for treatment of difficult chronic conditions, as ways to enhance the body's ability to resist disease, or as ways to mitigate the effect of treatments that insult the body (e.g., radiation or chemotherapy used for cancer treatment).

"Nutraceutical" is a new term used to describe medical foods, although its meaning is not as specific as it sounds. It has crept into the vocabulary of the press and general public to describe medical foods. The definition of the word is as follows: "A nutraceutical is a food or a naturally occurring food supplement thought to have beneficial effect on human health."

Medical nutritional or nutraceutical products encompass both enteral and parenteral dosage methods and combine multiple ingredients to achieve a clinically proven therapeutic effect. They are distinctly different from vitamins and supplements. The proven medical benefit is supported by clinical research with a patient population random enough and large enough to determine the benefits and shortcomings of a product. Products with clinically proven claims do not rely on anecdotal evidence for support.

Medical foods, in contrast to vitamins and supplements, have their claims closely regulated. Both the FDA and the Federal Trade Commission (FTC) monitor and challenge the claims made by manufacturers of this subclass of food products. The FTC becomes involved if a medical food is marketed to the public at large. The claims are not as expansive as drugs, and support the way the product, most likely in combination with other products, improves the patient's well-being. They are positioned in what is considered a gray area of pharmaceutical products between foods and drugs.

Packaging and labeling of medical foods are very specific. Labeling has been exempted from standard food and nutritional labeling requirements but still must meet a rigid and unique set of requirements that state definitively that it is a medical food. The label must state that the product is directed toward treating or managing a specific medical condition or medical disorder. A good example of this type of labeling and the type of product it describes is found on labels for foods formulated to be free of the amino acid phenylalanine for treatment of PKU. Examples of other chronic conditions treated with medical foods and labeled as such include anorexia, hypermetabolism, malabsorption, atrophy of various organs, impaired cell-mediated immunity, and anemia.

Medical foods must not be confused with foods that are part of a healthy diet or foods that are marketed to decrease the risk of disease. Examples of products promoted or classified as regular foods and considered part of a normal diet include fat-free foods, low-sodium foods, or foods marketed to promote lower cholesterol or a healthy heart. Medical foods used for weight loss are highly specialized food forms, including liquid diets that are used under the supervised care of a physician.

The FDA has seen a marked increase in the number of products being marketed as medical foods during the last 10 years. This is happening in response to greater awareness by the medical community regarding the role of nutrition in patient's well-being and as understanding of food's role in patient's recovery from disease, surgical intervention, or in metabolic conditions becomes better understood.

COMPOSITION AND FORMULATION OF MEDICAL NUTRITIONAL PRODUCTS

The basic idea behind any medical food is that a combination of traditional food products formulated in a unique way can treat or help the patient with a medical condition. It combines the idea of good nutrition with active or modified ingredients that produce a clinically substantiated outcome to a condition or disease. This type of thinking is new to the public at large but not new when you think of specialized products that address nutrition and therapeutic needs of specific patient groups.

Medical nutritional food products comprise the same ingredients as regular foods including protein, carbohydrate, fat, fiber, and water along with vitamins and minerals. The recommended daily allowance (RDA) of food is the basis for formulation of complete foods and is considered when specialized medical nutritional products are used to support a patient. Included in every hospital staff is a trained dietician to advise and assist with nutritional requirements of oral diets and the diets of special needs' patients, including patients supplemented or sustained by enteral feeding. Additional supplements may be needed when a disease or the medication used to treat a disease interferes with nutrient function and metabolism or if they decrease nutrient absorption or synthesis.

Formulas may be polymeric, that is, containing intact protein and other intact nutritional components, or they may be partially hydrolyzed by enzymatic action into smaller components of these molecules. Partially hydrolyzed formulas have unpleasant odors and taste.

Medical foods are chosen on the basis of the patient's absorptive or digestive capacity. The various components comprise the essential elements needed by a patient to sustain the body.

Protein sources are predominantly based on soy or casein. Protein may be intact or partially or completely hydrolyzed. Intact protein molecules are high-molecular-weight molecules, while hydrolyzed protein consists of smaller fragments called polypeptides. Enzymes produce the polypeptide from the intact protein. The enzyme breaks down (digests) the protein into smaller fragments called oligopeptides, and still smaller species called di- and tripeptide fragments.

Carbohydrates are the next component of enteral nutritional formulas. They are primarily derived from cornstarch or maltodextrin. Carbohydrates consist of starch, disaccharides, which may be sucrose (composed of glucose–fructose), maltose (composed of glucose–glucose), or lactose (composed of glucose–galactose), and other small sugars, along with monosaccharides such as glucose, dextrose, and fructose.

Another component of an enteral formula is fat (Fig. 1). Fat comprises saturated fatty acids, polyunsaturated fatty acids, and medium-chain triglycerides. Fat in enteral food formulas improves the palatability of the product and also improves the mouth feel and flavor of the product when taken orally. A variety of vegetable oils are the source of fatty acids in the formulas, with a key component being linoleic acid (Fig. 1). Fat calories are a key component of the total nutritional energy provided to the patient.

Vitamins and minerals are essential to the body and to the nutrition provided by an enteral formula. Enteral products are designed to meet the complete vitamin and micronutrient needs of the patient. Some disease-specific formulas are incomplete in their vitamin and mineral makeup when required to treat a specific metabolic condition of the patient. Disease-specific formulas are designed to treat specific organ dysfunction or metabolic stress. Fat-soluble

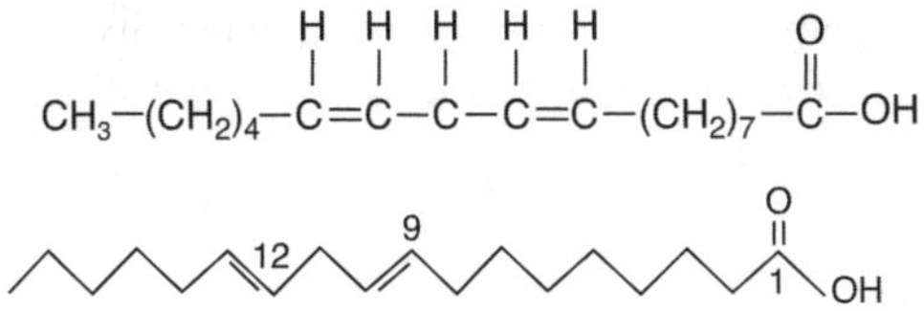

Linoleic Acid
cis, cis 9, 12 - octadecadienoic acid (IUPAC Name)
omega - 6 fatty acid
(Safflower Oil contains 78% linoleic acid)
N-6 unsaturated
$C_{18}H_{32}O_2$

Figure 1 Linoleic acid.

vitamins may be indicated for patients with specific vitamin deficiencies. An example would be vitamin K, a nutrient that is normally produced by the intestinal flora of the GI tract. This type of deficiency is rare, but most enteral formulas contain some amount of vitamin K.

The last two components of an enteral formula are fiber and water. Fiber is made up of insoluble fiber such as cellulose and hemicellulose and soluble fiber comprising pectin, mucilage, algal polysaccharide, and gum. The normal amount of fiber in a nutritional product is 5 to 14 gm/L, and the most commonly used source of fiber is soy polysaccharide.

The amount of water in an enteral product is expressed as water content or moisture content. The total quantity of water is expressed as milliliters of water per liter of formula or milliliters of water per 1000 milliliters of formula. Most enteral formulas range between 690 and 860 mL of water per 1000 mL of formula. Nutritional formulas for adults contain 10 cal/cc of formula or 1.5 to 2.0 cal/cc of liquid. This is a lot of nutrition in a small volume of product.

Medical foods are classified into four different and distinct categories:

1. Nutritionally complete products
2. Nutritionally incomplete products
3. Formulas for metabolic or genetic disorders
4. Oral rehydration solutions

Nutritionally Complete Products

Nutritionally complete products are exactly what the name says—products that deliver a diet sufficient in protein, fat, carbohydrates, vitamins, and minerals to sustain a patient receiving no other source of nutrition. The products may or may not contain fiber. The products are used to support patients with a wide variety of medical conditions. The products vary in composition from natural or whole food components, intact protein, long-chain fats, and complex carbohydrates to those with partially digested or partially reduced components such as medium-chain triglycerides, simple sugars, and amino acids. Caloric density, the amount of energy delivered by a standard amount of the medical nutritional product, is another variation in this class of products.

Nutritionally Incomplete Products

Nutritionally incomplete products are generally referred to as modular products. These products may contain only a single nutrient or a limited group of nutrients that are insufficient to maintain the patient by themselves. Modular components provide flexibility in the nutrition delivered to a patient and can, when combined, be formulated into a nutritionally complete product. The major problems with incomplete products are the training and education necessary to avoid microbial contamination from environmental factors when they are combined or mixed, the expense of multiple components, the potential for additional nutritional problems due to the lack or excess of specific nutrients in the blended formulas, and the lack of sufficient skilled labor in many institutions to mix the components in the proper order.

Formulas for Metabolic or Genetic Disorders

Formulas for metabolic or genetic disorders are made for patients with specific inborn errors in their metabolism, such as cystic fibrosis, propionic academia, PKU, or MSUD (3). Although these products compositionally are not complete in the normal sense, they are complete in providing all the nutrition needed by a person with one of these metabolic disorders. For infants and juveniles, they provide complete nutrition for growth and development while managing the metabolic or genetic disorder. Patients under this type of care normally see their histopathology change over time. These changes require modification of the modular formula in new and different directions and are one of the more difficult decisions a physician must make because the change may be in an unknown or unusual direction.

Oral Rehydration Solutions

Oral rehydration solutions are products used to rehydrate the body enterally with both water and electrolytes. The products are manufactured for infants and adults with varying amounts of sodium, potassium, citrate, dextrose, and sterile water. An analog to such products is found in parenteral solutions used to maintain or replace water in the body. Infants and adults suffering from severe diarrhea are treated with rehydration products to maintain electrolyte balance and hydration of the body. Probably the best consumer analog to this class of products is Gatorade®.

ENTERAL PACKAGING

Packaging of an enteral product in cans and glass bottles uses the same techniques and processes as consumer products in the same packages. The package must be able to withstand the sterilization or processing of the product discussed later in the chapter and must also be able to protect all of the food components from degradation over an extended period of time, typically one to three years, which is required to meet the needs of the health care distribution system. Packaging, processing, and the product itself are inextricably linked in providing enteral nutrition to critically ill patients.

Medical nutritional products typically come in two forms, powders or liquids. The types of packaging used for medical nutritional products include metal cans, glass bottles, plastic bottles (multilayer), plastic films (multilayer), paperboard composites, pouches, and laminated multilayer plastic structures.

MEDICAL FOOD ADMINISTRATION TO THE PATIENT

Medical foods are delivered to the patient using either enteral or parenteral routes into the body. Both forms used are complementary to each other; however, the preferred method of administering nutrition to a patient is through the enteral route using the GI tract. Oral feeding of a patient via the mouth is always best when possible, but when problems with swallowing, impairment from stroke, or the presence or absence of dysphagia, along with other problems prevent a patient from eating or drinking normally, enteral nutrition is used. Enteral nutrition is used to support patients when the GI tract is functioning properly. Many times you will hear doctors comment, "When the gut works and can be used safely, use it."

The majority of patients supported by oral means use medical foods. The medical foods supplied for oral intake are a form of a tube-fed product that is made appealing and tasty to the patient. Hospitals provide a variety of specialized diets and dieticians who work in conjunction with physicians to determine and support unique nutritional needs of patients. Nutrition for patients who can swallow or eat on their own is supplied in the form of liquid products, although some solid food may be included with a medical nutritional liquid. The liquid form of the nutritional product is used because it is easier for patients to drink than to eat, and liquid products are typically easier for a patient to digest. These products can supply the calories needed by the patient in a concentrated form when they cannot eat a sufficient volume of food. One drawback to liquid feeding is that many patients find it less satisfying and less filling than solid food.

Feeding patients usually progresses through three different stages. In the first stage, the patient eats or drinks a regular or modified diet; in the second stage, the regular or modified diet includes supplements; and in the third stage, tube feeding supports the patient who is unable to ingest food. Enteral nutrition provides the calories a patient needs when he/she cannot support his/her nutritional needs through oral intake of food products.

The term "enteral feeding," although applicable to oral feeding, typically refers to tube feeding. Oral feeding of a patient is always preferred, and tube feeding is used only when oral feeding is not possible. Tube feeding is instituted when the patient has trouble swallowing or when other physical problems prevent the patient from taking food orally. Enteral feeding involves administration of food products through a tube directly into the GI tract. It provides many benefits to the patient. Food in the GI tract promotes blood flow, secretion of hormones, and GI tract integrity; prevents the translocation of bacteria from the gut; and even in small amounts maintains the trophic or minimum nutritional level needed for several gastric hormones. Parenteral nutrition cannot supply glutamine, a key source of nutrition for the small intestine.

Tube Feeding

Enteral nutrition uses a tube placed in a specific portion of the GI tract to feed the patient. Enteral food products, when administered via a tube placed in a specific position of the GI tract, use a number of different medical devices such as pumps, tubes, endoscopes, and other surgical equipments to put in place and maintain the feeding system along with the specially designed and prepared food product. The packaging of the product is an integral part of the system along with all the related medical devices, and everything works in combination to deliver the product to the GI tract of the patient.

Tubes are placed into the GI tract of the patient in two ways, either via the nose (nasoenteric) or via a direct surgical incision in the abdomen and the stomach (4).

Nasogastric tubes are passed through the nose and placed in different portions of the patient's GI tract depending on need. Nasal tubes are only used for a short period of time, usually six weeks or less. The names associated with this type of tube feeding refer to their placement in the GI tract and include nasogastric, nasoduodenal, and nasojejunal tubes. One sidelight is prisoners refusing to eat are force-fed using nasogastric tubes.

The second method of tube placement is tube enterostomy. This is a surgical procedure that may be performed laparoscopically, percutaneously, or by a regular surgical operation. The tube is placed in the same position in the GI tract as with the nasoenteric method, and the terms used to describe these tubes are much the same. The most common placement of a tube in a patient via surgical means is called a PEG (percutaneous endoscopic gastrostomy). Other examples include esophagostomy, percutaneous endoscopic jejunostomy (PEJ), needle catheter jejunostomy (NCJ), operative laparoscopic gastrostomy, and operative laparoscopic jejunostomy.

The gastric feeding tube or "G-Tube" provides a long-term solution to feeding patients with conditions that prevent them from taking nutrition by normal (oral) means. Conditions that require tube feeding include stroke, esophageal atresia, and tracheoesophageal fistula. The G-tube is also used to prevent the occurrence of aspiration pneumonia. Feeding tubes surgically implanted are used to maintain the weight of patients who cannot take enough food orally to avoid malnutrition or weight loss. They can be used in reverse to empty the contents of a patient's stomach.

A jejunostomy tube is similar to a gastric tube but is placed in the jejunum, the portion of the small intestine between the duodenum and the ileum. They are used when the upper GI tract must be bypassed. These tubes differ from the standard G-tube in the use of a smaller diameter tube with a finer bore.

All of these components used for tube feeding must be packaged and sterilized. Radiation and gas sterilization using ethylene oxide are the two most common methods for sterilization and both require a package tailored to the device and the sterilization method. Packaging can range from a thermoform tray for the tube or a tray that is a complete kit that contains not only the feeding tube but also all the other items needed to implant the tube. These kits are designed as a one-time-use item.

Parenteral Nutrition

Parenteral nutrition is used to feed patients when the GI tract cannot be used safely. Parenteral nutritional products are considered drugs and are infused directly into a central vein or a peripheral vein of a patient. Parenteral nutrition is used for the same reasons as enteral nutrition, that is, to prevent the effects of malnutrition and to prevent or correct specific nutrient deficiencies in the patient.

Intradialytic parenteral therapy (IDPN) is a form of parenteral nutrition used with patients undergoing hemodialysis. The parenteral product is delivered during the dialysis treatment to retard or stop the effect of protein calorie malnutrition, which occurs in 25% to 40% of patients undergoing chronic dialysis.

For both parenteral and enteral nutrition, 2000 to 2200 cal/day is required to maintain body weight. The use of parenteral nutrition to supply less than 750 cal/day is considered supplemental.

Typical conditions that necessitate the use of parenteral nutrition include

- Crohn's disease;
- obstruction secondary to stricture or neoplasm of the esophagus or stomach;
- loss of the swallowing mechanism due to a central nervous system disorder, where the risk of aspiration is great (stroke);
- short bowel syndrome secondary to massive small bowel resection;
- malabsorption due to enterocolic, enterovesical, or enterocutaneous fistulas (parenteral nutrition being temporary until the fistula is repaired); and
- motility disorder (pseudo-obstruction).

PARENTERAL FORMULATIONS FOR INTRAVENOUS FEEDING

Parenteral nutrition comes in a number of forms. Each component is packaged separately and must be designed to be easily blended into a complete prescription determined by the physician. This is in contrast to enteral nutritional products, which have moved away from individual components to formulas that are specialized to treat diseases and conditions without the requirement of hospital, nursing home, or in some cases, family members mixing the products.

Parenteral products require good aseptic technique in their blending to avoid microbial contamination. Again, the design of the package components aids in minimizing microbial problems; however, the multiple components put together in a 1, 2, or 3 L bag for infusion present multiple opportunities for error. This is one area where new and unique package design is improving the existing product delivery systems. Many packages are similar to or the same as those used for drug products, so the possibility of mix-up or error is always a problem. Creative design and labeling can prevent errors in identification.

Fat emulsions are the primary source of calories in parenteral nutritional products. A total of 15% to 50% of the total caloric content of a parenteral nutritional product is delivered by fat. Fatty acid deficiency is a concern with this type of energy delivery, and 2% to 4% linoleic acid (Fig. 1) is included as part of the mix to avoid the condition.

Plasma triglycerides are used to determine the patient tolerance for the amount of fat being delivered. Patients with severe liver disease, liver dysfunction, or hyperlipidemia AIDS have problems with high fat and have a decreased ability to clear infused fat that must be treated cautiously with this component of a parenteral nutritional mixture.

The carbohydrate requirement of the parenteral nutritional mixture is another component that must be calculated by the physician. The body has a limited ability to process glucose, and excess glucose is converted to fat, which is an additional stress on the patient. Overfeeding with glucose can result in hyperglycemia, fatty liver, and weight gain. This is avoided by calculating the maximum carbohydrate tolerance rate using the following formula:

$$\text{Maximum carbohydrate (g/day)} = 5 \text{ mg carbohydrate/kg/min} \times \text{ideal body weight (IBW)} \times 1.44$$

$$\text{where } 1.44 = (60 \text{ min/hr} \times 24 \text{ hr/day})/(1000 \text{ mg/g})$$

Electrolytes are the next component of the parenteral nutritional formulation. The standard mix of electrolytes includes the following elements:

- Sodium
- Potassium
- Calcium
- Magnesium
- Acetate
- Gluconate
- Chloride

There are a number of considerations for electrolytes. Potassium is normally not included in standard electrolyte mixtures. Cramps may result from excessive potassium in the parenteral nutritional solution. Single electrolyte components are available to tailor the electrolyte composition to individual patients. Nonstandard electrolyte formulations must be formed with a balance of anions and cations.

A clinical pharmacist is consulted when designing nonstandard electrolyte formulas for individual patient needs. Another concern with electrolyte composition is the amount of potassium prescribed for patients with renal impairment, which requires caution and oversight of the physician.

Vitamin requirements of parenteral nutritional products are another part of their composition. A typical vitamin profile includes

- Vitamin A
- Vitamin D
- Vitamin E
- Vitamin C
- Thiamine (B_1)
- Riboflavin (B_2)
- Niacin (B_3)
- Pyridoxine (B_6)
- Pantothenic acid
- Folic acid
- Biotin
- Vitamin B_{12}

Manufacturers prepare a standard premix package of the vitamins in the above list, which is formulated to meet both the recommendations of the American Medical Association (AMA) and the FDA-recommended allowances of the vitamins. Vitamin C in the form of ascorbic acid may be increased in the formulation for wound healing during critical illness or to enhance immune system response. Vitamin premixes can be packaged as liquids or powders. In liquid form, the premix may have two phases, one phase for fat-soluble vitamins and another for water-soluble vitamins.

Trace elements are the final components determined by the doctor, dietician, and clinical pharmacist in a parenteral nutritional product. Trace elements are added to avoid a clinical deficiency or to treat individual deficiencies of one or more trace elements. A standard list of trace elements includes

- Zinc
- Copper
- Manganese
- Chromium
- Selenium
- Iodide

Iron is not part of a single trace element package and normally does not need to be added for patients on parenteral nutrition for a short period of time.

A parenteral form of iron (iron dextran) is used when needed as an injection product, but it has the potential to cause allergic reactions. Iron supplementation is normally part of enteral nutrition, and most patients have sufficient stores of iron in the body for short durations of parenteral feeding. Patients may have one or more trace elements increased to meet high metabolic needs. Zinc and chromium may be increased in patients with high GI fluid losses.

Infant Formulas

Infant formula provides nutrition to newborns when their mothers cannot provide adequate nutrition by breast-feeding. Infants born with allergies or partially developed digestive systems benefit from infant formulas that eliminate allergens until their bodies can handle them properly, and infant formulas based on partially digested ingredients permit the newborns with underdeveloped digestive systems to develop in a normal manner.

Infant formula products must meet a specific set of regulations. These foods provide the nutrition an infant needs to grow in a healthy manner. Infant formulas contain far more than milk and corn syrup that many people think are their ingredients. Trace minerals, vitamins, and ingredients that provide the small bodies with concentrated energy to sustain rapid growth during the first year of life are all part of the infant food formula delivered to the child. Infant formula should be used throughout the entire first year of an infant's life and should be used in combination with solid cereals and other foods to transition the infant from a liquid to solid diet.

The FDA defines infant formula as "a food which purports to be or is represented for special dietary use solely as a food for infants by reason of its simulation of human milk or its suitability as a complete or partial substitute for human milk" [FFDCA 201 (z)]. Infant formula is a food, and the laws and regulations pertaining to foods are applied to infant formula. A number of additional regulatory and statutory requirements also apply directly to infant formula because it is the sole source of nutrition to the most helpless part of our population during the child's critical period of growth and development. The additional requirements are part of section 412 of the FFDCA. The regulations that implement section 412 are part of 21 CFR 106 and 107 (5).

Within the FDA, a number of different offices work together to regulate infant formula products. Three different offices are responsible for infant formula regulation. The offices are The Center for Food Safety and Applied Nutrition (CFSAN, a division of the FDA); the Office of Nutritional Products, Labeling, and Dietary Supplements (ONPLDS); and the Office of Food Additive Safety (OFAS). Each individual office of the agency is responsible for oversight into different aspects of infant formula.

The ONPLDS reviews and evaluates product submissions to determine if an infant formula manufacturer has met the requirements under section 412 of the FFDCA. This office consults the OFAS, which reviews the suitability and safety of ingredients used in the manufacture of the formula. This office is also responsible for suitability of the packaging of the infant formula and reviews packaging materials, closure methods, and other issues related to the packaging. The packaging is required to meet all requirements of current good manufacturing practice (CGMP) during procurement, manufacture, and testing. The OFAS evaluates the safety of materials and substances that are used in or come in contact with the formula as outlined in sections 201 (s) and 409 of the FFDCA.

The components of an infant formula are similar to those used in all foods. The ingredients must be safe and lawful to use. The components of an infant formula, like those for all food ingredients, must meet generally recognized as safe (GRAS) requirements for infant formula just as food ingredients must also be recognized as GRAS. The FDA maintains a list of items that are identified as GRAS.

The ingredients must also be compliant with the FDA's food additive regulations [FFDCA 201 (s) and 409].

The FDA specifies nutrient requirements for infant formula. These requirements are found in section 412 (i) of the FFDCA and 21 CFR 107.100 (5). These specifications detail the minimum levels for 29 nutrients and the maximum amounts that may be used for 9 of those nutrients. If an infant formula is determined to be deficient in any of the nutrients or over-fortified (containing more than the maximum amount) in the nine nutrients with maximums, the product is considered adulterated and can be seized or removed from the market. The only exception to this requirement is "exempt infant formula," which is defined as "any infant formula which is represented and labeled for use by an infant who has an inborn error of metabolism or low birth weight, or who otherwise has an unusual medical or dietary problem" [FFDCA 412(h)(1)].

When a new substance is proposed by a manufacturer for inclusion in an infant formula, the FDA requires manufacturers to work with OFAS to review and resolve any safety issue before submitting the infant formula notification required by law 90 days before offering the infant formula for sale.

There are three key steps required for introduction of an infant formula into interstate commerce in the United States.

1. Registration—When a new infant formula is ready for commercial introduction, the manufacturer is required to register the name (manufacturer), the place of business, and all manufacturing locations of the product [FFDCA 412(c)(1).
2. Notification—The manufacturer or person(s) responsible for introduction of a new infant formula must submit to the FDA all information relative to its manufacture at least 90 days before the product is available for sale. The 90-day notification must include a listing quantitatively of the ingredients in the infant formula; descriptions of how the product is manufactured, including any changes in processing or any reformulation of an existing product; certification that the product will not be sold if it does not meet all the nutrient and quality requirements of the FFDCA; and documentation that the formula complies with CGMP and has appropriate quality control

procedures governing its manufacture [FFDCA 412 (d)(1)]. This notification includes all information relating to the packaging of the product.

Completion of steps 1 and 2 at the same time is stressed by the agency to smooth the progress of the product's review by the FDA.

3. Verification—The manufacturer or the person(s) responsible for introduction of the product into interstate commerce is required to submit written verification that summarizes test results that prove whether the formula complies with all requirements of the FFDCA. This submission must be done after the first full-scale manufacture of the product by the manufacturer and before the product is offered for sale. Typically, the results submitted to the FDA are those obtained from testing the first full-scale production lot of the product.

The FDA is also explicit about notification of what it calls a major change to a product. This is intended to avoid any deleterious changes to the product due to formulation change, processing changes, or packaging changes. The FDA wants to insure that a manufacturing or packaging change does not modify nutrients and nutrient availability. Examples of major changes are as follows:

- A new manufacturing plant
- A new production line
- The use or deployment of a significant new technology
- A change in the type of primary packaging being used for the product
- The introduction or addition of a new macronutrient (protein, fat, or carbohydrate)
- Addition of a new ingredient

In addition to all the requirements for marketing an infant formula, the FDA also monitors each batch or lot of infant formula. The monitoring program is carried out under the FFDCA section 412. The monitoring requires the following steps for compliance:

1. Manufacturers are required to test each manufactured lot of products for composition during production and throughout its stated shelf life.
2. The manufacturer must maintain records on production, testing, and distribution of each batch of formula and use CGMP and quality control procedures. This includes documentation of deviations in manufacturing or packaging and the testing used to determine the impact of the deviation and the data used to determine disposition of the deviated product.
3. In addition to the internal records, the FDA requires the manufacturer to maintain a record of all complaints it receives. These are reviewed periodically by the agency to determine or reveal possible hazards to health.

The FDA conducts yearly inspections of all production lines and manufacturing plants producing infant formula. The agency also inspects any new facility prior to or during early production runs and collects samples from all lines new or old for analysis of ingredients, labeling, and any other requirement of the FFDCA. If a product is found to be deficient or adulterated (hazard to human health) or if the labeling of the product is incorrect (misbranded), the FDA is authorized by the FFDCA to initiate a mandatory recall of the product.

PRENATAL NUTRITIONAL PRODUCTS

Nutritional products for the mother during pregnancy improve prenatal health and the health of the child. Nutritional products for mothers are an essential way to supplement diets deficient in key nutrients. This intervention means that the mother is healthier and more fit for birth and the infant is bigger and healthier at birth. These products are most widely used in Asia.

JUVENILE NUTRITIONAL PRODUCTS

Nutritional products also carry on beyond the first year. Children may need increased nutrition to meet unique needs, and a variety of products aimed at children between the ages of one and four years are available to help these young people thrive. These products include those formulated to address allergies or metabolic disorders.

MEDICAL FOODS: LEGISLATIVE OVERVIEW AND REGULATIONS

The Food, Drug, and Cosmetic Act (FD&C Act) of 1938 regarded medical foods as prescription drugs. This was to guarantee that these products would be used under the direction of a physician and prevent abuse by people promoting health benefits. This situation remained static until 1972 when the Congress revised the classification of products from "medical foods" to "special dietary foods." This change is cited in the Code of Federal Regulations (21 CFR 105.3) (5). The purpose behind this change was to enhance the opportunities for development of specialized products by both the drug and food industries. Another change was made to the code in 1973 to exempt medical foods from conventional nutrition food labeling. This change is found in 21 CFR 101.9(h) (5,6). This change moved the definition to "foods represented for use solely under medical supervision to meet nutritional requirements in specific medical conditions." This addressed the problems outlined in the discussion of foods containing or eliminating specific nutrients that create problems for the body or nutrients that create dangerous and life-threatening conditions.

The FDA tried in 1984 to publish a regulation that specifically addressed medical foods, but this was rejected by the Office of Management and Budget during a period when deregulation was in vogue and regulations were being eliminated or streamlined.

Following this rejection of further regulation, the FDA has used 21 CFR 105 to define and direct the use of medical foods (5). Unfortunately, this regulation only defines products for weight loss, hypoallergenic diets, and infant formulas. The lack of definition was further increased with the Proxmire Amendment to the FD&C Act (section 411) in 1976 that separated the regulation of vitamins and supplements from medical foods. This was done because the vitamins and supplements were available over-the-counter and did not require the supervision of a physician.

The most confusing piece of regulation for medical foods came in 1994 with the passage of the Dietary Supplement Health and Education Act (DSHEA). This Act expanded the definition of vitamins and mineral supplements to include extracts of natural products, metabolites (e.g., amino acids), botanicals, herbs, and proteins. It changed the name of this class of products to dietary supplements and permitted the use of claims that border on drug claims for disease. Examples of these claims include lowering cholesterol and increasing muscle mass and strength. These changes make these products pseudo-medical foods but release them from being defined as medical foods in the strict legal sense and permit people access to use them for self-diagnosis and self-treatment of medical and nutritional conditions.

Congress passed the strict legal definition of medical foods in 1988 as part of the Orphan Drug Act. The Orphan Drug Act refers to conditions or diseases that have fewer than 200,000 new cases per year. The first true definition of medical foods is found in 21 U.S. Code 360ee(b)(3). This definition restricted the definition of medical foods to foods that were administered or consumed via enteral means for the specific treatment or mitigation of a disease or medical condition, which has unique or distinctive nutritional requirements based on established scientific principles under the care of a physician, and the condition is established in the patient by medical evaluation.

This definition of medical foods was established in the Nutrition and Labeling Act of 1990 (21 U.S. Code 343) and is now the current authoritative definition. Medical foods were exempted from nutritional labeling in this Act in order that specific regulations would be developed to control medical foods. The CFR [21 CFR 101.9(j)(8)] contains the following definition:

> "A medical food is a food which is formulated to be consumed or administered enterally under the supervision of a physician and which is intended for the specific dietary management of a disease or condition for which distinctive nutritional requirements, based on recognized scientific principles, are established by medical evaluation. A food is subject to this exemption only if:
>
> (I) It is a specially formulated and processed product (as opposed to a naturally occurring foodstuff used in its natural state) for the partial or exclusive feeding of a patient by means of oral intake or enteral feeding by tube;

> (ii) It is intended for the dietary management of a patient who, because of therapeutic or chronic medical needs, has limited or impaired capacity to ingest, digest, absorb, or metabolize ordinary foodstuffs or certain nutrients, or who has other special medically determined nutrient requirements, the dietary management of which cannot be achieved by the modification of the normal diet alone;
> (iii) It provides nutritional support specifically modified for the management of the unique nutrient needs that result from the specific disease or condition, as determined by medical evaluation;
> (iv) It is intended to be used under medical supervision; and
> (v) It is intended only for a patient receiving active and ongoing medical supervision wherein the patient requires medical care on a recurring basis for, among other things, instruction on the use of the medical food."

INFANT FORMULA REGULATION

In the early 1970s, there was considerable concern regarding the development and regulation of infant formulas. At that time, a hypothesis was put forth that conjectured that infants with high salt intake were subject to hypertension (high blood pressure) in later life. This concern led to the establishment of regulation regarding the salt and chloride intake of an infant. The amount of salt in products was limited to 0.25%, and the regulation also asks for a voluntary reduction of salt in infant formulas and semisolid baby food. Unfortunately, this emphasis on salt led to a manufacturing error in soy-based infant formula that resulted in alkalosis and failure to thrive syndrome in 118 infants. Because of news reporting of this condition and public awareness of its consequences, the Congress passed the Infant Formula Act (FD&C Act, section 412). The portions of the CFR, which provide the implementing regulations for section 412 are as follows:

- 21 CFR part 106—Infant formula quality control procedures
- 21 CFR part 107—Infant formula
- 21 CFR part 110—Current good manufacturing practice in manufacturing, packing, or holding human food
- 21 CFR part 113—Thermally processed low-acid foods packaged in hermetically sealed containers.

The specific nutritional and compositional requirements for infant formulas are found in 21 CFR 107.10 wt. seq. Additional regulation contained in 21 CFR 107.50 (a) provides for exceptions to the overall standard and permits infant formulas without all the nutrients required in 107.10 when treating a variety of conditions such as low birth weight, inborn errors in metabolism, or other special disease conditions. These formulas are classified as exempt infant formulas by the Act. Manufacturers are required to submit information to the FDA on the composition and need for the product 90 days before the products' commercial introduction for sale.

Additional discussion of regulations for producing medical foods for enteral nutrition is found in the "Processing Authority" section of this chapter. This section details manufacturing regulations and expands the number of agencies and the number of regulatory requirements used to supervise the production of medical foods including infant formulas.

Regulation for the production of parenteral nutritional products is found in chapters 2 and 5 of this book. These chapters describe the basic requirements for injectable products.

MANUFACTURE OF INFANT FORMULA AND MEDICAL NUTRITIONAL PRODUCTS

Infant formulas and medical foods are produced using the same techniques as many commercial food products. Both processes rely on a term called "commercial sterilization," not absolute (complete) sterilization that is found with drug products. Absolute sterilization is performed on parenteral nutritionals. The focus of this chapter is on enteral nutrition and the manufacturing details and requirements for production of this class of product.

Commercial sterilization or commercial sterility means a condition created by heat by itself or in combination with other processes or ingredients that eliminate any microorganisms in the product from being capable of growing at temperatures above 50°F or 10°C while sealed in their container (2). This thermal processing of food eliminates all dangerous microorganisms but does not destroy one class of microbe, called thermophiles, which pose no significant threat to public health and which will not manifest under normal environmental conditions found in supply chain storage and distribution. Normal conditions mean shelf stable with no refrigeration under any condition found anywhere in the world and any unusual conditions that might be encountered in the distribution system. An extreme case may be a container or truck left in the sun for extended periods of time. Note that the product is not organism free; all the organisms are inactivated under normal conditions, but may under extreme conditions still be viable. Total destruction of microorganisms would cause degradation of the food to the point that it is inedible and unusable.

Retort processing (6) and aseptic processing (7) are the two predominant methods used to produce enteral nutritional products. Retort processing (Fig. 2) is one of the oldest ways of preparing food (6). Many people are familiar with this method of preservation from home canning done in home kitchens in many parts of the United States. Aseptic processing (Fig. 4) is a much newer and more complex method of achieving the same result. It was developed first for metal cans and glass

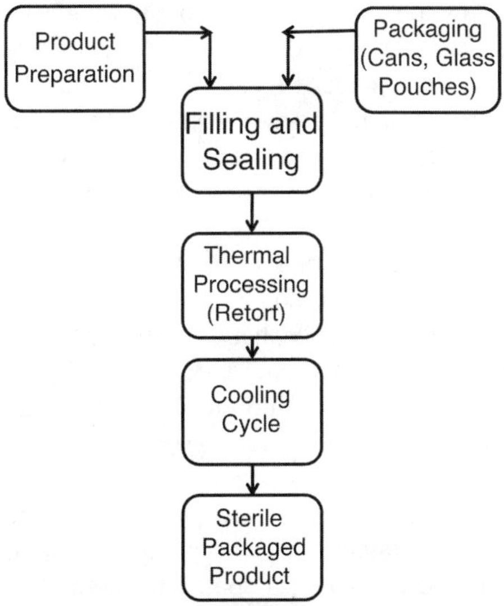

Figure 2 Block diagram of thermal (retort) processing and packaging system.

jars and later for all types of containers to permit the preservation of foods that could not withstand high heat for the extended period required by retort processing.

Aseptic processing's high-temperature short-time (HTST) or ultrahigh-temperature (UHT) preservation cycle is ideally suited for dairy products and products with ingredients sensitive to heat for extended periods of time.

Both preservation methods require the packaging to be an intimate part of the system. The package in retort processing (Fig. 2) is actually the container in which the food is cooked. The package in aseptic manufacturing operations is typically plastic and provides a means to deliver products with convenience features found only with plastic containers.

Retort Processing

Retort or heat processing of food is an old process. In a retort process, the product is heated to a temperature that is lethal to the harmful microorganisms it contains (6). The heat also inactivates the natural enzymes found in food, which would cause degradation in the sealed container. In retort processing (Fig. 2), the heat is applied to the product in each individual sealed container. In essence, each container is its own pressure pot or cooking vessel.

Originally, retort processing (Fig. 2) was done as a batch manufacturing process and for small-volume food producers it still is. Cans or glass jars are loaded into individual "baskets" or holders. These holders were then placed into pressure vessels and subjected to heat and pressure to raise the temperature above that of boiling water and permit the contents inside the container to heat to a point where all harmful microorganisms are killed and to a point where the enzyme action of the food is stopped or greatly retarded. The more elaborate systems provide for movement of the basket or holder to mix the product inside and reduce the total cycle time of heat processing. The movement of the basket creates a mixing action within the container that speeds heat transfer into the container and at the completion of the heat processing, out of the container during the cooldown phase of the operation.

During the sterilization cycle, a can or glass jar must withstand the internal pressure generated by the steam contained inside the container and later, after cooling, be strong enough to withstand the outside atmospheric pressure pushing against the vacuum formed inside the container. This requirement to withstand both pressure and vacuum inside the container is in addition to the strength the container needs to withstand the rigors of the distribution system, including stacking of pallets in a warehouse or truck.

Hot filling, heating the product to near boiling before placing it in the container, saturates the headspace of the container with water vapor prior to closing. After the processing step, when the container is cooled to room temperature, the water vapor in the headspace of the container condenses back into a liquid as part of the product, thus creating vacuum inside the container. This was the major reason that only glass and metal packages were available for most shelf-stable foods. They possessed the structural strength needed to withstand the positive and negative pressures created during the manufacture and the additional strength requirements of the distribution processes.

Originally plastics, both rigid bottles and flexible pouches, could not stand up to the rigorous temperature and pressure extremes placed on the container. New handling and processing techniques for plastic bottles, bowls, and pouches changed retort processing (Fig. 2) equipments by adding thermal process heat and pressure programming capabilities that balance the pressure inside and outside the container during the retort cycle. This balance combined with techniques to minimize headspace in plastic containers has permitted them to withstand retort processing just like that in metal cans and glass bottles. The containers have also grown more sophisticated with the advent of multilayer plastics that deliver low oxygen and moisture vapor transmission performance.

During the 1950s, two major improvements in retort processing were introduced. These innovations were the hydrostatic continuous process (Fig. 3) and the rotary continuous process.

The hydrostatic retort process (Fig. 3) consists of a tower with multiple sections. At the entrance and the exit of the hydrostatic operation is a column of water used as a pressure seal to permit other sections or "legs" of the process to maintain temperatures above 212°F. Normal retort processing uses maximum temperatures in the 250°F to 270°F (121–132°C) range.

Multiple cans are loaded in "sticks" or long lines of cans placed on their side in a conveyor that resembles a long tray. The conveyor moves up and down a number of times, (each up and down a leg of the process) within the hydrostatic

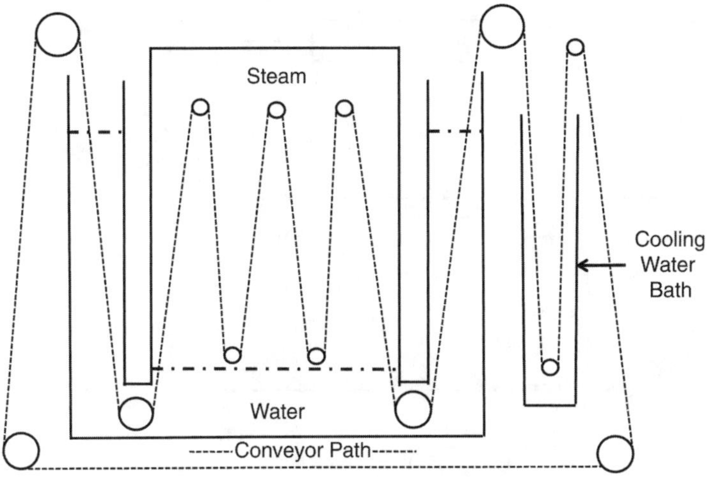

Figure 3 Internal diagram of a hydrostatic cooker.

retort (Fig. 3) to preheat, cook, and cool the containers before they emerge on the other side of the operation. The product in the packages receives little agitation. The thermal process must take into account the time it takes for heat penetration into the container and food to hold that temperature inside the container until the harmful microorganisms are killed or rendered inactive. This includes the time it takes for the heat to penetrate to the inside of food particulates within the container along with heating the liquid phase of the product. This aspect of thermal processing is covered in the "Thermobacteriology" section of this chapter.

Retort processing (Fig. 2) was the catalyst for research and development into describing and testing thermal processes and the mathematics and microbiology involved. The calculations and methods developed for retort processing are used for all thermally processed foods. The measurement methodology for microorganisms is used or referenced for new processes under development to increase or extend the methods of manufacturing used for food products.

Aseptic Processing

Aseptic processing and packaging of foods (Fig. 4) (7) is and continues to be one of the more dynamic areas of process and package development for consumer foods and nutraceutical products. It has opened the door to convenient, lightweight plastic containers. In hospitals, it has eliminated glass containers. It has permitted the substitution of lightweight, plastic containers into food, infant formula, and medical food product presentations replacing glass and metal.

Although the emphasis on the packaging used by aseptic systems is primarily plastic, it can be used for metal cans, glass bottles, pouches, and composite containers. In fact the first wide-scale use of aseptic packaging was based

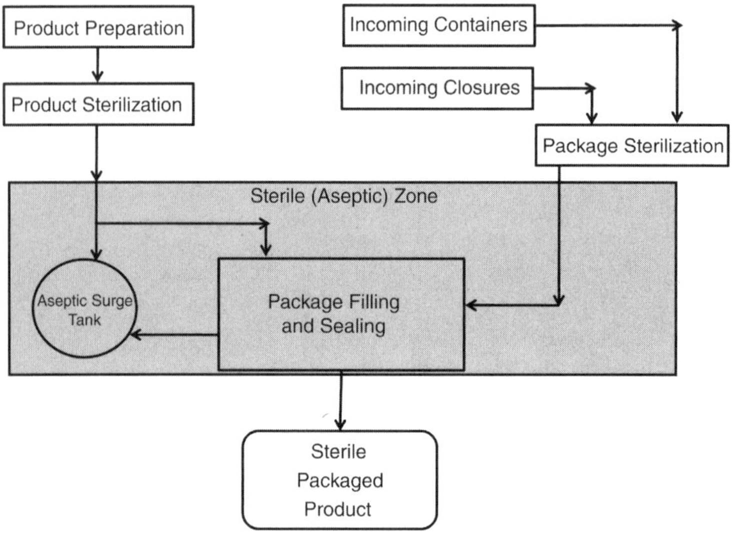

Figure 4 Block diagram of aseptic processing and packaging. *Note*: shaded area denotes the processes contained in the sterile zone.

on a metal can. One of the most popular and widely used containers in the world, Tetrapak®, is based on aseptic processing (Fig. 4) and filling. The Tetrapak package is a composite of paper, foil, and plastic, with simple and unique opening features that make it highly portable and widely accepted.

Aseptic processing and packaging (7) is the bringing together of product, package, and closure, each sterilized in separate operations in a sterile environment to fill and seal the package (7).

Aseptic processing (Fig. 4) in the United States began in 1940 when the aseptic canning system of Martin Dole Company was employed to produce milk products without subjecting them to a long-duration heat cycle. This next innovation was aseptic drums of products manufactured in the 1950s. Aseptic bulk storage systems followed in the 1960s and in the 1970s with aseptic bag-in-box systems and aseptic rail tankers becoming prevalent as methods for storing bulk food ingredients. In the 1980s, specifically 1981, the FDA approved the use of hydrogen peroxide as a sterilant for aseptic packages. This approval opened the way for containers to be sterilized without subjecting them to high-temperature steam or other high-temperature sterilization processes that were deleterious to the package.

The use of a chemical sterilant promoted the use of plastic bottles, plastic and composite laminates, and pouches in aseptic packaging. Heat is still required to dry and inactivate the hydrogen peroxide, making it difficult for heat-sensitive plastics with plastic memory [e.g., polyethylene terephthalate (PET) bottles] to be sterilized with this chemical. A second sterilant, peracetic acid (Fig. 5) or PAA (chemical names are peroxyacetic acid and ethaneperoxic acid), has been

Acetic Acid + Hydrogen Peroxide = Peracetic Acid + Water

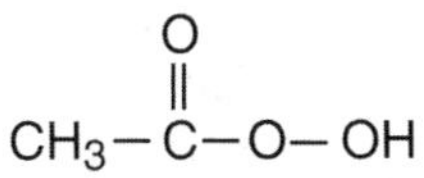

Peracetic Acid (PAA)
$C_2H_4O_3$

Figure 5 Peracetic acid.

submitted in a process filing and accepted by the FDA in 2007. Its approval opens the way for lower-temperature sterilization cycles for plastic containers and will permit wider use of PET as an aseptic packaging material.

The heating and cooling process used commercially for liquid product sterilization of aseptically prepared food or medical nutritional product takes place in a closed continuous system (Fig. 6). The closed system moves the product through highly efficient heat exchangers that heat and cool the product very rapidly to sterilize all the nutritional ingredients and other unique ingredients (e.g., vitamin fortification or unique trace metal combinations) that provide therapeutic benefit while retaining high levels of nutrient value and excellent sensory qualities for the patient or consumer.

The processes are called HTST and UHT sterilization processes. The letter abbreviations are often used when referring to or discussing the processes. Because the two processes shorten the amount of time an ingredient is subjected to degrading heat, the process yields a product that retains its nutritional benefits and its favorable appearance and flavor and sensory characteristics. The product also retains its natural color because degradation reactions and caramelizing reactions are held to a minimum.

COLD ASEPTIC STERILIZATION—ASEPTIC FILTRATION

Cold sterilization or filtration is another aseptic method used to sterilize clear solutions. It is used for medical and food applications. It is most often used for rehydration solutions, solutions designed to replenish and maintain the electrolyte and water balance within the body.

This method of sterilization consists of passing a clear liquid through a membrane filter with an opening or pore size small enough to remove the microorganisms without removing the beneficial ingredients contained in the solution (Table 1). The method becomes impractical for emulsions or solutions with dissolved ingredients equal in size or larger than the bacteria being removed. Table 1 lists the size of various organisms found in food and medical nutritional products and provides a guide about what can and cannot be filtered. The table also describes the various classes of microorganisms that must be rendered harmless in any food product.

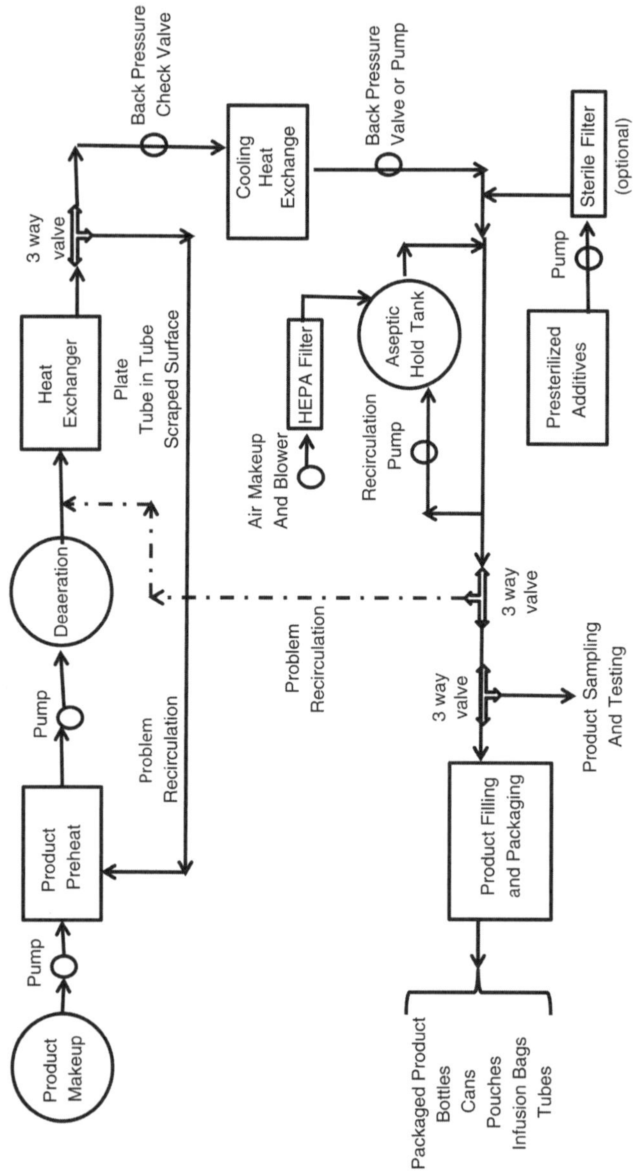

Figure 6 Typical aseptic processing and packaging schematic.

Table 1 General Microorganism Characteristics

Microorganism	Cell wall composition	Cell shape	Cell size
Virus	Protein and lipid	Isometric Rod Filamentous Pleomorphic Isometric head Attached to a rod-shaped tail	20–450 nm
Gram-negative bacteria	Polysaccharides Peptidoglycan (1–10%) Lipoprotein	Rod Vibrio spirilium Cocci	0.2–1 × 0.6–4 μm 1–2.5 μm
Gram-positive bacteria	Peptidoglycan (40–80%) Lipids Teichoic/teichuronic acids	Rods Cocci	0.3–2.2 × 1–14 μm 0.3–3.5 μm
Bacterial spore	Peptidoglycan Calcium-dipicolinate Protein	Oval Round	
Yeast	Mannans Glucans Chitin (small amount)	Spherical Oval Oblong	1–9 × 2–30 μm
Mold	Polysaccharides (chitin, cellulose) Glucans Glycoproteins	Filamentous	

ASEPTIC MANUFACTURING EQUIPMENT

Thermal treatment of food in aseptic systems uses a number of proven types of equipments called heat exchangers, widely accepted around the world, for the sterilization of food products. These systems include

indirect heating systems (heat exchangers)

1. Plate
2. Tubular
3. Scraped surface
4. Electrical (induction)

direct heating

1. Steam injection
2. Steam infusion

The type of product to be sterilized normally dictates the choice of heat exchanger or sterilization system. Each system has its advantages and disadvantages relative to the product attributes, ingredients, ability to clean, amount of product surface buildup in the equipment, and other factors all of which influence the effectiveness of heat transfer used for sterilizing the product.

All aseptic systems have difficulties with particulates. Particulates in products, such as diced meat or vegetables, make fast heat transfer into and out of the particle very difficult and are typically avoided in aseptic product formulation. Particulates when used in aseptic food processing must be extremely uniform and consistent in size and shape. An equipment producing extremely uniform particles required to fit within the thermal process parameters for the product is highly specialized. Particulates are handled with the equipment addressing the unique characteristics of the particulate in question. This specialization has been the most successful way of addressing the particulate sterilization problem. Medical nutritional products and infant formulas do not contain particulates; however, one way to enhance a medical nutritional product will be the introduction of particulates in, for example, a soup broth that a patient consumes orally.

Aseptic systems are continuous and closed. This means you cannot sample the product to determine its sterility during the system's operation. Aseptic processing (Fig. 4) is different from most other systems of food sterilization in its reliance on process controls and statistical methods. These methods focus on controlling the process in a repetitive manner to deliver a repeatable result. Computerization of controls has improved this capability and has improved real-time monitoring of process conditions during an aseptic sterilization operation. It has also brought with it other questions regarding validation of software and input/output validation of controls and alarms.

Recently, some exotic nondestructive test methods, notably NMR (nuclear magnetic resonance) imaging, have been used as a test of an aseptically manufactured product after packaging. Because the test looks for microbial changes in the product, it cannot be used in line at the time of product manufacture. It is employed as a secondary measure after the product is stored for a time sufficient for a microorganism to produce enzymes, a marker indicating a change that it can detect.

ASEPTIC PACKAGE STERILIZATION

Aseptic packaging provides the same functions found with all food packaging. These functions are containment, protection, and preservation of the food product. One additional function, not part of the physical, chemical, or mechanical properties of a container, is the package's ability to inform the consumer about its contents. This is referred to as labeling.

Package sterilization in aseptic systems uses two different mechanisms to achieve sterility of the package and the package closure before being filled. The first method uses heat, typically in the form of wet heat or wet steam, to inactivate spore-forming bacteria. Super-heated steam can be used, but it typically does not transfer its heat to the packaging very efficiently The second mechanism uses a chemical sterilant along with modest heat to inactivate the bacterial spores. Bacterial spores are the most resistant form of bacteria to all sterilization processes.

In the United States, the only sterilant approved through 2006 was hydrogen peroxide (H_2O_2).

Peracetic acid (Fig. 5) produced by reacting acetic acid with hydrogen peroxide (H_2O_2) underwent extensive evaluation and was accepted in an aseptic filling by the FDA as an alternate to hydrogen peroxide in 2007. Peracetic acid is an alternate sterilant needed by equipment and container manufacturers because it is effective in inactivating microorganisms at lower temperatures than hydrogen peroxide, and the residual materials from its breakdown are benign. Peracetic acid (Fig. 5) decomposes into acetic acid, water, and oxygen, all of which do not come under stringent residual standards. Hydrogen peroxide has requirement of no more than 0.5 ppm residual material remaining on the packaging after sterilization as defined by FDA regulation.

Outside the United States, the list of sterilants include hydrogen peroxide, hot air/steam, radiation, halogens, peracetic acid, citric acid, or combinations of these materials for sterilization of both the equipment used for manufacture and the packaging materials. The most important aspect in the choice of sterilant is its ability to kill microorganisms on surfaces of equipment and packaging materials. Care must be taken that the material properly "wets" or contacts all surfaces. Ideally, the sterilant has a low surface tension to wet the surface of the material being sterilized, is easily applied, leaves no residue, and is fast acting in its ability to kill spores.

What makes aseptic packaging different from traditional packaging methods is that the product and the package are continuously sterilized separately from each other. The two units, product and package, are brought together in a protected sterile chamber. The package is filled and hermetically sealed inside that sterile chamber to produce an extended shelf-life product that does not need refrigeration.

A number of methods are used to achieve package sterilization in aseptic systems:

- Mechanical processes
- Thermal processes
- Irradiation
- Chemical
- Combination processes
- Other processes

Mechanical Processes

Mechanical processes by themselves are normally not sufficient to completely sterilize a container. They are used for precleaning containers and reducing the microbial load of the container prior to sterilization. These processes include water rinsing/flushing, air rinsing or air blasting, and brushing (to remove dust or particulates).

Thermal Processes

Thermal processes are typically high-temperature processes. They are not compatible with most heat-sensitive plastics. The first commercial aseptic system, the dole process, used steam to sterilize a metal can, and the can end was used for packaging.

One unique or novel way to overcome the problems of using heat to sterilize plastic containers is to use the heat required to extrude the plastic during the sterilization process and then maintain sterility of the container from this point of manufacture through the aseptic filling and sealing operation.

Examples of thermal processes used for package sterilization include saturated steam, superheated steam, hot air, and mixtures of hot air and steam. These same processes may be employed to sterilize the aseptic equipment and the aseptic zone or sterile zone of the machine where product, package, and closure are brought together, filled, and sealed.

Irradiation Processes

Irradiation of packaging was first introduced into aseptic bulk container sterilization, notably bag sterilization for large volume bag in box products. The bag, in a bag and box product, is assembled into its final form and sealed prior to radiation exposure. The radiation sterilizes the bag, and its seal maintains sterility until it is opened and filled in the sterile zone of an aseptic packaging operation.

Irradiation can be used to sterilize heat-sensitive packaging materials, and in single use, containers can be employed in the same way the heat of extrusion is used in thermal processes. Again, after exposure the package is maintained in a sterile environment until after it is filled and sealed. Examples of the types of irradiation include ionizing radiation (β or γ rays), infrared radiation, and ultraviolet (UV) radiation.

Chemical Processes

Chemical processes are used more often than any other method to sterilize containers. They are fast, efficient, and easy to monitor and maintain. In the United States, the sterilant is limited to hydrogen peroxide and peracetic acid (Fig. 5), although other sterilants and combinations of chemicals are used outside the United States.

The container surfaces are exposed to the chemical sterilant as it enters the sterile zone of the packaging operation. After exposure to the chemical sterilant, the package is either rinsed or dried in the sterile zone of the machine to remove the sterilant by chemical breakdown into benign substances before filling and sealing.

The sealing material used to close the container follows the same steps and normally uses the same sterilant.

Chemical sterilization is often used by itself or in combination with heat or mechanical means to sterilize the packaging equipment and the sterile zone in an aseptic system.

Combination Processes

Combination processes use the same conditions and materials, heat and chemical sterilant, listed above to sterilize containers. Examples include hydrogen peroxide/heat, hydrogen peroxide/UV, ultrasound/hydrogen peroxide, peracetic acid/hydrogen peroxide, and peracetic acid/alcohol. None of these combination systems with the exception of hydrogen peroxide/heat is approved in the United States.

ASEPTIC PACKAGING SYSTEMS

Aseptic packaging systems combine the product, package, and closure in a sterile environment to manufacture a product. Commercial aseptic packaging systems and the equipment that produces the packages fall into six broad categories. The standard names of the systems are

1. fill and seal
2. erect, fill, and seal
3. form, fill, and seal
4. thermoform, fill, and seal
5. blow mold, fill, and seal
6. bulk storage and packaging

Fill and Seal

This type of aseptic packaging system uses glass or preformed containers. This would include thermoformed, injection-molded, or blow-molded plastic containers. The preformed containers are sterilized, filled, and sealed in a sterile environment that describes the classic aseptic operation.

Erect, Fill, and Seal

A container that is folded or mechanically manipulated at the packaging operation is opened or erected and then sterilized, filled, and sealed in a sterile environment. The best example of this type of container is the CombiBloc® container recognized as a type of a liquid multilayer composite box container.

Form, Fill, and Seal

This is by far the most familiar and widely used aseptic system. TetraPak is the leading example of this type of aseptic packaging. In these systems, a roll of container-making material is sterilized, formed, filled, and sealed in an aseptic environment. Sterilizing the flat material in a bath or heat tunnel simplifies the process and makes removal of residual sterilant prior to container forming much easier. Pouch systems and the TetraPak carton system are examples of this type of aseptic operation. Drug packaging uses single-layer form, fill, and seal technology to produce small-dose packages. The package is injection molded, or may be injection blow molded. A good example is a single-use eye drop container produced with no preservative using this packaging technique.

Thermoform, Fill, and Seal

This is the flat plastic roll stock equivalent of form, fill, and seal. A roll of extruded plastic container–forming material, made up of multiple layers of oxygen, moisture, or solvent barrier materials required for product protection, is sterilized via chemical means, or is sterilized using the heat required to thermoform it into the container shape, and then moved into the sterile zone for filling and sealing. This process requires a chemical sterilant if the plastic is forged or shaped at a temperature too low for complete sterilization of its surface.

The process uses the heat required to produce the plastic container and maintains the sterility of the container into the sterile zone of the filling and sealing equipment. Cups produced in this manner for liquid products normally remain in the plastic web through filling and sealing and are trimmed into individual units or bundle packs after they emerge as sealed containers from the sterile zone. Lid stock, the sealing material used to close the package is a multilayered composite structure of plastic and foil materials which is preprinted with the product labeling and is matched in size to the width of the thermoformed web used to produce the plastic containers. It may be sealed to each individual container, or it may only be sealed at the outside perimeter before emerging from the sterile zone. For equipments using the perimeter-sealing technique, each individual cup is then sealed before the container or bundle of containers is separated from the web.

Blow Mold, Fill, and Seal

This process relies on the heat of extrusion to sterilize the container. The plastic is extruded and blow molded using sterile air to force the plastic into the final container shape in the blow-molding cavity or die. The package may move immediately into the sterile zone for filling and closing, or the package may be resealed using excess plastic to reclose the bottle before the molds are released. This second bottle-producing method is used in a process called extrusion blow molding, where a parison or melted plastic tube emerges from the extruder and is placed in the center of the bottle-producing mold.

The bottle, which is sterile on the inside, is then introduced to a sterilization system that re-sterilizes the outside of the container or the neck of the container as it enters the sterile zone prior to opening or breaking the bottle seal. The totally sterile bottle then moves into the sterile zone where it is opened, filled, and sealed. The dome covering the neck of the bottle is trimmed to produce a bottle finish compatible with the closure used for the bottle. It may also receive a heat-sealed membrane or lid stock prior to application of the closure. The process can be used for flexible containers as well as bottles when a fitment or unique opening area or mechanism is part of blow molding a container.

Bulk Storage and Packaging

These systems are used during the manufacture of products that have unique or extended manufacturing operations or periods. They are typically used for concentrated intermediate products or products handled in bulk. Examples would be milk products that have completed pasteurization and homogenization as a starting point for multiple finished products. The final product is then customized with the addition of unique ingredients tailored to specific patient needs. This is one method that manufacturers use to generate products that do not require mixing or manipulation at the patient's bedside.

Bulk packaging includes large volume bag in box products, drums, tote tanks, and rail cars. Large storage tanks required to decouple product processing from aseptic packaging operations fall into this classification. The large tanks typically are sanitized using chemical disinfectants, iodophors, or bromine solutions, and then treated with heat to complete the sterilization process. Specially designed valves are used to attach these tanks and bulk units to aseptic units without breaking the sterility of the system.

Aseptic systems are also categorized by six distinct elements that occur within the closed system. These elements are the product, flow control, product heating, hold tube (the product being held at temperature), product cooling, and packaging. Each of these elements must work correctly to produce an acceptable result. Aseptic systems are extremely complex mechanically and electrically (controls and monitors the equipment) and require rigorous and meticulous attention to detail to operate properly.

Along with the elements of an aseptic system listed above that actually contact and act on the product, all equipment surfaces and the methods used to sterilize and maintain sterility of the equipment must be considered. This includes the positive pressure systems used to insulate the sterile parts of the machinery from environmental recontamination after sterilization.

Manufacturing plants typically use a positive pressure inside the building or in the room containing the aseptic equipment as a first layer of protection against airborne pathogens because untreated air is a source of contamination for the equipment and the product. Inside the processing plant, airborne contamination can originate from humans, raw materials used in manufacturing, dust, dirt, and anything that enters the facility. Aseptic systems use positive air pressure produced by filtration systems [high efficiency particulate air (HEPA)] throughout the equipment, and particularly in the sterile zone of the equipment to keep out molds, spores, bacteria, viruses, yeasts, and other normal airborne pathogens that would contaminate the system. All air entering the sterile zone where product filling and sealing take place must be treated to remove any contamination including microorganisms. This can be done with heat (incineration) or high-efficiency filtration (HEPA).

Laminar airflow is used within aseptic manufacturing equipment to further insulate the operation from contamination. Laminar flow (nonturbulent airflow) prevents mixing of the air in the system and eliminates potential low-pressure areas that could pull in external contaminants. The air in an aseptic system is typically changed more than 100 times per minute to maintain a sterile environment around the product and packaging and sealing operation and a positive pressure within all sterile parts of the equipment. In laminar airflow, this high flow rate keeps any particles produced by the machine airborne for removal. The positive pressure permits the movement of containers and lid stock into the sterilization sections of the equipment without bringing in environmental contaminants.

There are normally two air filtration systems employed to clean the air entering aseptic processing equipment (Fig. 6). These same systems and filters are also used for surge tanks and bulk storage tanks to maintain sterility as the tank fills or empties and while the tank is static (empty) or holding a processed product.

Filters are tested for effectiveness and are usually changed on a fixed schedule based on hours of operation to maintain the performance of the filtration system. UV systems are sometimes added to the filtration air as a second level of defense and are effective in open areas of plants around tanks, fillers, and CIP (clean in place) units.

Plant environments are constantly checked for airborne contaminants, as are the areas inside a manufacturing equipment. A device consisting of a membrane filter with a vacuum pump and a liquid or gel impingement device are employed. These devices collect the microorganisms and permit evaluation microscopically or through actual attempts to culture the trapped contaminants. The type of area being used for an aseptic operation may define the amount of contaminant. For drug installations, and food installations that mimic drug guidelines, class 100 or class 10,000 requirements found in the CGMP regulations are used. For food manufacturing, the literature suggests an environmental standard of 50 viable microorganisms or less for aseptic packaging areas.

BASIC PRINCIPLES OF THERMAL PROCESSING

Thermobacteriology

High temperature has been used as a method to kill or inactivate microorganisms for a very long time. Since the 1920s, thermal processes have been designed on the basis of data and characteristics that describe the thermal destruction of microorganisms. This information was originally applied to products in metal cans and glass containers (retort processing, Fig. 2) and then adapted to aseptic processing of products. During the 1950s, continuous retorts were developed for the sterilization of individual containers. This greatly speeded the process of producing canned food but required a complete understanding of the thermal process employed to make the food safe.

Aseptic processing began during the 1940s, as a method to process foods that underwent degradation or discoloration in traditional retort processes, and it required the same process understanding to assure safety.

The design of all thermal processes, both aseptic and retort, follows the same conventions. Generally, the most important microorganism studied and verified as killed by thermal processing is a bacterial spore. These entities are very resistant to heat, and the determination of how to inactivate them is key in attaining commercial sterility in food products. Thermal processes are asymptotic and can never achieve zero in the complete destruction of organisms. Instead a log reduction is used to describe the number of orders of magnitude the amount of the organism has been reduced to.

All thermal processes rely on a time/temperature relationship of heat exposure to inactivate or kill the microorganism. The reaction (destruction) of the microorganisms by heat obeys first-order reaction kinetics. This reaction is expressed by the following equation:

$$\frac{dc}{dt} = \frac{-2.303}{D}c \qquad \text{(Eq. A)}$$

where dc/dt = rate of destruction

D = decimal reduction time or the time for population to decrease by a factor of 10 at a given temperature.

c = instantaneous number of viable cells or spores.

2.303 = conversion factor for $\log_e$ to $\log_{10}$.

The D or destruction value is characteristic of the specific microorganism being processed, the medium it is contained in (e.g., solution pH, composition, and presence of oxygen) with the value expressed in a time constant base 10.

This equation can be integrated and rearranged into the following form:

$$\int_{c_i}^{c_f} \frac{dc}{c} = \int_0^t \frac{-2.303}{D}dt \qquad \text{(Eq. B)}$$

$c_i = t = 0$, which represents the initial number of spores

$c_f = t =$ the final number of spores

Integration of the left-hand side of the equation results in $ln\ (c_f/c_i)$

The problem with this equation is the right-hand side value because the time integral cannot be integrated in this form as the D value is not constant. The

D value is a dependent variable based on temperature, and temperature is based on time.

The thermal death time model for the temperature dependence of a microorganism is expressed by the following equation that describes the temperature dependence of the D value.

$$\log \frac{D}{D_R} = \frac{T_R - T}{z} \qquad \text{(Eq. C)}$$

Where D_R = the D value at T_R
z = temperature change in degrees Fahrenheit (°F) necessary for the D value to change by a factor of 10.

Solving this equation and substituting it in the previous equation produces the following result:

$$D_R \log \frac{c_i}{c_f} = \int_0^t \frac{1}{(T_r - T)/z}\, dt \qquad \text{(Eq. D)}$$

The log of c_i/c_f is the number of log cycle reductions the concentration of cells must go through to achieve the proper D value.

The FDA requires a 12-log reduction of *Clostridium botulinum* spores in nonacid or low-acid foods (pH $>$ 4.6). This is usually based on a temperature of 250°F (121.1°C) making D_R log (c_i/c_f) the thermal death time normally called F_0. This means that the time/temperature exposure of the thermal process results in an integrated effect of F minutes at 250°F.

The problem is that the right-hand side of this integral still cannot be integrated since temperature is dependent on time.

A process called the improved general method for thermal process calculation solves this problem. The term "$1/10^{(250 - T/z)}$" is expressed as the lethal rate. Values for lethal rate are determined from heat penetration experiments and then calculated for each time/temperature value. Lethal rate is then integrated over the process using graphical means or using a numerical integration procedure such as Simpson's rule or Trapezoid rule.

This was not an easy way to determine how to make these calculations to obtain a safe thermal process and a major improvement was needed. C.O. Ball in 1923 was able to make a major improvement on this method. Ball designed an equation that described the time/temperature profile of foods with sufficient accuracy to describe foods undergoing a thermal process. His insight and formula became the formula method of thermal process calculations. His description of the time/temperature relationship is exhibited in the formula:

$$\log(RT - T) = \log(j_h I_h) - \frac{t}{f_h} \qquad \text{(Eq. E)}$$

RT = heating medium temperature
$I_h = RT - T_{ih}$

$j_h = (RT - T_{pih})/(RT - T_{ih})$
F_h = time in minutes for $(RT - T)$ to change by a factor of 10
T_{ih} = initial temperature of the product
T_{pih} = simulated initial temperature of the product. This is obtained by extrapolating the linear portion of the semilog heating curve to corrected zero time.

Ball determined the factors j_h and f_h by conducting heat penetration experiments on food. He then differentiated the equation and substituted the result into the previous equation as an answer for d*t* in Eq. D. This, along with a few assumptions to simplify the process, resulted in the formula method used to calculate the total accumulated lethality of the thermal process. The same method is adaptable to any thermal process for food provided the *z* value is independent of temperature. This is problematic in aseptic processing because the *z* value is constant over a narrow range of temperatures that span approximately 70°F.

Heat Exchange/Heat Transfer

Heat transfer now becomes the next major consideration in retort and aseptic processes. In addition to this description of how a thermal process is determined, including the time it takes to kill or render inactive the microorganisms, a complete review of the heat transfer characteristics of the product and the process must be undertaken. The effect of the time/temperature relationship is independent of whether the product is in a closed container or passing through a heat exchanger (retort) or in an aseptic process heat exchanger and hold tube.

The time/temperature profile is specific to the process used because aseptic processes can use UHT or HTST heat treatments. This extreme heat, compared with the limitations in heating and cooling the product in its own container (retorting), drastically changes the time/temperature profile received by the product. This difference is the major advantage of aseptic processes because the HTST process results in less degradation of the food product and its nutrients, producing superior flavor, appearance, and quality. Dairy-based products are whiter (less browning) in color and better retain their flavor. Try tasting scalded milk sometime to understand this difference.

Heat transfer can be affected in either a batch or a continuous system. Both systems are effective in producing a satisfactory result.

In continuous aseptic processing systems (Fig. 6), the velocity of the product is a key determinant in the amount of heat treatment or the amount of sterilization it undergoes. The velocity of the product through the system, the mixing the product receives in the system, and the nature of the liquid itself are all factors that must be evaluated. Liquids can be a Newtonian fluid (linear viscosity profile), thixotropic (shear thinning—viscosity decreases with shear or stress on the fluid), or dilatent (shear thickening—viscosity increases with shear or stress on the fluid), and this physical characteristic of the liquid must be understood as part of the system design. Velocity also must produce laminar or nonturbulent flow through an aseptic system to produce the effect of plug flow. This means that a given volume of liquid moves through the system in a uniform manner, like a bucket being passed in a bucket brigade. This attribute permits the calculation of the flow behavior and permits an accurate determination of the holding time required in the heat exchanger for the liquid being processed. The portion of the aseptic system that accomplishes the thermal treatment of the product is called the hold tube. Each unit (bucket) of the product must receive and maintain the required thermal process temperature required to deactivate or destruct the harmful organism.

Following this heating and holding of the product at temperature, referred to as HTST or UHT processing, the product must be rapidly cooled to stop the degradation reactions of the food product. Cooling of the product is also done in a heat exchanger.

Following the cooling of the product, it moves directly to packaging or into an aseptic surge tank that balances the variations in processing and packaging within the system.

DEAERATION

Aseptic systems normally require a deaerator in the system. Deaeration is applied to the liquid prior to its introduction into the heat exchanger for sterilization. Air that becomes entrapped in the product is removed to extend shelf life. Shelf life is extended because removing the air removes the oxygen driver for oxidative reactions that occur at the high temperatures used to kill microorganisms in liquid food products. The deaerator is a vacuum vessel of some type through which the liquid flows. Deaeration can be tricky because many of the flavor constituents of a product are volatile, particularly at elevated temperatures. Elevated temperatures make the noncondensible gases less soluble in the liquid, hence the conundrum of temperature and vacuum is required to deaerate a product without affecting its sensory qualities. Deaeration is always operated at the highest temperature the product will tolerate without stripping away key flavor and aroma components. The higher temperature makes the noncondensible gases in the product less soluble and easier to remove.

Residual oxygen in a product is a concern with most plastic packaging. Plastic packaging materials all display some degree of permeability that leads to deterioration in product quality over time. Removing the oxygen prior to processing and packaging eliminates a source of oxygen in the product and extends packaging shelf life by reducing the product's exposure to oxygen to the rate it permeates the container. The finished product is also maintained at temperatures far below those used in processing, which also slows the oxidative degradation reaction.

ASEPTIC SURGE TANKS

Aseptic operations combine a number of different operations, any one of which can experience problems. Portions of systems can be upgraded to increase heat process throughput capacity, or different products because of their physical makeup or physical characteristics can be thermally processed or produced faster during the sterilization step. When the flow rate of sterile product is greater than the packaging capacity, a buffer or aseptic hold tank is built into the system.

Aseptic surge and, in some cases, hold tanks can range from as small as 100 gal to several 1000 gal in size. These tanks provide operational flexibility and operational capacity to aseptic manufacturing systems. An aseptic surge tank essentially decouples the processing and cooling section of the operation from the packaging operation and allows the two parts of the system to operate independently. Heat-sensitive ingredients, sterilized by other means, may be added to a product in the aseptic hold tank.

The tank operation is very simple. The product is pumped into a previously sterilized tank and maintained in a sterile environment with filtered air. The tank is always under positive pressure, that is, the pressure inside the tank is greater than the external or atmospheric pressure. By keeping the pressure in the tank above the pressure of the outside environment, microorganisms and other contaminants are barred from entry. When the product is pumped from the tank, a sterile gas or sterile air is used to maintain the positive pressure in the tank. The positive pressure in the tank is monitored throughout the filling and discharge cycles.

PROCESSING AUTHORITY

The term "processing authority" generally refers to an individual or organization, which through a combination of scientific knowledge and training along with actual field experience is capable of designing, developing, evaluating, and implementing scientific thermal processes. This individual or organization develops the sterilization criteria for a product, its packaging, and the equipment used to manufacture the product and proves that it meets all pertinent regulations and requirements set forth by the FDA, the USDA if meat or fish are involved, or in the case of dairy products, the FDA and Public Health Service Pasteurized Milk Ordinance (PMO) (8) typically enacted into state law and monitored by a state agency.

In addition to the product itself, processing authorities responsible for aseptic systems must be capable of judging multiple systems or factors, such as equipment sterilization, package sterilization, and maintenance of sterile package filling and sealing environment. These additional systems, all part of any aseptic operation, are not part of retort or canning process systems, which are also under the responsibility of a processing authority.

The FDA requires that anyone operating or involved in thermal processing operations attend an FDA-approved course on preservation technology. This requirement is contained in 21 CFR 108.35 (g) and 21 CFR 113.10 (6). The FDA sponsors multiple better process control schools during the course of a year. A number of universities involved in Food Science and the Food Processors Institute also sponsor the schools. Attendance at the school and a curriculum with a portion devoted to aseptic processing will fulfill the requirements for both retort (Fig. 2) and aseptic (Fig. 4) processing involvement. A number of the schools with advanced programs or significant pilot plant capabilities such as Purdue University and North Carolina State University offer advanced training in aseptic system management and operation. These and a number of others schools have aseptic pilot operations and permit hands-on experience with processing and packaging of aseptic products.

A food processor may have one processing authority responsible for everything or may split the responsibility for the product and place it with one group or individual while placing responsibility for the packaging system with a different processing authority—an individual or an organization. Regardless of whether the processing authority is working on one system or all systems in an operation, they must ensure that the system is designed, installed, and instrumented in a way that guarantees it will produce a safe food product. The processing authority is responsible for designing the thermal or chemical processes used in the system to achieve and maintain sterility and also must design the procedures and validation protocols necessary for biological confirmation of the system's performance and process effectiveness.

The processing authority is responsible for the "process filing" (the industry term used for the complete documentation package proving the system is safe and efficacious) with the FDA and any other agency involved with the product's safe manufacture. The process filing comprises all the data and information, including conclusions and summaries of results, which prove that the system is safe and efficacious. The FDA requires registration of food processing plants conducting aseptic operations and a detailed filing of the thermal processes and sterilization procedures before a product can be sold in interstate commerce. The plant where the product is manufactured is registered using FDA form 2541. The form used to file aseptic processes for low acid foods is 2541c. Form 2541a is used to file processes used for acidified aseptic foods.

The processing authority is held responsible by FDA regulations to develop and prove the adequacy of all parameters used in an aseptic system for sterilization of the product, equipment, and packages. These parameters must be robust and assure that commercial sterility of the final product is guaranteed. The regulations also detail record-keeping requirements for every operation, both aseptic and retort, and also require detailed procedures for evaluation of any process deviations that may occur during operation of the equipment and manufacture of the product. The regulations require any deviation in the manufacture of a food product, either retort or aseptic, be reviewed by a competent processing authority. The processing authority is the final internal arbiter that decides if a product is safe when the batch history includes a process deviation.

The FDA does not approve equipments or food processes per se, but it does exert authority over the equipment types, sterilants, and packaging used for aseptic products. This information is all part of the process filing presented to the agency to register the plant and the system. The FDA maintains an engineering staff that reviews the equipment design, control, and function as part of the review of a process filing. Many times the equipment manufacturer, the packaging supplier, and the sterilant supplier will be involved in the data development and submission to the FDA. If the FDA does not feel the information supplied is sufficient, they can and do request additional information. This is accomplished by returning the process filing forms to the manufacturing company, which mean the processing system is not registered and approved by the agency and the company is not permitted to place anything produced on the equipment to enter or be distributed in interstate commerce.

The FDA relies on periodic (typically yearly) inspection of food processing plants to monitor compliance to all regulatory requirements. Inspection frequency of individual food plants may vary significantly depending on the occurrence of potentially hazardous problems, the type of product being packed, and the availability of FDA personnel trained for field inspection.

The U.S. FDA/CFSAN Grade A Pasteurized Milk Ordinance

The Grade A PMO is another portion of the federal code that governs aseptic operations. It is a code of practice covering milk and various milk-based products. This code was revised in 1983 to include aseptically produced milk and milk products. The latest version of the code was issued in 2003. This revision, among other things, placed whey and whey protein requirements in the aseptic portion of the regulations. The Public Health Service and the FDA developed the code, and it sets minimum standards and requirements for production and processing of milk and milk-based products.

This set of regulations (Grade A PMO) is recommended to the FDA by the National Conference on Interstate Milk Shipments (NCIMS) (8), which comprises members from state and local public health or departments of agriculture or agencies. The board also has nonvoting members from the dairy industry with the FDA providing input. The FDA typically accepts the recommendations of the NCIMS and incorporates them into each revision of the PMO. Each individual state then adopts the PMO standards for regulation and control of milk production, processing, and packaging within the state. Enforcement in each state typically falls to the State Department of Agriculture or State Health Department. The PMO for each state is the minimum standard required by manufacturers, and additional more restrictive standards may be added to each state's adoption and incorporation of the PMO into its regulatory codes.

The PMO specifies and regulates the manufacturing plants, aseptic processing, pasteurization, packaging, labeling, examination, distribution, and sale of milk and milk-based products. The PMO is similar to the FDA's CGMPs, which list specific requirements for equipment setup, operation, controls, and instrumentation. The PMO requirements do not replace the FDA's CGMP requirements, they are in addition to those requirements. The PMO contains regulations that are much more specific than the FDA's with regard to equipment manufacturing, sanitary design, and construction standards. This is a continuation of the 3–A Sanitary Standards for dairy equipment that are part of this regulation.

The FDA and the Public Health Service cannot enforce the standards of the PMO unless the product from a facility enters into interstate commerce. It is up to each state to enforce the code within its borders.

The PMO and the FDA CGMP requirements have a number of conflicting areas. The PMO has been adopted and is administered by the states, not the Federal Government. The FDA Milk Safety Branch personnel and the state regulatory agencies have resolved conflicts between the two sets of regulations in a satisfactory and amicable manner. Examples of differences in the ordinance arise in the placement of instrumentation in the system and the way the sterilization is performed on parts of the backside of flow diversion valves in aseptic systems, to name a few.

For packaging specific sections, the PMO refers to package manufacturing (package supplier) plants, the containers they produce, and the microbial load that may be present in fabrication materials or finished containers. The code contains requirements for chemical, physical, bacteriological, and testing standards for all types of milk products including aseptically manufactured products.

USDA Requirements

The USDA is responsible for regulation of meat, fish, and poultry products within the United States. Manufacture of formulated meat products containing 3% raw meat or 2% cooked poultry is under the jurisdiction of the Food Safety and Inspection Service (FSIS) of the USDA.

Dating back to 1984 when the USDA issued "Guidelines for Aseptic Processing and Packaging Systems for Meat and Poultry Plants" (2), the USDA has taken a very different regulatory and enforcement direction compared with the FDA for monitoring of aseptic processing and packaging systems (Figs. 4 and 6). These regulations have been updated since the initial presentation but still present differences in approach between the two agencies.

The USDA regulations are found in 9 CFR 318 (9) and 9 CFR 381 (10) portions of the U.S. regulatory code. Initially these regulations required thermal process operating personnel and container closure technicians to take training beyond the FDA-mandated course. Most universities and schools offering training for food-processing employees modified the curriculum of their better process control schools to address this requirement.

The USDA has a difference in philosophy behind their regulations compared with the FDA. The USDA reviews and approves the equipment and procedures used in meat and poultry establishments. The USDA guidelines detail the types of information the FSIS requires to approve processing and packaging systems. This is in contrast to the FDA procedure where only an acceptance of a process filing takes place and no formal approval is issued. The aseptic guidelines of the USDA are supplementary to other meat and poultry plant and inspection regulations. These regulations also require the manufacturer to submit an acceptable quality control program that covers the operation and maintenance of the aseptic processing system (Fig. 6). The USDA also requires that all materials used in containers meet FDA approval standards and that the secondary and tertiary packaging adequately protects and maintains container integrity during distribution.

FSIS requires each batch or lot of materials must be retained and samples incubated and evaluated as part of the lot performance testing. FSIS relies on continuous in-plant inspection similar to, and in parallel with, the continuous inspection and grading of meat carried out in other packing plants.

STERILIZATION TECHNOLOGIES UNDER DEVELOPMENT

During the last 10 years, a number of new technologies were introduced to the list of sterilization systems. These include

ultrasonic waves
irradiation
pulsed light
ohmic heating
extreme high-pressure sterilization
sterilization treatments using the above technologies with moderate product heating

All of the technologies represent methods that may provide improvement in the way food is sterilized without degradation. Unfortunately, the organisms targeted for sterilization are most resistant to many of these forms of disruption. A number of the technologies have been around for a while and continue to show promise. Most of the problems surrounding the technologies are the cost to develop a pilot-scale facility and the amount of engineering and developmental effort required to prove the efficacy of the technology to the FDA.

The most novel of the above technologies is the pulsed light systems. This technology employs a burst of light covering the complete visible spectrum to achieve the sterilization effect. Laboratory scale systems move the product through a tube, restricted in diameter, to allow the energy from the light to penetrate though the liquid. Clear liquids and translucent or opaque liquids like milk have been successfully sterilized using this technique. The technique has also been combined with moderate heating of the product to improve its effectiveness.

It appears that these technologies will remain on the fringe for the immediate future. The more mainstream technologies of retort processing

(Fig. 2) and aseptic processing (Figs 4 and 6) are accepted and operational and meet all current requirements for preparing enteral nutritional medical foods, infant formulas, and foods in general.

FUTURE TRENDS

Increased attention to the role of medical foods is a given in the coming years. The possible benefits range from possible increases in IQ in infants through the use of specialized ingredients in infant formulas to rapid recovery from traumatic conditions and some diseases. The needs of patients undergoing radiation or chemotherapy for cancer are one area where patients suffer from loss of appetite. This loss of appetite is reflected in a general reduction in the bodies' well-being and manifests with the loss of weight in the patient. Products that are tolerated by patients and provide increased levels of nutrition required by the body to sustain itself and combat the harmful effects of treatment are needed.

Patients undergoing dialysis and patients suffering from diabetes while undergoing invasive or disruptive treatments require increased nutrition and nutrition easily tolerated by the body under stress. Both food and pharmaceutical companies are actively pursuing products in these areas.

Packaging will play a key role in many of these therapies. Products that provide convenience and are easily portable to fit today's lifestyles are needed. Patients may be undergoing radiation, chemotherapy, or dialysis while continuing to work. The ability of packaging to deliver medical foods to these patients and fit into the hectic nature of daily schedules will spur developments beyond existing packaging.

Packaging for patients with medical needs and suffering from arthritis or other physically restricting conditions will also benefit from medical foods that are convenient to open and use.

FURTHER READING

Code of Federal Regulations, 21 CFR 114. Acidified Foods.

Code of Federal Regulations, 21 CFR 178. Indirect Food Additives: Adjuvants, Production Aids, And Sanitizers.

Wikipedia. Total Parenteral Nutrition, page last modified August 1, 2006.

Schmidl MK, Labuza TP. Medical Foods. Minnesota: University of Minnesota, 2000.

Trujillo EB, Robinson MK, Jacobs DO. Nutrition feeding critically ill patients: current concepts. Crit Care Nurse 2001; 21(4):60–66.

REFERENCES

1. Bussell S, Donnelly K, Helton S, et al. Clinical Nutrition. A Resource Book for Delivering Enteral and Parenteral Nutrition for Adults. Washington: University of Washington, 1997.

2. Code of Federal Regulations 9 CFR 318.300 Entry into Official Establishments; Reinspection and Preparation of Products, Subpart G, Canning and Canned Products, Food Safety and Inspection Service, January 1, 2003.
3. Marcin MPH, James P, Kallas MD, et al. Metabolic and Nutritional Support of the Critically Ill Child. California: University of California, 1997.
4. Wikipedia. Total Parenteral Nutrition, page last modified July 30, 2006, retrieved from en.wikipedia.org/wiki/Feeding_tube.
5. Code of Federal Regulations 21 CFR 100 to 169, Chapter 1, Subchapter B-Food for Human Consumption and various section references. Revised April 1, 2006.
6. Code of Federal Regulations, 21 CFR 113. Thermally Processed Low-Acid Foods Packaged in Hermetically Sealed Containers.
7. James V, Chambers JV, Nelson PE, eds. Principles of Aseptic Processing and Packaging, 2nd ed. Washington D.C.: The Food Processors Institute, 1993.
8. U.S. Food and Drug Administration, Center for Food Safety and Applied Nutrition. National Conference on Interstate Milk Shipments (NCIMS) Model Documents, April 25, 2006.
9. Code of Federal Regulations, 9 CFR 318, Requirements for Food Processed in Meat and Poultry Plants. Entry Into Official Establishments; Reinspection and Preparation of Products.
10. Code of Federal Regulations, 9 CFR 381, Poultry Products Inspection Regulations. Poultry Products Inspection Regulations.

5

The Regulatory Environment

INTRODUCTION

Pharmaceutical products are among the most regulated products on the planet. Pharmaceutical packaging, particularly the labeling of pharmaceuticals, must meet a myriad of regulations and approvals from government agencies all over the world. The regulations essentially make the package a part of the product and place it under evaluation and review in much the same way the drug receives evaluation and review. Each product and package must face and pass rigorous approval requirements mandated in the United States by the United States Food and Drug Administration (FDA)-administered statute and by similar agencies in almost every country of the world. Typically, the countries require multiple stages of testing with multiple submissions of data with voluminous documentation to prove that the product and its packaging are safe, efficacious, and perform as the manufacturer claims. This scrutiny is for the public protection, and it places a considerable burden on the manufacturer to prove and document that a product works and meets all claims related to its performance. Regulations vary around the world, but for the most part, the varying processes require and use much the same information as required by the FDA. In this chapter, we will discuss and highlight what the FDA requires.

Packaging faces the same rigorous review process as the drug itself. Packaging is considered part of the drug, and this is stated clearly in the regulations as part of the complete descriptions and definitions used to define packaging as part of any drug submission. The packaging portion of the submission information is contained in the Chemical, Manufacturing, and Controls (CMC) section of New Drug Application (NDA) (1). The materials used to protect a product must pass numerous tests and produce data for documentation in the same way the active pharmaceutical ingredient (API) and the excipients blended with it must pass to support any claims for efficacy and disease treatment made by the manufacturer.

This chapter is designed to provide the reader with a general overview of the regulatory requirements and is not to be considered definitive in any way. It is a summary to help in understanding the demands placed on packaging; it is not a guide for how to put together a submission or how to qualify packaging. The company or individual preparing a regulatory submission to the FDA is responsible for meeting the requirements set by the agency for approval. These requirements are contained in the statutes and also in the FDA's interpretation of the statutes presented as guidance to their current interpretation of a specific statute for the drug or device in question. The process is interactive between the agency and the manufacturer, with multiple phases and steps required to move from the identification of an API to drug approval. The process does not stop after approval. Postlaunch surveillance by the manufacturer and the agency continues throughout a drug's life. Data are constantly gathered on adverse reactions, complaints of any type about the product, its packaging, its labeling, or any other issue considered significant and reported by a user. This information is periodically reviewed to determine if some long-term effect or reaction to the drug or device manifests itself in long-term use. The results can be both positive and negative. With a positive result, new claims or other extended uses of the drug will be added to the labeling. If the information is negative, additional warnings or restrictions for use may be added to the labeling.

The regulatory process follows the orderly sequence that all pharmaceutical products go through from initial discovery of a biologically active compound to its final approval as a pharmaceutical. It is built around increasing levels of review and evaluation each time a compound is subjected to animal or human interactions. A review of the various phases of this process is necessary to understand the regulatory environment in the United States and the rest of the world. These reviews coupled with the regulatory scrutiny any pharmaceutical or medical device undergoes are necessary to understand why it is so hard to get one compound through the review and approval process. In fact, the process and the odds of developing a new drug are so difficult and long that it is not unusual for a researcher at a pharmaceutical company to never work on an approved drug. The pharmaceutical company spends large amounts of money at each phase of this testing and deploys the majority of its R&D staff in the various phases of this product development review and approval process. It is interesting to note in the accompanying tables the large number of people who follow and study the development of a drug. This group of people are constantly looking for problems and issues with any new product. They are audited and reviewed in their methods and ability to monitor patients and develop data necessary for a NDA (1). It provides everyone with a measure of security that a drug has undergone significant review and permits everyone to understand just how complicated the human body really is when problems surface even after very diligent screening at each phase in the approval process.

Each level of review and testing raises the bar on requirements for approval (Table 1). Only one in five drugs entering clinical trials gains FDA

Table 1 R&D Spending by Function Pharmaceutical Research and Manufacturers of America Member Companies, 2003

Function	Dollars (in millions)	Share (%)
Prehuman/preclinical	10,983.3	31.9
Phase 1	2333.6	6.8
Phase II	3809.6	11.1
Phase III	8038.1	23.3
Approval	4145.4	12.0
Phase IV	3698.1	10.7
Uncategorized	1445.2	4.2
Total R&D	34,453.3	100.0

Source: Pharmaceutical Research and Manufacturers of America Annual Membership Survey, 2005.

approval. Many times a material or compound can complete the entire process and not be introduced because in the last phase of testing it is determined that the product is equal to or possibly less than equal to a product already on the market or perhaps only equal to an older generic product. This last situation is the worst possible outcome for a pharmaceutical company. They are faced with discontinuing work on a compound that has taken many years to understand and have a huge investment in research, development, and testing. Alternatively, the company may try to recoup some of this investment by introducing and promoting a product that is no better than something already in use. This decision complicated by the costs of scale-up and marketing is weighed against the size of market and the need for another competitor in a therapeutic area. A determination of whether the sunk costs (research and development spending) or at least a substantial portion of them can be recovered with a product equal to others on the market is extremely difficult. The Pharmaceutical Research and Manufacturers Association (PhARMA) estimates that only 3 out of every 10 drugs completing the approval process recover their research and development costs (2). Remember, it is estimated to take approximately $800 million (Table 1) to bring a drug to market, with the majority of that being spent on clinical testing needed to develop data necessary for approval of the product. This cost has increased from $138 million in the 1970s (2).

Stages in the Identification and Qualification of a Drug

The drug approval process follows the following steps:

1. Drug discovery
2. Preclinical testing
3. Investigational New Drug (IND) review
4. Clinical trials

5. FDA approval (or approval in other parts of the world)
6. Post-marketing surveillance
7. Phase IV studies

DRUG DISCOVERY

Drug discovery is just what the name states, researchers identify a disease or condition to study that may respond to pharmaceutical treatment. During the study of the biochemical causes or other initiators of a disease or genetic condition, numerous chemical compounds that may interfere or stop the chemical path of the ailment or malady are identified and screened. Small chemical molecules, both organic and inorganic, received the majority of screening during the 20th century, and as a result, it is somewhat unusual for a new small chemical molecule to be identified as a potential pharmaceutical product. Biotechnology and its promise to harness the power of much larger molecules have now become the primary area of investigation in pharmaceutical drug discovery. This technology has highlighted the tremendous power of highly specialized proteins for catalysis or chemical intervention. A good example of the power of these discoveries is in treatment of rheumatoid arthritis. Here a biomolecule, [e.g., etanercept (Embrel) or adalimumab (Humira)] interferes with the chemistry of an agent called tumor necrosis factor and has significantly improved treatment of a debilitating condition. It has also been found to remediate a number of other conditions all catalyzed by the same chemical pathway. This type of research constitutes one of the main areas of drug discovery.

Genetics, particularly the sequencing of the human genome, is another area of intense research that attempts to understand how the body works and what may be missing or added in people with specific genetic conditions. The compounds are screened using a technique called computational chemistry. This unique computer-based technique identifies all of the potential molecules or compounds that may interact with the chemistry of the disease. The problem with doing this is its complexity. For example, the number of combinations of 20 or so amino acids in a typical protein of 350 amino acids result in more possibilities than any laboratory would have time to test. The molecules that may show biologic activity in blocking or treating a chemical sequence of reactions that constitute a disease may only be one or two of the possible combinations. Using computer techniques, all the possible combinations of these amino acids can be cataloged and evaluated statistically for their possible interaction with the disease. The total number of possibilities is reduced to a manageable number of possibilities for investigation. From this computer-assisted identification technique, the scientists will create the molecules with promise and begin testing them in cell cultures or other surrogates for cells that permit a determination of potential activity. As a sidelight to this description of proteins, keep in mind that it is not just the amino acid sequences that present a major problem to development of a new drug. Proteins fold or can be pictured as odd-shaped entities where the polymer backbone folds and convolutes to present small, very specific active sites on the surface of a protein in one configuration of possible multiple folds that manifest in the ability of the protein to be active. The synthesis, in vitro testing, and other investigatory requirements and techniques mean that drug discovery takes many years to complete.

PRECLINICAL TESTING

Candidate pharmaceuticals identified in the drug discovery phase of the approval sequence move into a second stage of testing called preclinical testing or trials. This stage of testing determines if a potential candidate will go on to clinical trials. Testing is conducted in the laboratory and in animals to determine toxicology of the compound and its potential biologic activity. This stage of development also begins to determine the potential safety of a compound. During this stage, the chemical and biologic synthesis methods needed to produce the active ingredient are studied along with the compound itself. The purity of a compound, the determination if one or more isomers or protein configurations of the compound are active, the degree of their activity, the stability of the compound, and some general pharmacokinetics are all part of the investigation (Table 2).

Both laboratory and animal testing are used in this stage to determine the toxicity of a product. These studies begin to quantify the minimum and maximum dosage limits for the potential API. All of the testing is designed to gain a complete picture of the biologic activity of the compound and to determine the

Table 2 Research and Development Staff Assignments Domestic R&D Scientific, Professional, and Technical Staff Personnel by Function, Pharmaceutical Research and Manufacturers of America Member Companies, 2003

Function	Personnel	Share (%)
Prehuman/preclinical	27,042	34.9
Phase 1	54,216	7.0
Phase II	66,879	8.9
Phase III	116,316	21.1
Approval	45,604	7.2
Phase IV	17,940	10.3
Uncategorized	1874	2.4
Total R&D staff	**71,077**	**91.8**
Supported R&D non-staff	6382	8.2
Total R&D personnel	**77,459**	**100.0**

Source: Pharmaceutical Research and Manufacturers of America Annual Membership Survey, 2005.

safety and efficacy of the candidate compound. Synthesis of the compound through multiple chemical pathways or with multiple biotechnology techniques for manufacturing are evaluated to determine if it is feasible and possible to make the product or the API in the necessary quantities. These tests determine some of the problems and difficulties associated with producing the compound on a large scale. Scale is usually determined by the identification of the number of patients worldwide who suffer from the disease or condition. During the preclinical testing phase of a product, other key starting points for dosing, packaging, and dispensing are explored. The formulation of the product and the form in which it will be presented to the patient, for example, tablet, inhaler, and injection, are all part of this phase of development. Packaging of potential products for stability testing and for delivering the various potential dosage forms or dosage delivery methods is a key part of this part of development and discovery. The decisions made on packaging size, materials, protection, and the other properties the packaging must provide to the product are for the most part determined and locked in during this phase of development. Many times these "best guesses" are wrong, and packaging must be redone when information developed in stability or further clinical testing indicates that the assumptions about stability, dosing, or the patient's method of use made at this early stage of drug development were wrong. Preclinical testing is extremely rigorous because it provides the data and the scientific rationale and basis for engaging the regulatory process and the FDA. All information developed in the preclinical phase of the drug approval sequence is required to prepare the first documented regulatory submission of the compound to the FDA as an IND application.

INVESTIGATIONAL NEW DRUG REVIEW

After all the possible investigations into the safety and efficacy of a compound are completed, the next step is to begin testing in humans. The pharmaceutical company submits an IND application to the FDA for review. At this point, the molecule or compound changes legal status and is considered a drug that is subject to the Federal Food, Drug, and Cosmetic Act. This is the entrance of the product into the FDA's pharmaceutical regulatory system. The IND application is designed to protect patients from an unreasonable level of risk or danger in clinical trials. The IND contains all the toxicology studies of the drug in animals, the chemical characterization of the drug, the methods for manufacture of clinical quantities of the drug, the clinical packaging of the drug, and the data derived to support that the drug treats a specific condition or conditions. The IND must also contain the proposed clinical protocols and information about the investigators who will perform the clinical trials. The clinical protocols must be detailed to insure that the patients are not being subjected to any unnecessary risks. The information about the clinical investigators, usually doctors, is to assess that they are qualified to carry out the trials. Last but not least are commitments to obtain informed consent from the subjects exposed to the drug in the trial, the method or methods that will be used in the study for determination of the drug's effects by an institutional review board, and the commitment to adhere to the IND regulations by the company, university, or investigator undertaking the evaluation.

Packaging for an IND is contained in the CMC section of the IND. The background and information supplied describe the components and how they are assembled into a finished packaging system. The information outlines data developed that ensures the protection and preservation of the drug during clinical trials. The submission also contains any precautions necessary to maintain the product and the package throughout the clinical trial. Container closure system information and how it is submitted to the FDA are outlined in the FDA guidance for industry entitled *Content and Format of IND Applications for Phase I Studies of Drugs, Including Well-Characterized, Therapeutic, Biotechnology-Derived Products* (November 1995).

CLINICAL TRIALS

Following acceptance of the IND, clinical trials begin in humans. There are three phases to these trials, called, as one would expect, phase I, phase II, and phase III clinical trials. Clinical trials can last from 1 to 10 years. At times they may be shortened if a drug shows such remarkable promise in its performance that additional testing is not required or if the therapy is for an orphan disease or condition. An orphan disease or condition is one that affects a very small number of people and has no currently available treatment. These exceptions and the methods used to determine the exceptions are contained in the regulations. Each of the clinical trials is designed to learn more about the safety and effectiveness of the pharmaceutical product in question.

Phase I Clinical Trials

Phase I clinical trials are designed to establish the safety of the drug, the safe dosage range, and the biologic mechanism of action that the drug creates in human test subjects. This phase of the trial is conducted on 20 to 100 volunteers. This type of trial is to ascertain that no surprises are present in the reaction of a drug with humans. All the subjects in this phase of the trial are healthy and typically do not have the condition under evaluation. This could be considered a very controlled screening of the toxicity of the molecule.

Phase II Clinical Trials

Phase II clinical trials are where the study of the disease in humans really begins. The testing introduces placebo controls to determine if the drug's effectiveness is real or whether it is a figment of the patient's imagination. During this phase of testing, the type of subject for the trial is detailed in the protocol for the test. The trial consists of testing the drug in 100 to 500 volunteers with the disease. Along with testing for effectiveness of the product against the disease, continued data gathering on safety, the determination of side effects, the effectiveness of the dosage form, and if the product is used outside the clinical site, for example, in the home, the effectiveness of the packaging in delivering the product correctly are all included in the data required. Subjects are studied intensively to determine the proper dosage regimen and if this varies dramatically from the assumptions used or determined in the preclinical stability studies and phase I testing, and if the product presentation to the doctor or patient thought to be best initially for treatment is adequate or if a complete rework of the packaging is needed to deliver the drug properly to the patient. This rarely happens for tablets, but for other dosage forms, problems are identified and more common. This is the most common place in the package testing and development that requires major packaging changes. If compliance is a key to the drug's performance, the design of packaging to maximize doctor and patient's understanding of use of the product is studied in detail. Compliance with treatments that require long-term use is one of the most difficult problems when developing and administering a drug to patients outside a controlled setting such as the hospital.

Phase III Clinical Trials

Phase III clinical trials are the last major hurdle for a product to successfully complete on the way to regulatory approval. Phase III is a large-scale trial of 1000 to 5000 patients in hospitals, clinics, and physician offices that removes many of the preselection criteria of candidates in phase II and exposes the drug to a wide variety of patients with the disease. These trials produce the statistical evidence in the general population needed to prove drug effectiveness. The patients are typically a random selection of people with the disease; however, patients with certain risk factors discovered in earlier clinical trials are identified and excluded. These exclusions are detailed in the labeling used for the product. The trial is completely blinded to the people administering the trial. This means that the placebo and actual drug product look or behave alike in their physical characteristics, and the packaging for the placebo and the drug are the same so as not to disclose the nature of the package content. Packaging in phase III trials may be a continuation of the earlier product presentation but should, if possible, be the final package in which the drug will be introduced without the colorful graphics and other elements of final labeling. Because a large number of patients are needed in a phase III trial and the amount of study of each patient throughout the treatment regimen, whether they receive drug or placebo, is required to determine if the product performs properly, these trials take the longest amount of time and consume the greatest amount of resources in the development of a product.

FDA APPROVAL

At the completion of the phase III clinical trials, the developer prepares a complete NDA for the product and submits it to the FDA. Following receipt of the NDA, FDA scientists and advisory panels review all the data collected about the drug over the many years of testing. The scientists and committees focus on the safety, potential risks, and benefits a product has to offer, other alternative treatments for the disease, and any other factors pertinent to determining if the product's data justify its approval and release to the public. Numerous meetings take place throughout the process, and it is not unusual for the agency to request additional information or to submit a list of questions regarding the product for a company to answer. The requests may be for an expansion of the data developed, a query specific to a particular aspect of how the submission was prepared, how data was treated or summarized in the submission, or a specific question or specific testing requirement that expands specific parts of the information to further elucidate the performance and function of the drug. The CMC section of the documentation undergoes equally rigorous scrutiny regarding all aspects of manufacturing, validation, product protection, and packaging developed for the product along with a submission of the proposed labeling of the product.

Labeling and its claims can be contentious at this stage of review, and multiple changes may be required by the agency up to and including the final labeling they approve with the final approval of the NDA.

Companies continue long-term toxicity testing, evaluations of dosage forms, potential manufacturing methods, and evaluation of package design and performance during the initial stages of manufacture and during the first months the product is on the market.

A drug at this stage of approval has undergone an amazing gauntlet of tests and evaluation to finally be approved. It is one of approximately 10,000 compounds originally identified that actually performs as hoped to treat a disease. This complex process yields approval of only one in five drugs that enter clinical trials.

POST-MARKETING SURVEILLANCE AND PHASE IV STUDIES

After approval is granted, a drug continues to be studied throughout its useful life. Companies are required to constantly monitor the safety of products and determine how these products affect particular groups of patients who had a small statistical representation in the clinical trials, particularly in phase III trials.

The product safety monitoring by the companies, including a review of all complaints and reports of side effects, drug interactions, and adverse events must be reported to the FDA on a very rigid schedule. Companies are required to report and investigate adverse events quickly and maintain communication with the FDA on the investigation into the adverse reaction and its results. It is not unusual for a product to receive initial approval in phase III, but as the number of patients exposed to the drug grows, new and different safety and efficacy results emerge. This information must constantly be submitted to the agency and may precipitate required updating of the labeling for a product. From time to time, a product may be removed from the market on the basis of safety data derived in the post-marketing surveillance of the product by the manufacturer. Conversely, data derived from the continued monitoring of a product may result in the discovery that it is an effective treatment for other conditions beyond those for which it was originally approved. In this case, the labeling is updated to include the new claims, and the manufacturer will begin promoting the product for the new conditions. Manufacturers and the FDA work together to insure that the patient is receiving the best information and best data available about the product's performance and safety.

THE REGULATORY ARENA

In the United States, a number of federal agencies have jurisdiction over packaging for drugs and food. The primary agencies are the FDA and the Consumer Product Safety Commission (CPSC). With the recent rise of nutraceuticals or food products falling somewhere between a drug and a food, the United States Department of Agriculture (USDA) may also be involved. Congress has also passed pieces of legislation that applies to drug manufacture and packaging that are outside these agencies' areas of responsibility. These pieces of legislation and associated agencies include the Toxic Substance Control Act (TOSCA), the Occupational Safety and Health Administration (OSHA), the Resource Conservation and Recovery Act (RCRA), the Clean Air Act, and the Comprehensive Environmental Responsibility Compensation and Liability Act (CERCLA). The requirements and information needed to answer the provisions of the legislation for these other regulatory agencies apply broadly to the manufacture of all types of substances, not just pharmaceuticals, so they are excluded from this discussion.

Along with all the governmental agencies, a number of other references and their requirements must be considered in meeting the regulatory requirements for packaging any drug. In particular, the United States Pharmacopoeia (USP)-National Formulary (NF) must be consulted and complied with as part of developing or qualifying packaging for a product. The Pharmacopoeia and Formulary is mentioned specifically in legislation and carries the force of law in its requirements. This reference is recognized as definitive in the Food, Drug, and Cosmetic Act.

The regulatory process is a complex web of laws, guidance documents, and opinions that constitute an area of study unto itself. A general summary of major or key requirements, considerations, and other items that are part of packaging design and development provide a basic understanding of how regulatory requirements affect the design, development, and approval process for any drug. The regulatory process effects the packaging material selection process, the package manufacturing process, the pharmaceutical acceptance process for materials, the pharmaceutical packaging and labeling process, and the continuing scrutiny that manufacturing (including batch records) receives in the ongoing manufacture of drugs. It extends to the methods and controls for recording out-of-specification products or procedural hiccups in the manufacturing process that may determine, influence, or question if a particular lot of products are equivalent to all others. It includes requirements for testing and documentation of the investigation of what are called major and minor deviations in product manufacture that occur from time to time when things are made on an ongoing basis. It extends to the investigation of complaints from the field and whether these real or perceived shortcomings of a product recorded by the doctor or consumer can be traced to a problem in manufacturing or packaging. It also extends to the requirements and methods used to make major or minor alterations to a product over the course of their useful life. An example of packaging change over the product's life would be the qualification of an alternate plastic material used to manufacture a container for the product.

THE UNITED STATES FOOD AND DRUG ADMINISTRATION

The FDA is not a new agency in the U.S. Government. It traces its roots back to 1906 when problems with adulterated foods were relatively commonplace. The original problem focused on packaged meat, and was highlighted by Sinclair Lewis in his book entitled *The Jungle*. This book captured the public attention and described many problems that existed in meatpacking at the turn of the 20th century. The original act passed by Congress authorized the FDA to police interstate commerce for mislabeled and adulterated food and drugs. This is when some of the original "snake oil" remedies came under review, along with canned, frozen, and refrigerated meats. This is why many of those quaint old-time remedies marketed in the 19th century are no longer around.

It was not until 1938, when the Food, Drug, and Cosmetic Act was passed by Congress, that the FDA gained the authority to establish definitions and standards of identity for foods, drugs, and cosmetics. The agency was also given authority to evaluate and provide clearance to drugs after appropriate safety evaluations were completed. The law brought cosmetics under the agency's review for misbranding, mislabeling, or adulteration.

Over the next twenty years, the growing use of food additives prompted Congress to pass amendments to the act in 1958 that regulated their use. A part of these amendments was the first direct statements that outlined requirements about materials in contact with food or in the packaging of food that were not permitted to become part of the food. Food colorants were included as another amendment to the act in 1960. Congress continued to add to the power of the agency in 1962 after thalidomide was shown to cause severe birth defects. The fortunate thing for the United States was that the Agency had not approved the drug for use here. In Europe the product was approved and prescribed for women who were pregnant. When the drug was used in pregnancy, infants were horribly affected and were born with missing limbs. The drug caused this birth defect. The tragedy did touch a few in the United States; women who received prescriptions and took the drug in Europe, after return to home had babies with the same tragic deformities.

Following this tragedy and the public outcry that ensued, the FDA was given the power to base preclearance of drugs on both efficacy and safety. They were also instructed to use an affirmative clearance process that differed from their previous policy of a waiting period and non-objection procedure. This affirmative clearance power has evolved into the sophisticated drug trials and reviews now used for all drugs. Today's law prohibits the introduction of a drug or the method to deliver a drug into interstate commerce without following an application procedure first set forth in Section 505(b) of the 1938 act.

To implement all the provisions of this law, the FDA has published numerous regulations contained in the Code of Federal Regulations (CFR) published under Title 21 (3). These regulations detail testing, studies, data, packaging, manufacturing practices and processes, and clinical studies that must be conducted and submitted to the agency for review and approval. Major regulations for the data submissions, testing, and approving new products as INDs are contained in Section 312, the first major hurdle in the drug approval process. Section 314 details all the information necessary to file a NDA, the formal result of years of testing and clinical trials that prove that the pharmaceutical product is safe and efficacious for use. Drug labeling requirements are detailed in Section 201 of Title 21 of the CFR (3). Essentially, the 1938 act and later amendments to the act require the submission of full reports detailing all investigations conducted to study and show that a drug is safe and effective for use, a complete list of the materials used in the components of the drug, a full statement detailing the composition of the drug, and a full description of all the manufacturing methods, processing, procedures, and packaging of the drug. It also requires the manufacturer to submit samples of the drug and its components as required by the secretary (read FDA or agency) and all specimens of the labeling proposed for the drug. The labeling is very important because it contains all claims, warnings, and uses of the drug. These are scrutinized in the same way the chemical and clinical results are reviewed. The claims of efficacy used to describe the performance of the drug must be proven by the data submitted.

In addition to the laws and the specific citations of the law in the CFR, the agency also issues guidelines or guidance that defines and clarifies the latest thinking of the agency on the subject. The guidance explains, defines, or clarifies what the FDA will use as criteria for their standards in interpretation of the regulation and how they will evaluate what a manufacturer provides in a submission. These guidelines, which are not binding because they are not regulations or laws, are not enforceable either through administrative actions or through the courts, but provide insight into the agency's current thinking on a subject. The guidance documents provide information regarding design, production, manufacturing, and testing of regulated products that are consistent with the position of the part of the agency that will review and approve or reject an application. Guidance provides clarification but not definitive steps to follow or the specific requirements necessary to meet the FDA's needs. The agency always refers to alternative approaches that may satisfy the statute or regulations. Alternative approaches require review and discussion with the FDA to avoid a later determination that the approach(s) being considered are not correct or are inappropriate to meet the requirements of approval. These guidelines in separate sections define the manufacture of drugs, samples, analytical data, validation methods, stability, and packaging. The packaging guideline was published in the Federal Register in 1984 and finalized three years later in 1987. The latest version of this original guideline or guidance was published in May of 1999.

The complete series of guidelines and each individual document are titled and are referred to as "Guidance for Industry" published by the agency and the various subdivisions of the agency such as the Center for Drug Evaluation and Research (CDER) and the Center for Biologics Evaluation and Research (CBER). They represent the current thinking of the agency on various topics and are provided to help all parties, private and corporate, understand the rigor of the review process as determined by the agency and its subsections. These guidance documents are not part of the law and are updated and changed periodically. For packaging, the current guidance for industry is titled "Container Closure Systems for Packaging Human Drugs and Biologics," Chemistry, Manufacturing, and Controls Documentation. These guidance documents and the laws behind them are not easy or simple reading but are required reading for a packaging professional engaged in the development of new packaging for a new drug. There are also multiple study aids, formal classes, and seminars to help you understand how to provide the agency with the information needed to make the process move forward smoothly. The FDA always has the final word, and unfortunately, if one of these outside aids is in error or is misinterpreted, the person or company must go back and do what the agency specifies and requires in its review and approval process. Many groups have lost considerable time in the development and testing process for a new drug by not reviewing carefully the information and laws governing the item in development and relying on outside information that may be out-of-date or erroneous. Large pharmaceutical manufacturers maintain extensive Regulatory Affairs departments. This department may be part of a Medical Affairs department and may use this title as well. These professionals are responsible for determining what the agency requires and what must be done in developing and documenting a drug's data for submission. These individuals meet frequently with their counterparts in the FDA to clarify and understand how the agency may view something novel or different from similar processes used for other products. They also explain and review with the agency novel approaches or ideas the manufacturer or developer wants to use in the various trials.

A GENERAL OVERVIEW OF THE DRUG APPROVAL PROCESS

The FDA considers packaging to be part of the pharmaceutical product. Information regarding packaging must be submitted as part of the IND and NDA processes, and additional procedures are provided by the agency for Abbreviated New Drug Applications (ANDA). This is sometimes referred to as an amended NDA. The ANDA covers many of the changes in source of supply, location of manufacture, or substitution of materials required over the lifetime of a product and uses the same methods and standards in abbreviated form to prove equivalence to the original product described in the NDA-developed material or procedure. It can be quite extensive depending on the changes being undertaken.

Originally, the requirements for packaging were spelled out in the Guideline for Submitting Documentation for Packaging for Human Drugs and Biologics issued in February 1987 and a packaging policy statement issued in a letter to industry in June of 1995 from the Office of Generic Drugs. Currently, the latest information is contained in FDA Guidance for Industry titled "Container Closure Systems for Packaging Human Drugs and Biologics" issued in May of 1999. The subtitle of the document is Chemistry, Manufacturing, and Controls Documentation. This guidance is broken up into the following sections:

INTRODUCTION

BACKGROUND
Definitions
CGMP, CPSC, And USP Requirements on Containers and Closures
Additional Considerations

QUALIFICATION AND QUALITY CONTROL OF PACKAGING COMPONENTS
Introduction
General Considerations
Information That Should Be Submitted in Support of an Original Application for Any Drug Product.
Inhalation Drug Products
Drug Products for Injection and Ophthalmic Drug Products
Liquid-Based Oral and Topical Drug Products and Topical Delivery Systems
Solid Oral Dosage Forms and Powders for Reconstitution
Other Dosage Forms

POSTAPPROVAL PACKAGING CHANGES

TYPE III DRUG MASTER FILES
General Comments
Information in a Type III DMF

BULK CONTAINERS
Containers for Bulk Drug Substances
Containers for Bulk Drug Products

ATTACHMENT A
REGULATORY REQUIREMENTS

ATTACHMENT B
COMPLIANCE POLICY GUIDES THAT CONCERN PACKAGING

ATTACHMENT C
EXTRACTINON STUDIES

ATTACHMENT D
ABBREVIATIONS

ATTACHMENT E
REFERENCES

The Federal Food, Drug, and Cosmetic Act defines the role of packaging very clearly in Section 501 and mandates the need for adequate information on packaging materials. This section states, "A drug or device shall be deemed to be adulterated . . . if its container is composed, in whole or in part, of any poisonous or deleterious substance which may render the contents injurious to health" [Section 501(a) (4)] or "if it is a drug and the methods used in or the facilities or controls used for, its manufacture, processing, packing, or holding do not conform to or are not operated or administered in conformity with current good manufacturing practices (CGMP) to assure that such drug meets the requirements of this Act as to safety and has the identity and strength, and meets the quality and purity characteristics which it purports or is represented to possess" [Section 501(a) (2) (B)]. CGMP is discussed later in this chapter.

Section 502 of the act defines misbranded product when there are packaging omissions, and Section 505 (b) (2) (D) of the act describes, "An application shall include a full description of the methods used in, the manufacturing, processing and packing of such drug. This includes facilities and controls used in the packaging of a drug product."

THE DRUG PACKAGING APPROVAL PROCESS

What does it take to get a drug approved by the FDA? What is required of the packaging by the FDA for the drug approval? Packaging is considered part of the drug. This means that data surrounding the performance of the packaging and the drug together are required as part of the proof or data needed by the agency to properly review and approve a new drug. The regulations, which are summarized in the sections of the guidance detailed in the last section of this chapter, require that "full information. . .in sufficient detail to permit evaluation of the adequacy of the described methods of manufacture, processes, including packaging and the facilities and controls used to manufacture the drug preserve the identity, strength, quality, and purity of the drug" (2). The agency also requires information "with respect to the characteristics of the test methods employed for the container, closure or other components of the drug package to assure their suitability for the intended use" (2).

Samples of the finished packaging must accompany the NDA submission. The information regarding the package in many cases involves details about the materials and components of the package that are proprietary to the manufacture of the package or one of its components. This proprietary information requires protection so that it does not become publicly known. Proprietary information is regarded as a trade secret by the manufacturer, which they do not want to make available to their competition. It includes information that is not patented because it is deemed sensitive to the product or the process. The FDA has a procedure to protect this information and requires that the manufacturer establish with the agency a Drug Master File (DMF) as part of the submission process. The proprietary information is submitted directly by the manufacturer to the FDA. It is kept in the DMF that can be accessed by the agency only when it is authorized to do so by the submitter of the NDA. The background information held within the DMF only proves the ingredients in the packaging or the other materials are considered safe on the basis of toxicologic or other standards established for those materials. It may also contain information on manufacturing processes or closure processes that are new or proprietary in how they are employed with the product. Even if all the materials and components used for packaging a drug have been on the market and have been used with similar drug compounds, the manufacturer must still prove that the packaging will maintain the quality, strength, purity, and other properties of the drug as specified in Title 21.

Once a product and package are cleared by the agency for use, it becomes very expensive in time and testing to prove that another material, process, or package are equally safe for use. If manufacturers want to change a package or introduce a new package, they must submit a supplemental application called an ANDA for review. This application may not need to go through all the testing of the original NDA if equivalence can be demonstrated in various aspects of the application. An example of this would be using a second source of an approved USP material for the manufacture or packaging of the product. In the review of supplemental applications by the agency, the focus and area of proof typically centers on stability. Some changes may require prior approval before implementation, and some changes may only require detailed explanations in the submission of the annual report. The annual report summarizes all changes in packaging and labeling made to a product throughout the year. In most cases, it documents updates to labeling required by the FDA for various classes of drugs, or it updates a change and qualification of a packaging material from another

Table 3 Examples of Packaging Concerns for Common Classes of Drug Products

Degree of concern associated with the route of administration	Likelihood of packaging component–dosage form interaction: High	Medium	Low
Highest	Inhalation aerosols and solutions: injections and injectable suspensions[a]	Sterile powders and powders for injection; inhalation powders	
High	Ophthalmic solutions and suspensions; transdermal ointments and patches; nasal aerosols and sprays		
Low	Topical solutions and suspensions; topical and lingual aerosols; oral solutions and suspensions	Topical powders; oral powders	Oral tablets and oral (hard and soft gelatin) capsules

[a]For the purposes of this table, the term "suspension" is used to mean a mixture of two immiscible phases (e.g., solid in liquid or liquid in liquid). As such, it encompasses a wide variety of dosage forms such as creams, ointments, gels, and emulsions, as well as suspensions in the pharmaceutical sense.

source. Many times a material is changed or discontinued, and the manufacturer must qualify a similar material as a replacement. Since the qualification is the same as that detailed in the original submission, the FDA requires that the manufacturer carry out the same validation and testing protocol, including stability, before the substitute material is placed into production. Because the data and testing are proving equivalence, the agency may permit this to be reported at the end of the year as part of the summary of all changes made to a product. Contained in the Guidance for Packaging is a table (Table 3) that represents the level of scrutiny a packaging component will receive depending on the type of drug and how the package is used.

CURRENT GOOD MANUFACTURING PRACTICES

CGMPs (note that this may also be abbreviated as cGMP) is not a new invention of the FDA (5). CGMP has gained prominence in the past 20 years because much more emphasis is being placed on the total system used to manufacture and package drugs. The regulations covering CGMP are contained in Part 211 of Title 21 of the CFR. They discuss multiple aspects of manufacturing and have a number of references to packaging. They are also used to emphasize the validation aspects of manufacturing.

VALIDATION

To digress for a moment, some discussion of validation is necessary in today's regulatory climate. Validation is not a new concept. It stands for the proof needed to manufacture anything reliably within established specifications. It is the data needed to prove that the equipment, materials, and processes used to produce anything, in this case drug products, when controlled properly and sufficiently, will produce the same result over and over again. Validation is required as part of the regulations for pharmaceuticals and medical devices. This requirement is stated in 21CFR Section 210 and Section 211 (3) and the Good Manufacturing Practice Regulations for Medical Devices, 21CFR Section 820.

Validation and certification began in the aircraft and nuclear power industries. Each of these technologies can have cataclysmic consequences if something does not perform in the expected manner. Certifying an airplane engine to insure that it will continue to operate in torrential rain, hail, snow, ice, or after multiple bird ingestions answers questions everyone, not just regulators, ask about a new piece of equipment. The development of test methods, carrying out rigorous tests, documenting the tests, and providing data that proves without a doubt that the engine will continue to operate in bad weather or under adverse conditions are together an example of validation or certification.

The FDA has adopted a similar position regarding drug manufacturing and packaging. They do not tell a manufacturer how to prove that a drug and its package will perform as stated. They hold the manufacturer accountable for maintenance of quality, purity, strength, and efficacy of a product over its claimed shelf life.

Process validation is part of current good manufacturing practices (CGMP) for pharmaceuticals and medical devices (4). The FDA recognizes that because there is a great variety of products, both pharmaceuticals and medical devices, and a wide variety of equipments and materials including packaging, no guideline can cover all situations. What they do is provide the manufacturer a broad guideline of concepts and expected requirements that a manufacturer can use to prove that a product is properly produced. The FDA follows some basic principles regarding quality assurance of a product, which are as follows (4):

- Quality, safety, and effectiveness must be designed and built into a product.
- Quality cannot be inspected or tested into a finished product.
- Each step of the manufacturing and packaging process must be controlled to maximize the probability that the finished product meets all quality and design specifications.

These guidelines emphasize that the process and the materials used in a process must be the same each time the product is produced. Packaging components are no different from chemicals used to synthesize the pharmaceutical product and no different in their manufacturing process from any of the pieces of processing equipment that the pharmaceutical passes through or is held in during the manufacturing process. For a medical device, each device cannot be tested and modified to prove that it works as stated each time. A good process validation identifies key process variables that must be monitored and documented and when maintained directly influence the operation of the device in the same manner each time.

The packaging components produced to protect a pharmaceutical product must follow the same idea of validation. Validation develops the proof needed to assure a high degree of confidence in a process to produce to predetermined specifications defining the packaging article or component in a consistent manner. A validation protocol is a plan that states how validation will be conducted, the testing regimen and parameters, the product characteristics, production equipment, and decision points that define what is required to produce an acceptable result in the package testing. It must be prepared by the manufacturer to prove that the packaging used for a product is reproducible each time it is manufactured for the pharmaceutical manufacturing process. If the process is modified slightly, such as variations in temperature or pressure used during the extrusion and blow molding of plastic bottles, the acceptable range of conditions must be reviewed and tested. The minor modifications required for the process to produce acceptable containers must be part of the range of conditions tested and approved. A packaging validation protocol is developed in the same way it is developed for the drug. The validation protocol must be written and reviewed using sound scientific, engineering, and statistical principles that assure that the process and resulting component remain the same. The protocol contains the procedures and tests that will be used to establish the level of performance and will specify how the data is to be collected and reported. Numerous testing protocols are needed for validation of a process or a package and are specific to the three standard qualification requirements being measured, installation qualification (IQ), operational qualification (OQ), and performance qualification (PQ). The letter designations IQ, OQ, and PQ are the common terms used to identify which validation component is undergoing review and challenge. The formal names are used less often. The expected result of the testing, the pretesting judgment of the person(s) writing the protocol, is a major determinant of acceptability if an outside industry standard or test method does not exist. The protocol defines quantity, sampling, and tests needed to evaluate the package or performance parameter(s) of the packaging and proves that it meets or exceeds the required performance level. When a validation protocol is completed for one of the three components in the validation, IQ, OQ, or PQ, it contains the test results and the data behind the results required by the validation protocol indicating how the operation, component or package performed. A summary is also part of the document and explains the purpose of the specific protocol, what it was designed to challenge, and the results of that challenge against the expected result the protocol defined. Remember the validation protocol was developed, reviewed and approved prior to the testing as a reasonable test to challenge the process or the package and prove it was robust and met the needs of the product it was producing or protecting. All the validation protocols from each requirement of validation, IQ, OQ, and PQ must pass the defined protocol and contain adequate documentation including all test data for FDA review and critique.

A typical protocol is written after all initial development work is completed on the new product or the amended product. This development work establishes the general acceptable limits needed to protect the product with packaging. In the case of a plastic bottle, for instance, the type of material, the thickness of the material used in its construction, and the physical characteristics designed to seal the product in the package are essential parts of the data sets used in producing stability samples and all the product samples used in clinical trials.

Product produced in a single-cavity mold or a low-cavitation mold has a specification developed over time that attests to product performance and quality. When this product moves into large-scale production, the number of mold cavities used to produce bottles increases. A typical problem for the packaging engineer is to validate that the new mold tooling with multiple molding cavities produces bottles that are in all material respects the same as those produced by the single-cavity tool. Specifications established on the single-cavity mold must be duplicated for each cavity of a large multi-cavity mold. Varying process conditions, representative of worst-case low operating range, (example low temperature, low pressure, etc.) conditions, nominal (mid-range or normal) operating conditions and high range (example high temperature and high pressure) operating conditions and in this case mixtures of high and low operating conditions (example high temperature/low pressure) are built into the validation protocol to insure that the mold on its best and worst days of operation still produces bottles of acceptable quality for packaging the drug. The defined terms are sometimes referred to as the lower, nominal, and upper control limits for the process. Most protocols also require the testing to be done multiple times and extend to multiple batches or multiple production cycles to prove the process is reproducible on a day to day, or other repetitive basis.

ELECTRONIC DATA SUBMISSION
ELECTRONIC SPECIFICATION SYSTEMS
ELIMINATION OF PAPER RECORDS
21 CFR PART 11 ELECTRONIC RECORDS

Probably no other topic has been discussed, interpreted, or misinterpreted more than FDA's issuance of regulations regarding electronic records, electronic signatures, and databases containing the electronic records (6). Part 11, as these regulations are normally referred to, was issued in 1997 to provide all interested parties with FDA's acceptance criteria for electronic records including the use of electronic signatures. Electronic signatures permit departments in multiple geographic locations to review and approve a document without physically transferring the document to each location for signature or using another method to obtain legally binding signatures and time stamps on documents. This set of regulations for storing and approving digital documents made electronic files and signatures equivalent to paper records and handwritten signatures.

Part 11 also applies to electronic records submitted to the FDA even if the electronic records are not specifically identified in the FDA's regulations (§ 11.1). The FDA defines the underlying requirements for documentation set forth in the Federal Food, Drug, and Cosmetic Act and the Public Health Service Act (PHS Act), and other FDA regulations as predicate rules. Predicate rules [e.g., §§ 11.2(a), 11.2(b), 11.50, 11.70, 11.100, 11.200, and 11.300] that define the agency's requirements for paper records relating to pharmaceutical products are well known for hard copy paper records and documents. Part 11 regulations were designed from the outset to encourage and permit the adoption of electronic documentation in the pharmaceutical industry while maintaining the same safeguards already in place for paper documentation. The original regulations generated a significant amount of review and discussion between the agency, industry, and contractors regarding interpretation and implementation of the regulations. These ongoing discussions and questions about possible issues led the FDA to publish a compliance policy guide (CPG) 7153.17 and a number of guidance documents describing electronic records and electronic signatures validation, glossary of terms, time stamps, maintenance of records, and copies of electronic records. Even with this effort to clarify and answer the many questions raised regarding interpretation of the Part 11 requirements, problems continued. The problems or questions raised were that the regulations would restrict the use of electronic technology, increase the costs of compliance, and discourage innovation and the use of technical advances without permitting the obvious benefits these advances provide. Complicating the problem were questions about what validation is required on any electronic system, how old paper and new electronic audit trails could be merged, what were considered legacy systems, and many others. Further complicating the problem was the fact that 21 CFR Part 11 also contains provisions of the CGMP regulations (21 CFR 211), the Quality Systems regulations (21 CFR Part 820), and the Good Laboratory Practice for Nonclinical Laboratory Studies regulations (21 CFR Part 58) (6). Because of these concerns and, additionally, the FDA's major review and upgrade of CGMP, Part 11 will undergo some changes. It is anticipated that most of the underlying goals of the original regulation will be retained as the technology evolves. The key point will be insuring that all records are maintained and submitted in accordance with the predicate rules and that the records provide the same level of accessibility and understanding as defined in the predicate rules.

Later in this section on regulation, new systems and initiatives that will enlarge the scope of electronic records are described. The important item here is the need to improve safety for the public and permit companies to benefit from the best possible method for record keeping.

Electronic record keeping is designed to deal with a number of problems that are always inherent with paper records. These include

- access,
- location of records,
- maintenance of records in multiple locations,
- search capabilities for the records,
- archive and retrieval of records,
- sharing of records, and
- review and approval of records.

Conversion of records to an electronic format in a true database permits a company to gain the maximum access to the records and the maximum benefit from the information they contain.

Over the past 10 years, the FDA and a number of leading companies have placed a great deal of emphasis on developing true electronic specification and labeling systems or repositories. In some cases, the companies had legacy systems that met the Part 11 requirements. These systems were permitted provided they were not materially changed after the March 1997 regulation date. Companies saw the benefit of electronic files for storage and retrieval and have converted many records from paper to electronic documents held in these systems. The goal for the FDA and the companies is improved accuracy of specifications, engineered drawings, batch records, standard operating procedures (SOPs), labeling, artwork, and promotional materials wherever and whenever they are used. Electronic systems permit reviewers in multiple locations to view and approve a document electronically, archive the document in a more accessible and reliable form when compared with microfiche of the paper records, and provide the FDA auditor or investigator with fast complete access to all records necessary to safeguard the public health. Many companies have begun the conversion from multiple paper and electronic systems to one electronic system and one electronic digital repository for all information. Think about the number of records generated on a single product in multiple research centers or multiple manufacturing locations. What is the best way to maintain this information and to update it in a controlled and systematic manner? This is the question that prompted Part 11 when electronic systems were needed and had the capability to replace paper systems. Real time electronic systems capable of maintaining a proper audit trail (sometimes called the paper trail) and capable of being available at multiple locations worldwide 24 hours a day, 7 days a week, were much more efficient and offered improved safety over systems that relied on the preparation, dissemination, and constant updating of paper records in each location. These improvements along with improved information sharing and improved productivity were some of the drivers behind the development, validation, and adoption of electronic systems. The information disseminated over the company's own intranet or over a secure Internet system permits the maximum use and maximum benefit to be derived from developmental data and information, manufacturing information, packaging specifications and labeling, and promotional literature.

CHANGE CONTROL

A goal of all regional or global pharmaceutical and biologic companies with multiple administrative, research, and manufacturing locations is to make specifications and records qualified during the development of a new product available, easily usable, and, most important, searchable by everyone inside the company. Products are constantly being developed, approved, updated, and revised. This process requires all records contained in a file system, either electronic or paper, to maintain revision control and to accurately establish the time and date any change is made, who authorized or approved the change, and when the change is effective. This type of control on all records mandates a disciplined and documented method for change control that is reliable and accurate. For paper records, the methods were documented in SOPs. The FDA requires the same for electronic records (7). How electronic records are to be developed and maintained are left to the companies or department developing the system. They are required to write the SOPs that establish how records move from an existing system to another, or how the records transition to a new electronic system, with most companies emphasizing minimizing the cost and effort required by converting paper records to electronic records.

Documentation for change control must also be defined in a set of SOPs that are part of any change control process. This means that each time a specification or label is revised for whatever reason, the previous version of the record is archived and the newest version of the record is put into use. Records used in the development and maintenance of products must be archived in a way that provides a simple paper or electronic trail that is available for review at any time. The FDA requires a historical archive and the capability to search development and product records. The benefits to a company developing an electronic system to accomplish this task are multifaceted. The most obvious example is a searchable database of information that can be configured for multiple purposes worldwide and can be referenced by anyone in the company anywhere in the world to review and use. The old saying "don't re-invent the wheel" may actually become a reality within a company with a well-designed, well-engineered, and easy-to-use electronic system. Even in small-scale implementations in document, heavy departments like packaging have the potential to save large amounts of money and time by providing information, data, test procedures, specifications, and labeling to individuals located outside the department and possibly in remote locations around the world. Many times, the person searching for needed information, for example, a specification for a specific type of packaging, would recreate the information because it was faster and easier than attempting to find it in existing records.

A database available over the company intranet or secured Internet connection is always up-to-date. It can be structured to contain not only the original records regarding a component specification or label text and graphics but also all subsequent changes to the records for anything used in a product's manufacture.

This ability to consult and reuse information previously developed allows the maximum benefit of the data to be available to all who are searching for the knowledge it contains. Examples of candidate records include raw material specifications, subassembly specifications, and packaging specifications for components used in any location anywhere in the world that manufacture the product. When multiple vendors are qualified to produce a component, it means that a company has the ability to cross-reference information or use substitutes without a laborious determination of what is qualified and what is not.

The goal of Part 11 has always been to make these benefits possible. It was also the first step toward paperwork reduction and the ability of companies to submit information electronically to the agency. Going forward, the FDA has multiple initiatives to encourage and improve how information is submitted and structured for use as electronic submissions and documents.

STRUCTURED PRODUCT LABELING: ENTERPRISE CONTENT MANAGEMENT, DIGITAL ASSET MANAGEMENT

In the previous section regarding Part 11 common specifications, labeling and engineered drawings are to be retained in a database with the same capabilities provided by a hard copy or paper file system containing information about a product. The newer systems called Enterprise Content Management (ECR) and Digital Asset Management (DAM) utilize advanced electronic data management techniques that bring together much larger and broader portions of a company's complete information infrastructure. A true ECR system permits the parsing and use of digital information components that have multiple presentations and uses. DAM is a subset of a true ECR system. It is the secure database that holds all digital assets of a product such as pictures, drawings, audio, or video files. These can be time consuming and costly to create. Making them searchable, transportable, and available to a broad group of departments, suppliers, and users is a very valuable capability.

The ECR and DAM systems permit digital assets like labeling text and graphics developed for product packaging to feed multiple secondary uses for the same information. An example would be the use of a product insert produced for printing and inclusion in a product being electronically accessed by and used by the company or product website that makes the information available for doctor, pharmacist, or patient review. These systems replace the need to update single product or multiple product (same product family) information dossiers one document at a time. The chance for error when tens or hundreds of documents require modification or change is greatly reduced.

Using eXtensible Markup Language (XML) formats and information tagged with XML-specific information, a document, web page, or specification that requires change can be updated simply and quickly. The change of information in the specification or labeling is modified once in the XML-tagged data or text. The electronic system has the capability to search and find all documents that are part of the change and to automatically substitute the new information in all similarly tagged text or graphics whenever and wherever they are used (read multiple plants, offices, sales locations, websites, headquarters, marketing, etc.) A good example is a Master document that is referenced by all other subdocuments pertaining to a product. In countries granting complete approval all references are immediately updated along with the website used by doctors and healthcare professionals for reference. In countries in different stages of review and approval the change would only happen on claims or warnings already approved. Fast, easy updating of graphics and artwork, and the ability to add or modify claims for product use are difficult if the information is scattered or held in multiple company locations. The ability of a company to understand and manage a complete picture of their specifications, manufacturing processes, and packaging (both structural and labeling) is not a simple task and few companies have truly mastered this challenge using paper records. ECR and DAM technologies offer the promise to make this difficult job of data synchronization one they can control and manage with a high degree of accuracy.

Information contained in the component specifications and labeling can be rearranged to generate batch records, testing records, and other documents needed to track and sustain product manufacturing. This information and the SOPs for its use are one way to improve CGMP. The packaging bill of materials (PBOM) is the one place where all of this information comes together in a concise form for each individual product. The PBOM identifies the product number, the NDC number, and all the physical characteristics of the finished product including all the labeling needed to produce the product. Along with the physical description of all of the components and subassemblies, it also contains how the product is bundled, case packed, palletized, and shipped. This information is necessary in many areas and different departments of a company including procurement, manufacturing, finance, regulatory affairs, quality assurance, sales, marketing, distribution, and, ultimately, the customer for the product.

Now what does this have to do with Regulatory Affairs and the FDA? The FDA is in the midst of changing and revising standards for the electronic transmission of information, and the rules regarding how the information is maintained within a company. This change being undertaken is to move away from PDF file formats to true electronic information interchange. This is referred to as structured product labeling (SPL). It is a logical extension of Part 11 rules discussed earlier.

The European Agency for the Evaluation of Medicinal Products (EMEA) (8) and the European Federation of Pharmaceutical Industries and Associations (EFPIA) (8) have a similar initiative under way under the heading of Product Information Management (PIM). PIM's goal is to provide a secure method of electronic submission of product information to the various relevant authorities in the European Union (8).

The FDA is working with an ANSI-accredited standards development organization named Health Level Seven (HL7) and other interested parties to develop the technology for exchanging information between computer systems. This set of electronic standards known as Clinical Data Architecture (CDA) permits information to be exchanged using XML. The same standard is also being reviewed for use as the Electronic Health Record (EHR). The FDA, under the HL7 initiative, has adapted the CDA into a standard identified as SPL. This standard was put in place during the autumn of 2005, and replaced PDF files as an acceptable method for submission. SPL has a number of advantages over the older PDF file format standard. The new standard provides the following advantages:

- SPL permits the exchange of information between computer systems in ways that cannot be accomplished using PDF formats.
- SPL lets individuals compare text and specific data elements.
- SPL can be used to exchange information needed for other submissions such as drug listing. This improves efficiency and complies with the paperwork reduction act by eliminating redundant data collection and multiple submissions of the same data.
- SPL makes full use of the XML format and information tagged in XML to easily exchange and to make the process far more efficient for both the FDA and the manufacturer compared with PDF documents. For example, documents prepared and formatted in XML would only require the submission of the labeling or data elements that change, not the complete document. It also permits updating of all places where the information appears, as compared with Hyper Text Markup Language (HTML), where each document or root document would require that the information be changed everywhere it appears and then reissued.

SPL has put into place a framework to make all submissions to the agency electronic. This is extremely important when you consider the volume of documents, data, analysis, and other supporting information required for an IND, particularly for an NDA, and, to a different degree, by an ANDA.

This change is extremely important to anyone involved in packaging. Typically, the systems that contain and manage direct product labeling are under the control of the packaging group, or the packaging group is a major contributor and user of the repository system. A packaging department is responsible for the packaging component specifications, the engineered drawings that represent the component, the artwork and dielines that define the labeling text and graphics on each piece of packaging, and, most importantly, the PBOM. These items are constantly updated and refined. They constitute the majority of the working documents in a specification system used for manufacture and distribution of a product. A completely electronic database of all of these components permits procurement, quality assurance, manufacturing, and distribution to know exactly what the product is and what it contains.

The Finished Product Bill of Materials [PBOM, which may consist of separate product and package Bills of Material (BOM) or as one combined Bill of Material] is used by marketing and sales for costing and selling. Distribution uses the specifications to define if a product is hazardous, if it requires special handling, as well as the weight and cube of the product to define its transportation requirements. This set of documents and the information they contain provide everything one needs to know about the product packaging from its most basic definition all the way through the pallet level and, in some cases, the truck or containerized load level. Accurate data is provided to the shipper of a product for safe handling, loading, shipment, and delivery of the product.

THE UNITED STATES PHARMACOPEIA-NATIONAL FORMULARY

The United States Pharmacopeial Convention is a unique publisher. This organization publishes the United States Pharmacopeia. This unique volume of information has been in use for a long time. It was first published in December 1820, and as you read this description, you will realize that it was a very unusual and different book for its time. It contained formulas, in essence, preparation instructions, for 217 known cures. These 217 "drug" preparations were comprised of commonly known ingredients like plant roots, barks, and herbs, or more truly chemical substances such as sulfur or calcium carbonate (limestone) that could be combined as directed (formulated) to produce a product with therapeutic effect. The text was updated periodically during the 19th century as new "medicines" became known. The book remained in much the same format until 1880, when it expanded to begin including a list of product standards. This text was published in ten-year intervals from 1880 until 1942. As the pace of information development increased, the interval between updates was reduced to five years between 1942 and 2000. The USP-NF became an annual publication in 2002.

A separate group, The American Pharmaceutical Association, began the publication of the NF in 1888. Its first title was The NF of Unofficial Preparations. This title was changed to the NF in 1906 when the Food and Drug act was passed by Congress. That act referenced both USP and NF standards for therapeutic preparations. The NF was acquired by USP in 1975. The United States Pharmacopeial Convention began publishing the compendium as the USP-NF.

The United States Pharmacopeial Convention is a nonprofit organization located and incorporated in the District of Columbia.

The compendia are organized first with monographs for APIs and preparations. Dietary supplements and related monograph information for excipients used in drug formulation appear in the NF section of the reference. These are cross-referenced to the USP, and sometimes the USP also contains an excipient monograph.

The USP-NF is recognized in the Food Drug and Cosmetic Act in the United States, making the information it contains carry the force of law. Outside the United States, it is also accepted and used as a standard reference for pharmaceuticals by many countries, and many of these countries also use the information and standards as a legal reference for pharmaceuticals. The United States act uses the term "official compendium" to mean the USP. It also means the NF and the Homeopathic Pharmacopeia of the United States. The FDA can and does use the USP-NF as standards for assessing adulterated or misbranded product. The standards are used by the FDA to exclude products from the US market and to remove products from the marketplace if they fail to meet the provisions of a USP monograph or standard. This includes the test methods behind the standards.

The USP has established requirements for containers that are described in the drug monographs contained in the USP-NF. The information is found in the "General Notices of Requirements (Preservation, Packaging, Storage, and Labeling)" section of the USP. Material requirements used in the construction of the container are included in the "General Chapters" of the USP-NF. The USP uses terminology that is different from the terms normally used to describe packaging. For example, when describing packaging for capsules or tablets, the design characteristics of the container may be stated as tight, well closed, or light resistant. Materials for construction of a container for tablets and capsules are rarely mentioned. This changes when injectable products are described in the USP-NF. Here the materials for single- or multi-dose containers are specified. For example, "Preserve in single- or multiple-dose containers, preferably of type I glass, protected from light." The USP should always be consulted for background on material and general product protection requirements for a class of drugs. It provides a starting point for development of packaging, and it will be consulted or considered by reviewers of a NDA or an ANDA.

A firm may choose not to use the USP procedures to demonstrate compliance. When doing so, the firm must develop a solid rationale that describes and presents the FDA with information indicating that the procedure or method used by the company provides all the information needed to characterize the drug or package and that the end point of the unique procedure still brings the drug to some equivalence in testing against the USP procedures. This approach requires close collaboration with the FDA. The FDA will determine if they are in agreement with the proposed excursion from normal USP procedures and will comment on whether they will accept the alternate approach. If a company chooses to follow this path without prereview and comment by the FDA, the development work or the data developed may be rejected as inadequate. The FDA will determine if a firm is compliant or noncompliant with the regulations after consulting and comparing the company methods with the established testing standards for the drug indicated in the USP procedures and standards. The FDA can and sometimes does publish procedures that differ from the USP if the USP procedure has been superceded by something better or if a USP procedure does not provide all the information the agency wants or needs. The USP has taken a position "that allows for the use of different procedures in a monograph, depending on the route of synthesis, dosage form performance, and other factors" (4).

THE UNITED STATES PHARMACOPEIA DICTIONARY

The USP also publishes a dictionary that contains the drug names of all products sold in the United States. The hard copy version can be a little behind at times, but the online electronic version is always up-to-date. The organization strives to be as complete and universal as possible in listing all products. The dictionary includes names adopted for drugs in the United States, official USP-NF names and nonproprietary and brand names for the drugs. It also contains chemical names, chemical formulas, molecular formulas, molecular weights, graphic formulas, CAS registry numbers, code designations, drug manufacturers, and the pharmacologic and therapeutic categories. The dictionary provides an accurate reference for finding information commonly used in product labeling and product inserts.

CONSUMER PRODUCT SAFETY COMMISSION

Another agency that is of primary importance to pharmaceutical packaging is the CPSC. This agency is responsible for administering and enforcing the Poison Prevention Packaging Act of 1970. The act stipulates the performance level for packaging used with hazardous household substances to prevent and protect children from handling, using, or ingesting these substances. It is designed to prevent personal injury or death by a child who could, with natural curiosity, gain access to a dangerous substance. Drug products, including over-the-counter (OTC) products, are subject to the act. This includes oral prescription products, including products in clinical trials and outpatient trials. OTC products containing aspirin, acetaminophen, diphenhydramine, liquid methyl salicylate, ibuprofen, loperamide, lidocaine, dibucaine, naproxen, iron, or ketoprofen require child-resistant and special packaging to comply with the act.

The regulations that define this packaging are contained in 15 USC 1471(2) (4), 16 CFR 1700.1(b) (4), and 21 CFR 310.3(1). These regulations establish the performance standards and test methods the agency uses to determine if a package design or construction is child resistant and adult use effective. This last provision is important. An adult must be able to access the drug in the package when the package includes barriers to entry for children. A common complaint from adults concerns packaging that requires some dexterity and strength to open when that dexterity or strength is beyond their capability. These standards apply to reclosable and non-reclosable packaging systems. Examples of a non-reclosable packaging system are a tablet packaged in a unit-dose blister or a tablet packaged in a pouch.

The requirements of the act define a number of circumstances where child-resistant packaging is not needed. These include bulk packages for products that will be repackaged by the pharmacist (16 CFR 1701.1) and products that are dispensed in a health care institution such as a hospital or nursing home. Hospitals are required to use child-resistant packaging for medications dispensed to patients when leaving the institution. A sample of product provided to physicians that they provide to patients is not required to be child resistant.

OTC products also have an exemption from child-resistant packaging. Manufacturers or third-party packagers may supply one size of package without child-resistant packaging provided that other sizes of child-resistant packaging are also supplied. Any OTC product that does not include child-resistant packaging must use special labeling that highlights this difference (16 CFR 1700.5).

The act (16 CFR 1702) also includes procedures to petition the CPSC for exemptions from the requirements. These exemptions are granted when the CPSC finds that a product is not required to protect a child from serious injury or if the special packaging is not feasible, practicable, or appropriate for the product. Some examples of prescription products have received exemptions including oral contraceptives in mnemonic packages, powdered colestipol, and medroxyprogesterone acetate. One standard regarding exemption to the child-resistant packaging regulations is the need of a manufacturer to prove that the product is not harmful to a child weighing less than 25 pounds.

SUMMARY

The regulations surrounding drugs and their packaging is extensive and covered by a number of different federal and state agencies. The same state of affairs is also present within the European Union. Anyone developing packaging for drug or device products must consult the regulations and the agencies charged with administration of the regulations to determine what is required. It is a complex and multifaceted question that surrounds every drug and medical device.

FURTHER READING

U.S. Pharmacopeia. Mission and Preface. Rockville, MD: U.S. Pharmacopeia; USP USP28-NF23, 2005:xi.

Department of Health and Human Services, U.S. Food and Drug Administration. Guidance for Industry: CGMPs (Pharmaceutical CGMPs for the 21st century).

Department of Health and Human Services, U.S. Food and Drug Administration. Guidance for Industry: Providing Regulatory Submissions in Electronic Format—Control of Labeling, Electronic Submissions, CDER, CBER, April 2005.

Federal Register, Proposed Changes to the CGMP Regulations About Validation, Out-of-Specification Finding, etc., May 3, 1996.

Department of Health and Human Services, U.S. Food and Drug Administration. Guidance for Industry: General Principles of Software Validation, Final Guidance for Industry and FDA Staff, January 11, 2002.

REFERENCES

1. Department of Health and Human Services, U.S. Food and Drug Administration, CDER Drug Applications. New drug application process. Available at: www.fda.gov/cder/regulatory/applications/nda.htm.
2. FDA Title 21 of the Code of Federal Regulations (CFR), 314.1 (c) (8).
3. United States Code of Federal Regulations

    a. 15 USC 1471 (2) (4)
    b. 16 CFR 1700
    c. 21 CFR 310.3 (1)
    d. 21 CFR 211
    e. 21 CFR 210

4. U.S. Food and Drug Administration, Center for Drug Evaluation and Research, . Guideline on General Principles of Process Validation, May 1987, reprinted February 1993.
5. Federal Register, Amendment to the Current Good Manufacturing Practice Regulations for Finished Pharmaceuticals, Companion Document to the Direct Rule, December 4, 2007.
6. U.S. Food and Drug Administration. Guidance for Industry: Part 11 Electronic Records; Electronic Signatures—Scope and Application, August 2003.
7. Federal Register, Revision of Certain Labeling Controls, July 29, 1997.
8. More information on Product Information Management in Europe can be found at www.emea.eu.int or www.efpia.org Web sites.

6

Pharmaceutical Packaging Materials

INTRODUCTION

Packaging begins with the material selection. Material selection drives the choice and type of packaging equipment and, most importantly, the final package performance. Packaging performance is the primary criteria used to define the success of a package. The choice of material for a package drives all other choices about the product's appearance and consumer attributes. It influences and many times determines how a product is manufactured, filled, sterilized, labeled, bundled, distributed, and presented to the customer. It can influence where a customer looks for a package in a retail store, how the customer uses the product in the home, and how a hospital, nursing home, or retailer handles a product through their inventory and distribution systems.

The hardest and most difficult choices a packaging engineer must make are the selection of the material(s) used to package a new product. Existing products or products of a similar class or type many times mimic the packaging used on the first marketed product, even if newer materials and alternative material fabrication, manufacturing, or converting techniques offer significant advantages. The best opportunity for upgrading and improving the materials used in pharmaceutical packaging happens when a new molecule or biologic entity is identified and new packaging is designed to bring the product through regulatory approval and ultimately to market. As materials, customer needs, package fabrication machinery, package-filling equipment, sealing equipment, and distribution methods change, the packaging engineer is presented with a unique opportunity. They must survey a complex and changing backdrop of choices to design a package and choose materials that fulfill multiple requirements of the value chain to produce a package that meets all company and regulatory requirements along with the expectations of the final end user in the health care system. The choice of material and the type of package that can be made from that material becomes a critical decision very early in product and package development. It is a decision most often made before all the characterizing information about the new drug molecule or biologic entity is established or fully understood. It is a packaging decision driven by inputs from research, manufacturing, and marketing that presume enough is known about the molecule's potential to identify its reactivity, what is needed for product protection, where the product will be manufactured, and, in many cases, the size and type of market for the drug. These early product and package decisions, based on the early identification of a biologic or chemical activity, carry multiple implications about how a product is supplied and ultimately shapes many of the doctor and consumer perceptions about the product. This decision can and does stretch over many years, representing a very large investment in development and qualification of the product. It is rare, given the cost of stability testing and other qualification tests, that a product after Food and Drug Administration (FDA) approval gets a makeover in its packaging. Typically, it receives cosmetic improvements in graphics and secondary packaging presentations, but unless there exists a major commercial advantage in making a change, the pharmaceutical product, even in an over-the-counter (OTC) form, will remain in the same packaging material and type of package throughout its manufactured life.

Materials used for all types of packaging, not just pharmaceutical packaging, are beginning to change. The pace and direction of the change, away from traditional packaging materials like glass, metal, and older plastics like polyvinyl chloride (PVC) to newer and more efficient materials, particularly crystalline and biologically derived plastics, open opportunities and present new challenges to the packaging professional. The change in materials is markedly different from packaging changes in the past. Its speed and direction separate modern day packaging and the global reach of today's markets from previous packaging designs and material choices tailored for regional markets or individual countries. Global reach can mean that different materials are available and needed in different regions or markets, different environmental requirements for packaging exist and present unusual market needs, or consumer expectations about how a drug product is packaged are different.

Pharmaceutical packaging materials are and have always been slow to change. The factors behind the slow and tedious process required to change packaging include a long data driven qualification and stability testing protocol and the cost and time involved to produce this data. The FDA has streamlined their informational and qualification requirements in recent years to permit simpler substitutions of similar grades of plastic resin used to package product [example high-density polyethylene (HDPE) from a different supplier] or to move the manufacturing or packaging location of a product. The relatively small volume (manufactured unit volumes) quantities of pharmaceutical packages compared with food or beverage products also slow down the need for innovation. Comparatively small-scale manufacturing, the capital required for manufacturing capacity, and the somewhat indifference to consumer and customer needs by pharmaceutical manufacturers contributed to and created the circumstances that hindered fast adoption and changes to newer and sometimes improved materials. Caution and safety are watchwords of the pharmaceutical industry and are engrained in the multiple scientific and engineering departments of a pharmaceutical manufacturer. Thus, when a new material becomes available and is proposed for a new product, the packaging engineer begins a qualification journey that is questioned and scrutinized by other scientific disciplines within the company that do not share the urgency or commitment to change that packaging and marketing advocate. The development and qualification of a new material many times require a parallel development of packaging machinery, sterilization processes, or both, which is difficult to justify unless substantial product or manufacturing improvement is the end result. A package system changes, that is, all the manufacturing equipment and related facilities of the components needed to produce the finished package are difficult to master, expensive, and extremely resource-intensive to qualify, particularly in terms of human resources from a variety of scientific and engineering disciplines within the pharmaceutical company.

An overview of the traditional primary packaging materials, glass, metal, plastic, and composites, including the composite structures found in flexible films, is a starting point. To a lesser extent, other materials like rubber and elastomers along with some of the newer composite materials and biologically derived polymer materials need exposition. Paper, if used for a primary packaging material in pharmaceutical packaging, is always converted and processed with something else and is not the true product contact material; in this chapter, paper will be considered a composite material in the discussion of packaging materials.

Many of the materials reviewed in this chapter have been used for a long time. Change, even in plastic materials, comes slowly and requires extensive development and testing. Pharmaceutical companies for the most part choose to use existing materials for similar pharmaceutical products and have not searched for alternatives unless a new need required a change. The older materials are safe and have a good history of performance. This position is now changing as new drug products and new biologic products demand higher-performance packaging materials. Slowly, manufacturers are introducing newer plastics that provide improved performance properties, and these materials such as polyethylene terephthalate (PET) or polypropylene (PP) are slowly replacing HDPE and PVC.

Penetration of very new polymers like the biologically derived polylactic acid (PLA) is still for the future. Today, the ideas of sustainable packaging are considered a prime packaging goal, and sustainable packaging for OTC products will be one of the principal drivers of change. Environmental awareness and the need for companies to be good corporate citizens will also be factors in the evaluation, qualification, and use of environmentally derived polymer materials and their use in pharmaceutical packaging.

Changes in sterilization techniques, the introduction of large biomolecules, and consumer demand for convenient OTC products are driving changes from the older traditional materials like glass, metal, and some plastics, polyolefins primarily, to newer, established, and common materials in food and beverage packaging. The idea of recycling and sustainability is other aspect of packaging materials. Global awareness of "Planet Earth" and the implications and concerns about packaging on sustainability are just beginning.

GLASS PHARMACEUTICAL PACKAGING

Glass has always been the traditional gold standard for pharmaceutical packaging. Its physical characteristics of clarity and impermeability coupled with its inert chemical nature when exposed to organic or inorganic liquids and solids always made it the standard starting point for package development.

That is not to say that glass works for every package situation. Glass is not totally inert. Glass, depending on its composition, can impart hydroxyl ions to a solution; some of the chemicals used in its manufacture are leachable, but understanding glass as a material, and its material limitations, permits the package engineer to mitigate the problems. Glass was and may still be the most widely used material in pharmaceutical packaging worldwide, but plastics and composites are displacing it.

GLASS COMPOSITION

Glass is a simple material that has a long history. The chemistry is straightforward and has been known for millennia. Glass making is a chemical process of melting inorganic oxides, primarily silicon dioxide (SiO_2) with alkali and alkaline earth oxides that react and combine with the silicon to produce a hard clear substance that is very inert. The addition of alkali and alkaline earth oxide materials in the makeup of glass change and improve the physical properties of glass while lowering the overall melting point of its liquid state to make it easier to fabricate into containers.

The chemical materials used to make glass are common materials with layman names of sand, soda ash, and limestone. These materials are readily available and easily obtained from naturally occurring mineral deposits found all over the world.

Sand, or, more properly, SiO_2, is blended with soda ash and limestone and then heated to produce glass. The latter two materials are naturally occurring carbonates, sodium carbonate or soda ash (Na_2CO_3) and calcium carbonate or limestone ($CaCO_3$).

Traces of other materials are added to impart clarity and hardness. Lead (Pb) provides brilliance and clarity but makes glass relatively soft and alumina (Al_2O_3) is a common addition to increase hardness and durability. Other trace materials are used to reduce seeds and blisters in glass, for example, arsenic trioxide (As_2O_3) and sodium sulfate (Na_2SO_4). The materials used in making glass are heated and melted at temperatures that can be as high as 1600°C to convert the molten mixture of silicon and inorganic oxides into glass.

Glass is recognized as an inert material used to package and contain both strong acid and strong bases (alkaline) along with all types of organic and inorganic liquids and solvents. One problem with glass as a pharmaceutical packaging material is measurable chemical reactions of the material with a number of materials, most commonly water. Distilled water stored in soda lime or flint glass will pick up 10 to 15 ppm of sodium hydroxide (NaOH) along with traces of the other materials used in making the glass. The sodium in the glass is loosely combined with the silicon and is leached from the glass surface by water. The other trace materials used in glass making can also be leached to lesser degrees.

Glass stored in high temperature and high humidity conditions or high temperature and humidity fluctuations can undergo a physical change called blooming. Blooming, or leaching, is a physical change in which salts in the glass "bleed out" or more properly migrate and accumulate on the surface of the glass.

Glass manufacturers employ a number of different surface treatment methods to change or modify the surface chemistry of glass to reduce leaching. Soaking glass in heated water or a dilute acid solution will remove most of the surface-leachable salts.

Glass can also be surface treated to make it more resistant to attack by acids. The reaction of glass with a sulfur compound will form Na_2SO_4 on the glass surface, making it more resistant to water or acid interaction. It will not make the glass resistant to alkaline solutions.

The fact that inorganic salts are available for chemical reaction and present on or easily leached from a glass surface by a wide variety of liquid or solid materials makes normal glass unsuitable for a large percentage of pharmaceutical packaging.

High-quality glass used for pharmaceutical packaging is designated as type I glass in the United States pharmacopeia (USP) (1) and is substantially more resistant to surface attack. Normal soda lime glass with surface treatments targeted or designed for specific pharmaceutical applications can and are used for packaging drugs but not in the most demanding applications. The addition of 6% to 10% boron to glass to form borosilicate glass reduces the leaching action of water to 0.5 ppm in a one-year period. Boron, added as an alternative to some of the earth oxides in the form of boron oxide (B_2O_3), reduces the melt viscosity of the material for fabrication to a manageable range, even though this glass still requires the highest processing temperatures for package manufacture.

Glass containing boron in the amounts used to produce the most inert glass compositions or chemistries reduces leaching by alkalis while greatly reducing surface leaching by acid solutions. Borosilicate glass is about 10 times more resistant to acids than soda lime glass found in typical consumer packaging. Boron-containing glass is more heat resistant and durable than other forms of glass and displays higher melting point and significantly smaller coefficient of thermal expansion. This is why this material is cited in the USP as the highest-quality choice for packaging pharmaceutical solutions, especially acidic solutions that have a potential to interact with the packaging and possibly change in stability or performance because of the interaction. The major drawbacks to this type of glass are cost, difficulties inherent with its fabrication, and a somewhat limited supply because of the small, specialized demand.

The reason glass is clear is the material's chemical structure after cooling. Glass remains an amorphous material after cooling. Glass is stabilized to remain clear with the addition of aluminum oxide (Al_2O_3) and lead oxide (PbO_2). These materials help prevent devitrification, a slow crystallization process that can, over time, reduce glass clarity and turn it cloudy.

If borosilicate glass is cooled too quickly, the material will not have time to relieve strains (internal forces in the glass) developed during casting or molding. These strains make borosilicate glass mores susceptible to breakage and must be relieved by annealing (long/slow temperature change cooling after molding) immediately after container fabrication. Annealing is needed with soda lime glass and other types of glass to relieve the stresses and strains induced by molding in manufacture. Stresses and strains are found in all glasses, including flat or plate glass.

Glass is typically annealed or relieved by time/temperature treatment in ovens, called lehrs, after an article is manufactured ("formed" or "fashioned" are other terms used as descriptors in the glass industry). Controlling the time and programing the amount of heat a package receives during various stages of cooling process remove the strains and potential weak points from the glass. Stresses and strains and tension or compression from the molding process are relieved as the glass flows or moves to a lower physical energy state. The need for this process and its control cannot be overemphasized because improperly treated glass will always remain brittle and prone to breakage.

Color is another important characteristic that can be added to glass. Colored glass is produced by the addition of other inorganic materials not mentioned earlier. Colored glass is common in both food and pharmaceutical applications. Light, particularly energetic ultraviolet (UV) light, can initiate and sustain chemical changes in a product, thus the need for protection of food or drug from light energy. In food, many vitamins, flavor components, and other nutritional ingredients are susceptible to breakdown by light. The USP defines the amount of light transmission that pharmaceutical products may receive and the wavelengths of light that must be blocked.

The addition of a number of different materials to glass can color it and can provide screening of the contents from specific wavelengths of light. A good example is the addition of iron oxide (Fe_2O_3) or manganese oxide (MnO_2) along with sulfur to produce amber-shaded glass that blocks all light at wavelengths below 450 μm.

Glass can be produced in a wide range of colors that offer different light transmission and aesthetic characteristics. Common colors of glass are produced with the addition of chromium and cobalt oxides to create green and blue colors, respectively. The addition of selenium will produce glass with red or ruby color.

Table 1 Glass Pigment Colors

Compounds	Colors
Iron oxides	Greens, browns
Manganese oxides	Deep amber, amethyst
Cobalt oxide	Deep blue
Gold chloride	Ruby red
Selenium compounds	Reds
Carbon oxides	Amber/brown
Mix of manganese, cobalt, iron	Black
Antimony oxides	White
Uranium oxides	Yellow green
Sulfur compounds	Amber/brown
Copper compounds	Light blue, red
Tin compounds	White
Lead with antimony	Yellow

A complete table of materials that create different colors in glass is shown below (Table 1). Many materials produce colors but not light-blocking characteristics at the energetic wavelengths of light that cause chemical degradation. Notably, the uranium oxide that creates yellow/green glass also creates glass that glows in the dark and is not suitable for any packaging application. For pharmaceutical packaging, only materials that remain inert and block light in accordance with USP requirements are used in colored glass for drug packaging.

TYPES OF GLASS USED FOR PHARMACEUTICAL PACKAGING

There are four types of glass used in pharmaceutical containers. The glass performance grades or "Types" used in pharmaceutical packaging are defined precisely in the USP as type I, type II, type III, and NP glass (1). Type I glass is borosilicate glass. Type II glass is very high-quality soda lime glass. The last two are lower grades of soda lime glass and approximate glass found in packaging food and other consumer products. Their performance characteristics are listed in the USP as type III glass and NP or nonparenteral glass. Both type III glass and NP glass are acceptable for food packaging. NP glass has a highest specification for leachable components in USP's standardized test. This difference makes its use and the development of testing to prove its suitability for a drug product more problematic. NP glass must be proven safe and acceptable for use in pharmaceutical packaging but really is closer to a general or generic grade of glass. It is one of the USP standards but not widely used in pharmaceutical packaging. Two limited applications for type NP glass are in packaging oral or topical products. It is not a material normally used in primary drug packaging.

Table 2 USP Designated Glass Types for Pharmaceuticals

Glass type	Material	Type of test	Size[a]	Alkalinity, mL[b]
Type I	Highly resistant borosilicate glass	Powdered glass	All	1.0
Type II	Treated soda-lime glass	Water attack	100 mL or less	0.7[c]
			Over 100 mL	0.2[c]
Type III	Soda-lime glass	Powdered glass	All	8.5
Nonparenteral (NP) oral or topical products	General purpose soda-lime glass	Powdered glass	All	15.0

[a]Size indicates the overflow capacity of the container.
[b]Maximum amount of 0.02 N sulfuric acid required to neutralize water autoclaved in contact with 10 g of powdered glass.
[c]Type II glass is tested with 100 mL of water autoclaved in contact with the treated surface of the glass.

USP Type I Glass

USP type I (Table 2) glass is the most inert glass used for pharmaceutical packaging. It is borosilicate glass with approximate composition of 80% SiO_2 and 10% B_2O_3. It still contains Al_2O_3 and sodium oxide (Na_2O) in smaller amounts for the properties these materials impart to the finished glass. Borosilicate glass typically does not contain arsenic or antimony. Borosilicate glass has the lowest coefficient of thermal expansion and is the least likely to crack or break when subjected to sudden temperature changes. This provides durability and resistance to breakage during severe sterilization cycles required in manufacturing many pharmaceutical products. Type I glass is necessary for solutions that are slightly acidic. Acid solutions dissolve the various oxides in glass, causing a rise in the solution's pH. This change in pH may alter the efficacy of the drug, or it may change and reduce its shelf life or stability.

Type I glass is the highest-defined or highest-specified quality level for glass used in the packaging of pharmaceutical products, and it is the most expensive. It has the lowest-specified limits for leachable materials defined by the USP. The glass is primarily used in ampules and vials for liquid parenteral products.

USP type I glass is first converted into tubing and then into ampules, vials, and small volume bottles for pharmaceutical packaging. The conversion process is discussed in chapter 8, in the section describing glass container fabrication. The other types of glass (soda lime) use a more traditional method of manufacture.

The use of USP type I glass in pharmaceutical packaging requires understanding of the need for its extremely high-performance characteristics. These characteristics cannot be obtained with other types of glass. The combination of inert chemical properties and impermeability are the two primary reasons for using this material. Other glass grades and other materials such as multilayer plastic materials cannot match these specific properties of borosilicate glass.

When chosen as the glass material for a packaging application, the decision is based on chemical characteristics of the ingredient(s) in the drug product and the type of protection their chemistries require. Atmospheric oxygen protection and no interaction of the package with the API and carrier or diluent required by the active molecule make up the primary reasons for choosing type I glass. Real time and accelerated stability testing at ambient and elevated temperatures is used to confirm the high level of performance of the package with the drug. It provides a gas-impermeable inert environment inside the package needed for hard-to-hold, highly reactive drug products.

USP Type II Glass

Type II glass (Table 2), sometimes called soda lime glass, is the next grade or level of performance described in the USP. This glass does not contain boron and does not possess the properties of type I glass. This glass is sometimes referred to as treated soda lime glass or dealkalized soda lime glass. As would be expected from the type definitions in the USP, it is more resistant to leaching than type III glass, but less resistant than type I.

The glass itself is made with the same ingredients and same processes as standard glass for packaging. The glass is made more resistant to leaching than normal soda lime glass by treating it with sulfur oxide (SO_2). This process converts the surface oxides in the glass to soluble compounds that can be washed away with warm or hot water and or dilute acid solutions. This glass after surface treatment is limited to one heat sterilization cycle and only one use as a package. The glass cannot be cleaned and autoclaved for reuse in dispensing liquid products. Repeated heat cycles will cause the soluble oxides to migrate or diffuse to the surface of the glass, negating the surface treatment.

Type II glass is much easier to fabricate into bottles and other glass packaging because it has a lower melting point than borosilicate or type I glass. It is suitable for solutions that can be buffered to maintain pH below 7. The oxides in glass are labile, that is, free to move, as described and observed by their diffusion or blooming characteristics. These oxides are more easily leached by base solutions ($pH > 7$).

Type II glass is seen as a lower-cost alternative to type I material. It can be fabricated at lower temperatures than type I glass, making manufacturing much easier. Bottle manufacture uses the same high-speed, high-volume equipment to make both food and pharmaceutical packaging.

USP Type III Glass

USP type III glass (Table 2) is untreated or standard soda lime glass. No surface treatment and no prerinse are used in its preparation prior to filling. Normally, this grade of glass material is used in pharmaceutical packaging for anhydrous liquids and dry products. The USP does specify the amount of leachable material permitted in their controlled test procedure, but the level is relatively high compared with type I and type II glasses. Depending on the nature of the product packaged, this type of glass can and is used for parenteral products following indications by testing that the product does not react with or leach out any of the glass contaminants. For larger volume containers, greater than 100 mL, this type of glass is suitable for use if testing determines that the amount of leachable interaction is low and acceptable to product stability. It is the lowest cost of the USP grades of glass.

Type III glass is analogous with glass used in food packaging. Because it is used in applications with larger volumes of product, normally more than 100 mL, the high volume of product to surface area dilutes and limits the amount of leachable contamination to a low level.

Type III glass provides a package cost standard for a pharmaceutical product equal to glass food and plastic containers. "Nutraceuticals," the generic term used to describe a food product with enhanced characteristics that are not quite drugs, would consider this material as a starting point for packaging if plastics could not provide the product protection required.

USP DESIGNATION NP GLASS

The USP designates a lowest or minimum level of quality for glass that is called type NP or simply nonparenteral glass (Table 2). Again, the USP sets a limit for the amount of leachable oxide from the glass, but the limit is very high. This glass would typically contaminate small volume parenterals and make them unusable, but it is satisfactory for topical products like creams or lotions and for oral products like mouthwash. Normally, the volume of product packed in NP glass exceeds 100 mL.

GLASS AS A PHARMACEUTICAL PACKAGING MATERIAL

Glass has advantages and disadvantages associated with its choice as a pharmaceutical-packaging material. Probably, the two best characteristics of glass are its resistance to chemical attack by almost all liquids except hydrofluoric acid (HF) and strong caustics along with its impermeability. Glass being impermeable prevents any volatile ingredients from escaping and prevents any environmental gases, primarily oxygen, from entering the container. Glass disadvantages include its brittleness and weight. Glass brittleness is a problem that translates into glass breakage and the tendency of glass to break into numerous fragments. Even when care is taken to prepare glass to break by scoring or other techniques that thins the glass, in a container designed to be broken such as an ampule, the glass can shatter into fine fragments that may be ingested with the drug. Glass has a high density (2–2.5 g/cc), which in combination with its brittle nature means that containers must be fabricated with thick walls to achieve adequate durability. The thick walls make the resulting product heavy and increase transportation costs. This is a disadvantage compared with plastic and metal containers. A short list of the advantages and disadvantages of glass as a package is as follows:

Advantages

- Compression strength (permits stacking in distribution)
- Material strength to permit hot filling and retorting
- Heat resistance (can be autoclaved and sterilized with heat treatments)
- Impermeable to gases
- Inert (most inert material of all drug packaging materials)
- Clarity (contents easily viewed without opening)
- Easily cleaned and sterilized
- Fabricated into multiple sizes and shapes
- Technology for filling, sealing, and labeling is mature.
- Consumers everywhere are familiar with the package.
- Widespread availability (except type I glass)

Disadvantages

- High density—high weight (high transportation costs)
- Brittleness—easily breakable (broken glass can contaminate ampule products designed to break)
- Slower and more costly to fabricate than metal or plastic

One way to determine if glass is the best choice in a packaging application is to consider and evaluate the need for two properties glass provides, inertness and impermeability. Glass containers are used when an inert surface and complete gas barrier protection are the primary requirements for protecting a product.

The pace of change from glass to other pharmaceutical packaging materials has been slow; cost and cautious approach to packaging change have been the hardest hurdles to overcome. No plastic container can match the impermeability of glass in gas barrier properties; however, plastics have slowly displaced glass used in pharmaceutical packaging over the past 25 years. Improved plastic packaging performance, its lower cost, and its easier manufacturing coupled with a reduction in the number of suppliers of pharmaceutical grade glass are some of the reasons behind the trend to prefer plastic. A consumer preference for plastic also contributes to this trend.

For older generic drugs and for many drugs that were tested and qualified in glass containers, the cost of qualifying a new container material is rarely justified. Glass will continue as a packaging material for pharmaceuticals, but it is no longer the only material capable of packaging hard-to-hold products. Glass use will not expand, but the majority of glass containers, bottles, vials, and ampules will not be changed without some strong outside influence that favors another material.

The only exception to the last statement is the conversion of prescription products that were originally supplied in glass to OTC pharmaceutical products that require plastic to match consumer expectations. This change of older products for consumer reasons is largely complete.

METAL PHARMACEUTICAL PACKAGING

Aluminum, tinplate, and steel are the three primary metals used to make metal cans. Metal cans have been used for over 200 years and date back to Napoleon. The general was looking for a better way of preserving food and making it highly portable for his Grand Armée. A gentleman by the name of Nicholas Appert, in 1809, claimed a prize offered by Napoleon for packaging and preserving food to make it portable, and viola! The glass bottle and metal can were off and running. Glass was used more extensively in France, while metal cans were developed in England. Appert's achievement is particularly noteworthy when one realizes that Louis Pasteur did not explain the mechanism for food spoilage that Appert's canning process overcame, for another 60 years. Cans are one of the most widely used containers for packaging food and beverage products.

Pharmaceutical packaging only uses two of the three primary metals for primary packaging: aluminum and tinplate. Steel is normally not used even when coated with an inert lining of plastic or multiple layers of thermoset organic coatings. It is used for bulk materials in the form of drums, but it is not for primary packages. Tinplate is really a steel composite material that uses a steel core coated with tin.

Aluminum and tinplate are not limited to cans and are materials used to make tubes and pouches. Either these two materials when used for can manufacture require a great deal of processing and manufacturing before they become suitable as a finished package for pharmaceutical or foods products.

Almost all metal cans need to coat or paint the metal with an organic lining to separate the product from bare metal. There are few minor exceptions, but almost every can used today requires an inert thermoset coating on the inside to protect the product and metal and a variety of coatings on the outside to label (can makers call it decorating) and protect the can.

A wide variety of organic coatings were first to be adapted to seal or insulate the metal from the product. Before coatings, cans relied on zinc or tin surface coating to protect the product from iron in the steel and to prevent or retard corrosion. As more products were sealed in cans and as trial and error knowledge about can performance identified shortcomings, the need for coatings to improve performance became obvious. For food, consumers demanded packages that delivered appealing products. For drugs the organic coating used in production of the metal container had to be similar in performance, that is in its ability to separate the metal from the product, to the performance delivered by glass bottles. An alternative to organic coatings in cans is the use of polyolefin liners. Other methods are also available that encapsulate the metal between layers of plastic to take the place of coatings and make these materials suitable for food and drug packaging. These processes include extrusion coating and adhesion coating of polyolefin film directly on the metal surface.

Both metals are produced in many different soft and hard alloys that impart unique physical performance properties, particularly strength and surface hardness, to the metal package. The processed metals range in hardness and ductility from very hard and brittle to very soft and malleable. Very soft and malleable forms of the materials permit metals to create a dead fold, material's ability to be shaped mechanically with no memory or to spring back to its original shape; this property is valuable when flexible pouches or tubes are crimped closed or when crimping a collar on a vial with an elastomeric seal. Harder alloys and stronger alloys provide the strength and durability needed for cans and metal closures for cans, bottles, and aerosol packages.

The harder more durable forms of the aluminum and tinplate permit the metals to be worked or mechanically shaped into a package with very thin container walls or container cross sections. This reduces the weight of the finished package and uses very efficiently two physical properties of metal: strength and ductility. A good example of this quest for reduced container weight is the evolution of two-piece aluminum cans over the past 25 years. Originally, 30 to 35 lb of aluminum were used to produce 1000 aluminum cans; today, the amount of metal used is closer to 25 lb.

Tinplate

Tinplate, or more properly, steel coated with a thin deposit of tin on the surface of a sheet of steel, was the first material used for manufacturing cans. Cans were fabricated one at a time and each joint was soldered individually by hand. After the can was filled with product, the lid was crimped or mechanically attached to the can and soldered in place to complete the manufacturing process. The can was then processed to preserve the filled contents. "Processed" or "processing" is a term that refers to heat treating or retorting a can to sterilize its contents.

The can-making process was mechanized and automated during the first years of the 20th century. This automation made widespread use of cans possible, and the resulting improvements in quality and consistency of can performance created the metal can industry.

The popular name or term for the package "tin can" and the use of the term "tinfoil" are both misnomers. The material used to produce a can is properly called tinplate and is a composite structure of steel and tin. Steel is rolled into a thin sheet and then electrolytically coated with tin. Originally, the steel was dipped in molten tin. The layer of tin deposited on the surface of the steel by electrolytic coating is extremely thin and measures approximately 1/1000-in thick on each side of a sheet or coil of can-producing metal.

Originally, tin provided the means for making and sealing the can. Tin provides a surface coating that can be soldered with an alloy lead and tin, which, when heated, flows into the mechanical junctions or seams to fill any voids in the mechanical bond. The original manufacturing process for cans was a hand-soldering operation for much of the 19th century.

One product benefit from using tin was that it provided a stabilizing material for green vegetables and other vegetable products packaged inside. This same stabilizing effect was seen in some early drug packaging when tin salts contributed to maintaining the stability of products. One downside was the introduction of lead from tin coating and solder into the product.

During the last 25 years, welded tinplate cans have replaced soldered tinplate. Welding the container side seam instead of forming a folded mechanical seam and then sealing it with solder eliminated lead exposure to workers in the manufacturing operation and to the consumer.

Soldering of metal ends to the two ends of a cylinder was replaced at the beginning of the 20th century with a mechanical process called "double seaming." A rolled "double seam," which combined a mechanical seal with an elastomeric rubber compound, achieved the same hermetic or impermeable closure of the can previously achieved with solder.

Pharmaceutical products use the tinplate can primarily in aerosol packaging, although some tinplate is used in foil structures for tubes. Most tinplate "tinfoil" tubes are used for topical ointments.

Can Coatings for Tinplate and Aluminum Cans

Another advance in metal cans came with the addition of organic coatings to the inside and outside of container. Outside coatings permitted the can to be decorated or lithographed with artwork and information about the product and the manufacturer. Interior coatings on the metal surface of the can insulated and protected the product from reaction with the metal used to produce the container.

Almost all polymer types, acrylic, polyester, vinyl, alkyd, and others, are used to produce coatings for metal cans.

Aluminum

Aluminum has become the can-making metal of choice for food, beverage, and pharmaceutical packaging. Cans, closures (metal can ends), tubes, pouches, and foils are made from aluminum in combination with organic coating materials or as one layer in a multilayer composite material.

The organic coating materials are similar to or the same materials used to coat tinplate cans and tubes. These materials provide additional physical properties to the aluminum that separate and insulate the metal in the container from the product. Coatings are also used to protect and label the outside surface of an aluminum container in high-speed can manufacturing.

Aluminum is produced by electrolysis to reduce bauxite ore, which contains a high concentration of aluminum in its oxide form, to the primary metal. The ore undergoing electrolysis and reduction is in its molten state. While the aluminum is molten, it is combined or alloyed with small amounts of silicon, copper, magnesium, manganese, and iron. These trace additives increase strength and improve other physical properties of the finished metal. The alloyed material is cast into ingots and later mechanically rolled into the coil form that is the starting point for the manufacture of aluminum packaging.

The material, rolled to a very thin cross section (metal thickness), must undergo a secondary heat treatment called "annealing" to adjust its hardness properties for fabrication. Unannealed aluminum is hard and brittle. Annealing permits the metal manufacturer to adjust the hardness and brittleness of the finished material to meet the needs of the package. A harder or more brittle alloy that works well for easy open ends used on cans would be completely unsuitable for a tube or pouch application.

Aluminum, alloyed and annealed to a softer more ductile state and then rolled to extremely thin thicknesses (e.g., 0.0002 in) is used as one layer in a laminated material. The aluminum layer of a laminated composite material provides the gas or oxygen barrier and the light barrier in multilayer plastic and metal composite structure. Laminated material is found in lidstock for plastic containers, pouches, and flexible packaging.

Aluminum rolled into thin foils is a defect called "pinholing" that appears as extremely small holes invisible to the eye in what appears to be a solid sheet of metal. When the metal is rolled and stretched into extremely thin foils, the oxides and other impurities in the metal move to the surface and tear, creating these extremely small pinholes that are difficult, if not impossible, to identify by visual inspection. Specifying a minimum thickness of material for a specific alloy is one way to eliminate pinholes in a material. A second method is to use online inspection equipment to scan the thin sheet continuously with detectors designed for identifying pinholes during the rolling process and before the material undergoes expensive conversion into packaging materials. When specifying foil for any package, not just a pharmaceutical package, a limit is set on foil thickness to ensure that an oxygen barrier is present and that it is not compromised in its barrier protection by pinholes.

METALS AS PHARMACEUTICAL PACKAGING MATERIALS

Metals have a number of properties that produce strengths and weaknesses in the construction of a variety of containers and laminates. A partial list of the positive and negative attributes is as follows:

Advantages

- Material strength (capable of withstanding internal pressure in aerosol containers)
- Shatterproof
- Impermeable to gases
- Light barrier (opaque, this is both advantageous and disadvantageous)
- Lightweight (due to the strength of the material in thin cross sections)

- High heat transmission (metals conduct heat well, approximately 100 times better than glass and 400 times better than plastics)
- Mature manufacturing methods
- Malleability, the materials can be tailored in hardness and flexibility to the container
- Dead fold capability (only material with the strength and durability to act as the overcap on vials with elastomeric closure)
- Low weight of finished package (a consequence of the high strength of the material)
- Exterior decoration (both aluminum and tinplate can be highly decorated)
- Tamper evidence (breaking a metal seal cannot be reversed)

Disadvantages

- Potential interaction with product (the metal must be coated or insulated from the product)
- Limited shelf life (liquids)
- Container weight compared to glass (aluminum containers with density of approximately 2.7 can compete well with plastics; tinplate containers with density over 8 cannot compete against plastics)
- Cost to produce in small unit volumes (this is both advantageous and disadvantageous, depending on container specification)
- Difficulty to produce small volume containers
- Primarily targeted to food products

Metal alloys, designed for specific package applications, are first rolled or fabricated to a specific gauge or thickness used to produce the metal container. This metal then proceeds through a number of specialized steps that fabricates the metal into cans, closures, or pouches. During the fabrication process interior organic coatings, which insulate the metal from the product, is required for most rigid metal cans. The exterior is also coated to label or decorate the finished container.

Pouches are made by combining a the metal layer (foil) with other layers of plastic and or paper to create laminated structures, a sandwich of materials, that are fabricated into pouches that tailor package performance to product attributes. Exterior decoration or printing are part of the preparation of one of the individual laminate layers or is part of the laminating process. All are performed to produce a highly specialized container with properties designed for and specific to the intended product being packaged.

General specifications for cans, tubes, and pouches are available from manufacturers, trade associations, primary metal suppliers, and material converters. These general specifications have developed from practical applications over many years and provide a basis for comparing comparable containers, which is used as a starting point for container specification or a starting point for further refinement of a metal container to a specific product or class of products. The availability of this core group of specifications provides a packaging engineer all the information needed to begin specification development or as a starting point to adapt an existing container to the product being packaged. This group of specifications also ensures that the container developed and specified can be produced reliably.

The strength and high degree of ductility of metals permit them to be thinned to a degree not possible with glass. This reduces the total weight of the container. The use of chemical coatings, extrusion encapsulation with thermoplastic resins, extrusion lamination, or adhesive lamination enhances and broadens the capability of a metal container.

Extrusion lamination is the extrusion of the molten plastic directly onto the metal surface or at the interface of the metal and another layer of material. Adhesive lamination is the combination of a plastic film bonded with an adhesive to the surface of the metal. These hybrid material combinations permit metals to provide a combination of properties in the hybrid or composite materials that enhance the performance attributes and strengths of separate material components.

Metals are used as one of multiple layers in a laminate for blister and strip packages. When metal is used for the complete package, the blister portion of the package is thermoformed at low temperature or cold formed. This means that the metal is shaped by a die without any heat treatment of the metal. The ductility of the metal and the organic coating on both sides provide the physical properties needed for shaping or forming with simple mechanical force.

The metal in combination with a coating or plastic laminate provides specific protection and by itself or in combination with other materials provides the capability to use a form of heat sealing, induction sealing, or radio-frequency sealing needed to close and seal the package.

Metal composites are used as liners in plastic bottle closures (cap) to form a heat-sealable barrier impermeable to oxygen. It also becomes a visual and mechanical tamper-evident device for the consumer. The same layer of metal in a closure liner may be used to combine a two-step sealing and capping process into a one-step process. Metal can be induction heated, this in turn melts the plastic layer on the metal liner and on the lip of a plastic container, creating a welded bond between the two.

Aerosol Cans

Aerosol cans are unique applications for pharmaceutical containers. They deliver drugs for asthma and other inhalation products, sterile gases used as anesthetics, and gas uncontaminated with other atmospheric gases to support the eye during ophthalmic surgery.

The application of antiseptics to wounds and the delivery of topical ointments are some additional uses of aerosols made possible by metal cans.

Metal packages, particularly tinplate packages, are competing with plastic and composite materials, making their choice for packaging a cost/performance/consumer preference trade-off. Metal tubes are competing with plastic multilayer tubes in the same way. This is an example of a standard package being challenged for a number of different reasons, and in some cases, a decision being made to change from one material to another. Metal cans and packages are durable, low-cost containers that will remain part of the pharmaceutical packaging mix for the foreseeable future.

PLASTIC PHARMACEUTICAL PACKAGING

Plastics Overview and Definition

Plastics are the fastest-growing material used in pharmaceutical as well as food packaging (Table 3). Plastics are replacing metal and glass containers for both food and pharmaceutical end uses. The variety of packages and package

Table 3 Common Packaging Plastics

Homopolymer	Abbreviation/ symbol	Copolymer	Abbreviation/ symbol
Epoxy	EP	Acrylonitrile/butadiene/ styrene	ABS
Polyamide	PA	Ethylene/ethyl acetate	E/EA
Polyacrylonitrile	PAN	Ethylene/propylene	E/P
Polybutylene acrylate	PBA	Ethylene/vinyl acetate	EVA
Polybutylene terephthalate	PBT	Ethylene/vinyl alcohol	EVOH
Polycarbonate	PC	Linear low-density polyethylene	LLDPE
Polyethylene	PE	Polyethylene terephthalate glycol modified	PETG
High density	HDPE	Vinyl chloride/ethylene	VC/E
Low density	LDPE	Vinyl chloride/ethylene/ methyl acrylate	VC/E/MA
Polyethylene terephthalate	PET	Vinyl chloride/vinyl acetate	VC/VA
		Vinyl chloride/ vinylidene chloride	VC/VD
Polyethylene naphthalate	PEN		
Polymethyl methacrylate	PMMA		
Polypropylene	PP		
Polystyrene	PS		
Polytetrafluoroethylene	PTFE		
Polyurethane	PU		
Polyvinyl acetate	PVA		
Polyvinyl alcohol	PVOH		
Polyvinyl chloride	PVC		
Polyvinylidene chloride	PVDC		

components made from plastic is broad and increasing. A quick list of plastic packages would include bottles, thermoformed trays, pouches, blister packs, bottle closures, laminates, and nonwoven materials such as Tyvek®. The reason behind this sweeping change is simple; plastic offers more flexibility and more ability to tailor the properties of the package to the needs of the product.

Polymers and plastics are both macromolecules of repeating units. The chemical composition of the repeating units (called monomers) combines to form the macromolecule or polymer and create the physical properties of the material (2).

Plastics are normally referred to or considered a subgroup of polymer materials. Plastics are quite different from rubbers, adhesives, and coatings that are also called plastics or plastic-in-chemical compositions. The main distinction of this subclass of polymers is their ability in final or finished form to flow or move without chemical change (3). They possess the ability to be molded and shaped when heat and/or pressure is applied. This makes them a unique subclass of polymers. The ability to shape these materials with heat and pressure sets them apart from other materials and other polymers that contain repeating units but cannot melt or flow with the application of heat and pressure.

Plastics are easily shaped into containers and all types of packaging by heating them above the plastic glass transition temperature (T_g) and then subjecting the hot pliable material to pressure or other mechanical forces that move the plastic into a new shape that is retained when the stress is removed and the material cools. In blow molding, hot plastic is literally frozen in place by a cold or chilled mold into which it is forced by mechanical pressure (4).

Another name for these types of formable polymers is thermoplastics. The problem with this distinction is that polymer materials may belong to multiple groups of materials, some of which display plastic properties and some of which do not. Some plastics are defined or designated thermoplastics and others are defined as thermosets.

Thermoplastics, as the name implies, can be softened and shaped multiple times with the introduction of heat and mechanical force. Thermoset materials can only be shaped once; these materials cross-link and form irreversible bonds between reactive sites on the molecules' long chain. This reaction can only be reversed with the breaking of chemical bonds within the polymer chains.

Thermoset materials, a good example being vulcanized rubber, have excellent chemical resistance and mechanical properties but normally are not made directly into packaging. They have been used in the past as one of the first generation dual ovenable containers (plates) for frozen dinners, but were supplanted by a thermoplastic polymer [crystallized PET (CPET)] that was less expensive and easier to produce. Thermosets in packaging are most often found in metal can coatings and anywhere a cross-linked thermoset material provides property enhancements. Thermoset materials, once reacted, form a structure that cannot be reformed without breaking the chemical bonds of the material.

Introduction to Plastics

Plastics play a large and important role in packaging of pharmaceutical products (Table 3). Plastic materials can be the primary packaging material, the material that contacts the product being protected often referred to as the primary package, or they can be used in other parts of the package all the way to the pallet and the shrink or stretch wrap film around the pallet. They can be part of a composite structure of materials or an adjunct to another material to improve its properties. They can be used as the sealant, adhesive, or coating material between a wide variety of metal and paper materials.

Before any discussion of plastic packaging is undertaken, a basic understanding of the polymers used to create the plastic package, their physical properties, their chemical properties, and their limitations require review and examination. Plastic processing technologies are another factor in packaging and is reviewed in chapter 8 "Container Fabrication."

Plastics are essential part of our lives. The breadth and depth of products made from plastic spans a list from the clothes one wears to consumer products too numerous to mention, to industrial products one never thought of, to defense materials crucial to the modern military, to medical devices and implants, and to packaging of the food one eats. Plastic plays a key and pivotal role in making modern life possible.

The materials display an extremely wide range of chemical and physical properties. The clever use of plastic materials in combination with other materials, creating symbiotic hybrids, is expanding their acceptance into high-performance products, including electronics and medical products few would have envisioned until recently.

All types of packaging for every conceivable product can and are made from a variety of plastics. Beyond packaging, think of fibers for clothes, carbon fiber structures for aircraft components, rotors used for helicopters, medical devices, including those implanted in the body, and an incredible array of consumer, industrial, and household products that are made possible by plastics. Nothing else comes close to the versatility of these materials. Plastics are a standard material in pharmaceutical packaging used for bottles, blisters, labels, and even the pallets used for transportation.

A Plastic Primer

To begin, let us start with the basic definition of a polymer. The dictionary defines a polymer as any of a class of natural or synthetic substances composed of very large molecules that are multiples or repeating units of simpler chemical units called monomers.

The word polymer is derived from the Greek word "meros":

Poly = many   meros = parts

A polymer can be organic or inorganic, but for introduction, the following definition will be used:

> A polymer is a long chain molecule consisting of a backbone or main strand with or without repeating side chain groups along the main strand or backbone.

Polymers are formed from monomers, which are the smallest unique individual unit in a polymer chain.

Monomer + (monomer)$_n$ ⟹ Polymer

Monomer ⟹ Polymer ⟹ Repeating unit

C=C —C—C—C—C—C—C—C— —(C—C)$_n$—

Ethylene ⟹ Linear polyethylene (Fig. 1)

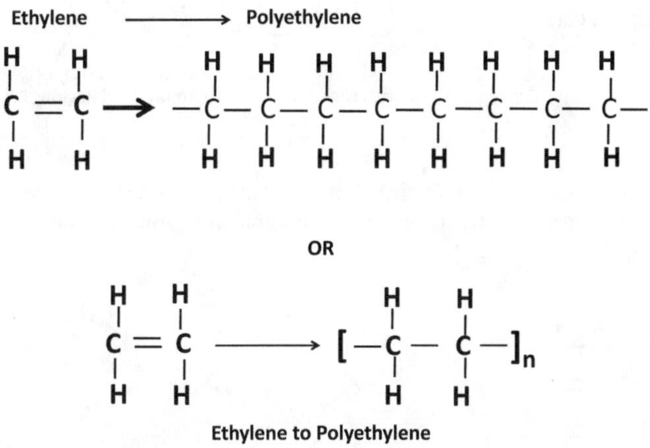

Figure 1 Ethylene to linear polyethylene.

The polymer chain can have four distinct features (Fig. 2). The first is an unbranched or straight chain molecule (polymer). The second is a branched chain, where multiple side groups are present along the molecule backbone (Fig. 3). Note in the figure representation the important point is placement not the size of the side chain groups which can vary in size and can be quite large in number of monomer units. The third is a cross-linked polymer (Figs. 2 and 4). The fourth can be a crystalline or semi-crystalline polymer (Fig. 2).

The polymer chain is not truly a straight line because if folds and curls on itself. A good mental picture of a polymer is a bowl of spaghetti with the individual strands representing the multiple strands of the polymer.

The smallest unit of a polymer is called a monomer. This is the repeating unit that attaches to other similar molecules. The total number of monomer parts

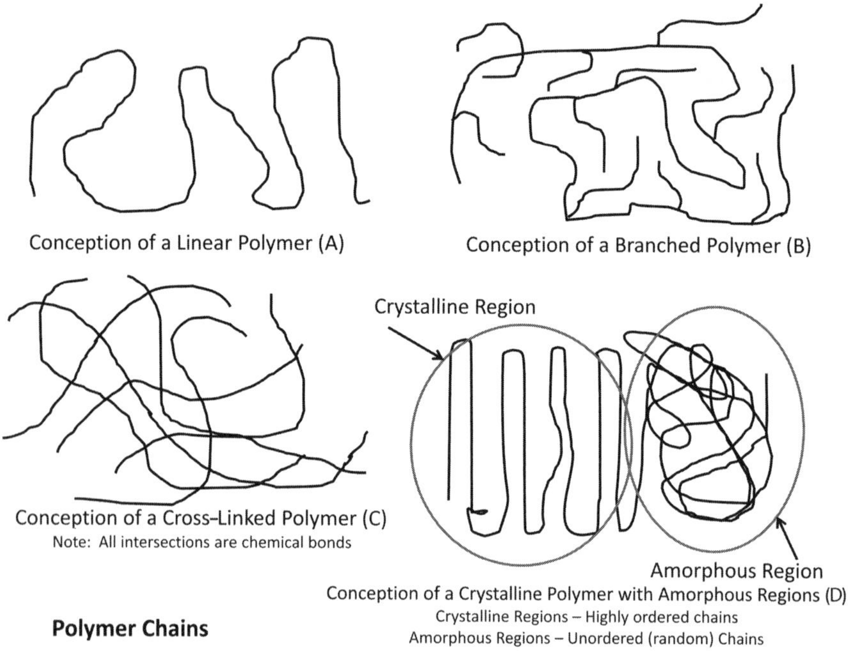

Figure 2 Examples of the three polymer chains. (**A**) Straight chain polymer, (**B**) branched polymer, (**C**) cross-linked polymer, and (**D**) crystalline and amorphous regions of a polymer chain.

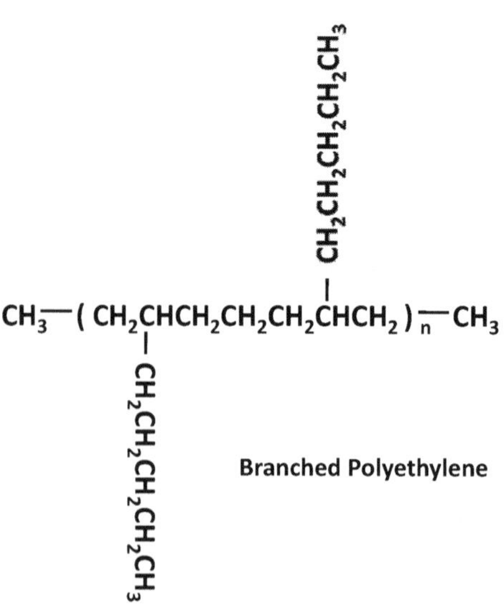

Figure 3 Branched polyethylene.

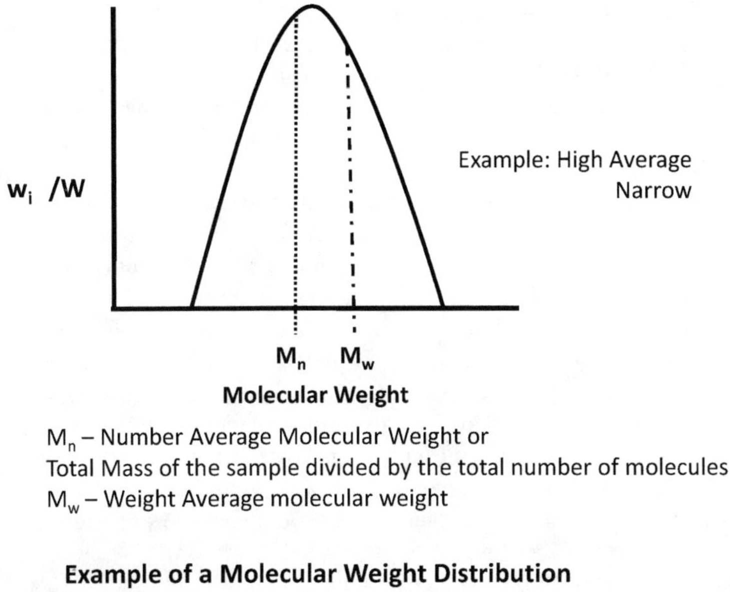

Figure 4 Molecular weight distribution.

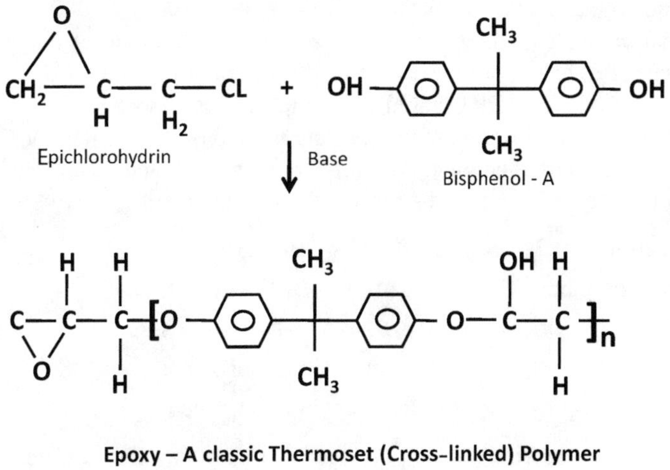

Figure 5 Representation of a cross-linked polymer.

in a polymer adds up to the molecular weight of the polymer (this includes the monomers branched from or not directly part of the polymer backbone). The molecular weight and the total length of a polymer chain are important chemical characteristics of a plastic material and contribute significantly to polymer performance in packaging (Fig. 5).

Since a thermoplastic polymer is not a single unique molecule, and since a single thermoplastic item contains many individual polymer strands of different lengths or different molecular weights, an understanding of this concept is needed to understand how the material and the finished package behave. One idea about molecular weight of a polymer is to consider the average of all the strands a given sample contains. Another way to characterize or visualize this for a thermoplastic material is to consider the statistical distribution of the molecular weights contained in all the polymer chains.

These ideas do not work for thermoset materials because the entire polymer structure can be considered one molecule (Fig. 2). The reactions between the polymer and monomer thermoset components or reactive sites built into the polymer chain create one very large molecule that irreversibly links all the smaller monomer components and multiple polymer chains together (Figs. 2 and 4).

For thermoplastic materials, molecular weight is usually represented as a bell-shaped distribution curve, with the range of the molecule or molecular weight represented by the amount of any given chain length characterized by each individual data point between the upper and lower limits or range of the curve (Fig. 5). This distribution is important for packaging, because the lower molecular weight components may be leached from the package to the drug or may absorb components of the drug (absorption and adsorption).

Monomers sometimes bind together in units of two or three, and these are referred to as dimers and trimers. They act the same way a monomer does; the difference is that the majority of the monomer units are usually found in one of these forms instead of the individual form.

Polymers can be made of the same unit repeating over and over as is the case of polyethylene (PE) or PP, or they can be made with multiple monomers to produce a copolymer. Nylon and PET would be examples of copolymers (Fig. 6).

Typically, two or more different monomers are used to produce a polymer. These materials are known as copolymers (multiple monomers A and B) and can come in a variety of forms (Fig. 6). The following are examples of copolymers:

Alternating copolymer	A-B-A-B-A-B-A-B
Block copolymer	A-A-A-B-B-B-A-A-A-B-B-B
Random copolymer	A-B-B-A-A-B-A-B-B-A-A-A
Amorphous copolymer	random chain arrangement
Crystalline copolymer	ordered chain arrangement

Understanding the descriptions of polymers is important. The different polymer types require different methods of manufacture that have implications about how the polymer is used in pharmaceutical packaging.

Polymer Descriptions

Addition Polymers

The first type of polymer is the addition polymer (Fig. 7). One chemical species reacts with a second chemical species to form a new and larger compound. These polymers are most often formed by reacting unsaturated monomers, building

A----ABABABABABABAB----B
Alternating Copolymer

A---ABAABBBABBBBABBAAAABAAB----ABBAAA
Random Copolymer

A----AAAAAABBB----BBBAAAAAAABBBBBB----AAAAAA
Blocked Copolymer

A---AAAAAA----AAAAAA----AAAAAA----AAAAAA
B B B B
B B B B
B B B B
B B B B
Graft Copolymer

Note: Capital Letters represent different monomers

Types of Copolymers

Figure 6 Examples of each polymer type.

$$M + M + M = P_n$$
$$P_n + M \rightarrow P_{n+1}$$

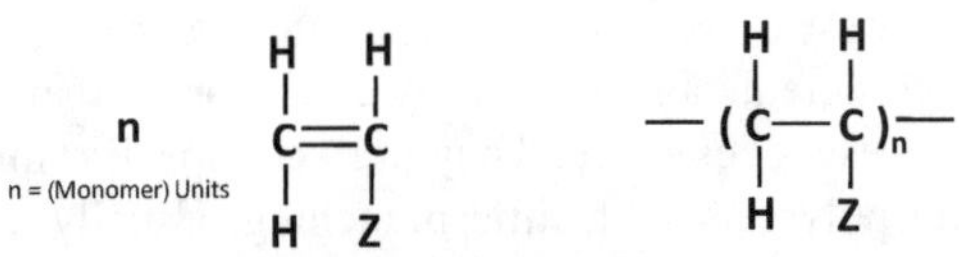

Pn = (Polymer) with Monomer Units
M = A (Monomer) Unit
Pn+1 = (Polymer) increasing by one monomer Unit

Addition Polymerization

Figure 7 Example of an addition polymer.

blocks containing one double bond, which combine to form the polymer chain. Addition polymers can also be formed with monomers containing multiple double bonds or even triple bonds; as each double bond opens, a new monomer unit is added without producing any by-products (Fig. 7).

Polymerization begins with the formation of a free radical and then the addition of the first monomer unit (Fig. 8). The free radical can be generated by interaction with a peroxide, azo compound or hydroperoxide, which may be called the "initiator." PE is an excellent example of an addition polymer (Fig. 7). It is produced using the addition process at what is considered low pressure

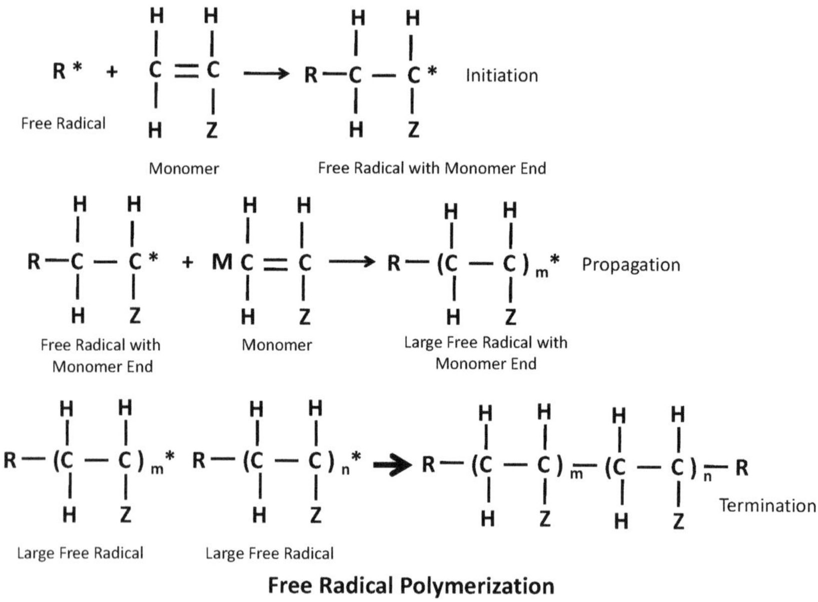

Figure 8 Free radical propagation.

(300 psi) and temperatures of 125°C to 250°C. Temperature is extremely critical to the PE-manufacturing process and is controlled to yield the type of polymer structure desired.

For HDPE, a Ziegler–Natta catalyst is used to propagate and control the polymerization reaction. These catalysts create a lower-energy condition on their surface where the polymerization process takes place. They are very specific in their actions and produce long chain polymers with little branching. Usually a small amount of branching is desired for processing (conversion into the package) of the polymer, and this is achieved by the introduction of a small quantity of hexene to the reaction.

The second method used to produce addition polymers is ionic polymerization. The ionic nature of the polymerization comes from the use of ionic intermediates that interact with the monomer to form the free radical. The reaction then proceeds in much the same way that free radical polymerization proceeds for free radical intermediates. Ionic polymerization is used to produce block copolymers and a few other specialty polymers that require unique control of how the components are put together. It is not a method widely used to produce polymer materials and is limited primarily to specialty applications outside of packaging.

Condensation Polymers

The second method for producing a polymer is a chemical reaction called condensation. One chemical species reacts with another to form the polymer with a second residual material (a by-product) formed in the process. Condensation polymers are made with monomers that have functional groups consisting of acids, alcohols, and amines (Fig. 9). These functional groups appear in pairs as diacids, dialcohols, and diamines (Fig. 10).

A condensation reaction follows a step process. In the first step, an ester or amide is formed from the diol or diacid, or from the diacid and a diamine, respectively (Fig. 10). The first step of this reaction process is the most rapid, using up the monomer very quickly (Fig. 11). The second step in the reaction is the growth of the polymer chain that takes place as smaller segments of monomers and short chain fragments that contain multiple monomer units are formed and combined into the final polymer (Fig. 11). The short chain or polymer backbone fragments are some of the reaction products formed in step 1. As the reaction proceeds, water or alcohol is formed as by-product and eliminated from the reaction. Water is the most common by-product of this type of polymerization. The reaction of a diacid and a dialcohol produces polyester with water as

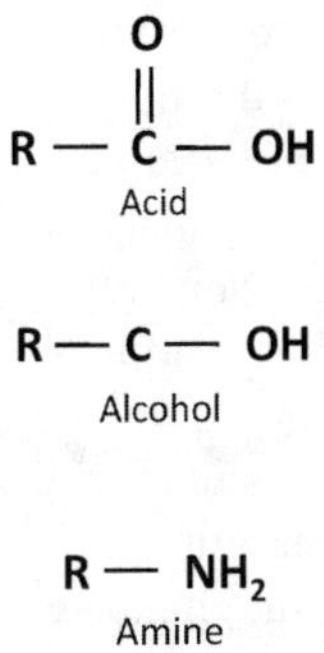

Figure 9 Examples of an organic acid, alcohol, and amine. "R" represents the organic molecule attached to the functional group.

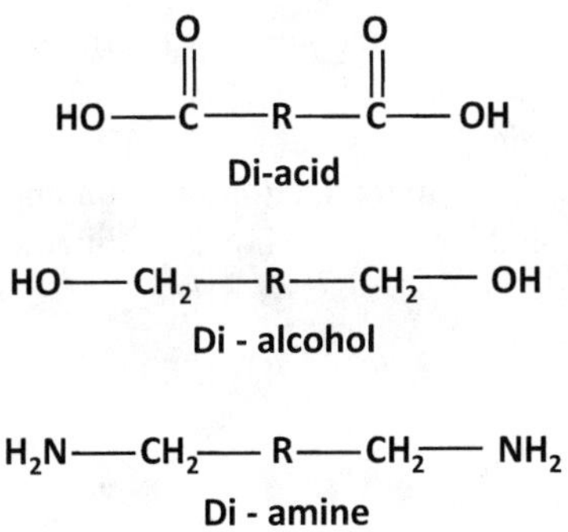

Figure 10 Examples of a diacid, a dialcohol (diol), and a diamine.

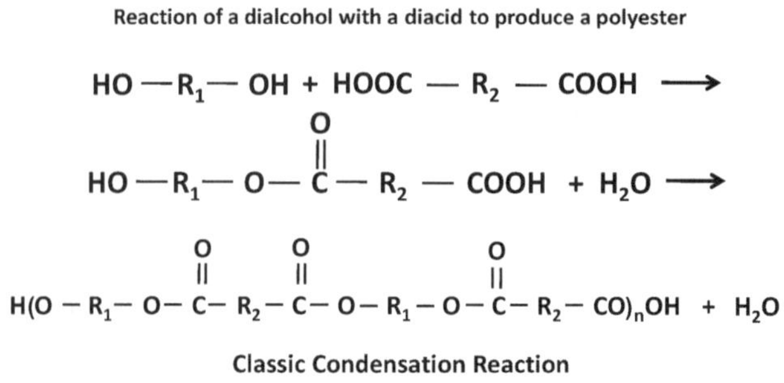

Figure 11 Example of a classic condensation reaction producing a polyester.

the residual by-product of the reaction. In the formation of an amide, the reaction takes place between a diacid and diamine.

For the second reaction, a common amide polymer, familiar by its trade name Nylon, is produced. The water by-product must be extracted from the polymer. This is done by the physical change of the materials; as the polymer grows, it becomes insoluble in water, which is also produced by the reaction. A good example of this type of condensation polymer reaction is the synthesis of PET. Terephthalic acid or dimethyl terephthalate is reacted with ethylene glycol to produce PET and water (Fig. 11). Water is constantly removed by distillation during the polymerization process.

Common polymers produced by the condensation process include PET, polyethylene naphthalate (PEN), and polycarbonate (PC). Condensation polymers are normally not considered copolymers. This is because the intermediates, esters and amides, are identical and form the polymer. The reactions to form these materials and their structures are covered later in this chapter.

Classes of Polymers

There are two classes of polymers: thermosets and thermoplastics. Thermoset polymers consist of polymers that cross-link to produce large chains. Once the chemical reaction is complete, the material becomes nonmeltable. Nonmeltable means that the polymer has actually completed the polymerization reaction and now cannot be changed without actually breaking the molecular bonds within the polymer.

Examples of thermoset polymers:

- Urea formaldehyde
- Epoxies
- Urethanes
- Unsaturated polyesters
- Rubbers

Thermoplastics are the other type of polymers that are generally referred to as plastics. These materials have completed the polymer reaction but maintain the ability to melt and become liquid at elevated temperatures. The hot liquid or semisolid polymer is formed into different shapes by a wide variety of plastic and packaging-manufacturing methods. The materials may be linear or branched polymer chain materials. Thermoplastic polymers may be amorphous or crystalline materials or combinations of both forms (Fig. 2).

Examples of the different types or classes of thermoplastic polymers:

- Amorphous materials—PVC
- Crystalline—PE (HDPE)
- Crystallizable—PET

What determines a polymer's physical properties? This is a key question for the chemist and packaging professional. By knowing some of the basic chemistry behind different types of polymers, the materials can be grouped into classes whose physical properties provide the protection required for the active pharmaceutical ingredient. A short list of properties that determine a polymer's performance properties include chemical composition, polarity or bond arrangements, chain size or length, chain structure, and crystalline or noncrystalline arrangement of the polymer in question.

The first item in this list, chemical composition, is a good starting point. Polymers contain carbon and hydrogen in combination with a wide variety of other elements such as oxygen, nitrogen, and chlorine. The chemical composition determines many of the properties of the polymer. Chemical composition consists of the atoms used to make the polymer molecule and the number or arrangement of the atoms in the polymer chain.

The monomers chosen for reaction, the reaction conditions used to make the polymer, and the control of molecular weight during the reaction contribute to the physical properties of the polymer. The chain structure or the orientation of the molecules and subgroups of molecules on the polymer chain also make major contributions to physical properties (Fig. 2).

The crystallinity, degree (percentage) of crystallinity, in a plastic material refers to the orientation of the molecules in relation to each other, and this makes a contribution to physical properties. The mechanical forces and heat history of the manufacturing processes used to make a polymer also relate and contribute to physical properties (Table 3). The chemical structure of a polymer generally determines the following properties:

- Chemical reactivity
- Density
- Diffusion characteristics
- Friction or lubricity
- Melting and softening points (sometimes referred to as T_g)

- Permeability
- Solubility
- Thermal properties

Determinants of a Polymer's Properties

A polymer can be modified in many ways to change its properties. Modification of any polymer is the reason it fits into so many different parts of our everyday lives. It is hard to imagine everyday life without polymers and plastics. Their widespread use is one of the things that make modern life possible.

One of the easiest ways to change the properties of a polymer is to modify its molecular weight. Typically, increasing molecular weight will improve the properties of a polymer. The improvement in properties includes physical properties like elongation, flexibility, and chemical resistance and, possibly, barrier properties (5). The trouble with increasing the molecular weight of thermoplastic resins is that it makes the plastic much more difficult to melt and process (manufacture into packaging). These larger molecules have more difficult time moving by each other in a melted state and passing through a constriction in the flow stream of the liquid material.

The long molecular chains of a polymer contribute to the following properties:

- Elasticity
- Melting temperature
- Strength
- Creep, or very slow viscous properties and stress relaxation
- Viscosity

Another way of modifying a polymer, which could be considered a way of changing its molecular weight, is to create or eliminate the branching characteristics of subgroups or monomers along the chain. An example of multiple side chain branching is linear low-density polyethylene (LLDPE). The use of monomers like butene, hexene, and octene create the branching; this is the reason that LLDPE is a copolymer, not a homopolymer. An example of a long straight chain molecule is HDPE.

Another way of modifying a polymer is to change the monomer content or type of monomers used. The change can have profound effects on chemical properties. An example is the difference between PET and PEN, which use two different but somewhat similar monomers. PEN displays enhanced oxygen barrier properties and resistance to UV light compared with PET. The change in a monomer can also affect the transparency of a material and can alter its crystallinity. Crystalline change is apparent to the eye because the differences in density between the amorphous and crystalline regions of a polymer cause light scattering that the eye sees as whiteness. Other ways to measure crystallinity is by measuring the impact resistance or density of the material.

Varying the proportion of monomers in a polymer can also produce significant changes in properties. It may alter the crystallinity of the polymer and may contribute to a change in a wide variety of physical properties.

Polymers are also modified or changed in their characteristics by the addition and incorporation of additives in their structure. Additives can range from inorganic fillers like talc or calcium carbonate that provide structural stiffness and cost reduction to organic stabilizers that improve UV resistance, thermal properties, or act as antioxidants.

For many plastics, the addition of lubricants is necessary to process the material. PVC relies on the addition of plasticizers, large organic molecules, typically a phthalate, to improve flexibility and other physical properties.

Pigments, colorants, and dyes are other common groups of materials used to modify a polymer. These can be inorganic or organic and along with the color produced can also provide improvements in other properties like thermal stability.

Glass and fiberglass strands along with other fibers may add considerable strength to a polymer, while impact modifiers, clarifiers, and stiffing agents can all contribute to improved properties needed in package design.

CHEMICAL ATTRIBUTES OF POLYMERS

The unique properties of polymers can be attributed to a number of different attributes that define the material. Some of these chemical attributes include:

- Chemical bonding
- Chemical resistance properties
- Viscoelastic behavior
- Molecular shape
- Flexing, mobility, and stiffness
- Crystallization
- Barrier properties

A brief overview of the chemistry and forces involved in this list of attributes will provide another basis for understanding polymer behavior, and will provide background understanding of the chemical and physical properties of polymers. Understanding inter- and intramolecular interactions in plastic materials provides a basis for why certain plastics perform the way they do.

Chemical Bonding

The type of chemical bond, the force that holds the individual atoms together is part of the construction of a polymer. The range of bonding can be both interatomic and intermolecular, and both play a role in how the very large polymer molecules behave. The interatomic forces are the bonds that hold the atoms of the polymer together forming the molecules. These bonds connect atoms like carbon, oxygen, nitrogen, and silicone to make a monomer or polymer molecule. In polymers, we find two types of atomic bonds, covalent bonds, and ionic bonds. Ionic bonds are very rare and only found in unusual applications. They are not found in common polymers. Understanding the basics of chemical bonds contained in a polymer chain has a profound effect on the way the material behaves.

The first bond is the covalent bond. This type of chemical bond occurs between two similar elements and can be referred to as nonpolar covalent bonds. Covalent means the sharing of electrons to form a stable bond. In a covalent carbon bond, electron orbitals overlap as each atom donates or shares one electron to the bond. The most common covalent bond is the one between two carbon atoms; however, many other atoms also form covalent bonds. Individual pairs of atoms form stable molecular bonds by the equal sharing of electrons, a good example being the atoms of oxygen and nitrogen combining to form the molecules (O_2) and (N_2).

Another type of covalent bond can form between atoms that are close to each other on the periodic table of elements. These materials share electrons with a slight imbalance in the sharing. These bonds are called polar covalent. The most electronegative element will get the greater share of the electrons forming the bond, and the more electronegative element will assume a slightly "negative" charge. Examples of polar covalent bonds are carbon/hydrogen bonds or carbon/oxygen bonds.

- $C^{\delta+}$—$H^{\delta-}$
- $C^{\delta+}$—$O^{\delta-}$

A bond that forms between two atoms where one atom completely donates an electron(s) to the other is called an ionic bond. The best example of a bond of this type is found in table salt or sodium chloride.

- $Na^{+}Cl^{-}$

Ionic bonds are rarely found in plastics. A class of polymers, called ionomers, used for heat-sealing materials contains sodium and zinc atoms in the side chains to neutralize carboxylic acid groups, and because of their presence, these ions display ionic bonding in the polymer chains and between the polymer chains. They are not the primary bonds of the polymer, and the ionic name is very limited. They should not be confused with a standard polymer.

Chemical properties in polymers can be viewed as performance characteristics and used to describe and contrast performance under a given set of conditions with that of another material. Some of the common ways to describe chemical properties are:

- Weatherability
- Gas barrier
- Water barrier
- Solvent resistance
- Stress crack resistance

Weatherability

Weatherability of a polymer material is not something of concern to a packaging application. It does highlight how some materials behave on the basis of their chemical makeup. Weatherability is a measure of the reactive chemical groups contained in the molecule. Unsaturated bonds (C=C) and reactive chemical linkages such as carbonyl groups provide a location for attack by environmental chemicals like water, with energy for the chemical reaction and breakdown supplied by sunlight.

Gas Barrier–Water Barrier

Permeation of plastic films and plastic materials by oxygen and water vapor are always a concern in choosing and specifying a plastic material to protect a drug (Table 4). All plastic materials, even those coated with high barrier materials, still exhibit some degree of permeability compared with glass or metal (5,6). For a gas or water to permeate a plastic, it must be soluble in the material and able to diffuse through it. Crystallization of the polymer, or the inclusion of bulky chemical groups that hinder the movement of gas and water molecules are ways to improve the barrier properties of a plastic.

Gas and water barrier properties are extremely important to the packaging of products. Determining the polymer properties as a gas or moisture barrier is fundamental to making the correct decision regarding package composition. The physical property is most dependent on the polarity of the polymer and the presence of crystallinity in the polymer structure.

Solvent Resistance

Solvent resistance is another important attribute. Solvent resistance in polymers is dependent on the polarity of the polymer and polymer chain length or molecular weight. In general, the effect of solvents on a particular polymer decreases with increased molecular weight, chain branching, and crystallinity.

Table 4 Relationship of Polymer Characteristics to Barrier Characteristics

Polymer characteristic	Change in characteristic	Effect on permeability
Density	Increase	Decrease
Molecular weight	Increase	No major effect
Crystallinity	Increase	Decrease
Cross-linking	Increase	Decrease
Plasticizer content	Increase	Increase
Orientation	Increase	Decrease
Humidity	Increase	Increase
Filler content	Increase	Decrease

Chemical attack causes a number of problems with packaging consequences. The most prevalent problem is absorption of the solvent, but other problems include partial dissolution, polymer plasticizing, and actual chemical reaction. These changes may not lead to a complete failure of the package but will result in reduced mechanical performance or another physical change in the package or its contents.

Environmental Stress Cracking

Stress crack resistance is a combination of the polymers' chemical makeup and the physical interaction of the polymer with another chemical. Environmental stress cracking can be defined as cracking or crazing of a polymer when exposed to solvents or aggressive chemicals while under tensile stress (7). This is a big concern in medical plastics because crazes caused by a solvent can create microfractures or microcracks in the material, resulting in reduced mechanical integrity. Both conditions, stress and solvent, must be present for the failure to be labeled environmental stress cracking.

Molecular Shape and Intramolecular Forces

The extremely large size of a polymer molecule and the interaction of the atoms in the chain with adjacent molecules create a number of significant effects on polymer's physical properties. Because the molecules are so large, there are a huge number of these interactions both within the same molecule and with adjacent molecules. These intermolecular and intramolecular (different parts of the same molecule) interactions or forces are far weaker than the primary bonds holding the atoms together and are typically referred to as secondary forces. They may also be referred to as secondary bonds. These forces are primarily responsible for the physical characteristics of the material. There are a number of different secondary forces to consider, namely, van der Waals forces, hydrogen bonds, dispersion forces, dipole forces, and induction forces. These forces are highly sensitive to the distance between the molecules that affect the force. The distance between the molecules for these forces to manifest is between 3 and 5 Å (angstroms). The strength of a secondary force decreases proportionally by the sixth power (10^6) of the distance separating the molecules.

Polymers can be polar or nonpolar molecules. The polarity of a polymer is determined by a combination of the three-dimensional shape that all polymers display and the type of molecular or electronic bond (polar, nonpolar, or ionic) present in the polymer. Along with the properties conferred on the polymer chain by the type of chemical bond(s) it contains, the ability of a polymer to dissolve in various chemicals or the ability of a polymer to absorb or contain materials in the interstices or voids of the folded chain is determined by the intermolecular forces that are the basis of a polymer's crystallinity and its polarity.

PE can easily absorb and transmit nonpolar gases such as oxygen (O_2) and carbon dioxide (CO_2), and the nonpolar nature of the polymer is the main reason that it cannot act as an oxygen or gas barrier in packaging. Contrast PE with a crystalline polymer with polar groups like PET. Polar polymer molecules like PET will easily absorb water or other polar molecules and will slowly lose liquids stored in the sealed container. Water or a polar molecule is absorbed and is soluble in the polymer and slowly diffuses through and evaporates from the inside of the package by moving through the polymer walls. An old PET water bottle or liquid-containing bottle will deform and the level of liquid will be noticeably below a normal fill level because of this property.

Another characteristic of a polymer determined by its molecular shape is the melting point of the polymer. The three-dimensional shape of a polymer determines how tightly packed the chains are within the molecule. This packing of the chain or the ability of the chain to fold back upon itself along with the ability of the chain to pack closely with the chain of an adjacent polymer molecule determines the crystalline network of the polymer. In general, the more crystalline a polymer, the higher the polymer's melting point. Polymers can also have different regions contained within the polymer chain. The attractions of the polymer chain within the same molecule or with the adjacent molecule are also determinants in the melting point of a polymer.

Finally, the type of molecule within the polymer can have a major bearing on the stiffness and density of a polymer. Benzene rings are a good example; the p orbitals above and below the ring tend to align with each other, somewhat like a stack of plates. This stack of plates tends to be much stiffer and stronger than a more random arrangement found inPE. If a polymer has large polar molecules attached to the chain, the attraction between these polar molecules can increase stiffness or rigidity within the polymer. Large chemical groups, such as amide or imide groups can also increase the strength or stiffness of a polymer by making it hard for the molecules to move past each other, and in some ways form a small lock and key arrangement on a molecular level.

The polymer properties just described lead to another set of characteristics that combine on the micro and macro level to affect polymer mobility and stiffness. They also play a big part in how flexible a polymer material may be. The two characteristics are microconformation and macroconformation. These two characteristics, influenced by chemical bonds and the relationships between the bonds, produce properties of polymer stiffness and mobility.

Microconformation refers to the interaction of the atoms and molecules within a polymer. These include atom-to-atom conformation, atom-to-atom single bond rotational energy, electrostatic interactions, repulsive and attractive (van der Waals) forces, potential energy inhibiting internal rotation, and hydrogen bonding.

Macroconformation refers to the larger molecular interactions and characteristics. These include polymer chain-to-chain conformation, intermolecular energy variations, amorphous content, chain folding, and chain extension.

Microconformation is at the atom-to-atom level. When a chemical formula is written on a page, the bond between two atoms, for example, two carbon atoms, is written as C–C. This does not reflect all the degrees of freedom that this bond can display. Thus, conformation is more about the geometric arrangement of the atoms in the chain utilizing the principle of free rotation about chemical bonds. One must always think of polymer chains in three dimensions with some bending, offsets, or rotations around bonds.

The most stable arrangement of bonds around a carbon atom bonded to another carbon atom is with the three remaining bonds on each atom displaced by 180°, forming a staggered trans conformation. Think of two 3-legged stools stacked seat to seat, one on top of the other with a pole between the seats. The pole between the seats represents the carbon-to-carbon bond, the seats of the stool, the carbon atoms, and the three legs represent the three remaining bonds. This is the most stable of the bond arrangements. Other arrangements are common and will be discussed as necessary.

Electrostatic interactions and hydrogen bonding are straightforward principles that cause atoms to attract or repulse each other. This kind of force can profoundly affect a polymer's ability to bend or flex and its overall stiffness.

Macroconformation refers to the structure of polymers beyond the atomic level. It is influenced by microconformation and by a number of other factors. Polymer chain-to-chain conformation refers to how each of the polymer chains fit or arranges with the next molecule. The best way to think of this is to picture a bowl of spaghetti and think of all the random noodles as the polymer chains; this is one form of polymer chain-to-chain macroconformation. Another way to think of this is to think of more orderly arrangements like stacks of plates or stacks of lumber. These represent how multiple polymer molecules can be arranged. This becomes important when one sees irregularities in an orderly structure, like a plate slightly out of alignment with all the other plates in the stack. These minor variations can produce effects and interactions with materials being packaged that create major differences in package performance.

Another idea or mental picture of polymers is amorphous content. PP is one of the best polymer examples to help one understand amorphous content. PP can exist in three forms, atactic, isotactic, and syndiotactic. In each of these forms, the $-CH_3$ group can be placed in different relationships along the polymer backbone. Isotactic PP has the $-CH_3$ group always on the same side of the chain in a regular order. Syndiotactic PP has the group alternating in a regular order on either side of the polymer chain or backbone. Atactic PP has the group placed randomly on either side of the polymer chain with no regularity or order. Atactic PP is amorphous and has little commercial value. The other two forms of PP are much more valuable and constitute the homopolymer and copolymer forms of the material.

Chain folding is another macroconformation constituent of polymers and describes how the long polymer backbone can bend back on itself to create a molecular state of the lowest energy. Chain folding can influence stiffness and shear fracture in a material.

Viscoelastic Behavior

Viscoelastic behavior describes how polymers can deform when subjected to stress and temperatures near their T_gs. All polymers display a viscous component, which is the ability to move or flow even in the solid form. This ability to move means a polymer is not rigid like a piece of steel, but is more like frozen molasses. The T_g can be described as the point where the polymer begins to change from a rigid solid to a flowable solid, more like putty, which can be shaped and moved easily. Depending on the polymer's molecular weight and some of the atom-to-atom and strand-to-strand characteristics, as the polymer approaches its T_g, the polymer begins to behave like molasses or other high-viscosity liquids that slowly flow to the lowest potential energy point, given enough time. This means the polymer may over time acquire a permanent distortion or shape change through fluid flow.

A more scientific explanation of this change is that the micro-Brownian motions of chain segments within the polymer begin to unfreeze. This unfreezing involves torsional oscillation and/or rotations around the bonds in the backbone of the polymer involving 2 to approximately 60 carbon atoms. Most crystalline polymers have amorphous regions, and this description fits the change in that portion of the molecule. The crystalline portions of the polymer have a specific energy or heat of fusion, and this energy is proportional to the percentage crystallinity in the polymer. This is the second factor that contributes to polymer movement. There are other relaxation temperatures and transitions present in a polymer, but without making the subject more complicated than needed for this description of viscoelastic behavior, they can be considered part of the T_g.

Understanding this property is important in understanding how plastic closures work. Plastic closures subjected to a heat sterilization cycle will flow during the high-temperature period of the sterilization. This flow is measured by comparing the amount of force (torque) required to remove the closure before the heat cycle to the amount of force required to remove the closure after sterilization. Comparing the initial torque of a plastic container closure to the same container after sterilization shows a significant reduction in the amount of force required to remove the container closure. The change is due to the polymer relaxing or "flowing" away from the area of stress to a more relaxed conformation that is significantly lower in mechanical energy. The same change also takes place at lower temperatures; it just takes much longer to become evident.

Polymers also exhibit elastic behavior. This is the ability of materials like rubber to stretch and return to their original shape. During this process, the polymer chains are able to slip by each other and unfold, allowing the material to expand. Most of this expansion and subsequent contraction is the polymer being placed in a higher-energy state (the force required to stretch the band) and then relaxing back to a position of lowest potential energy.

Viscoelastic behavior of a plastic material is the combination of both of these characteristics. The polymer can stretch because of a stress or force applied, and it can also flow in response to the same stress. The plastic can flow more readily at higher temperatures, and the point where the polymer flows is called its T_g. The plastic material can deform immediately to take on the shape of the surfaces applying the force. Think of a package picture in your mind, the gasket material being squeezed or stressed from the top by the closure, and the force of the downward pressure causing the material to spread out or conform to the lip of the bottle and the underside of the cap. Then think of the material passing through a heat sterilization step. The increased temperature permits the material to move or flow easily, so the plastic takes on the permanent shape of the area in which it was confined. Since it is able to flow and rearrange itself, it usually shrinks slightly in volume, reducing the total force on the material.

In this discussion of the viscoelastic nature of polymers, another important parameter is time. Increasing the temperature causes the deformation to accelerate, so temperature is a key to the rate of change of the material.

This means that time–temperature relationships follow Arrhenius behavior, and this temperature-induced acceleration of time permits the packaging engineer to test and predict the final performance of a package at the end of shelf life. The time–temperature superposition or a position of increased energy being available to the system provides the basis for accelerated testing. Guidelines for accelerated testing of materials are found in ASTM and ISO standards used to test all types and varieties of materials. The problem with accelerated testing is the limits imposed on the temperature or other conditions of the test on the basis of limits contained in the physical properties of the material itself. Higher temperatures cannot induce immediate permanent deformation. It takes time for a high-viscosity liquid like molasses to flow from a jar even at elevated temperatures. The elevated temperatures make the molasses flow more like water, but it retains a measurable time component that is not instantaneous. Higher temperatures cannot instantly change the polymer structure whether it is crystalline or amorphous, but it can speedup a change in the polymer in a much shorter period of time. When the accelerated testing temperature exceeds the T_g of the material, it permits the polymer to change shape under stress from its own weight or the weight of its contents, making the evaluation worthless in terms of expected real-world performance at room temperatures or slightly elevated temperatures.

Using the Arrhenius equation, potential time/temperature aging regimens can be devised; however, they require judgment in the selection of the test times and temperatures to represent a reasonable simulation of material performance. Normally, accelerated aging of a package does not exceed 70% to 80% of the degradation temperature of the contents it holds, and most often this is considered extreme. Actual temperatures for testing are specified in the various standards and normally run in the 30° to 45°C range.

Physical Properties of Polymers

Polymers display many different physical properties (8). Some of the common properties that are used to describe how a polymer may behave are stiffness, hardness, melting or thermal characteristics, impact strength, and resistance-aging characteristics. Stiffness of a polymer can be modified in a number of ways. Stiffness is a result of a number of molecular interactions caused by the type of chain and the molecules in the chain. Normally, increasing the aromatic content of a polymer, adding conjugated bonds to the backbone, or adding cyclic groups to the polymer can increase the stiffness of the material. The use of bulky side groups is another method to increase the stiffness of the material. Polar molecules or polar bonding in the material, which result in electrostatic interaction within the polymer chain, such as when different polar areas of the chain fold back and align with each other, create electrostatic forces that cause the material to become much stiffer. Finally, crystallinity in the polymer can provide the basis for its stiffness.

Hardness, sometimes confused with stiffness, is another characteristic of a material. Hardness is the ability of the material to withstand a direct force. Think of how rubber deforms versus how a harder material like glass, or a crystalline polymer like PET, or an amorphous PC resists this direct deformation (9).

The melting characteristics of a polymer are another way to describe the physical properties of a polymer. Picture in your mind common sealing wax sold for home canning and household use. This material is very low in molecular weight but will help you to better understand how a long-chain polymer such as PE behaves. PE is a very high-molecular-weight version of sealing wax. It can have a range of viscosities depending on temperature. When molded, its flow characteristics (viscosity) are modified to match the equipment used to shape the item by adjusting the temperature of the material. If you melt wax and pour it into a mold to make a candle, you have an understanding of the way PE is molded. A wide range in the amount of melting in a polymer is the result of the amorphous nature of the polymer.

The melting characteristics of a crystalline polymer are very different. When dealing with a crystalline material and observing its melting characteristics, one notices that the material will remain at one temperature until enough energy (heat of fusion) has been supplied to break all the crystalline bonds in the material. Only after the crystalline bonds have been energized to a point that permits the molecules to move freely does the material absorb more heat and flow easily. Most polymers contain both amorphous and crystalline regions, so their melting characteristics are proportional to the amount of each physical form in the material.

Aging of polymers is another physical characteristic. The easiest way to understand aging is through personal experience. Think of a clear tape used to repair paper cuts and tears, and remember how you have observed yellowing of the material over a period of time. Materials may also become cloudy or brittle as they age. All of these characteristics are the result of the plastic molecules moving (viscoelastic behavior) and chemical or molecular change within the molecule. Polymers can slowly degrade through oxidation to lower-molecular weights (weatherability) or shorter chain lengths. This happens when a bond strained by folding receives enough energy to break, or the polymer receives enough energy from the environment to separate the polymer chain.

Branched molecules that contain multiple groups display this tendency toward partial chain separation and separation of the branched groups. Decreasing molecular weight is a sign of aging in a polymer, and lower molecular weight results in a decrease in most physical properties.

Polymers can crystallize over time, a characteristic mentioned multiple times in this chapter. Crystallinity typically comes about because it is a lower energy state or the lowest energy state for the molecules. Molecules respond to stress or potential energy by finding the lowest energy state, and orderly arrangements in crystalline structures allow for the closest packing of the molecules and the lowest energy state. This is called a reduction in "free volume" within the polymer. The rearrangement to a regular pattern found in a crystal results in the material becoming increasingly brittle and sets up the possibility of material failure along a plane in the crystal.

Polymers can become softer or change in structure through the absorption of gases and solvents. This includes water from inside a liquid container. Plastics with highly crystalline characteristics can absorb moisture during a steam sterilization cycle and become cloudy or display white lines indicative of environmental stress cracking. These physical characteristics are examples of accelerated aging. The cloudy portions of the polymer are regions of increased crystallinity. The cloudy appearance is the result of larger groups of molecules, crystals, having a different density than the amorphous material surrounding them. These larger molecules scatter light and appear milky or cloudy to our eyes.

The same characteristic can influence impact resistance and the overall strength of the polymer. The ability of the chains to move and stretch within a molecule provides the basis for impact strength. As materials become more crystalline and regular in their arrangement, the molecules become more resistant to movement and more prone to shatter or break upon impact. Chain mobility is significantly reduced in a crystalline structure, and the size and shape of the crystals, particularly crystals forming between regions of one chain and another, greatly influence the physical characteristics of a material.

Most of these characteristics of aging in a polymer can be explained as Arrhenius behavior within the molecule.

TEMPERATURE DEPENDENCE ON REACTION RATES

Temperature can have a significant influence on reaction rates, and this dependence is used to predict how quickly a material will react at a given temperature. One way to use this dependence is to use the Arrhenius equation as a method of predicting how fast something will age at a given temperature. By calculating how fast something ages at room temperature, and then substituting with a higher temperature, the time required to reach the same end state is decreased. An empirical observation is that many reactions, including predictions of polymer aging, have rate constants that follow the Arrhenius equation,

$$K = Ae^{\frac{-Ea}{RT}}$$

A is the pre exponential factor.
E_a is the activation energy.
R is the gas constant.

This brief background about polymers is by no means complete. It is intended to provide a very brief overview of how and why different plastics behave as observed in packages and how to work through the process required in choosing a plastic material to packaging a product.

PLASTICS AS DRUG PACKAGING MATERIALS

Plastic have become the most widely used material for food and drug packaging. It has found its way into packaging of countless consumer products. Plastics have rapidly replaced glass and metal in many food and beverage packaging applications. In pharmaceutical packaging and, particularly, drug packaging, this rate of conversion has been much slower.

The reasons behind the slow rate of change are significant. The most common reason and the one most often cited is the potential for adverse health consequences if a drug interacts or in some way is changed by its packaging. A second reason for the slow change is the relatively small unit volumes of plastic containers used for drugs when compared for food and beverage products. This means food and beverage manufacturers tend to focus more on packaging innovation than drug manufacturers, and these two consumer-directed industries have provided the initial developments needed to interest companies and consumers in the material or package benefits and eventually move the pharmaceutical companies to convert or install new packaging systems. PET bottles are a good example of this transition from consumer products to pharmaceutical products.

Estimates for the number of plastic containers used by pharmaceutical manufacturers are significant and growing. Examples of some of the conversions from glass and metal to plastic include

- Intravenous solutions and premixed nutritional products
- The bottles used for OTC pharmaceutical products
- The bottles used for tablets and other solid dosage forms
- Bottles for liquid pharmaceutical products
- Bottle closures
- Laminated pouches
- Blister packs

- Strip packs
- Tubes for ointments

With the exception of aerosol containers and parenteral vials, all other types of pharmaceutical packaging are slowly moving to plastic packaging.

Plastics have a number of inherent advantages over traditional packaging materials, and these advantages have led to plastic assuming the dominant position in drug packaging. A list of these attributes includes

- Consumer preference for plastic
- Density differences with traditional materials resulting in lighter weight packages
    - Plastic densities 1–1.5 g/cc
    - Glass 2–2.5 g/cc
    - Aluminum 2.7 g/cc
    - Tinplate 8.5 g/cc
- Shatterproof
- Clear and opaque (plastic can be either, or it can be translucent)
- Heat sealing
- Decoration
- Thin films with high strength and toughness
- Easy handling
- Design freedom.
- Potential of lower cost

Many of the advantages listed above translate into very significant container attributes to the manufacturer and the consumer. Density is a good example of a property that translates into multiple packaging attributes.

Density Differences/Consumer Preference for Plastic/Easy Handling

Plastic containers, particularly those made from plastic films, are much lighter than glass or metal. This lightweight characteristic means that the container is less expensive to manufacture because the package contains less material than that in its denser alternative. Generally a rigid plastic container, such as an HDPE or PET bottle, are less expensive than their glass counterparts by as much as 20%, and a container made from a plastic film will be several times less expensive.

The lightweight nature of plastic compared with glass and metal make the container easier for the consumer to handle. This makes transportation and dispensing simpler, and makes possible individual dose packages in the form of blisters and pouches. Plastic is shatterproof, a big advantage over glass when the bottle or pouch transits the distribution environment or is handled by a consumer in a bathroom or in an environment where dropping the container would cause it to shatter.

Plastics can be heat-sealed. This produces a hermetically sealed package that the consumer does not need a tool to open and eliminates the problems of a seal made from fused glass.

Design Freedom

Plastic offers a wide range of design possibilities that may be more difficult to obtain with metal or glass.

Plastics can be easily shaped and sealed. This permits the packaging engineer the freedom to create complex shapes and permits the incorporation of administration aids, such as squeezable droppers, thermoformed cups, bottles that dispense measured doses, and premeasured amounts of product ready for mixing and dosing.

Pouches are a unique form of packaging available to the package engineer. They permit the creation of lightweight, low-cost packages, which can be tailored to specific needs through the incorporation of different plastic and metal films that contribute unique properties to the finished composite. They also permit the manufacturer to provide a superior package when a small unit volume is involved.

Aluminum is the only other material made into thin films or foils. Aluminum is more expensive than plastic and inferior in providing the strength and toughness required for a finished package. Foil is used in packaging drugs and in blister applications, but it is always laminated to a plastic film to provide and enhance properties it lacks.

The disadvantages of plastic, some of which cannot be overcome by design or the combination of multiple materials, limit the penetration of plastic into a number of niches of drug packaging. A list of plastic disadvantages that cannot be easily overcome for specific packaging properties is as follows:

- Chemical inertness (no plastic can match type I glass)
- Stress cracking (the presence of alcohols, organic acids, and many oils can cause a plastic package to crack and fail over time)
- Resistance to heat (glass and metal can withstand higher temperatures)
- Resistance to light (glass and metal are not changed by long-term light exposure, even high-energy light)
- Resistance to oxygen (glass and metal are impermeable to environmental oxygen)
- Leaching of low-molecular-weight polymer fragments by a drug or solvent

PLASTIC DISADVANTAGES

Chemical Inertness/Stress Cracking/Additives/Electrical Properties

No plastic material can match type I glass for impermeability and chemical inertness. Fluorocarbons can match glass in being chemically inert but not in gas impermeability. Fluorocarbons are much more expensive than glass. This point highlights why many different materials are needed for packaging pharmaceuticals.

Stress cracking in plastics can be a major problem. Environmental stress cracking is caused by the presence of alcohols, organic acids, ethers, and many natural and synthetic oils along with mechanical stress induced in the package during molding. This mode of deterioration can reduce performance and produce package failures. Polymer tendrils (very thin threads of polymers) may be the only portions of the polymer bridging and holding together a microcrack. These will break easily and may compromise the integrity of the package and its sterile barrier.

Plastics' resistance to light, oxygen, solvents, and heat does not match that of glass or, in selective cases, metal. The enhanced resistance properties required must be provided by the addition of additives or surface treatments to the plastic. These additives and treatments can be leached from the plastic with solvents and, in some cases, by the drug ingredients. The low-molecular-weight polymer pieces, contained in any plastic and characterized as one part of a bell-shaped molecular weight distribution curve, also are susceptible to leaching by solvents and the drug.

The FDA and the USP define pharmaceutical grade polymers. These polymers do not contain certain additives and are restricted in the amount of residual catalyst they contain. Catalyst is removed and reused in the polymerization reaction, but a small amount remains in the finished polymer. These residual amounts of catalyst are limited and "pharmaceutical grade" polymers meet FDA guidelines for the amount they contain.

The resins are also restricted in the amount of extractables they produce in standard hexane and water extractable testing procedures.

COMMON PLASTIC PHARMACEUTICAL PACKAGING MATERIALS

There are a number of plastic materials commonly associated with pharmaceutical packaging. These materials range in molecular weight from 10,000 to approximately 1,000,000 (1 million), the same range for almost all commercially useful polymers.

Polyethylene Polymers

PE polymers represent the most widely used packaging plastics (Fig. 12). PE and PP polymers are often referred to as polyolefins. The olefin name was originally used

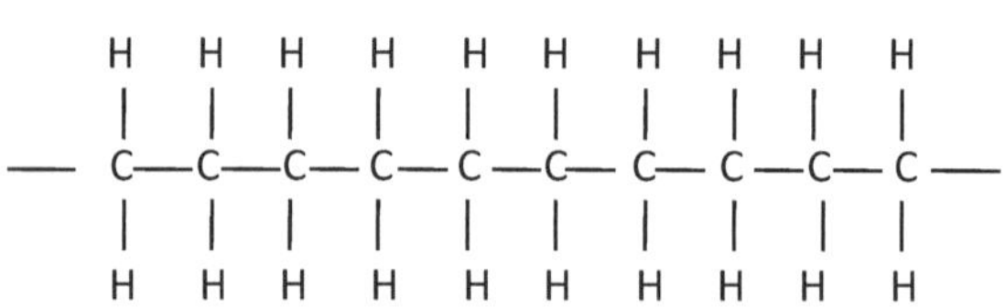

Polyethylene

Figure 12 Polyethylene.

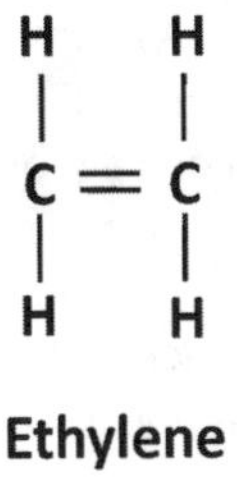

Figure 13 Ethylene.

Table 5 Comparison of HDPE, LDPE, LLDPE

Property	HDPE	LDPE	LLDPE
Density g/cc	0.945–0.967	0.91–0.925	0.916–0.940
Tg		−120°C	
Tensile strength, Kpsi	3.0–7.5	1.2–2.5	
Tensile modulus	125	20–40	
Haze	25–50	4–10	
Water vapor transmission g-mil/ 100 in^2/day@100°F and 90% RH	0.3–0.65	1.2	

with alkenes and means oil forming. In the plastics industry, the term "polyolefin" is used to describe both ethylene (Fig. 13) and propylene (Fig. 26) polymers.

PE comes in a number of forms. It is divided into different groups on the basis of densities of the different materials. A list of the density differences of the three grades of PE used in pharmaceutical packaging (Table 5) is provided.

The four common forms of PE are

- ULDPE—ultra low-density polyethylene
- LLDPE—linear low-density polyethylene
- LDPE—low-density polyethylene
- HDPE—high-density polyethylene

ULDPE is excluded from the discussion of pharmaceutical packaging materials because of its high level of extractables in hexane.

PE is a collection of addition polymers that can be linear or branched and either homopolymer or copolymer (Fig. 2). PE was first introduced in the 1950s as a packaging material, and quickly became a staple for packaging food (Fig. 12). LDPE is the most widely used member of the PE or polyolefin family of polymers used as packaging materials. After PE's introduction in the 1950s, it moved into wide commercial use as film, molded containers, and closures.

PE copolymers are also addition polymers that substitute comonomers such as propene, butene, hexene, or octene for ethylene. The comonomers may be compounds with polar functional groups such as methyl acrylate (MA), ethyl acrylate (EA), acrylic acid (AA), or vinyl acetate (VA). A normal distinction between homopolymer and copolymer blurs when the molar percentage of comonomer is less than 10%. These polymers may be referred to as either homopolymer or copolymer.

HDPE, LDPE, and LLDPE are all used in pharmaceutical packaging. These materials provide most of the properties required for drug packaging at the lowest cost.

PE is a long-chain polymer with the minimum useful length of repeating monomer units of 1000. The repeating or monomer units are $-CH_2-$ units derived from the polymerization of ethylene (Fig. 13). General molecular weight ranges for the various grades of PE are

Medium-molecular weight	<110,000
High-molecular weight	110,000–250,000
Very high-molecular weight	250,000–3,500,000
Ultra high-molecular weight	>3,500,000

Molecular weight and density must not be confused or substituted for each other when referring to different grades of PE. They are different properties.

The different grades of PE come from modifications in manufacturing conditions that affect the finished polymer. The choice of catalyst also plays a significant role in the grade of finished polymer.

Linear PE is a highly crystalline polymer. This crystallinity, which ranges from 70% to 90%, is a result of the small size of the repeating units or pendant groups that produce a high degree of stereoregularity. Creating branches in the polymer structure through the use of different alkenes and different reaction conditions reduces the amount of crystallinity because the branches create irregularities in the backbone of the molecule and prevent the packing and folding that linear PE exhibits.

High-Density Polyethylene

HDPE is produced by the polymerization of ethylene at low pressures and temperatures using a coordination catalyst. This catalyst under these conditions produces a polymer molecule with few branches and side chains. This is linear PE with a high degree of crystallinity and a limited number of branches and side chains. Limited branching accounts for its high (relative) density.

The material is thermoplastic and milky in appearance. HDPE is not only used for pharmaceutical bottles but also has broad application in a variety of household products ranging from foods like milk to strong household chemicals like bleach. Bottles are produced from the material using both injection molding and extrusion blow molding processes. HDPE is manufactured into film for packaging using both blown and cast film processes.

The crystallinity found in HDPE produces materials with good moisture barrier properties, chemical resistance, and opacity (translucence).

HDPE is prone to environmental stress cracking.

Low-Density Polyethylene

LDPE is produced with a different catalyst and a different set of reaction conditions from those used with HDPE. LDPE is produced using a free radical catalyst and reaction conditions of high pressure and high temperatures (approximately 300°F). LDPE is a polymer consisting of many branches and many side chains that hinder the molecule packing or the strands orienting with a minimum of space between them. The density of the material is between 0.91 and 0.93 g/cc, and the crystallinity of the polymer ranges from 40% to 60%. LDPE is a low-cost packaging material and is the most widely used plastic in the world.

A slight variation of LDPE is sometimes referred too or referenced as a separate polymer called medium-density polyethylene (MDPE). This material is the high-density end of the range of LDPE resins. It is slightly stiffer and somewhat less permeable than LDPE materials found at the lower and middle sections of the polymers density range.

The majority of LDPE is used in films; in fact, within the United States, approximately 55% of LDPE is made into films of less than a 12 mil or 300 μm in thickness. LDPE is an easy material to process and can be manufactured as blown or cast film; it can be injection or blow molded, and it can be used as an extrusion coating on a variety of plastic and paper substrates. In applications requiring more strength, the material can be produced in thicker sections for thermoforming containers, or it can be blow molded by either injection or extrusion blow molding to produce bottles (8,10). It is also widely used as an extrusion coating to provide a heat-seal medium or layer on other substrates.

Linear Low-Density Polyethylene

LLDPE is the third widely used variation of PE (Figs. 14 and 15). LLDPE is produced using the same conditions as those used for HDPE. The variation

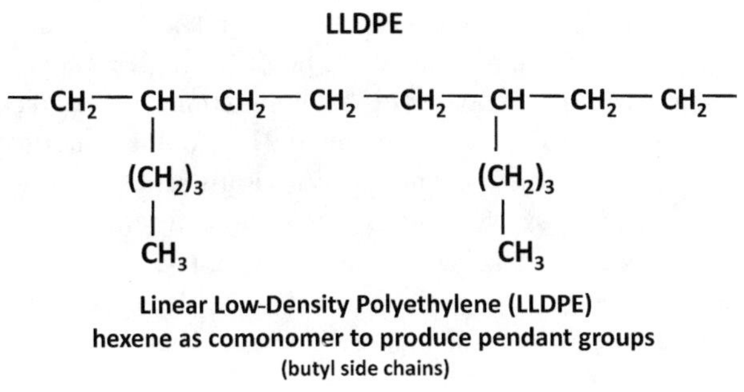

Figure 14 LLDPE structure using hexene as the monomer to produce the pendant groups (butyl side chains).

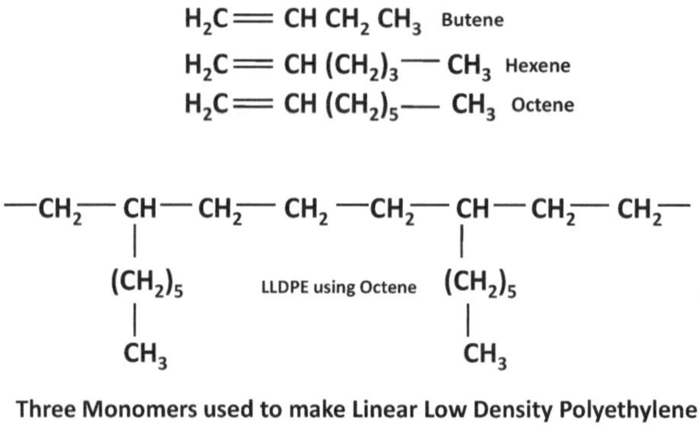

Figure 15 LLDPE, the three monomers used to make it, and octene represented in this example to produce the pendant groups (hexyl side chains).

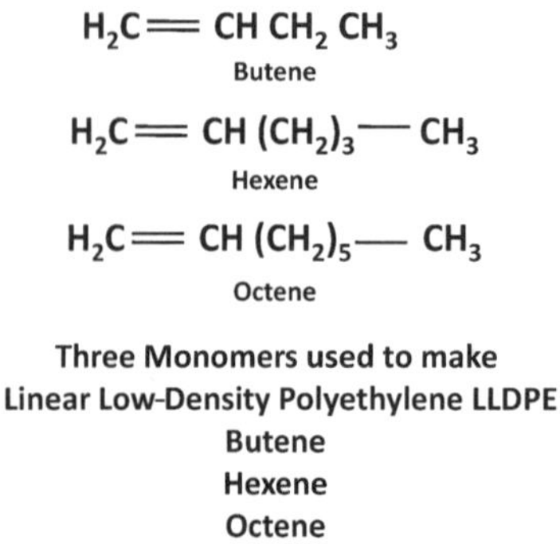

Figure 16 The three monomers used to make LLDPE, butene, hexene, and octene.

comes from the introduction of one or more of three different comonomers, butene, hexene, or octene (Figs. 15 and 16). At pressures around 300 psi, the slightly altered reaction conditions with a stereospecific-modified catalyst produce a form of PE that has very limited side chains or very short-branched pendant groups. The amount of the comonomer added to produce LLDPE ranges from 1% to 10% on a molar basis. If higher levels of comonomer are added during production of LLDPE, pushing the density below 0.91 g/cc, the material displays a high level of hexane extractables that are beyond limits sanctioned by the FDA. The extractables, which are low-molecular-weight fragments of polymer produce off-flavor, odors, and a leachable material in the packaging that can further oxidize. The limited branching, which modifies the density of the material is the reason behind the use of the name linear in describing this polymer.

LLDPE combines the strength and toughness of HDPE with the clarity of LDPE. It is not exactly the same as either of these two materials in properties but is a good compromise between the two. This material rivals the ionomer resins in producing very strong heat seals at low temperatures while maintaining a property called hot tack. Hot tack is high adhesive strength at temperatures close to the melting point of the polymer, allowing the seal to be strong and effective while the melted plastic cools.

The three forms of PE offer major differences in properties that pharmaceutical manufacturer's value in the packaging of their products. All three forms of PE, particularly HDPE and LDPE, are compatible with a wide variety of drug products and are recognized as a standard packaging plastic by the FDA (Table 5). Any use of a PE polymer to package a drug must be well documented as part or the development of the CMC section of the NDA describing packaging (Table 5). The testing and qualification of a product in this material must answer questions about permeability and absorption of ingredients from the drug's makeup. It also must be demonstrated that the material does not sorb the drug product.

HDPE is stronger and stiffer than LDPE; it is also milkier in appearance and less clear. HDPE is more resistant to chemicals, solvents, and oils, and it is less permeable to gases than LDPE. HDPE can be autoclaved but cannot be retorted as a container for liquid drug products. HDPE's strength, stiffness, and moisture barrier are the reason it is widely used as the material of choice for bottles packaging solid dosage forms. The containers lack the clarity of LDPE, but this is not a drawback. The majority of bottles are pigmented to reduce or eliminate light transmission and improve label contrast and clarity.

LDPE on the other hand possesses more clarity, is more flexible, and can be stretched more easily than HDPE. These properties make LDPE the material of choice for squeeze bottles. Clarity permits the patient or consumer to see the amount of liquid remaining in the bottle.

The restrictions on using HDPE, LDPE, and LLDPE in drug packaging are directly related to the permeability of the materials. A list of the restrictions and the reasons behind the restrictions are detailed below:

POLYETHYLENE RESTRICTIONS IN DRUG PACKAGING

1. Permeable to oxygen They cannot be used to package oxygen-sensitive products.
2. Poor odor barrier. Storage of a drug product next to an odoriferous substance or product may result in absorption of the volatile components and a degradation or contamination of the product.
3. High permeability to halogens. This prevents their use as a package for solutions containing chlorine, bromine, or iodine.
4. Oil softening and permeability. A number of natural oils, such as castor oil, soften PE, and a number of essential oils used in pharmaceutical

formulation for flavoring or aroma, such as coconut oil or oil of peppermint, are permeable through the material.
5. Poor resistance to strong oxidizing acids.
6. Tendency to environmental stress cracking. Solvent can induce stress cracking in PE materials when mechanical stress is present. This can be offset by using high-molecular-weight grades of the materials. If exposure of the material is to only one of the conditions, stress or chemical solvent, it does not produce environmental stress cracking failure.
7. PE will sorb other materials. It will sorb steroids, bactericides (e.g., benzalkonium chloride, benzyl alcohol, phenylethyl alcohol), and alkaloids [e.g., hyoscine (scopolamine), g-strophanthin or ouabain (a cardiac glycoside), pilocarpine]. PE will also sorb vegetable oils and coloring from tinctures. The sorption tendency is lowest in HDPE and can be mitigated to some degree by molecular weight.

OTHER ETHYLENE POLYMERS

Ethylene is also polymerized with other unsaturated comonomers to produce ethylene copolymers. The resulting polymers have a wide range of properties adapting them to multiple end uses in packaging and in medical applications. The applications include a wide variety of packaging applications such as adhesives in coextrusions, heat-seal and cold-seal formulation components, and components of chemical coatings. Cyclic olefin copolymers are used as moisture and oxygen barrier material. In medical devices, the plastics can be primary structural materials, modifiers, and, in the case of cyclic olefin copolymers, optically clear materials.

Ethylene Vinyl Acetate

Ethylene vinyl acetate (EVA) (Fig. 17) is a random copolymer with properties dependent on the amount of VA (Fig. 18) used in the reaction and the molecular weight of the polymer. Ethylene acetate polymers are produced by reacting ethylene with VA monomers to form a random copolymer.

EVA copolymers exhibit toughness, flexibility, and good heat-seal characteristics. The molecule is polar and its introduction into the backbone of branched ethylene copolymers results in increasing density and lower crystallinity. VA is added to ethylene copolymers in a range of 5% to 50%, but for most food, medical, and related packaging applications the ratio is 5% to 20%. The introduction of the polar molecule also results in improved flexibility, better barrier performance, and a wider heat-sealing range. At the 50% level of introduction, EVA becomes an amorphous polymer. EVA polymers are processed at relatively low temperatures because of their low thermal stability and

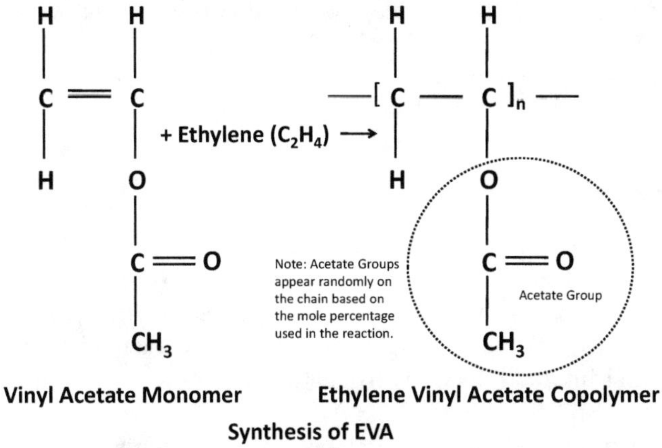

Figure 17 Vinyl acetate copolymer with acid group side chain highlighted.

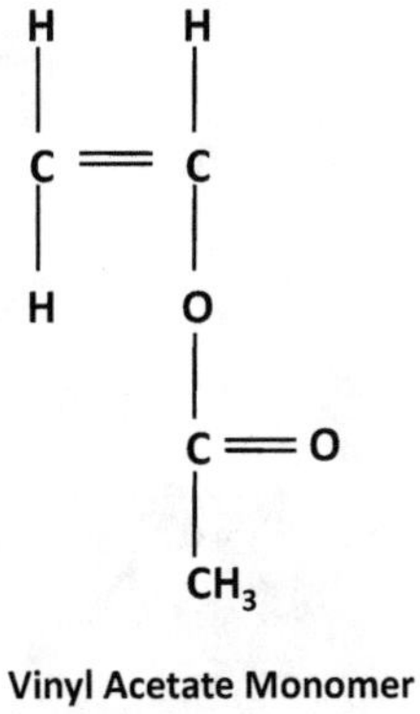

Figure 18 Vinyl acetate monomer.

low melting point. This characteristic explains some of their improved strength and toughness at low temperatures.

The acetate group is relatively large and bulky, and its inclusion in the copolymer creates random irregularities in the structure that reduce crystallinity. The group also increases the intramolecular forces within the polymer. The decrease in crystallinity would indicate a decrease in density; however, the addition of the oxygen atoms with their increased mass offset the decrease in density.

These changes in physical properties also produce increased clarity, improved low-temperature flexibility, and improved impact strength of the polymer.

EVA copolymers also are used as extrudable adhesives in multilayer plastic structures. They also are coextruded with PET and biaxially oriented PP to produce a heat-sealable layer on one or both sides of these polymers.

Ethylene Acrylic Acid

When ethylene is reacted with AA (Fig. 19), another useful copolymer of ethylene is produced. This polymer is ethylene acrylic acid (EAA) (Fig. 20).

EAA is a polymer similar to LDPE, but superior to it in adhesion, hot tack, and strength. These property improvements are created by the increased intermolecular interactions of hydrogen bonds within the polymer molecule.

This polymer changes in the same ways as EVA with the introduction of increasing amounts of AA. Increased amounts of AA result in decreasing crystallinity. Decreased crystallinity improves clarity and reduces heat-sealing temperatures. The increased polarity in the molecule results in increased adhesion strength.

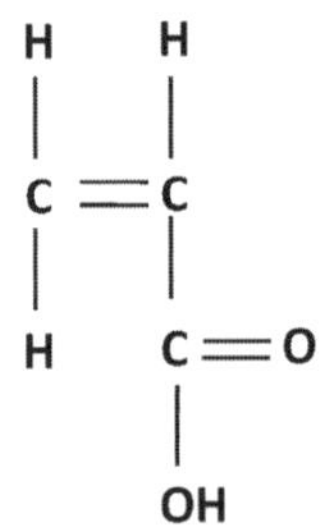

Figure 19 Acrylic acid monomer.

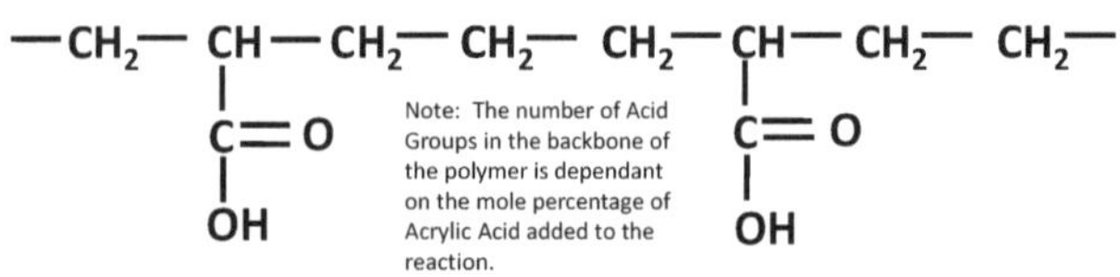

Figure 20 Ethylene acrylic acid.

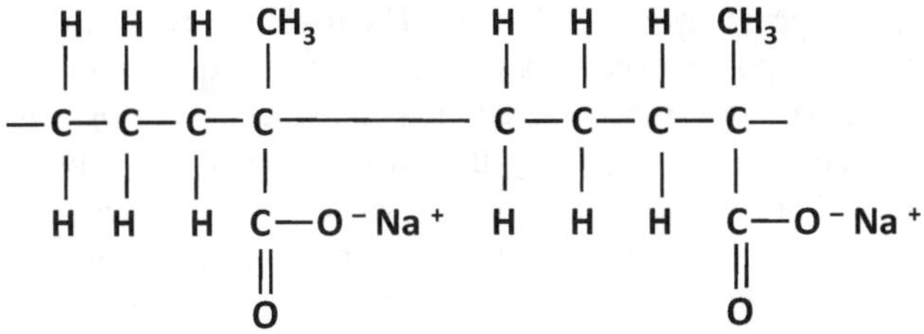

Ionomer
Ethylene Methacrylic Acid (EMAA) Neutralized with Sodium

Figure 21 Ionomer structure—ethylene methacrylic sodium acrylate.

EAA (Fig. 20) is most often used in pharmaceutical packaging as a film in flexible packaging. It is also used as an extrusion coating on plastics, paperboard, composite cans, and certain types of tubes, most notably toothpaste tubes. EAA is sanctioned by the FDA at up to 25% acrylic acid content for use in food. In medical applications, this amount is problematic and must be proven safe as part of the packaging qualification and validation.

EAA polymers are the starting point for the production of ionomers.

Ionomers

DuPont developed ionomers through the work of R.W. Rees and D. Vaughn in 1965 (11). They are marketed under the trade name Surlyn®. Ionomers (Fig. 21) are primarily used as a heat-seal coating on nylon, PET, LDPE, polyvinylidene chloride (PVDC), oriented polypropylene (OPP), and aluminum substrates. These materials produce excellent hermetic seals with outstanding hot tack characteristics. Their hot tack performance permits the seal to remain strong and resist any type of breach while still hot, immediately after the pressure, and heat applied to produce the seal are removed. Hot tack maintains the hermetic seal during cooling and hardening of the plastic resulting in excellent seals with minimum failures. Ionomers also display high resistance to puncture and retain their impact resistance to –90°C.

Ionomers are made by polymerizing ethylene with ethylene methacrylic acid and then treating the carboxyl groups with sodium or zinc compounds to replace the hydrogen atoms (Fig. 21). This can be accomplished in two ways. The first is to add a sodium or zinc compound to the high-pressure polymerization reaction. The second method is to partially neutralize the acid copolymer with a sodium or zinc compound as a second step after polymerization of the copolymer.

Ionic bonds produce unique effects in the polymer. The presence of polar molecules creates random cross-link-like ionic bonds between the different polymer chains. These bonds improve polymer performance, making their behavior similar to high-molecular-weight polymers. The ionic bonds are easily disrupted by heat, making the polymer process (extrusion or injection molding) in the same way as conventional thermoplastic ethylene polymers. The covalent bonds in the polymer behave in the same way they do in other thermoplastic polymers and are more robust than the ionic bonds that produce a synergistic effect with regard to performance. The ionic bonds disrupt the formation of crystallization within the polymer and also inhibit the formation of spherulites, which are crystalline regions within a polymer structure.

The sodium form of ionomers produces oil resistance, hot tack, and improved optical properties (Fig. 21). Modification of the polymer with zinc improves the polymer's resistance to water and improves its adhesion properties to aluminum.

Ionomers are the polymers most often used in critical extrusion coating or extrusion laminating applications in films. The ionic bonds also impart elongational viscosity to the polymer film, resulting in resistance to pinholes. The chemical bonds of the metal ions also enhance the puncture resistance of PE materials. Their high cost (3× to 5× of LDPE) tends to limit their use to demanding applications.

Ethylene Vinyl Alcohol

Ethylene vinyl alcohol (EVOH) (Fig. 22) is another of the ethylene polymers useful in pharmaceutical and food packaging. It is used primarily as a barrier to oxygen, flavors, and odors in a wide variety of packaging structures. It was the first material to compete with and find uses as a substitute for Saran® or PVDC polymers when a high oxygen barrier was required. It remains a preferred choice as a barrier material when it can be protected from exposure to moisture.

EVOH is not produced from polyvinyl alcohol (PVOH) (Fig. 23), as one would expect. It is produced by the controlled hydrolysis of EVA copolymer

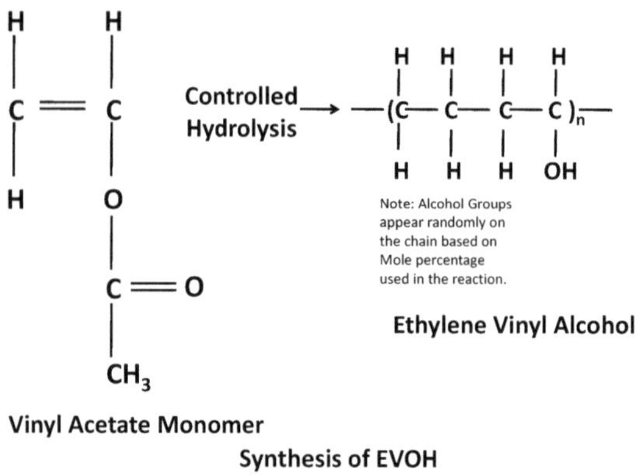

Figure 22 Ethylene vinyl acetate copolymer.

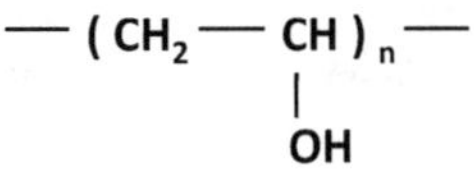

Polyvinyl Alcohol (PVOH)

Figure 23 Polyvinyl alcohol.

(Fig. 22). The VA group is transformed to vinyl alcohol during the controlled introduction of water. The –OH groups are very polar and increase the intermolecular forces, while the ethylene backbone of the polymer maintains the thermoplastic characteristics of molecular mobility. This polymer is highly crystalline, and the crystalline structure contributes to the high barrier properties of the material.

Unfortunately, the –OH groups are also hydrophilic, making the polymer and its barrier properties susceptible to degradation if the material is "wet" or water saturated. The barrier properties recover, as the polymer dries, but this characteristic must be minimized when the polymer is used as an oxygen or odor barrier in packaging. This loss of barrier when saturated is a characteristic that led to the coextrusion of EVOH with other materials (Fig. 24). The coextrusion buries or hides the layer of EVOH between layers of polymers with good water vapor resistance. EVOH does not adhere well to most of these nonpolar polymers so a tie or adhesive layer must be used to bond the multiple plastic layers together. The ability of a polymer with good water vapor resistance, like PP, is not enough to protect EVOH from loss of oxygen barrier properties after retorting, or sterilization, using water and steam at temperatures in excess of 250°F (121°C). Work by the American Can Corporation resulted in an effective polymer modification to improve retort performance. The incorporation of a desiccant in the tie layers of the polymer sandwich are used for demanding applications to minimize the amount of moisture reaching and compromising the

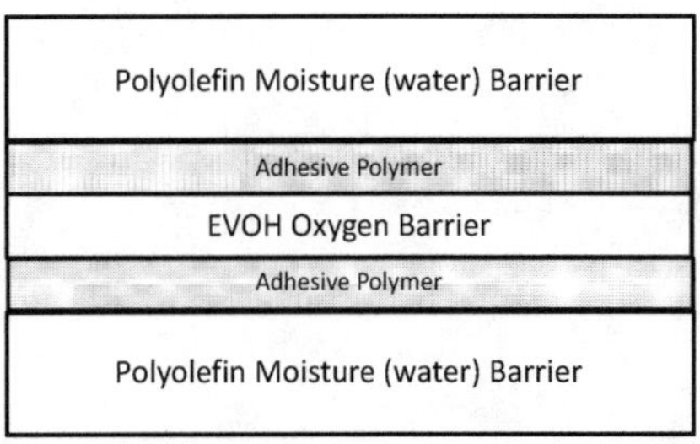

Multilayer Polymer Structure (coextrusion) with EVOH Oxygen Barrier

Figure 24 Diagram of a typical multilayer structure.

barrier layer of EVOH. The desiccant absorbs the moisture that penetrates the water-resistant polymer before it reaches the oxygen barrier layer during the heat and pressure of the retort sterilization process and preserves to a great degree the ability of EVOH to remain a barrier, resistant to the migration of oxygen, flavors, or odors through the packaging structure. This effect is not absolute, and testing will reveal the degree to which it protects the barrier and the time it takes for the barrier to recover to its full performance level. Keep in mind that the barrier will always be in some state of equilibrium with moisture if the contents of the package are liquid, and the structural layer of the material in the package has any permeability to the liquid.

The EVOH polymer (Fig. 22) is produced using between 27 and 48 mole% of ethylene. The lower the amount of ethylene, the better the barrier properties of the dry polymer, and conversely the more difficult it is to process the polymer. EVOH can be blow molded, injection molded, thermoformed, and extruded in films.

Polyvinyl Alcohol

PVOH (Fig. 24) is very similar to EVOH with a number of interesting twists. The polymer is water soluble, making it unsuitable for barrier uses, even though the dry polymer is an excellent oxygen barrier.

PVOH is not produced from PVOH in a standard addition polymerization (Fig. 25). Vinyl alcohol monomer is unstable and cannot be polymerized. The polymer is made from the hydrolysis of PVA.

The polymer is amorphous and atactic. The material will crystallize when subjected to orientation. Both the –OH and –H groups on the polymer are considered isomorphous groups, and this pendant group does not interfere with crystallization. The –OH and –H groups provide the mechanism for extremely strong hydrogen bonding within the polymer. This bonding is so strong that the polymer cannot be melt-processed in the same manner as other thermoplastics.

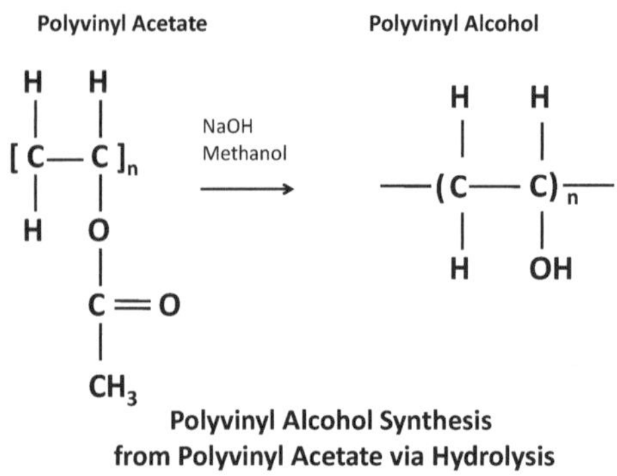

Figure 25 Polyvinyl alcohol synthesis.

The melting point of the polymer is raised above the decomposition point of the polymer by the hydrogen bonding within the polymer structure. Even though the crystalline polymer is a lower energy state of the material, stress is required on the polymer to overcome the hydrogen bonds and permit the polymer to crystallize.

Because of the extreme difficulty in processing PVOH, the polymer has limited uses. The interesting twists on using this polymer result in its combination of barrier properties and its water solubility. Films produced from the polymer and fabricated into pouches are excellent for containing highly toxic substances. The pouches are protected from moisture by the secondary packaging that is the moisture barrier. When the packaged product is needed, the pouch is removed from the container and placed in a solvent or water, dissolving both the plastic package and the toxic substance without exposing a person to touching or mixing the toxic material. A more mundane application of the polymer is in bags used to handle linen in a hospital. The linen is placed in a bag and the bag is placed directly into the washing machine. This eliminates potential exposure of hospital personnel to infectious waste.

PVOH is also biodegradable, so materials packaged in the polymer can be released in a controlled manner by exposing the material to sunlight or moisture.

Polypropylene

PP (Fig. 26) is similar to PE but much more complex in its structure. PP has a number of advantages over PE, with the most significant advantage for packaging being its resistance to higher temperatures. PP also has better resistance to grease and oil, is a better odor barrier, has fewer tendencies to absorb certain ingredients, and normally contains a lower quantity of additives in its final form.

Propylene monomer (Fig. 27) is normally polymerized in a hydrocarbon solvent at 200 psi using a Ziegler–Natta coordination catalyst or a metallocene catalyst (Fig. 28). This produces isotactic PP, the most common commercial grade of PP (Fig. 29). Reaction conditions are modified to produce syndiotactic PP (Fig. 29). The atactic version of the molecule has little or no commercial value (Fig. 29).

PP is also produced under the same conditions with the addition of between 1.5% and 7% ethylene monomer by weight. The polymer structure

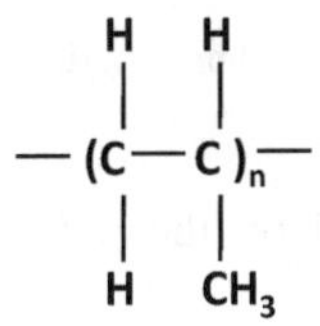

Figure 26 Polypropylene structure.

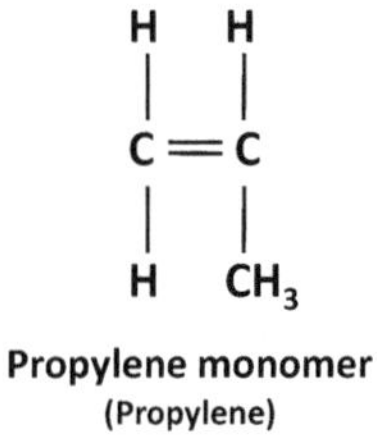

Figure 27 Propylene monomer.

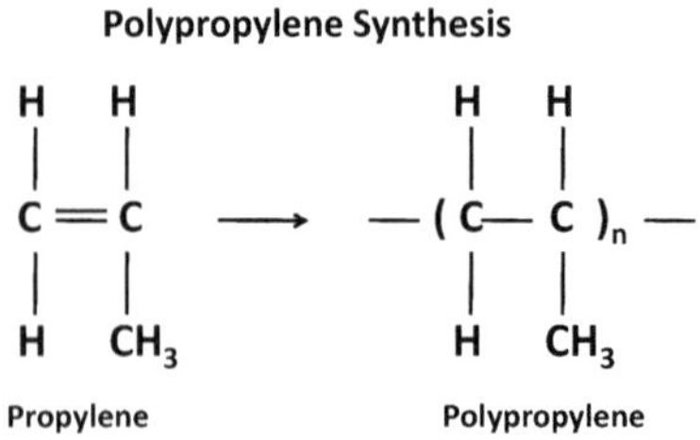

Figure 28 Polypropylene synthesis.

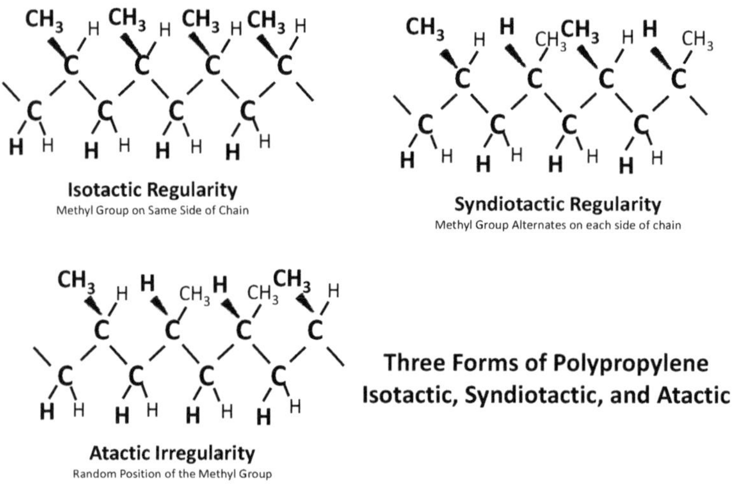

Figure 29 Three types of polypropylene polymers: isotactic, syndiotactic, and atactic.

remains the same and varies in the random fashion introduced by the polymerization of the ethylene groups in the backbone.

The molecular weight of PP averages between 200,000 and 600,000. A broad distribution across the molecular weight range enhances injection molding characteristics and permits easier processing of the polymer. PP can undergo oxidation that results in chain scission and a reduction in molecular weight. Antioxidants are a common additive in PP resins to combat this tendency.

PP also has a tendency to pick up a static charge. This is also true of PE, and the addition of an antistatic agent that permits the dissipation of static charges is common. The amount of additive for both purposes is generally lower than that found in PE materials.

The addition of ethylene to the polymer composition results in a lower crystallinity, better clarity, improved flexibility, and a lower melting point than standard isotactic PP. The polymer with this modification also displays improved low-temperature impact strength and better toughness and resistance to puncture.

As noted earlier, the presence of the methyl group ($-CH_3$) on the propylene chain permits three different forms of PP to be produced. The three variations of the same polymer result from the position of the methyl group in relation to the backbone of the polymer. The group can be on one side of the chain, it can alternate in a regular pattern from side to side along the chain, or it can be randomly oriented along the backbone chain. The three forms of PP are named isotactic, syndiotactic, and atactic PP, respectively (Fig. 29).

PP has gained favor in pharmaceutical packaging primarily because of its heat resistance and excellent moisture barrier characteristics. The higher heat resistance of the material permits its use in high heat sterilization processes. Its ability as a packaging material to be retorted or heat processed as a container for liquid and solid products set it apart from PE polymers that soften or deform at normal sterilization temperatures. This higher heat-resistant polymer is used in thermoformed trays and blow molded bottles for drug and medical device applications. PP can form hinges in packages that resist cracking after repeated flexing, making it a material of choice for clamshells.

PP film is normally oriented (stretched or pulled in one or two axial directions) to enhance it mechanical properties. The oriented film can be used as a shrink-wrap. The polymer film when heated reverts back to its previous configuration. OPP film is difficult to heat seal because of orientation and is made with the PP/PE copolymer or is modified with PE/PP copolymers that can be coated or coextruded with PP to permit heat sealing over a broad temperature range. These materials have a lower melting point and lower modulus. They can be applied by coextrusion or by coating of a PP substrate.

PP has good clarity and in the oriented film form produces a clear material that shrinks to produce a good overwrap film for bundling product.

A number of additives called clarifiers have been developed for PP to permit it to compete with PET in bottles as a lower-cost alternative material. Bottles produced with these chemical modifiers rival the clarity of PET when filled but lack the surface gloss and true clarity of PET.

PP is also produced as unoriented sheet for blister packing. It is an alternative to PVC and other plastics used in blister packs of tablets and liquids.

PP has another unusual characteristic. It can be formed using a process called solid phase pressure forming (SPPF). Because the polymer is crystalline, its melting point is very sharply defined; however, the crystalline nature of the material and the effects of the methyl groups on its structure produce unique properties that permit the polymer chains to retain strength with a great deal of mobility over a wide temperature range above the materials softening point. This property makes it mobile enough to be forced mechanically into a container shape far below the melting point of the polymer. This results in strong containers that contain a great deal of residual strain. Heating a PP container that has undergone SPPF will result in the container reverting to the shape of the material before forming. The containers are somewhat stiffer and stronger than those produced by more conventional thermoforming close to the melting point of the material.

Catalyst Background for Ethylene and Propylene Polymers

Manufacture of PP along with the manufacture of PE has undergone a major change in the past 10 years with the introduction of metallocene catalysts. These catalysts produce higher yields of polymer with less catalyst required per pound of monomer and permit manufacture of a much more selective (narrower) molecular weight range in the finished polymer by varying reaction conditions. The catalysts also produce polymers with better stereoregularity, or polymers with less variation in branching or three-dimensional irregularities.

Metallocene catalysts are not new. They were first discovered in 1954 and were used to make PE in 1957. FINA and Exxon both began touting and commercially developing these materials in the 1980s. Actual commercialization did not take place until the 1990s. The catalyst actually consists of two different compounds, cocatalysts, to further increase their activity.

The reason for more selective molecular weight range and improved stereoregularity is found in the differences between Ziegler–Natta catalysts and metallocenes. Ziegler–Natta catalysts have multiple reaction sites on the catalyst particles. The multiple sites (three) on the catalyst produce different results when incorporating the comonomer into the polymer. Thus, the same catalyst can produce high molecular weight and low branching at one of the three sites while still producing low-molecular-weight and high-branching sections at another site on the catalyst particle. The third reaction site on the particle would produce medium molecular weight with moderate branching.

Metallocenes are very different. Metallocenes contain only one site that is active. This is referred to as a single site catalyst. Because they only have one type of active site on the catalyst particle with only one type of geometry, the polymer produced by these catalysts is very regular and incorporates the comonomer into the polymer in the same molar proportion it is added to the reaction.

The regularity of the polymer results in improved molecular weight control and better control of molecular weight distribution. The regularity permits the incorporation of higher amounts of comonomer into the polymer and reduces the amount of low-molecular-weight species. This change in polymer structure improves tensile and tear strength while reducing the total amount of low-molecular-weight extractables. These polymers have a softer feel than the same material produced with the Ziegler–Natta catalysts.

Another interesting property of metallocene catalysts is their ability to produce polymers not possible with Ziegler–Natta catalysts. Long-chain α-olefins can be incorporated into the polymer backbone to produce the effect of long-chain branching. This property translates into improved heat sealing while maintaining tight molecular weight control to avoid excessive extractables.

When first introduced into widespread commercial production in the late 1980s and 1990s, these catalysts provided the resin manufacturer with the ability to tailor the polymer to the customer's needs and requirements. The problem with this capability of producing multiple grades of the same material means significant problems in inventory management. Thus far metallocene catalysts produce materials that cover broad ranges of end use applications. It is possible that a very large end user (multiple rail cars per week or month) could get a polymer tailored specifically to their needs.

It is worth noting that PE and PP were the first materials produced by ionic addition polymerization on a commercial scale. The coordination catalysts, named Ziegler–Natta catalysts for the work of Karl Ziegler (Max Planc Institute) and Giulio Natta (Polytechnic Institute of Milan), revolutionized polymerization by permitting control of the polymerization process to a degree not previously attainable. It also permitted the polymerization of PP in the three different configurations based on the position of the methyl group. For their work, Ziegler and Natta received the Nobel Prize in Chemistry in 1963.

Polyvinyl Chloride

PVC, next to HDPE, is the most widely used plastic in pharmaceutical packaging (Figs. 30 and 31). It was first used in intravenous (IV) bags to replace glass bottles for blood products and for intravenous (IV) glucose and saline solutions. PVC possesses good fabrication flexibility, clarity, and low cost. It also has a long history in pharmaceutical packaging and as such is familiar across a wide range of science and manufacturing disciplines within pharmaceutical companies.

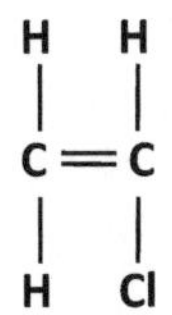

Vinyl Chloride

Figure 30 Vinyl chloride monomer.

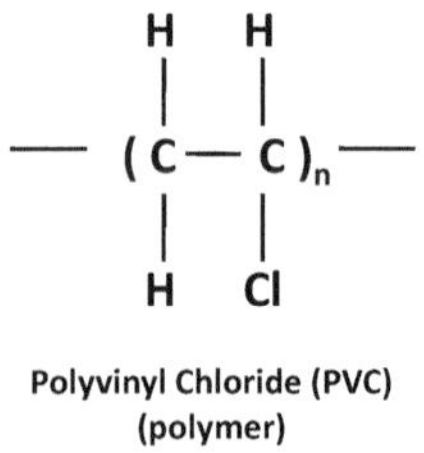

Figure 31 Polyvinyl chloride polymer.

PVC is manufactured using a free radical process at moderate pressures and temperatures (Fig. 32). A suspension process is the most favored for commercial production of PVC for packaging, with the remainder of the polymer produced by either emulsion or solution processes. PVC is a thermoplastic homopolymer.

PVC in its unmodified state (unplasticized) is very clear and stiff. It has low water vapor transmission (WVTR) somewhat comparable to that in LDPE. The material, because of the chlorine in its structure, is approximately 30% more dense than PE. Because the melting point and the decomposition point of the polymer are very close together, unmodified PVC is difficult to process. Stabilizers, primarily octyl tin compounds, are used to mitigate decomposition. Decomposition of PVC produces HCl, a very strong acid. Decomposition can occur at temperatures as low as 100°C. Stabilizers are used in PVC to make it viable for the manufacture of food and pharmaceutical packaging.

PVC (Fig. 31) displays a number of other characteristics that make it an excellent material for blister packs and for bottles. These characteristics include

- High flexural strength
- Chemical resistance

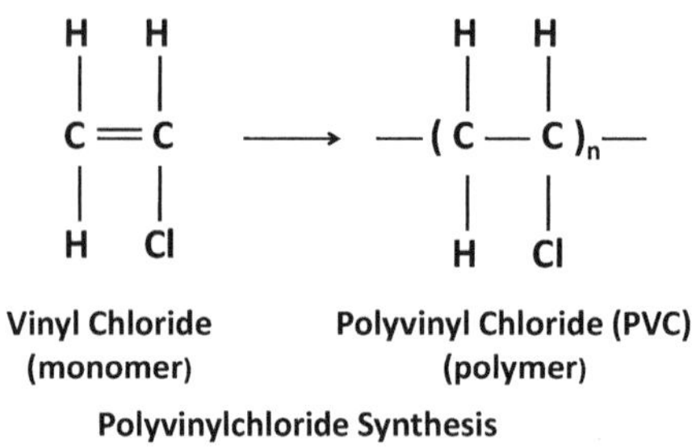

Figure 32 Polyvinyl chloride synthesis.

- Low permeability to oils, fats, and flavorings
- Easy coloring or tinting
- Low cost

PVC through the 1990s had virtually a 100% share of the market for the plastic component of blister films. PP and PE copolymer have begun to cut into this market share by providing a lower-cost alternative to PVC and an alternative that is chlorine free, an important environmental attribute particularly in a number of European countries and worldwide markets.

Films of PVC (Fig. 31) can be laminated with high barrier plastics such as PVDC or polychlorotrifluoroethylene (PCTFE), widely known by its trade name ACLAR®, to improve the blister's oxygen and moisture barrier (WVTR) properties.

PVC provides better clarity than HDPE and is used in bottles as an alternative when this property is desired. For opaque bottles, PVC absorbs fewer flavors and components, is a better odor barrier than HDPE, and presents a glossier appearance on the store shelf that is appealing to consumers in OTC applications.

One drawback to PVC is the wide array of additives used to modify the properties of the finished plastic. PVC without plasticizers has a relatively high T_g and is hard to manufacture into packaging. Plasticizer permits the formulation or compounding of PVC to produce a wide variety of physical properties and permits the packaging engineer to specify or dial in a very specific set of properties based on additive composition.

The polymer is miscible in a wide variety of additives and plasticizers. The highly polar nature of the PVC molecule and polymer give it an affinity for a wide variety of plasticizers and additives. The plasticizers increase the volume within the polymer matrix, producing an effect that permits the material to flow easily on a molecular level. Lubricants are also added to compounded PVC (PVC with plasticizer) to improve its ability to be formed into packaging. The ability to flow at low temperature without decomposition makes PVC adaptable to a wide variety of molding and film-making processes. PVC materials can range from stiff materials found in bottles and thermoformed packaging to soft and very flexible films.

PVC with small amounts of plasticizer displays good barrier characteristics. PVC with a mid-range amount of plasticizer added can produce a film with moderate barrier properties that retains its excellent resistance to fats and oils, making it a good material to package meat and other products containing fat and oil by-products. There is a wide range of plasticizers and other additives used to formulate PVC that are acceptable for both food and pharmaceutical packaging. PVC is also compounded with a number of impact modifiers. These materials modify the toughness of the material and enhance its ability to withstand a hard impact without cracking or breaking.

PVC (Fig. 31) has been dogged by intense scrutiny on health and environmental issues. This has led to regulations in countries and regions of the world banning the use of PVC or making it difficult to qualify for new packaging applications.

The first problems for PVC surfaced in the 1970s over the amount of residual vinyl chloride monomer in the polymer (Fig. 32). Workers involved in the manufacture of PVC developed a rare type of cancer that was linked to the vinyl chloride monomer. Thus, vinyl chloride was proven a carcinogen under conditions found in the bulk polymer manufacturing and bulk polymer packaging processes. During the polymerization process not all vinyl chloride monomer is converted into polymer. The monomer may remain trapped or unreacted in the polymer matrix. To eliminate the retention of vinyl chloride monomer the resin is subjected to multiple applications of strong vacuum to remove any residual monomer. Currently, the level of vinyl chloride monomer found in resins used to produce packaging is less than 10 ppb (parts per billion).

The second question surrounding PVC comes from the environment. The disposal of PVC raises a number of questions about safety. The most serious problems identified occur when the material is incinerated. Burning PVC produces HCl gas. This material in combination with carbon degradation products produced by incineration leads to the formation of chlorinated dioxins. These toxic chemicals are considered dangerous to the environment and are regulated in many locations.

A second environmental concern is the introduction of chlorine and the resulting chlorinated organic compounds to the environment.

The concerns over these two items and others relating to halogens in the environment have given PVC a very bad image in Europe and other parts of the world. As a result, companies have been substituting PET, PP, and cyclic polyolefin copolymer for PVC to avoid the negative environmental concerns. These materials provide the same performance characteristics as PVC when correctly adapted to the end use, without the perceived environmental problems so sensitive to many environmental organizations.

Polyvinylidene Chloride Copolymers

PVDC (Fig. 33) copolymers are known more by their trade or common name Saran that is a trademark of the Dow Chemical Corporation. PVDC can be produced as a homopolymer but has little commercial value in this state. The pure polymer decomposes at 205°C, while its melting point is in the range of 388°C to 401°C. PVDC produces dangerous HCl upon decomposition in much the same way that PVC decomposes. PVDC is impossible to melt process in its pure form.

PVDC is modified with comonomers such as vinyl chloride, various acrylates, but most often methyl acrylates, and vinyl nitrile to modify the polymer and decrease its melting point. The addition of the monomers reduces

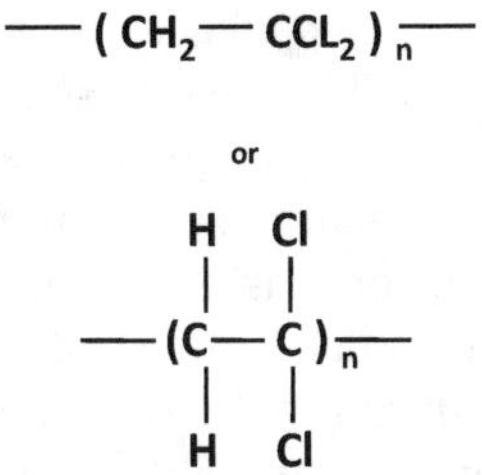

Polyvinylidene Chloride (PVDC)

Figure 33 Polyvinylidene chloride structure.

the melting point of the polymer below the decomposition temperature to a range of 140°C to 175°C, making the production of films and the coextrusion of the polymer as a barrier layer in packaging possible. The polymer normally contains heat stabilizers and plasticizers in a range of 2% to 10% to further improve its melt processing characteristics. Common plasticizers include dibutyl sebacate or diisobutyl adipate. The molecular weight range of most commercial PVDC polymers is 65,000 to 150,000.

PVDC was the first widespread barrier material used in packaging before the introduction of EVOH. The material is not only an excellent aroma and gas barrier but is also an excellent barrier to most organic liquids and water. The addition of comonomers to reduce the crystallinity of the polymer, making melt processing possible, also reduces the barrier properties of the material and increases its permeability.

The incorporation of comonomers inhibits the crystallization of the polymer and modifies the solubility characteristics of the polymer, permitting varying degrees of solubility in organic solvents. The most common monomers used to modify vinylidene chloride are vinyl chloride and methyl acrylate, and these modifiers make extrudable resins from PVDC possible. When the material is used to make solvent-based coatings for treatment of films and other plastics to impart barrier characteristics, the most common monomers used are acrylonitrile, methyl methacrylate, and methacrylonitrile. PVDC is supplied in a powder form, not in the form of resin pellets.

PVDC is one of the most effective barrier materials used widely in packaging. The material has excellent barrier properties to flavors, odors, gases, and moisture. The gas barrier it provides against oxygen is the most common reason it is found in pharmaceutical and food packaging.

PVDC is coextruded in multilayer structures to add barrier properties to the package in the same way EVOH is coextruded as part of a multilayer barrier structure. The expense of PVDC makes it uneconomical to use as a structural layer of the container, and coextruding it with less expensive materials make the overall container cost acceptable.

Coatings using PVDC as one of its components are applied to paper, cellophane, plastic films, and rigid plastic containers. The coatings add barrier properties to the substrate and in the case of paper and paperboard also impart grease resistance. Commercial coatings using this material most often use a latex base for PVDC resin dispersion into a film-forming medium. A dispersion grade of the PVDC polymer used in coatings reduces the need for and the amount of plasticizer and additives included during polymerization compared with the amounts of these materials used to make extrudable grades of the material.

Fluoropolymers

Fluoropolymers have become very familiar to most people under the trade name Teflon®, a registered trademark of Dupont (Fig. 34). In pharmaceutical packaging, an equally familiar name, Aclar®, a registered trademark of Honeywell Inc., is used to designate a modified fluoropolymer material primarily used for blister materials.

The chemical structure of these two materials is different, but they both derive superior properties through the inclusion of fluorine in the molecule. Teflon is polytetrafluoroethylene (PTFE) (Fig. 34). This is a very crystalline polymer that is extremely inert and has a very low coefficient of friction and excellent barrier properties. PTFE has a very low T_g of –100°C and a melt temperature of 327°C. Even at temperatures significantly above the polymers' melt temperature, the material retains a high viscosity, making it difficult to process into plastic components or shapes.

PTFE is not used in packaging, but is used in packaging equipment, most notably as a coating for heat-seal surfaces. This material permits the sealing of plastic without buildup of partially melted plastic on the sealing equipment. It is also found in equipment to lower friction between components, to protect

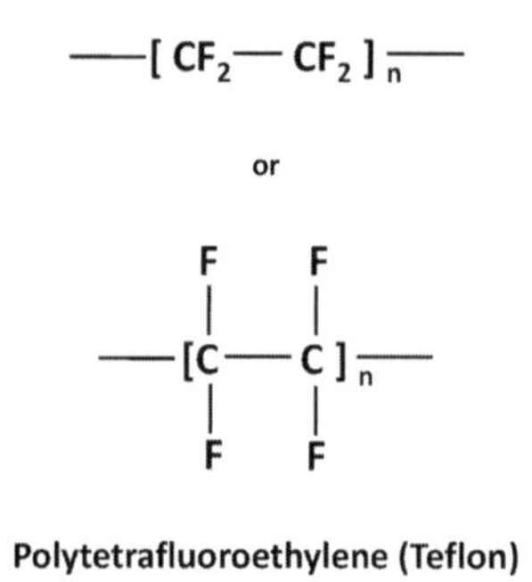

Figure 34 Structure of polytetrafluoroethylene.

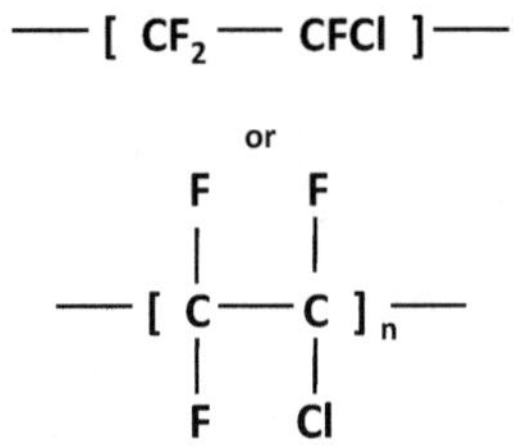

Polychlorotrifluoroethylene ACLAR®

Figure 35 Structure of polychlorotrifluoroethylene.

components from aggressive materials, or as an inert material that can contact tissue.

Aclar is PCTFE (Fig. 35). This material incorporates a chlorine atom into the polymer structure and is further modified by the addition of comonomer. These two modifications limit the crystallinity of the polymer and make the material semicrystalline. The chlorine atom in the structure is random in its placement along the backbone of the polymer. The material exhibits a T_g of 45°C and a melt temperature of 190°C. This material has found wide acceptance in pharmaceutical packaging as a water or moisture barrier material primarily in blister packaging. This material displayed the best moisture vapor transmission rate (MVTR) performance of any plastic film until competing materials such as cyclic polyolefin copolymers and aluminum cold-form blisters were introduced. PCTFE is also a good gas barrier, but the material is not often used in applications that take advantage of this property.

Polystyrene

Polystyrene (Fig. 36) is a plastic that has found many uses in packaging, including pharmaceutical packaging. The most common use of polystyrene is in

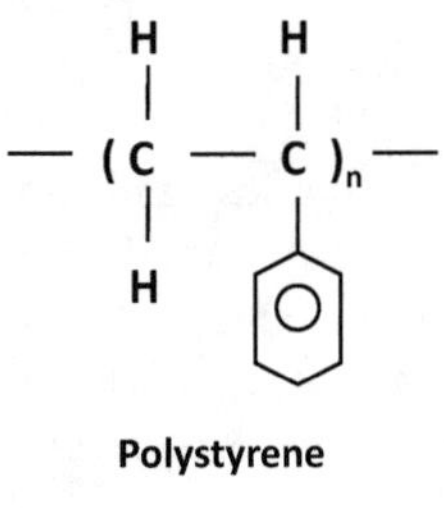

Polystyrene

Figure 36 Polystyrene.

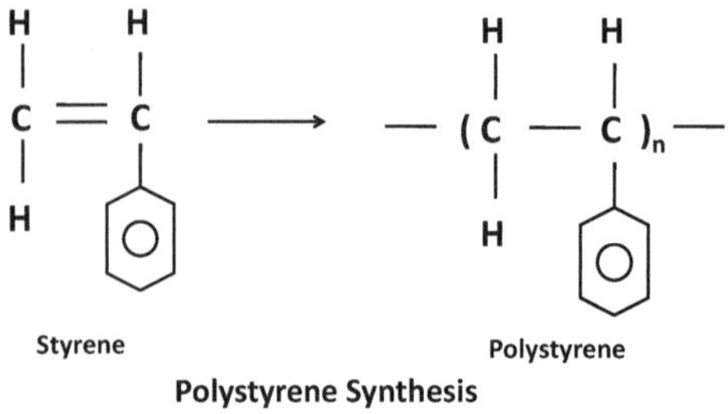

Figure 37 Polystyrene synthesis.

bottles for tablets and capsules. These are the bottles that are typically dispensed by the pharmacist when he or she fills a prescription. Polystyrene is also used for sample collection where its clarity permits the nurse or technician to easily determine the amount of fluid collected.

Polystyrene is a low-cost plastic. The polymer is atactic and cannot crystallize. This seems like a contradiction until you examine the structure of the monomer and note the benzene ring. The ring in the polymer structure resists rotation of the chain, and this chemical constituent and its effect make the polymer a stiff and brittle material (Fig. 37). Polystyrene has a density higher than PE or PP at 1.05 g/cm^3. Because the polymer is amorphous, it does not have a sharp melting point. This is observed in gradual softening of the material over a wide range of temperatures. The T_g of the material is 74°C to 105°C. Polystyrene softens at relatively low temperatures, making it unsuitable for pharmaceutical packaging requiring heat resistance. The material will flow like a liquid at 100°C (212°F). The material will also flow under stress, making it easy to thermoform or extrude. The material has poor WVTR characteristics and is a poor gas barrier. Compared with that of HDPE, the material has only one-tenth the moisture barrier and one-third the oxygen barrier properties.

Polystyrene is found in three different forms as a packaging material. The forms are as follows:

1. Crystal polystyrene (also called K resin, a play on the crystal word)
2. High-impact polystyrene (HIPS)
3. Polystyrene foam

Crystal polystyrene is a clear material found in cups, bottles, and any application where clarity and rigidity at normal temperature is required. The material is brittle and will break when subjected to a drop or sudden stress. Polystyrene materials are subject to brittle failure when subjected to sudden impacts and relatively low strains, as low as 3%. Crystal polystyrene is an atactic amorphous polymer that cannot crystallize.

HIPS is a graft copolymer. Polystyrene is modified with the addition of butadiene rubber. The addition of the rubber makes the polymer opaque and

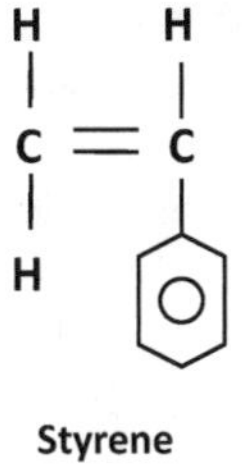

Figure 38 Styrene.

eliminates its gloss but greatly improves its impact resistance. HIPS is most often found in containers produced by thermoforming.

Polystyrene foam is standard polystyrene that has been modified during processing with a blowing agent to create a cellular material structure. Blowing agents typically used to create the cellular finished structure are carbon dioxide or a hydrocarbon gas. The cell structure can be either open or closed depending on the blowing agent and the conditions used to produce the polymer foam. Polystyrene foam is used as an insulating material for heat-sensitive products and as a cushioning agent. The foamed structure reduces the density of the material and reduces the material cost on a cost per package basis. Thermoformed trays for a variety of surgical supplies and other medical products use polystyrene foam.

OTHER STYRENE-MODIFIED COPOLYMERS

Styrene (Fig. 38) can be modified with a variety of other monomers and polymers to produce a chemical combination in the polymer that achieves multiple performance characteristics at costs not available with any one of the constituents.

As was noted earlier, HIPS is a graft copolymer of styrene and butadiene. Other polymers with packaging application include blocked copolymers. Blocked copolymers of styrene and butadiene can be sterilized by both γ-irradiation and ethylene oxide gas. The material is tough, shatter resistant, and somewhat transparent. All these properties are dependent on the size and ratio of the various blocks used in manufacturing the polymer. These materials are easily fabricated, particularly by thermoforming, making them a good choice for a wide variety of secondary packaging applications such as trays, holders, and other noncontact packaging applications.

Styrene–butadiene may also be blended with acrylonitrile polymers, producing acrylonitrile–butadiene–styrene (ABS) plastics. This blend is an attempt to overcome a number of physical shortcomings inherent in styrene or styrene–butadiene copolymers. The shortcoming of all styrenic copolymers in packaging applications is the need for HIP. Accidental impacts from dropping a container or the residual stress in the molded styrenic lead to environmental stress cracking or crazing and cracking in a finished package or assembly, all of which are unacceptable. Styrene and butadiene also have poor weatherability characteristics, and although this is not a prime concern in packaging, it needs to be noted as one of the reasons acrylic monomers are added to styrenic copolymer blends. ABS plastics attempt to overcome the shortcomings of styrene primarily and the shortcomings of butadiene to produce materials with a wide variety of uses. Each of the different materials contributes a different set of properties:

- Acrylonitrile—provides chemical resistance, higher temperature capability, weatherability, and resistance to creep.
- Butadiene—prevents crack formation, can store energy to resist impact failure, and through yield deformation minimizes crack propagation.
- Styrene—provides low cost, high modulus of elasticity, and excellent thermal-processing characteristics.

Styrene–butadiene polymers may also be modified with olefin and acrylate monomers to further enhance specific properties, particularly impact resistance. These materials are relatively low cost, and the same properties can be obtained at lower cost from other polymers suitable for packaging.

Polyamides (Nylon)

The first polyamide material introduced to the public was nylon (Fig. 39). It was the first revolutionary polymer created by Wallace Carothers of Dupont in the late 1930s. Nylons are condensation polymers produced by reacting two different monomers, a dibasic acid and a diamine. As the two monomers react, water is the by-product of the reaction the same by-product observed in polyester reactions.

Nylons are typically clear materials that are used widely in engineering applications where their excellent mechanical properties make them suitable for a wide variety of applications over a wide temperature range.

In packaging, polyamide polymers provide good puncture resistance, temperature stability, impact strength, chemical resistance, and good barrier properties to gas, oil, and aromas. Nylon is a hydrophilic polymer that can absorb

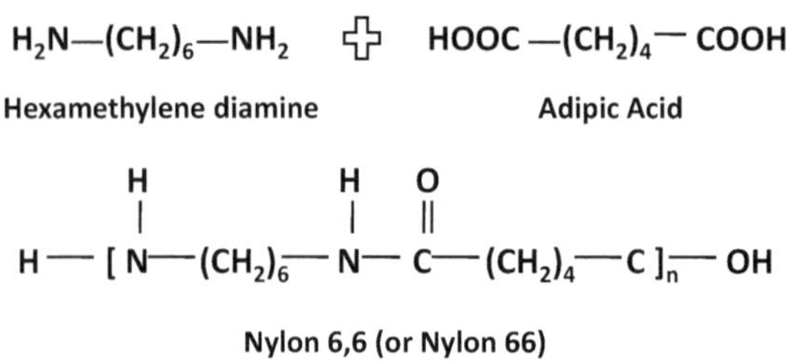

Figure 39 Nylon produced from a dibasic acid and a diamine.

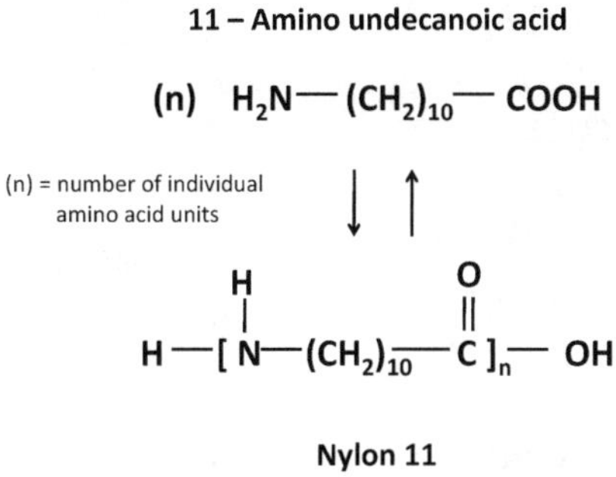

Figure 40 Nylon produced from the condensation of amino acids.

as much as 6% to 8% of its weight in water. As nylon absorbs water, its barrier characteristics decrease. When nylon is employed as an oxygen barrier material, it is used in a coextrusion (Fig. 24), usually with a polyolefin on both sides of the nylon layer to protect it from moisture.

The convention for naming nylon is based on the number of carbon atoms contained in each of the monomers. In nylons produced by the condensation of a diamine and a dibasic acid, the number of carbon atoms in the diamine is always first. To illustrate this, hexamethylenediamine is reacted with adipic acid to produce nylon 6,6 or nylon 66 (Fig. 39).

When nylon is produced by the condensation of amino acids (Fig. 40), the naming convention contains only one number. For example, the nylon produced from 11-amino undecanoic acid is named nylon 11. Amino acids contain both amine and acid functional groups.

Nylon is widely used as a film whose properties are modified by controlling the amount of crystallinity in the film and by controlling the rate at which the film is cooled or quenched. The faster the film is cooled, the lower the amount of crystallinity. The decrease in crystallinity produces a film that can be thermoformed more easily and a film more transparent than one of higher crystallinity. Nylon film can be cast or oriented. Biaxially oriented nylon film displays good mechanical properties and good barrier characteristics. Blow molding of nylon for containers also produces this orientation effect in the container structure.

Pharmaceutical end uses for nylon include disposable medical device products where nylon is a structural component and as a coextruded barrier layer in a multilayer structure. The nylon layer provides the gas barrier.

Polyester

Polyesters are another condensation polymer used widely in pharmaceutical, food, and beverage packaging. They are produced by the reaction of a diacid or a diester with a glycol (a material containing two –OH groups or alcohol groups on

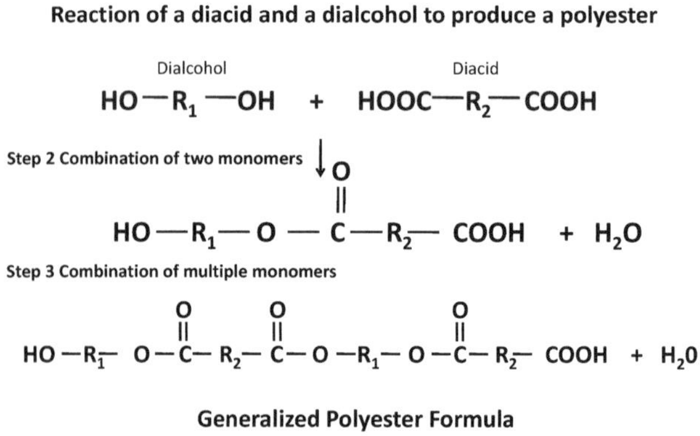

Figure 41 Generalized polyester formula.

the monomer) (Fig. 41). The polymer consists of a series of ester linkages in its chain. They can be both thermoplastic and thermosetting based on their chemical components. The most common polyester used in packaging is PET. It is primarily used in water and soft drink bottles because of its strength, clarity, and CO_2 barrier properties. Other polyesters common to pharmaceutical packaging include PEN and glycol-modified polyesters (PETG).

Polyethylene Terephthalate

PET (Fig. 42) has become the most widely used material for carbonated soft drinks, water, liquor, or distilled spirits and in pharmaceutical packaging for cough syrups and custom containers for a wide variety of liquid products.

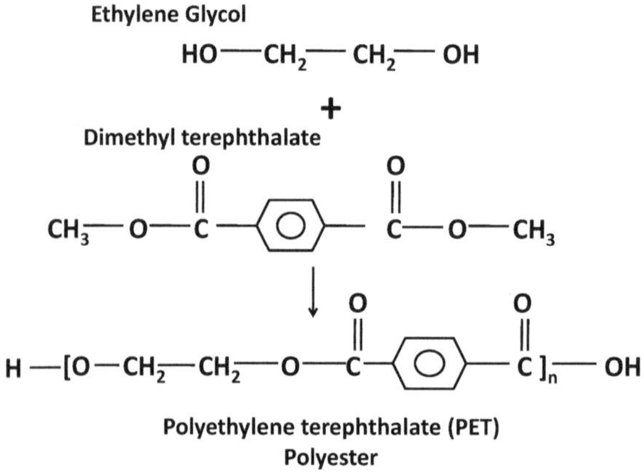

Figure 42 Reaction of ethylene glycol and dimethyl terephthalate to produce polyethylene terephthalate.

Reacting ethylene glycol monomer with either terephthalic acid or dimethyl terephthalate produces PET (Fig. 42). The condensation reaction produces PET and water or PET and methanol depending on whether the acid or ester material is used as the comonomer. The polymer as produced via normal condensation reaction is too low in molecular weight for meaningful use in packaging. A second-stage manufacturing process called solid stating is employed to increase the polymers molecular weight through the use of high temperature and vacuum to continue the polymerization reaction. The high temperature promotes continued reaction, while the vacuum removes the water produced in the reaction. Water at high temperature can reverse the reaction and attack the growing polymer chains, reducing their molecular weight. This process, which is time consuming, is being phased out with newer processes that produce the higher-molecular-weight grades of PET directly in the reactor. The newer processes also replace antimony catalysts with titanium catalysts.

Processing or converting PET into packaging requires understanding of the polymers' properties. The polymer has very low melt strength. This means the material cannot be heated beyond its T_g and formed into a container because the glass transition and the melting point of the polymer are very close together and sharply defined. An example would be an attempt to extrusion blow mold PET. As the material is heated and begins to flow, it loses its intermolecular strength. This loss of strength means the material exiting the head of the extruder would be like water, or at best more like hot syrup, and would not maintain enough strength to be stretched and placed in a mold and blown to a desired shape. Specialty grades of PET, which utilize comonomers and additives, have been produced to make the polymer capable of standard melt processing, but these modifications are expensive and these materials remain laboratory curiosities.

PET is typically produced into bottles in a two-stage process. In the first stage, the material is extruded and injection molded into a shape called a preform. Injection molding overcomes the problem of melt strength and capitalizes on it. The natural flow of the polymer in a very low viscosity state permit the material to flow and fill the mold quickly before the cooling process reduces the polymers mobility. The injection molded preform contains the threads used by the closure on one end and a test tube–looking shape below the threads. The PET in the preform is cooled into this shape and then either immediately moved to another station that blow molds the finished bottle, or the preform is stored or shipped to a location where it is reheated and then mechanically stretched and blow molded into a bottle. Almost all soda bottles and all high-volume custom bottles are produced this way in what is described as a two-stage molding process. It is also referred to as reheat stretch blow molding, or injection.

PET films and sheets are produced by extrusion of the hot material onto a cooling or quenching roll. Even though the material has little melt strength, it flows naturally by gravity onto the supporting surface of the roll and begins cooling and gaining mechanical strength. For biaxially oriented film the partially cooled material is mechanically stretched in two directions. The first is in the same direction it exits the extruder, with the takeoff slowly increasing in speed compared with the speed of the cooling roll. The second direction of orientation is 90° opposed to the first direction of stretching and takes place concurrently with the stretching going on in the machine direction. Orientation produces a strong tough film.

PET is typical of many types of polyester that are sensitive to hydrolysis and can depolymerize if water is present at elevated temperatures during its processing into containers. Water, a natural by-product of the condensation reaction, acts as the reactant causing chain break and depolymerization when it is present in large amounts in the melted polymer during the extrusion steps in plastic container manufacture. Its presence reverses the effects of solid stating and reduces molecular weight quickly and significantly. PET pellets being prepared for extrusion must be dried to extremely low moisture levels, less than 0.005%, to avoid a reduction in the final molecular weight of the polymer and to avoid the reduction in performance caused by lower-molecular-weight polymer.

Most PET used in bottle manufacture also contains a small amount of comonomer. The comonomer is present to retard or resist the formation of spherulites or small crystals that detract from the clarity of the finished bottle.

PET Physical Forms in Packaging

PET is unique in that all three physical states of the material are used in packaging. The three forms are microcrystallized PET in bottles, amorphous PET (called APET) in clamshells and thermoformed applications, and CPET used for dual ovenable products. As mentioned earlier, PET is found in packaging for food, distilled spirits, soft drinks, pharmaceuticals, medical nutritionals, and OTC pharmaceutical products.

Clear Microcrystallized PET Bottles

The first form, microcrystallized PET, is found in bottles. As the bottle is blow molded, it is biaxially oriented, and this process forms microcrystals in the bottle sidewalls to enhance gas barrier and other physical properties while remaining crystal clear. These same properties have contributed to the rapid rise in the use of PET bottles for "custom" applications, which include a wide variety of pharmaceutical applications.

PET is the material of choice to replace PVC bottles, and conversion to PET or the initial application of a PET bottle in a new product results in a container with greatly improved performance properties compared with PVC. The barrier in PET bottles is very good but not equal to a coextruded barrier structure using EVOH and is not adequate for the most stringent applications particularly in small sizes. In small sizes, the surface-to-volume ratio becomes a limiting factor in the polymers barrier performance characteristics. Surface-to-volume ratio means the smaller the bottle the larger the surface area of the container compared with the volume of liquid it contains. Different products including pharmaceutical products with gas sensitivity or pressurized products have greatly reduced shelf life because the large surface volume permits rapid ingress or egress of a relatively large amount of the gas compared with the amount of the product. Coatings and blends have been the favored routes to enhance barrier properties of PET; however, all these materials degrade the performance of PET when it is recycled or make the material unsuitable for recycling. This problem is a prime consideration when one considers the sustainability of packaging and the design of a new container that has a potentially large unit volume with the capability to contaminate a large amount of recycled material.

Amorphous PET

APET is another form of PET that has arisen to take advantage of the clarity and strength of PET and to address recycling issues and legislation in parts of the world that prohibits or discourages the use of PVC or materials containing halogens. Most APET is found in thermoformed containers. The material is modified slightly by copolymerization of a monomer that does not crystallize and resists the natural crystallization that is a characteristic of this polyester. APET is manufactured using standard bottle grade PET. Even small amounts of crystallization create haze in the finished package.

Crystallized PET

CPET has found a unique niche in packaging. By crystallizing PET, its heat resistance can be increased substantially to be better than 500°F. This combined with the good low-temperature performance of the polymer makes PET a low-cost material for containers that must withstand sterilization temperatures, or, in the case of consumer products, are reheated in conventional cooking (baking) ovens. CPET replaced aluminum for frozen foods and in "TV Dinners" during the mid-1980s through work done by Campbell Soup Company. The polymer is transparent to microwaves and ideal for use in a microwave oven; it is relatively low cost, and because of its crystallized temperature performance, TV-dinner customers who heat their meals in conventional ovens can use it. This work resulted in the development of an entirely new class of plastic container. Once introduced, the high heat-resistant characteristics of the material and the ability of the material to be thermoformed have seen its adoption into pharmaceutical uses where autoclave temperatures and other high-temperature steam sterilization processes are employed. For medical devices, it is used as trays with a porous membrane seal, usually Tyvek or its equivalent in sterilization operations.

CPET is always a nucleated material. The addition of a nucleating agent provides the "seeds" needed for rapid formation of the relatively large crystals found in its structure. PET can be crystallized in two thermal directions, the first by reheating from ambient temperature to a point above its glass transition where crystals begin to form and the second by inducing crystallization as the material cools from the melted state.

PET Films

PET films are used in a wide variety of pharmaceutical packaging and food packaging. Biaxial orientation of the film when it is manufactured induces crystallinity into the film, making it heat resistant. These materials are best known by the trade name Mylar®. The film serves as the structural layer for clear sterilizable pouches and boil-in-bag applications. It is also laminated or coextruded as one layer of pouch material construction to provide a puncture-resistant layer in pouch material.

Glycol-Modified Polyester

A common modification to PET is the introduction of an additional glycol and additional diacid or both into polymerization to modify the finished properties of the polymer. These materials reduce the polymer's crystallinity and increase its melt strength, making it possible to process the materials on more conventional plastic-manufacturing equipment. It also makes thermoforming much easier. The modifications also increase the polymers' impact resistance by reducing the amount of crystallinity in the structure.

For medical and other food packaging the most common PET-modified material is PETG. PETG is a copolymer that introduces cyclohexane dimethanol into the monomer mix with ethylene glycol and terephthalic acid. The resulting polymer is clear and colorless with improved melt strength and impact resistance. This material is commonly thermoformed into trays and clamshells for medical devices. The material can be sterilized with ethylene oxide and γ-radiation. The material is far easier to close and seal with a film by heat sealing when compared with PET and, particularly, CPET.

One problem with the polymer modification by this glycol comes from recycling. PETG has a much lower softening point than that of PET and does not require drying before processing. When the material is in the PET recycling stream, it can soften and cause large agglomerates in the dryer used to remove water prior to extrusion.

Polyethylene Naphthalate

PEN (Fig. 43) is another polymer in the polyester group with pharmaceutical packaging capability. The material is another condensation polymer similar to all polyester materials. The material is produced by a condensation reaction of ethylene glycol and naphthalate dicarboxylate (NDC). PEN is a material approved by the FDA with much better barrier properties, tensile strength, and flexural modulus than that of PET. This material offers a significant

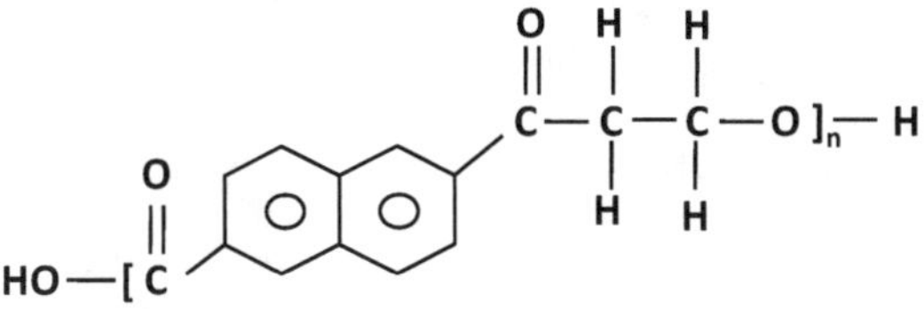

Figure 43 Polyethylene naphthalate (PEN).

improvement in barrier properties to water and oxygen in the order of four to five times better than PET. It has half (150%) the flexural modulus and is 35% higher in tensile strength. Other properties of PEN that are significant improvements over PET include it ability to block UV light and its resistance to UV degradation. It has a greater resistance to hydrolysis than PET. The major problem with wide-scale adoption and use of PEN is its price, which ranges between three and four times that of PET.

PET and PEN can be blended in an extruder, and the resulting material is a highbred blend that enhances the properties of both materials, because the transesterification reaction continues to take place in the extruder. Because PEN is expensive, extensive work was done to determine if blending it with PET could make useful materials. PET and PEN blends/copolymers fall into two groups. The low and high material blends containing either 85% PET or 85% PEN, respectively, are the only combinations with useful packaging properties. The intermediate blends of these two materials cannot crystallize and have poor performance properties when compared with the two high-percentage blends. The two materials, PET and PEN, are immiscible homopolymers that require the transesterification reaction in the extruder along with some specialized mixing techniques to produce a uniform blend that can be molded into useful packaging.

The cost of the improved properties, even at the relatively low levels (15% or less) of PEN, has limited the use of the polymer. PEN and PET have very different stretching or free-blowing characteristics. This difference means that to produce acceptable PET/PEN bottles specialized preform and bottle tooling is required. This added capital cost, the higher cost of the material, and the lack of specific markets, including pharmaceutical packaging, have kept this promising material from large volume use and acceptance.

Polycarbonate

PC (Fig. 44) is an amorphous polymer with excellent clarity that can be easily processed by injection molding, thermoforming, extrusion, and blow molding. The polymer is relatively heavy, having a density of 1.2 g/cm^3, is very rigid, and possesses good impact strength and dimensional stability, heat resistance, and

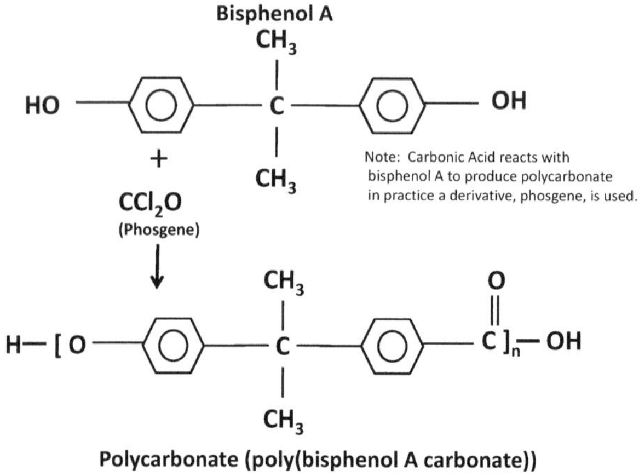

Figure 44 Poly bisphenol-A carbonate.

reasonable low-temperature performance. The polymer is primarily comprised of *bis*-phenol-A as the only building block. In fact, the polymer can be referred to as poly (bisphenol-A carbonate).

PC is resistant to alcohols, water, aliphatic hydrocarbons, and dilute solutions of ethanol. Alkalis, acetone, and other more polar functional solvents attack it.

PC is FDA approved for food contact. PC has been used for cells (analytical sample holders) in diagnostic instruments, including blood analyzers to provide a disposable container with clarity and chemical properties that can be placed in spectroscopic analysis instruments. PC is capable of withstanding sterilization by γ-irradiation, electron beam, and autoclave. This makes it an ideal and tough substance used in medical devices, particularly ones that require some limited reuse. It also makes PC films good materials for packaging medical devices and other pharmaceutical items.

PC in pharmaceutical packaging is limited primarily by its cost.

Polyurethane

Polyurethanes (Fig. 45) are highly specialized plastics made from the reaction of a diisocyanate with a glycol(s) to form polyurethane.

These polymers are very specialized in their use in pharmaceutical packaging. These materials, which possess a number of properties produced by the wide variety of glycols and diisocyanates available for polymer synthesis, find their way into specialized and demanding pharmaceutical packaging applications. If the reactants produce CO_2 as part of the polymerization reaction, polyurethane foam will be produced. The foam can be a flexible open-celled material or a closed-cell rigid material depending on the processing conditions.

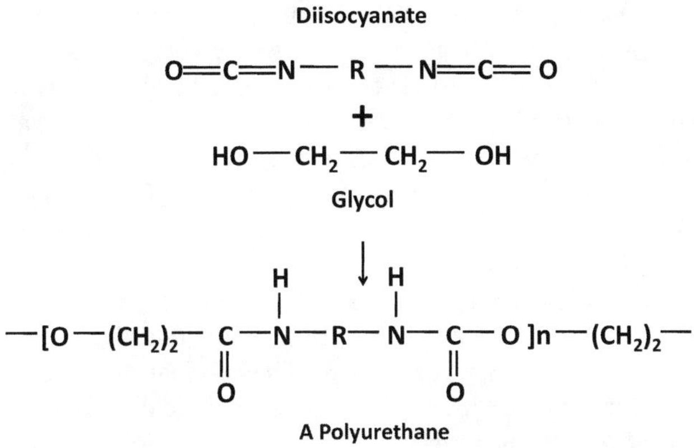

Figure 45 The reaction of a diisocyanate and a glycol to produce a polyurethane.

These materials have been used as a replacement for cotton dunnage in bottles because they resist the absorption of moisture in the same way as cotton. Polyurethanes are also used as films with varying degrees of permeability to the active compounds used in transdermal drug delivery.

Acrylonitrile Polymers

Acrylonitrile polymers (Fig. 46) are a specialized group of materials. The acrylonitrile polymer itself is not suitable for fabrication into packaging because it cannot be melt processed. The polymer degrades at 220°C, which is below its T_g, but the polymer retains too much stiffness and resistance to movement to be fabricated at temperatures below the degradation value.

Acrylonitrile has very strong intermolecular forces resulting from the very polar carbon–nitrogen bond in its structure. It has excellent gas barrier properties. To overcome the problems with processing, the material is polymerized with other materials.

The most common modifications of acrylonitrile are with styrene and with various ratios of methyl methacrylate and butadiene.

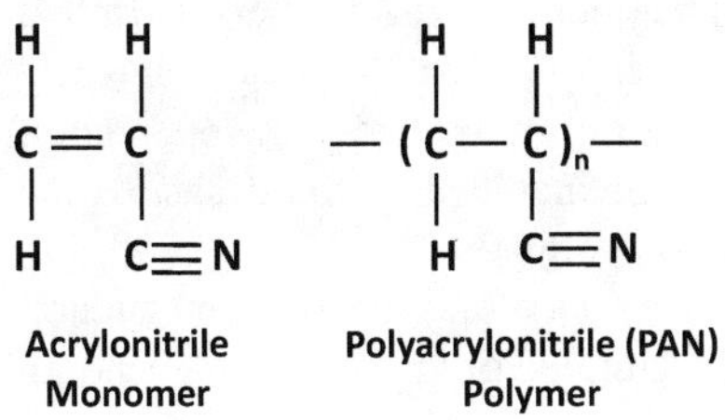

Figure 46 Polyacrylonitrile (PAN).

Styrene acrylonitrile material, usually produced in a weight ratio of 3 to 1, respectively, results in a material with heat resistance, gloss, good tensile and flexural strength, and chemical resistance. The material is found in bottles, overcaps, closures, and spray nozzles. The material does not have good gas barrier characteristics because of the high percentage of styrene in its structure. In order to improve gas barrier of this type of copolymer, the percentage of acrylonitrile would need to be increased to a 70% level, with the remainder being styrene.

Probably the most important acrylonitrile material is Barex®, a material produced by BP Chemicals. This material is a terpolymer of acrylonitrile, methyl methacrylate, and butadiene, with high nitrile content. The acrylonitrile content is in a 75/25 ratio to the methyl methacrylate used to make part of the polymer, which is then polymerized onto a nitrile rubber backbone. This material has excellent barrier properties and at one point during the 1980s was a prime contender for use in carbonated soft drink bottles. It was superseded by PET because of FDA review and concerns that were later removed.

These materials can be blow molded or injection molded, and they can be extruded into film and sheet materials. They are approved by the FDA for direct food contact and can be used in pharmaceutical applications in the same way as all approved food contact materials. They are found in many rigid containers for chemicals, cosmetics, and spices. The material can be sterilized by γ-radiation or ethylene oxide gas, two primary methods of sterilizing pharmaceutical and medical device products.

Rubbers and Elastomers

Elastomers are a group of polymers usually referred to as rubber (Fig. 47). In fact, they are highly formulated materials that may contain 2 to 10 different raw materials to achieve the properties desired for the specific pharmaceutical end use application (12). As a general definition, elastomers are polymers that can be stretched a minimum of twice their unstretched length and then return to their original length when the force is removed. Almost all other materials, glass, metal, and other polymers can only be stretched over a much more limited range or not at all.

Elastomers are primarily used as stoppers, the name the pharmaceutical industry uses for the closures on parenteral containers. Elastomers permit a hypodermic needle to enter a container, and when the needle is removed, reseal the container. They also mold or conform to small irregularities in the opening of a glass parenteral container, permitting the container to be closed and sealed tightly. The property that makes this possible is compressibility.

Elastomers are materials made up of a variety of synthetic and natural rubber compounds. They may be classified as saturated or unsaturated depending on the frequency of the double-bond content along the polymer chain. Unsaturation determines their physical and chemical properties. Materials that are

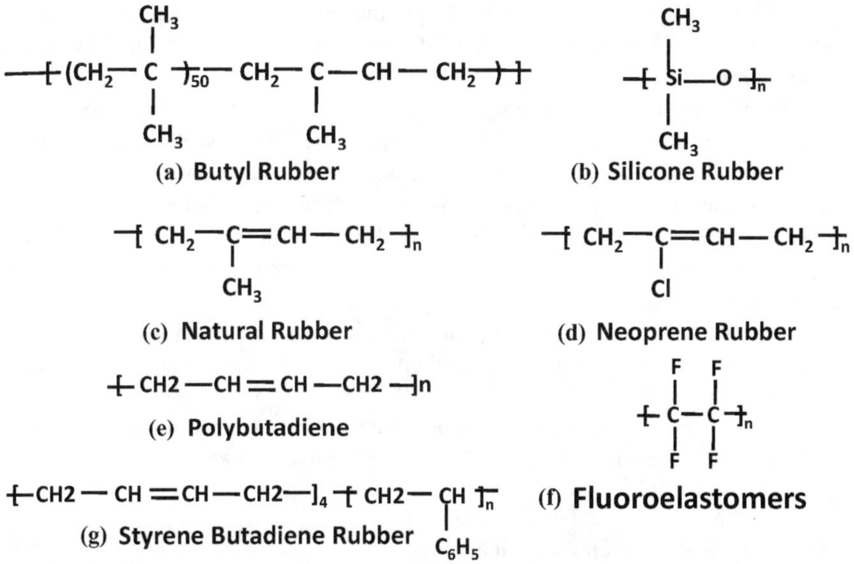

Figure 47 Common rubbers used in pharmaceutical packaging. (**a**) Butyl rubber; (**b**) silicone rubber; (**c**) natural rubber; (**d**) neoprene rubber; (**e**) polybutadiene; (**f**) fluoroelastomers, and (**g**) styrene butadiene rubber.

highly unsaturated have a more "rubbery" feel or characteristic. The unsaturated elastomeric material has rubberlike mechanical properties but loses its resistance to oils, solvents, and water. For pharmaceutical packaging, elastomers typically possess the following properties:

- Coring resistance (the ability to resist fragmentation when penetrated by a needle)
- Solvent resistance
- Resistance to radiation and ozone
- Resistance to interaction with the packaged components
- Impermeability to gas and moisture
- Resilience

Elastomers used in pharmaceutical applications are typically classified as natural or synthetic. Natural elastomers contain rubber extracted from rubber trees; synthetic rubbers are polymers derived from petrochemicals (Fig. 47).

Common rubbers used in pharmaceutical packaging are as follows:

- Butyl rubber
- Chlorobutyl rubber
- Natural rubber
- Silicone rubber

There are a number of other rubbers that may be used in pharmaceutical packaging. The four materials in this list represent most of these materials used as parenteral stoppers.

Butyl rubber and chlorobutyl rubbers have the majority share (~80%) of the parenteral closure market. These materials offer the best resistance to permeation by oxygen and water vapor. These materials are not used alone; for example, when the stopper must withstand multiple penetrations by a needle, natural rubber is included in the chlorobutyl formulation to resist coring. When a drug contains mineral oil, a nitrile rubber or neoprene rubber is part of the formulation.

Silicone rubbers are used in pharmaceuticals but have limited application. They are prone to tearing, making them unsuitable for most mechanically demanding applications.

Elastomers are formulated products that contain a large number of other substances to make them useable. A list of these agents includes

- Vulcanizing agents (sulfur, peroxides)
- Antioxidants (phenols, amines)
- Cure accelerators (amines, thiazoles)
- Activators (zinc oxide, stearic acid)
- Plasticizers (phthalates)
- Lubricants (oils)
- Fillers (carbon black, silicates)
- Pigments (inorganic oxides, titanium dioxide, carbon black)

All these materials must be considered when formulating or choosing a rubber material as a stopper for a parenteral product. These materials fall under the same FDA regulations (21 CFR 175, 177, 178, 182, 184, 185) as all other packaging materials. One point to highlight here is the choice of color in these components. Many drugs when packaged for ophthalmic use are color-coded using the closure or the stopper to help guide the physician and to avoid errors. Producing the proper color, while using all the other ingredients needed to make the material functional, is sometimes difficult.

Most rubber materials are cross-linked products. The materials listed above are mixed in a roll mill or some other type of mechanical mixer that breaks the basic components into small fragments and produces a uniform dispersion of all the ingredients. The mixed material is placed in a heated mold, where heat and pressure promote polymer cross-linking and "cure" the rubber. Radiation, both γ and electron beam, can be used for the curing process, but their application is rare. The curing process creates bonds in three dimensions that produce the required chemical and mechanical properties. The mix that is placed in a mold to produce a stopper or packaging component is a viscous liquid, soluble, and inelastic. Through the curing process it becomes the strong and tough rubber material needed for parenteral closures and other packaging components.

Following molding the parts are trimmed and washed to remove residual materials that may have bloomed to the surface during molding.

Elastomers may be surface treated with chlorine to create a shiny glaze on the surface, or they may be coated with a variety of other materials, most often silicon oils, to reduce their coefficient of friction. Elastomers may undergo a variety of extraction techniques to eliminate residual materials. This is most often done by autoclaving the rubber part.

Elastomers behave like other plastics in that they are not totally inert. They all display some degree of permeation and some degree of sorption that may cause a problem with the drug product being packaged. The drug product may have the ability to leach residuals and low-molecular-weight fragments from the elastomer.

The most common way permeation of these materials is overcome is through the use of very thick cross sections to eliminate the ingress of oxygen into a container.

SUMMARY

The variety of pharmaceutical packaging materials is extremely large. Through the proper choice of material, driven by an understanding of the drug and the end use requirements, the packaging engineer has a wide variety of options available.

Plastics are the preferred materials for new products, but metal and glass will always be needed to provide protection for products that interact with plastics.

The large selection of plastic materials and the equivalence of many of the materials open the opportunity to take packaging in new directions. The material choices permit the design and inclusion of new package features not available today. They may present the consumer with options that could not be obtained in traditional packages. They also make possible multiple component products that must be mixed at the time of use. All these opportunities make an understanding of how to properly use materials to package a new product a very exciting field.

Biological products are just beginning to become large-scale pharmaceutical products. These drugs have in the past relied on more traditional materials, primarily glass, to provide protection. As more understanding of these materials and their inherent physical properties are understood, the advantages of plastic packaging will be incorporated in their presentation.

FURTHER READING

The National Formulary, NF 23. Prepared by the Council of Experts and Published by the Board of Trustees. Rockville, MD: United States Pharmacopeial Convention Inc., 2005.

REFERENCES

1. The United States Pharmacopeia, USP 28. Prepared by the Council of Experts and Published by the Board of Trustees. Rockville, MD: United States Pharmacopeial Convention Inc., 2005.

2. Gruenwald G. Plastics—How Structure Determines Properties. Munich, Vienna, New York, Barcelona: Hanser Publishers, 1993 ISBN 3-446-16520-7.
3. Hernandez RJ, Selke SEM, Culter JD. Plastics Packaging—Properties, Processing, Applications, and Regulations. Cincinnati, Munich: Hanser Gardner Publications, Inc., 2000 ISBN 1-56990-303-4 (Hanser Gardner) ISBN 3-446-21404-6 (Hanser).
4. Berins ML, ed. SPI Plastics Engineering Handbook of the Society of the Plastic Industry, 5th ed. New York: Van Nostrand Reinhold–Chapman & Hall, 1991.
5. Crank J. Mathematics of Diffusion. London: Clarendon Press, 1975.
6. Vieth WR. Diffusion in and Through Polymers, Principles and Applications. Munich, Vienna, New York, Barcelona: Carl Hanser Verlag, Hanser Publishers, 1991.
7. Moskala EJ, Melanie J. Evaluating Environmental Stress Cracking of Medical Plastics, Medical Plastics and Biomaterials Magazine. Original publication date 1998. Available at: www.devicelink.com/mpb/archive/98/05/001.html.
8. Progelhof RC, Throne JL. Polymer Engineering Principles, Properties, Processes, Tests for Design. Munich, Vienna, New York, Cincinnati: SPE Books Hanser/Gardner Publications, Inc, 1993.
9. Information from Dow Plastics, Chemical resistance of CALIBRE* POLYCARBONATE. Available at: www.gallinusa.com/pdfs/polycarb.chemicalresistance.pdf.
10. Dominghaus Hans. Plastics for Engineers: Materials, Properties, Applications, Munich, Vienna, New York, Barcelona: Carl Hanser Verlag, Hanser Publishers, 1993.
11. Rees RW, Vaughan DJ. Polymer Preparation. American Chemical Society. Division of Polymer Chemistry. 1965; 6:287–295.
12. Jenkins WA, Osborn KR. Packaging Drugs and Pharmaceuticals. Lancaster Basel: Technomic Publishing Inc., 1993.

7

Medical Device Packaging

INTRODUCTION

Medical devices are an extremely broad category of pharmaceutical equipment that requires a very broad range of packaging materials and processes. Medical devices come in all sizes, ranging from the very large to the very small. They span room-filling imaging devices for examination and diagnosis of diseases or conditions within the body to small highly specialized implantable devices that repair or replace malfunctioning parts of the body. Devices may cure diseases, repair chronic conditions, or supplement and, in many cases, replace body parts that can no longer manage a particular body function or are simply worn out. Medical devices may be hybrids, combining a pharmaceutical drug product with a delivery device to release a drug in a specific area of the body or the device may be a radioactive implant designed to deliver a specific dose of radiation to cancerous tissue.

Without a doubt, medical devices span a wide array of ideas to improve the human condition and are the most varied assortment of regulated objects in the medical field. They are the product of many creative entrepreneurs and companies that see how to use a machine, a manufactured mechanical device, an electrical device, or some combination of technologies to examine or fix someone. Tongue depressors, contact lenses, magnetic resonance imaging (MRI) scanners, pacemakers, stents (including new drug-eluting models), replacement joints, and X rays are just a few examples of the breath of ideas this category covers.

Medical devices have become commonplace in our accepted methods for analyzing and treating disease and chronic conditions. Today, no one questions the need for a pacemaker to regulate and improve function for the heart, dialysis units for supporting failed kidneys, or stents used to hold open clogged or restricted arteries or veins. Everyone accepts the need for an X ray to determine if a bone is broken, and far more advanced MRI, positron emission tomography (PET), and radiologic imaging for noninvasive (nonsurgical) examination and determination of disease. Blood analysis is commonplace and not only determines if a specific disease is present, it also provides a complete overview of a patient's health. Medical devices are the home of clever inventors, engineers, entrepreneurs, and others who create, design, and market unique and novel equipment for medical use. They range from academics and industrial researchers to businesspersons who find, develop, and bring to market equipment or equipment hybrids that provide significant benefit to patients.

Medical device packaging is complicated because it covers a broad breath and depth of requirements using a wide variety of packaging forms. Just as the array of devices is broad and varied so too is the packaging designed and developed to protect it. Medical device packaging must accommodate a system to sterilize both the product and package, maintain that sterility through distribution, and provide easy access to the product with maintenance of sterility directly to the operating room or treatment area for the patient. It is common to find packaging that permits a medical device to be opened and moved into a sterile operating room while maintaining sterility of both environment and device.

Medical device packaging uses every form of packaging available. Probably the most common medical device package is the pouch, but primary packaging in bottles, cans, clamshells, and thermoformed plastics are all commonplace. These packaging forms are used for everything from the instruments in the operating room and treatment kits containing consumables (all the pieces needed for a procedure that are discarded after use) for testing and treatment to implantable devices that remain in the body. Pouches are used for surgical drapes, and all the different disposable items (scalpels, needles, swabs, sponges, sutures, etc.), provided by an equipment manufacturer as a complete combination of everything needed to perform a surgical procedure. They may be a complete sterile "kit" used for each patient when connecting the patient to the dialysis unit. The graph in Figure 1 (1) illustrates just how broad the range of products requiring packaging is.

Medical devices require a broad range of packaging skills. The range extends from simple pouches for surgical devices or instruments to packaging for devices that include sensitive optical and electronic systems. Packaging assumes a major role in the development of a device and must be part of the design criteria used to regulate how a device evolves in the development process. This is of primary importance when the device must undergo any one of a number of sterilization techniques. In devices using multiple components, the components may undergo different sterilization techniques and require different packaging to meet the needs of each of the components and to provide a finished group of components that make the device work.

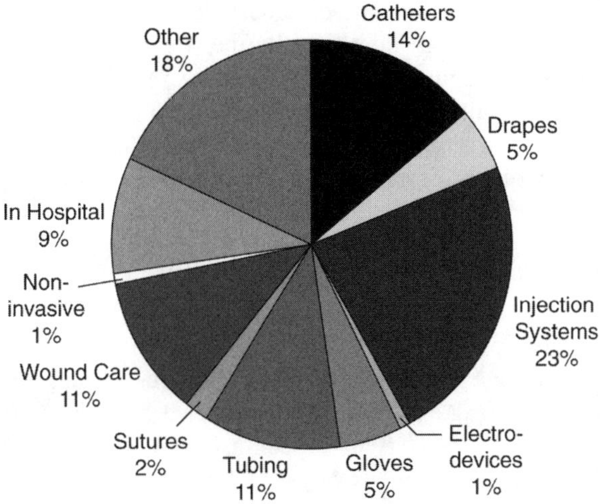

Figure 1 Graph of medical device packaging segments by end use market. *Source*: From Ref. 1.

REGULATION OF MEDICAL DEVICES

Medical devices in the United States are regulated by the Food and Drug Administration's (FDA) Center for Devices and Radiological Health (CDRH). In Europe, the European Union Council regulates medical devices.

The FDA's regulations covering medical devices is found in the Code of Federal Regulations 21 CFR parts 800 to 1299. The European Union's Medical Device Directive (MDD) is found as (93/42/EEC) published in 1993 that lists the "Essential Requirements" for medical devices. The methods and regulations regarding medical devices must be considered for both organizations, with in some cases the addition other of requirements unique to Asia, Japan, or a specific country. A general overview of these different systems and regulations and the use of standard test protocols to satisfy the packaging requirements are discussed in a later section of this chapter to understand how medical devices are placed in worldwide commerce.

FDA's CDRH regulates firms that manufacture, repackage, relabel, or import medical devices sold in the United States. The CDRH also regulates radiation-emitting products, both for medical and nonmedical end uses; examples of these products include X-ray systems, lasers, microwave emitters (including microwave ovens), televisions, and ultrasound equipment.

Medical devices in the United States are classified as class I, class II, or class III type devices. Regulatory control increases for each class of medical device, with class I being the least regulated and class III being the most regulated. Each classification defines the regulatory requirements for a general device type (2,3). Because the devices are so varied in their makeup and use, the classifications define broad characteristics that must be met for each level of classification and do not attempt to regulate specific devices (4,5).

Both regulatory bodies promulgate regulations that are similar to each other's and permit the use of standard testing for packaging and transportation qualification testing of a medical device for sale in their respective regions.

MEDICAL DEVICE DEFINITIONS AND TESTING STANDARDS

Medical device packaging is regarded as extremely critical to the success of the device and its ability to deliver reproducible outcomes to patients. Medical devices, because of their nature, face challenges not found in drug packaging. Some of the challenges include damage due to vibration caused by a truck or airplane transporting the device. There is a lot of confusion in the industry around the definition and testing requirements required for medical devices. Medical devices are also described in the National Formulary (NF), the United States Pharmacopeia (USP), and these two volumes must also be consulted, and any requirements they contain must be satisfied as part of the approval process.

The definition of a medical device in the Food, Drug, and Cosmetic Act is "an instrument, apparatus, implement, machine, contrivance, implant, in vitro reagent, or other similar or related article, including a component part, or accessory which is: recognized in the official National Formulary, or the USP, or any supplement to them, intended for use in the diagnosis of disease or other conditions, or in the cure, mitigation, treatment, or prevention of disease, in humans or other animals, or intended to affect the structure of any function of the body of humans or other animals, and which does not achieve any of its primary intended purposes though chemical action within or on the body of humans or other animals and which is not dependent upon being metabolized for the achievement of any of its primary intended purposes."

This extensive sentence/paragraph captures the breath and depth of products called medical devices, while separating them from drugs and the chemical actions they use for treatment of disease. Beyond the definition the FDA has placed on medical devices distributed in the United States are a number of General Controls along with pre-marketing (6,7) and post-market regulatory controls. The General Controls include (8)

1. Pre-market Notification 510 (k) (21 CFR Part 807 subpart E) unless exempt, or Pre-market Approval (PMA) (21 CFR Part 814)
2. Establishment Registration (21 CFR 807)
3. Medical Device Listing (21 CFR Part 807)
4. Quality System (QS) Regulation/Good Manufacturing Practices (GMP) (21 CFR Part 820) (9)
5. Labeling Requirements (21 CFR Part 801) (2)
6. Medical Device Reporting (21 CFR Part 803)

The first items on the list are device and device classification specific (5). There are three FDA classifications of medical devices: class I, class II, and class III (8). The FDA determines the class of every device, and this classification establishes the regulatory control assigned to the device (4,5). This control indicates the level of concern the FDA feels is necessary for qualification of the device for it to be marketed legally in the United States. As the risk increases, the regulatory control placed on the device increases. Basic definitions and background on each classification (5) are

Class I: Class I devices are defined as presenting minimal harm to the user. These devices are simple in design, manufacture, and carry a long history of safe use. Examples of devices in this category are hand-held surgical instruments, arm slings, tongue depressors, and other simple aids that we take for granted and probably did not realize that they were categorized as medical devices. These devices are the least complicated and require only cursory oversight. Their failure or a problem with a device like this poses little risk to the patient. Devices can and are exempted from the regulations regarding pre-market notification and possibly are exempt from current good manufacturing practice (CGMP) regulation (9,10).

Class II: Class II devices come under a set of requirements referred to as Special Controls. These special controls are much more specific concerning safety and effectiveness and have FDA Guidance Documents available that address questions or interpretation of the requirements for test methods, material standards, manufacturing standards, and other pertinent points needed to provide assured safety and effectiveness of the device. Examples of some of the items included in the Special Controls are

- Special labeling requirements
- Mandatory performance standards (International and United States)
- Post-market surveillance,
- FDA medical device–specific guidance

Class II devices require pre-market notification and the submission and clearance of a 510 (k) form describing the device. Examples of class II devices include X-ray systems, pumps, gas analyzers, surgical drapes, and physiologic monitors.

A very limited number of class II devices are exempted from the regulations. Background information on exempted class II devices is found in the device regulations (21 CFR 862–892).

Class III medical devices sustain or support life and have the most stringent regulations and controls. Their requirements go far beyond the General Controls and Special Controls of the two lower classes, and these controls make up part of the performance criteria for a device of this class. The FDA requires a PMA of a class III device before marketing the device and offering it for sale in the United States class III devices because they sustain or support life or are needed to prevent impairment of life present a very high risk to the patient of injury or illness and hence receive the extreme scrutiny and control. Examples of a range of class III medical devices include stents, heart valves, dialysis machines, silicone gel breast implants, and implanted cerebella stimulators.

510 (k) Pre-market Notification

A few class I and almost all class II medical devices receive market approval through the submission and review of the 510 (k) Pre-market Notification by the FDA (6,7,10). The 510 (k) identifies the characteristics and attributes of the new medical device compared with similar medical devices with similar intended use, and currently on the market in the United States. The current device already approved and legally marketed is referred to as the "predicate" device.

A 510 (k) requires the following information as defined by 21 CFR 807.87:

- Submitters name and address, contact person, telephone number, fax number, and representative or consultant, if applicable
- Trade or proprietary name of the device, its common or normal name or classification, and class of the device (class I, class II, or class III)
- Name and address of the manufacturing, packaging, and sterilization facilities and the FDA registration number of each facility
- All actions taken to comply with Special Controls requirements
- Proposed labels, labeling, and advertisements that describe the device, its intended use, and the directions for its use
- A 510 (k) Summary or 510 (k) Statement regarding the device
- For class III devices, a class III summary and a class III certification
- Engineering drawings of the device and photographs of the device
- Identification of similar devices currently marketed that are claimed as equivalent devices. This includes the labeling of the claimed devices and a description of the claimed device's medical use
- Comparison statement of similarities and differences to the marketed device
- Performance data for the modified device that shows the effects and consequences of the new device. This includes all data that confirms performance including bench, animal, and clinical data gathered during development
- Sterilization method and other information regarding sterilization of the device, if applicable
- Data detailing the development, verification, and validation of software used by the device
- Complete design data and review of the hardware development

- References to and information required as found in specific Guidance Documents
- Kit Certification Statement [required for 510 (k) submission of kit components used with the device]
- A statement attesting the truthfulness and accuracy of the information contained in the 510 (k)

Upon receipt of the 510 (k) document by the FDA a formal review of the device and the information supplied gets underway. The FDA's review of a 510 (k) requires anywhere from 30 days to 90 or more days depending on the novelty and complexity of the device. As devices become more complex, the FDA review process increases in length and scope. It is not unusual to receive questions from the FDA during the review, requesting clarification or additional data about specific parts of the device's operation or performance, or regarding substantiation of claims attached to the device. The agency is very thorough in its investigation of new and novel devices even if the device or a similar device has received a class III certification. This review also takes place for any device manufactured outside the United States and exported to the United States.

Pre-market Approval of a Medical Device

A PMA is needed when a medical device is complicated in its function and requires significant medical or scientific review of its safety and effectiveness. This applies to any device that presents significant risk to a patient. Almost all class III medical devices require a PMA on the basis of the definition of class III devices, and some class II devices may require this scrutiny.

The contents of a PMA are defined by the Federal Food, Drug, and Cosmetic Act Section 515 (c) (11) and in 21 CFR Part 814. Key points required in a PMA are specified by this legislation, and the supporting regulations are as follows:

- All reports and other information gathered or used in the development of the device, including any outside published information that should be reasonably known by the applicant. This information should detail all the investigations that show the device is safe and effective. The agency also requires a complete summary of the information in the application that permits the reader to gain a general understanding of the data and information in the application
- A complete disclosure of all components, ingredients, principles of operation, and properties of the device
- A complete description of the manufacturing methods and controls used to produce the device. This includes the facilities, controls, processing, packaging, installation, and sterilization methods used to produce the device. This requirement essentially details the CGMP for the device

- References to performance standards that would be applicable to any aspect of the device if it were in class II. This includes all information that shows the device meets these performance standards and any additional information necessary to allow for and justify a deviation from the standards.
- Samples of the device, its components, pictures, and anything the secretary (Agency head) may reasonably require unless the submission of the physical samples would be impracticable or unduly burdensome. This requirement may be met by submitting information specifying the location of one or more of the devices that are available for examination and testing
- Specimens of all labeling for the device
- Any other information the FDA may request. If necessary, FDA will obtain the concurrence of the appropriate FDA advisory committee before requesting additional information
- Periodic updates of the information filed regarding safety and effectiveness of the device that may effect the device's application, or may affect the statement of contraindications, warnings, precautions, and adverse reactions in the draft labeling of the device

The above points are an abbreviated summary of the requirements for a PMA, and detailed information should be obtained directly from the regulations and the Guidance Documents issued by FDA on the arrangement and content of a PMA.

By statute, the FDA has a requirement to review a PMA application within 180 days. If the device is complicated, is new or novel, does not fit into an established classification, or if it is not similar to an already approved device, this review process will require more than 180 days. The FDA may set up an advisory board to review and determine the merits and shortcomings of the device and use their recommendations in granting approval to market the medical device. This is similar to the setup of advisory panels for the review of drugs.

The FDA will carry out a complete facility inspection, facility audit, and verification (validation) of all manufacturing processes, packaging, and systems before PMA approval.

All of these requirements and the need to test and gain understanding of a complex device will stretch the approval process beyond the 180 days indicated in the statute.

Good Manufacturing Compliance (CGMP)

QS regulations found in 21 CFR 820 (11,12) are applied to the device and to the manufacturing establishment producing the device. The QS is required for the design, manufacture, packaging, labeling, storage, installation, and servicing of any finished medical device that goes into commercial sale and use in the United

States (9). The QS regulations are very similar to many ISO standards, including ISO 9001:1994 with added FDA-specific requirements (12).

QSs and their application are required to cover

1. Quality organization (QA department) management and organization
2. Device design
3. Buildings and equipment
4. Handling of components including purchasing
5. All controls, production, and process applied to the device
6. All packaging and labeling controls
7. Methods used for device evaluation
8. Distribution of the device
9. Installation of the device at the customers site
10. A system for tabulating and handling complaints
11. A system for servicing the device throughout its life cycle
12. All records detailing information covered in points 1 through 11

The FDA audits most facilities every two years for compliance to CGMP standards (9). The audit for compliance and the inspections they entail are based on prior audits, potential device risks, recalls of devices, and FDA-based initiatives that may significantly affect one of the device classifications.

Establishment Registration

The FDA requires any establishment that produces or distributes medical devices sold in the United States to register with the agency (13). The requirements for registration are found in 21 CFR 807 and are verified and updated annually.

Foreign establishments that manufacture, prepare, propagate, compound, or process a medical device outside the United States for shipment to or import into the United States must register as well. They must also provide the FDA with the name of the United States agent representing their establishment. They must also provide the FDA with a list of all devices they are exporting or expect to export to the United States. As a point of clarification, compounding for a medical device means production of a material that aids device use, but does not chemically react with the body. An example of these types of devices would be contact lens wetting solution or lubricants used to aid insertion or function of the device.

All establishments should complete listing a new device offered into commercial distribution in the United States within 30 days of its being offered for sale (14).

Device listing is updated when marketing a device is discontinued, a new classification of device is placed in commercial distribution, or if the manufacture and distribution of a discontinued device is restarted (14).

Medical Device Reporting

Medical devices are monitored after introduction into the marketplace. The FDA requires firms who receive complaints regarding malfunctions, serious injuries, or deaths associated with a device to notify the FDA of the incident. Medical device reporting is covered in 21 CFR 803 and requires the following:

- Written procedure for medical device reporting
- Files for each medical device reporting event
- Individual reports covering all adverse incidents
- Reports of remedial action taken to prevent unreasonable risk or substantial harm to the public health (5 days)
- Medical device reports of deaths, serious injuries, and malfunction attributed to a medical device (30 days)
- Baseline performance reports for the medical device

HARMONIZATION OF STANDARDS FOR TERMINALLY STERILIZED MEDICAL DEVICE PACKAGING—UNITED STATES AND EUROPE

Approval of packaging for terminally sterilized medical devices until 2006 was difficult and time consuming for devices sold in Europe and the United States. Two standards, ISO 11607 (3) and EN 868-1 (15), represented two different methods of approach and approval of a medical device (16). These two standards were originally published in 1997, with the ISO standard updated in 2000 (16,17).

A development cycle for two working groups, one for European Regulations and one for ISO standards, began in the early 1990s to develop a standard protocol for the approval of medical devices. The two groups worked in parallel and each published the original ISO 11607 and EN 868-1 standards, respectively, in 1997. The ISO working group charged with developing a new ISO standard for medical device packaging began with the intent to use both the ISO and EN background and standards to produce a new ISO standard that would be applicable for both the United States and Europe. The goal was one global document that represented the same requirements for both continents. Unfortunately, for a wide variety of reasons, the ISO working group decided to complete and publish the 11607 standard without harmonization with the European EN 868-1 standard. At the same time, the European Committee for Standardization (CEN), the working group producing the EN standard, chose to move forward with its EN 868-1 standard and a number of accompanying vertical standards to meet their goals. There was an understanding between the groups that both documents would be standardized at a later date. This harmonization was understood to take place in 2002 when the ISO standard would be up for its periodic five-year review. The ISO working group also produced a Guidance Document as a companion document to its standard to help users understand the ISO requirements. The problems with the two documents were the areas covered and the approach the two standards took to the problem of approving medical devices. The ISO 11607 standard (ANSI/AAMI/ISO 11607) spoke to terminally sterilized medical device packaging, while the EN 868-1 document addressed packaging materials and systems for medical devices that are sterilized.

It was quickly apparent that the ISO standard was not accepted in Europe. The ISO standard was updated in 2000 to ISO 11607:2000 to try to address the problem. Manufacturers outside of Europe were asked if they met and complied with EN 868-1 even if they met the requirements of ISO 11607:2000. The reason the two documents could not be reconciled was in their approach to the qualification and testing problem. The ISO document followed the final package for criteria regarding package materials selection, package forming, and package sealing, all followed by references to the final package. The document was mute to the package development process. The EN 868-1 document had its own set of problems. It did not address requirements for package design qualification, stability testing, or process validation, and many considered the document difficult to read and understand.

Because of these problems between the documents, a decision was made in 2002 to harmonize the two. The original idea for the revision and harmonization was the creation of two parts to the final standard with Part 1 covering materials and package design and Part 2 covering package assembly and validation. As the working group charged to develop the new standard quickly realized one of the biggest problems with the two documents was a lack of standard terminology. This problem was the major deficiency in resolving many of the problems between the two documents. There were a number of terms that meant different things to different people, including terms like primary package, secondary package, sterile package, barrier package, shipping package, and others that were used poorly and inconsistently. The problem was resolved when the working group decided to develop four key definitions that would be used throughout the standard. The four definitions are

1. Sterile Barrier System (SBS). The minimum packaging that prevents ingress of microorganisms and allows aseptic presentation at the point of use.
2. Preformed SBS. The SBS that is supplied partially assembled for filling and final closure or sealing, e.g., pouches, bags, and open reusable containers.
3. Protective Packaging. The packaging configuration designed to prevent damage to the SBS and its contents from the time of their assembly until the point of use.
4. Packaging System. The combination of the SBS and the protective packaging.

The two-part document structure contained other significant upgrades to the standard to address both the European and United States concerns about the former documents. The best part of the change was that the two parts of the standard now followed the normal order and methods used in the packaging development process. The new standard worked through material selection, package design, testing, process development, and validation in the order just listed. The other big change was the elimination of ambiguity about fulfilling all requirements. The standard is very clear about how to evaluate packaging and that all provisions and requirements must be met.

Along with the new standards two annexes were also published. Annex A provides guidance and overview of medical device packaging. It is extremely helpful to individuals who are new to packaging or for individuals who are not familiar with packaging as an engineering discipline. Annex B lists and describes all the test methods required for both parts of the standard and indicates how they can be used to demonstrate compliance to the standard.

In the old ISO standard, and to some degree in some of the European standards, testing information was part of the various sections of the documents, making them difficult to identify and locate. The new test methods along with the improved compliance and applicability comments are also categorized by the performance they gauge. Examples of some of the categories listed in Annex B include seal strength, performance testing, package integrity, and accelerated aging.

The standard was broken into two parts, ISO 11607-1 and ISO 11607-2. Part 1 of the standard covers the general packaging requirements, material selection, preformed package SBSs, and the design and development requirements for packaging systems. It also lists all the information to be provided as substantiation for the requirements. These points are in addition to the four critical definitions.

The General Requirements Section of Part 1 discusses QSs and sampling including test methods and proper documentation requirements for data developed. The Materials Section is extremely comprehensive in the breath of information it covers. The Materials Section covers physical and chemical properties, cleanliness, compatibility, microbial barrier properties, including biological and toxicological attributes. It also touches sterilization compatibility, labeling systems, and includes storage and distribution requirements of materials that maintain the sterile barrier of the package. The design and development sections provide an overview of general requirements for the design of packaging as a system and not as the combination of many components. It discusses many of the requirements to be considered in the package design process, and for package system performance testing including testing of stability and the sterile barrier. One-section details the information required for the sterile barrier material and the requirements for materials and information related to the materials and semi-finished packages described as a preformed SBS.

Part 2 of the standard expands and upgrades the requirements for package validation. This document works hand in glove with Part 1 and provides the flow for finished package qualification and validation in a normal package development project. Before the document was published, the EN standards did not require manufacturers to consider process control and process validation. This new emphasis is highlighted in the introduction to the standard and reinforces the FDA demands for improved process design and validation along with CGMPs.

One interesting point about the standard is that it also applies to all entities involved with medical devices including hospitals, health care facilities, reprocessers, and reconditioners, which mean it is applicable wherever a medical device is packaged and sterilized. It also forces for the first time secondary packaging component manufacturers to validate and document their processes used to produce a preformed packaging component. As an example, pouches produced and sealed on three sides and then supplied to a medical device manufacturer must undergo the design, development, and validation process required by the standard.

Another key point about the standards is found in Section 5 of Part 2. This section standardizes terminology by requiring an Installation Qualification (IQ), an Operational Qualification (OQ), and a Performance Qualification (PQ). It is very plain that the qualifications are done in this order. Validations can use earlier IQ and OQ qualifications, but they require an explanation and rationale that links them to the sterile barrier manufacturing processes being used. Any worst-case scenarios must provide documentation and a justification.

Another key feature of this section is that IQ critical process parameters must be identified, defined, controlled, and monitored. The OQ standard requires that the process parameters be challenged to guarantee that these process conditions will produce a packaging system that meets all defined requirements. This part of the section is directed at establishing upper and lower control limits for all key variables in the process. Some overview is also provided regarding forming and assembly of packages.

The final portion of the qualification, the PQ, is very specific on a number of points. It requires that the process parameters defined in the OQ be reviewed and confirmed as part of the qualification. These parameters must then be controlled and monitored throughout the PQ. It permits the use of actual or simulated product for testing system performance. These requirements force process control and thus prove process capability, process repeatability, and reproducibility. This portion of the standard also discusses package assembly, sterile fluid path packaging, and considerations for reusable SBSs. One point to remember is that the packaging may act as a holder or a dispenser for the device, and that this must be considered and included as part of the qualification of the device and its packaging.

AN OVERVIEW OF A PACKAGE VALIDATION

A medical device must reach a patient ready to perform its function. The packaging of the medical device must insure the integrity of the device, shield and protect it from mechanical damage, insure the sterile barrier is intact, and in some cases the package must act as a dispenser, a holder, or a fixture for the device. This is why the packaging must meet the same rigorous proof of performance required by a validation procedure. The critical acceptance criteria for any medical device package validation protocol must be package integrity. Regulatory authorities recognize that the packaging process and packaging performance, particularly the sterile barrier aspects of packaging, are nearly as important as the device itself in delivering the device for safe use (18). The packaging protects and keeps the device safe through its manufacture, distribution, transportation, and storage. The extensive review and scrutiny that both domestic and foreign regulatory bodies attach to the packaging of a medical device were the prime drivers for developing the ISO 11607 Part 1 and Part 2 standards.

Maintenance of sterility is the primary concern of most agencies and is the most common defect created by exposing a medical device to general conditions found in the distribution chain including handling, dropping, or possible mishandling. The vibration of packages by over the road vehicles and airplanes can significantly damage the device and the packaging. Common defects created by shipping and handling include slits, cuts, pinholes, tears, fractured thermoform clamshells, crushing, and deformation that may call the integrity of the package or the packaging into question. The performance of adequate and comprehensive testing on the packaging of a medical device comes as a surprise to many new medical device firms. Some are not aware of the standards and their importance to the FDA and the European Community in approving a medical device.

Packaging development should parallel the product development process. It needs to begin almost immediately after the basic concept or first model of a device is conceived. These prototypes can be used to guide and aid in the development process of the packaging for the device. This use of early concept models saves time and money and provides confidence that the packaging validation will be successful. This concept permits testing of key elements in the packaging, such as seal strength (ASTM F88), integrity (ASTM F2096), and the absence of pinholing before committing to the final package concept. Materials can be tested for tear resistance and integrity on the basis of the needs identified in the development of packaging and the testing of that packaging with the prototypes of the device. Design changes in both the packaging and the equipment, such as eliminating sharp edges identified in package testing of earlier prototypes that could tear the packaging can all be done and prequalified as part of the parallel design effort. Pre-shipment tests (ASTM D 4169 and ISTA 2A) done in the laboratory highlight package performance in distribution and further improve the capability of the packaging during the project development phase and provide real performance data that is directly applicable to the validation testing and documentation. This early corrective action made on iterative versions of the prototypes and the packaging of those prototypes makes fast and efficient validation of the final device and packaging systems flow quickly and provides the developers with confidence that they have identified any major problems that could throw project timing into disarray.

After all the regulatory discussion and interpretation discussed earlier in this chapter and some of the background behind the need for testing during the development of a medical device, it is useful to outline a typical plan needed for completing a package validation. This example is intended to provide a very generalized path for meeting the ISO standards. It is not definitive but will act as a general guide.

There are a number of items in the standards that must be addressed to complete package validation. Each of these elements is highlighted along with a generalized flow diagram that helps conceptualize just how a package qualification flows. Each part of the qualification plan requires the preparation of a number of protocols that describe the actions taken, the result expected, and define pass or fail criteria for the package system or package element being evaluated.

Writing a protocol is very important. The protocol is the roadmap that describes the purpose, scope, responsibilities, test parameters, production equipment, including how production and packaging equipment is operated, and the acceptance criteria for a successful test outcome. Planning and preparation are hallmarks of a good validation, and it begins with a good understanding of the validation goal in a well-written easy-to-understand protocol. As one progresses through the validation procedure, the concept requires that the performance factors stated in each protocol used for the IQ, OQ, and PQ be met. If a protocol is not met, or in other words the system fails the challenge testing, a review and explanation must be prepared, including an investigation and documentation of what was found, what are the causes for failing the protocol, and what are the next steps forward. If the failure were in a critical packaging component, an example would be a barrier material; the worst case would require a complete redesign and development of an alternative material or process that overcomes the deficiency found in challenging the original system.

All reviewing agencies in all parts of the world are becoming more cognizant of the need that the equipment used to produce the package and test the package are also validated. The FDA in its QS regulation has required validation for processes that cannot be completely verified, an example being package integrity. Manufacturers or packagers of medical devices must perform a formal IQ, OQ, and PQ on their package design and materials at standard operating conditions. ISO 11607 requires that packages be produced at the lower control limit of the process parameters for performance testing. Without knowing the capability of the equipment used and formally establishing this capability and its reproducibility through validation, a manufacturer can only guess at what the upper and lower control ranges are for the equipment being used. Without validation of the packaging equipment, it is impossible to ensure that a controlled process is producing quality packages. The FDA has published a definition for validation in its Guidance on General Principles of Process Validation: "establishing by objective evidence that the process, under anticipated conditions, including worst case conditions, consistently produces a product which meets all predetermined requirements (and specifications)."

MAJOR ELEMENTS OF A PACKAGE VALIDATION

There are a number of key elements in a package validation. The first step is defining the plan objectives and writing a protocol that describes what needs to be done for the complete validation. This describes all parts of the validation and provides an outline and summary of each point in the validation. This can be prepared as a separate document for each of the individual validations, IQ, OQ, and PQ, or it can be one document that encompasses all of the three validations.

If it is done separately, a summary document detailing the links and substantiating the relationship of each of the validations will be required, so doing a comprehensive plan and then breaking it down into smaller sections detailing the validation of different phases of the qualification is the best way to prepare the protocols. The key is to have a defined purpose and objective with conceptually sound performance testing that proves or disproves the ability of the materials, package, and process to deliver a safe and effective package. One key point to remember is that medical devices that use the same packaging system may be grouped and a representative sample of packages using the same package and packaging system, possibly the largest and smallest package, can be used for the challenge to represent the entire family of devices packaged that way. A general description of the elements needed in the overall validation protocol include

Plan Objectives

- Purpose
- Scope
- References
- Description/Definition of Materials
- Description/Definition of Equipment
- Description of Samples
- Preparation of Samples
- Test Procedures including Sterilization Methods and Testing
- Acceptance Criteria
- Documentation
- General Test Plan

After preparation of an overall plan that discusses and highlights all the different steps, procedures, testing requirements, along with all the materials, equipment, subassemblies, and other items, it may be difficult to communicate how this will produce a validated system. At this point, the key steps, usually the bullet points above, can be worked into a flow chart that describes how the validation will proceed and what will be done in the various steps of the validation process. The chart is useful in communicating the plan across all the different departments and disciplines within a company and with any outside companies or laboratories that will participate in the validation. It permits the validation team and management to measure and report progress. Validation is

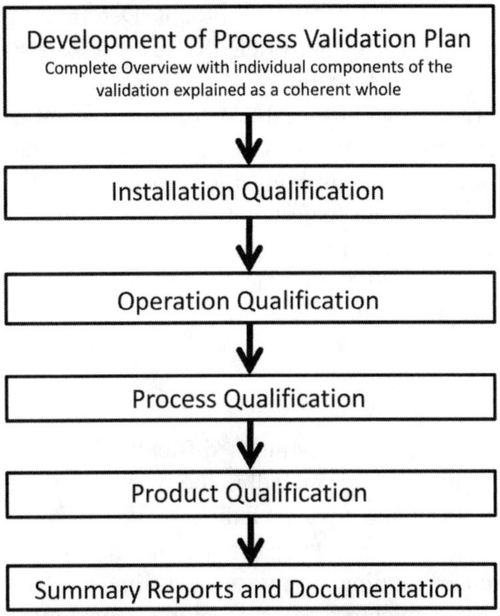

Figure 2 Generalized validation flow chart.

required before products can be submitted to the FDA for review (class II and class III). Delays in completing the necessary documentation of a device costs money in lost sales and may delay a patient from receiving the best possible analysis of their disease or treatment of their condition. The elements contained in a flow chart to track progress are listed below (Fig. 2). They typically carry some chronological reference regarding the amount of time it takes to complete each item:

Flow Chart for a Medical Device Validation Protocol (Fig. 2):

a. Permits Conceptualization of the Plan and Sequential Steps in the Plan
b. Provides a Method of Tracking and Measuring Progress
c. Permits Understanding of Validations with Multiple Segments contributing to the finished packaging system
d. Key parameters may be included to define pass/fail or minimum acceptance requirements

VALIDATION TESTING, PROCESS SAMPLING, AND VALIDATION REPORTING

A number of key concepts such as sampling of a process and reporting the results, package integrity testing, distribution simulation and testing, accelerated aging, including the environmental conditions used for doing these tests, are another part of writing and executing a good validation protocol. Determining the independent and dependant variables to test is often the hardest part of putting a protocol together. This requires understanding of the system and an understanding of the key variables (independent variables) that are not influenced or determined by others (dependent variables). An example for an independent variable would be the melting and sealing temperature needed for the material being sealed. This will be fixed by the material choice, and testing of equipment in the process must be designed to affirm that the minimum and the maximum temperatures applied to the package by the packaging equipment are suitable for the material to perform as specified.

Sample Size Testing

Determining the appropriate size of a sample for testing is difficult. There are many different factors that weigh into this decision including the type of test, the difficulty of the test, the cost to perform the test, including the cost of the materials or devices, and the risk factors also described as the confidence intervals needed to ensure proper understanding of the process limits. Many times lot sizes vary considerably when producing a device, making it even more difficult to choose a sample size. This can be overcome using reliability statistics that also provide an acceptable confidence interval. The minimum confidence interval permitted is 95%, but it can range as high as 98% depending on the needs or requirements of the device manufacturer.

The normal method for passing statistical testing of this type is zero failures.

The way to accomplish this is to make a declaration on the basis of the sample size in question. This statement would read the failure rate at 95% confidence interval, with a sample of n is x%. For example, if zero defects are found in a sample of 100 devices, the failure rate is 3.6%. Is this level of reliability acceptable? There are no black and white answers to that question. It is up to the company or individual developing the data to make a determination of its adequacy.

Test Methods

A number of test methods found in the ASTM catalog are used for package testing of medical devices. The list below provides the packaging engineer a sampling of some of the tests.

- Seal Strength ASTM F 88-06
- Seal Strength/Peel–Instron Testing
- Burst Testing ASTM F 1140 (unrestrained) and ASTM F2054 (restrained)
- Package Leak Testing ASTM F2096-04 Bubble Leak Test and ASTM F 1929 Dye Penetration

- Other ASTM Package Leak Tests
    - F2227-02
    - F2228-02
    - F 2338-04
    - F2391-05
    - F 2095-01

Many manufacturers have the mistaken idea that there exists an absolute minimum seal strength standard. The minimum required seal strength value must be established during the package validations as the one that establishes and maintains package integrity. The seal must be continuous and homogeneous, which means it provides package integrity, this type of visual or tested result on a seal does not prove the seal is adequate, for all the distribution challenges a package must negotiate before it reaches the patient. ISO 11607 requires that the upper and lower limits of seal strength be determined as critical seal process variables and these values must be demonstrated through testing as suitable for the intended purpose of packaging the device.

Distribution Testing

After a package is produced, it must be tested and proven to survive the transportation and distribution environment. This means the package must protect the product through all the drops, vibrations, and other stresses normally associated with transportation and storage. Laboratory simulation of these rigors has become accepted practice in the last five years and is eliminating the time consuming, costly, and many times inaccurate practice of actual shipping. Shipping a product from point a to point b and then examining or testing it for damage is only anecdotal evidence to performance of the package system with too many distribution stress factors left unknown. Examples of problems in ship tests include questions like whether the truck or train always takes the same route this shipment took, whether there were variations in road conditions, whether the package would always receive the same handling. Testing to answer these questions usually involves the following procedures:

- Drop Testing
  Drop testing is just what the name implies and is carried out from a specified height on all sides of a package and on the corners.
- Compression Testing
  A compression test squeezes the package in a manner similar to it being placed on the bottom of a stack or at the bottom of a pallet.
- Loose Load Vibration and Shock Testing
  Loose load vibration and shock testing subjects the package to being vibrated at a frequency and bounced multiple times. This simulates typical truck shipment.

- Random Vibration Profile Testing
  A random vibration profile subjects the package to the multiple frequencies it would undergo during shipment.

By using ASTM D4169-05 Distribution Cycle 13 and the ISTA (International Safe Transit Association) 2A pre-shipment test protocols, a worst-case scenario can be applied to the package, and passing these test methods provide a high degree of assurance that the package will withstand the handling and distribution environment. The two standards cited are for specific package sizes and conditions, and their applicability to a specific situation must be determined on a case-by-case basis.

Accelerated Aging

The European Union Directives require that all sterile medical devices must have and display an expiration date. This means that documented evidence is required for all medical devices to substantiate an expiration date. This can be developed in two ways. The first is real time testing to affix the date. The second is using accelerated aging to prove an expiration date.

Accelerated aging is based on a thermodynamic temperature coefficient that states for every 10°C rise in temperature the chemical reaction rate will double. This formula refers to rate kinetics of a single chemical reaction. Using this method as the age testing method for packages with multiple interactions and potential problems requires some caution and understanding. It also means this accelerated method for aging has limits in what conditions can be used to predict aging. Although this formula will work at high temperatures, any accelerated aging test using a temperature greater than 65°C cannot be justified by any rationale. There is no condition in a controlled supply chain that would subject a device to greater temperatures for extended periods of time, and temperatures greater than 65°C may cause melting or deformation of plastics and other materials used in device construction. Remember the device and the packaging are to survive accelerated aging and destroying the device to determine if the packaging will survive has little value.

The testing duration for an accelerated aging study can be developed from the following formulas:

$$\text{Accelerated Aging Rate} = \text{Q10}$$

$$= 2\left(\frac{\text{elevated temperature} - \text{ambient temperature}}{10}\right)$$

Where Q10 = 2, and ambient temperature is 23°C or 16°C.

$$\text{Accelerated Aging Time Duration} = \frac{\text{Desired Real Time Aging}}{\text{Accelerated Aging Rate}}$$

Example for 55°C Temperature Test

$$\text{Accelerated Aging Rate } (Q10) = 2\left(\frac{55 - 23}{10}\right)$$

Accelerated Aging Time Duration = 365/9.19 = 39.7 Days

Accelerated Aging Time Duration is normally rounded up to indicate that 40 days of testing would equal one year at ambient conditions of 23°C.

Testing a product under these accelerated conditions and determining the sterile barrier remained intact would mean for every 40 days of elevated temperature testing the manufacturer could claim one year of shelf life in the expiration dating of the device and package. Different agencies may restrict this extrapolation to a maximum of two years and require any dating beyond two years be proven with real time data.

Similar testing can be applied to the device itself, but one must use extreme caution to ensure that the temperature is defensible. This testing is also useful if a product undergoes a short-time temperature deviation from recommended storage conditions. This data would permit the manufacturer to determine if the device and its packaging were still functional after experiencing a higher than normal temperature exposure.

ISO STANDARDS

ISO, The International Organization for Standardization, is a worldwide federation of national standards regulatory bodies or agencies. These bodies, which contribute people and expertise to the various technical committees of ISO, are the mechanism used to develop and periodically update standards. This standard consists of multiple parts that define and describe the criteria required to test a medical device for use.

REFERENCES

1. Allied Development Corp. Packaging Strategies January 31, 2006.
2. Food and Drug Administration (FDA), Center for Devices and Radiological Health, Device Advice, Labeling Requirements, (www.fda.gov/cdrh/devadvice/33.html), April 24, 2003.
3. International Organization for Standardization (ISO) 11607. Packaging for Terminally Sterilized Medical Devices, Part 1 and Part 2. Geneva: International Organization for Standardization, 2006.
4. Food and Drug Administration (FDA), Center for Devices and Radiological Health, Device Advice Device Classification Panels (www.fda.gov/cdrh/devadvice/3131.html), June 10, 2003.

5. Food and Drug Administration (FDA), Center for Devices and Radiological Health, Device Advice, Classify Your Medical Device, (www.fda.gov/crdh/devadvice/313 .html), August 4, 2004.
6. Code of Federal Regulations Title 21 8, 21 CFR 814 Subchapter H Medical Devices Premarket Approval of Medical Devices.
7. Code of Federal Regulations Title 21 8, 21 CFR 807 Premarket Notification 510K.
8. Food and Drug Administration (FDA), Center for Devices and Radiological Health, Device Advice Device Classes, (www.fda.gov/cdrh/devadvice/3132.html), November 21, 2002.
9. Food and Drug Administration (FDA), Center for Devices and Radiological Health, Device Advice, Good Manufacturing Practices (GMP)/Quality System (QS) Regulation, (www.fda.gov/crdh/devadvice/32.html), January 28, 2004.
10. Food and Drug Administration (FDA), Center for Devices and Radiological Health, Device Advice Class I / II Exemptions, Class I/II Devices Exempt from 510(k) and class I Devices Exempt from GMPs(www.fda.gov/cdrh/devadvice/3133.html), February 9, 2000.
11. Code of Federal Regulations, Title 21 (pt 820), 21 CFR 820.
12. Food and Drug Administration (FDA) Guidance – Design Control Guidance for Medical Device Manufacturers, This Guidance relates to FDA 21 CFR 820.30 and Sub-clause 4.4 of ISO 9001, March 11, 1997.
13. Food and Drug Administration (FDA) Center for Devices and Radiological Health, Device Advice, Establishment Registration (www.fda.gov/cdrh/devadvice/341.html), August 7, 2006.
14. Food and Drug Administration (FDA) Center for Devices and Radiological Health, Device Advice, Medical Device Listing (www.fda.gov/cdrh/devadvice/342.html), July 17, 2006.
15. EN 868-1. Packaging Materials and Systems for Medical Devices Which Are To Be Sterilized, General Requirements and Test Methods. Brussels: European Committee for Standardization, 1999.
16. International Organization for Standardization (ISO) 11607. Packaging for Terminally Sterilized Medical Devices. Geneva: International Organization for Standardization, 2000.
17. Food and Drug Administration (FDA), Modernization Act of 1997: Guidance for the recognition and use of consensus standards. Federal RegisterFebruary 25, 1998; 63 (37 notices): 9561–9569. Available at: www.fda.gov/cdrh/modact/fr0225af.html.
18. Nolan PJ. Common mistakes in validating package systems, packaging business dot Com. News Release. May 22, 2006.

8

Container Fabrication

INTRODUCTION

Choosing how to fabricate a container is a hard decision and is as important as choosing the container material. Container fabrication equipment is specific and dependent on the packaging material chosen, even though a material can be fabricated into a container in many different ways. Selection of fabrication equipment and the process it uses to shape the material create the look, feel, physical shape, and style and determine to a large measure the performance of a container. A material may be able to do many different things and provide many different physical attributes to a package, but how the material is fabricated defines precisely the performance limits of the material and container and establishes the physical attributes of the package. Plastics and composite materials manufactured into packages by different methods are good examples of the differences in performance. Blow molding (1–4) will produce plastic bottles with one set of characteristics, while thermoforming (1,2,5,6) or pouch making will use the same material and produce an entirely different container.

The material and the method of fabrication are very symbiotic and require tailoring the material performance to the fabrication process. For example, metal can be a hard or soft alloy, but the final determination of which hard or soft alloy to use is dictated by the container fabrication process. The fabrication process changes the raw material into a finished package with all the features and attributes the material and process can deliver to the finished container. A consumer's perception and expectations about a container's performance is a result of how the container was made. The strength, durability, cost, and convenience features as well as other consumer attributes come from the choice of container material and fabrication method.

Glass and metal containers provide a specific set of performance attributes and capabilities that are well established with manufacturers and consumers. Glass bottles and metal cans are familiar packaging to all consumers.

Modifications in opening features, shape, stacking features, labeling, and other physical attributes of the package are based on the fabrication of the container and show the ingenuity and diversity available in fabricating a material. These package improvements and modifications highlight how the packaging has evolved and improved to become more consumer friendly. Examples of some of the improvements are vacuum buttons for metal closures on glass bottles and jars, the increased use of easy open ends, and both stay-on tab for beverage cans and full panel easy open ends for food and pharmaceuticals. Closures that are plastic tamper evident, closures for sports drinks, and closures that permit easy access and use of pharmaceuticals have become commonplace as part of this evolution in the manufacture of all types of packaging.

Plastic and flexible containers likewise have improved and evolved as manufacturing techniques, and new fabrication technologies were introduced. When the package manufacturing technique is chosen, the filling and closing technology used to fill and then seal the package is determined. This means the package construction must follow a set of well-established specifications to produce a packaging component that performs the same way millions of times to provide the robust and repeatable performance characteristics the packaged product needs for protection and consumers require and expect every time they purchase a packaged product.

Pharmaceutical packaging is different from food or beverage packaging. The size of a manufacturing lot and the total number of containers needed for individual products is far less than food or beverage products. For example, blister packaging has a predominant position in pharmaceutical packaging and must be considered as a form, fill, and seal manufacturing technology, but the total number of any one blister is far less than the smallest nationally distributed food or beverage. This is different from high-volume packaging. Another example of blister packaging is that all parts of the container fabrication are coupled. In food and beverage packaging, the manufacturing of packaging components is decoupled. The container and the closure are produced in a process that is removed from the filling and sealing of the container. This difference permits high rates of speed in food- and beverage-filling operations, with package components fed to a filler and to the closing equipment. In pharmaceutical blister packaging, the blister cavity that holds the tablet is thermoformed from plastic sheet and then moves forward for filling with a tablet or capsule and sealing with a lidding material.

Nutraceuticals, which fall between food and pharmaceuticals in composition and regulation, most often use food packaging techniques and technologies for enteral products and pharmaceutical techniques for parenteral products.

Each material, and the way it is fabricated into a package, is unique. This chapter discusses most of the common container fabrication methods for glass, metal, plastic, and composite materials. There are many more methods of making containers. New and specialized methods create products for unique niches in the marketplace. Innovation and entrepreneurship are always introducing new and better ways of making packages. It is a fascinating combination of material science with engineering. It provides an understanding of the large capital investments and the highly specialized nature of equipment used to make packages.

GLASS CONTAINERS

Glass can be made into a container in a number of different ways. Pharmaceutical packaging uses two basic methods for making glass packaging. The first process is blow molding, the standard method for making glass bottles for centuries. The second process starts with glass tubing for fabrication of small bottles, ampules, and vials.

Blow molding, both as a free form and with a mold, is as old as glassmaking. Free form blow molding is what a glassmaker does when he makes bottles one at a time by taking a gob of molten glass on a metal tube and blowing and shaping the glass into a bottle. The glass is not constrained in any way as the glass blower expands the gob of glass into a bottle, while parts of its shape are produced with a number of different hand tools used to shape specific parts of the bottle.

Blow molding or expanding the gob of glass inside a mold was an offshoot of this original technique. It was automated early in the 20th century by Michael Owens of Libby Glass in Toledo, Ohio to create the high-volume glass bottle and glass container industry that is known today. Glass is made with many different materials that create different manufacturing characteristics and require specialized fabrication equipment to produce bottles. In consumer products, the variations in materials produce everything from works of art to normal everyday containers.

Pharmaceutical containers place more demands on glass containers than those of food products, and USP type I and type II glasses are prime examples of unique materials that require a highly specialized manufacturing process to produce pharmaceutical vials, ampules, and bottles.

Glass has been the default standard for pharmaceutical packaging for most of the last century and has evolved into a highly automated and highly controlled set of manufacturing processes.

Blow Molding of Glass Containers

Glass is produced in a ceramic-lined furnace, where a number of earth alkalis are melted and mixed to form the product we call glass. (2) The process is a bulk-batching process of the raw materials followed by a continuous melting process. Bulk materials are batched and fed into a furnace that continuously melts the raw materials, mixes them, removes impurities, and then delivers molten glass to fabrication equipment for forming into bottles. Glassmaking only shuts down when a ceramic furnace lining has deteriorated and requires reconditioning after long periods of operation.

Molten glass is approximately 1500°C (2600°F) in the ceramic firebrick-lined furnace used to produce raw glass. Glass raw materials melt and slowly move through a horizontal furnace that mixes and chemically reacts the materials until the molten glass mixture is uniform. Mixing of the materials, particularly the molten materials, results from convection currents and mechanical flow that develops as the glass moves through the horizontal furnace process. The uniform glass mixture found at the end of the furnace flows into a refining chamber through an opening below the level of glass in the furnace. Keeping the exit opening below the surface of the glass removes impurities that float on the molten surface of the glass. The impurities are materials contained in the raw materials that cannot be removed easily in the raw state. The impurities float on the surface of the molten material and are skimmed off and removed above the refining chamber.

Molten glass in a refining chamber is cooled to approximately 1000°C to 1100°C (2000°F) and maintained at this temperature for container manufacturing. The reduction in temperature is necessary to increase viscosity of the molten liquid and produce a material that can be handled through the remainder of the process. From the refining chamber, sometimes called the forehearth, glass flows into a hemispherical bowl with an orifice and multiple openings in its bottom. As the glass flows out of the openings, it is cut by rotating knives into "gobs," the term used to describe small pieces of highly viscous molten glass, and then fed through a delivery system into a glass blowing machine that forms finished glass bottles. The gobs travel on rails as individual units, and each gob represents an individual bottle (Fig. 1).

Gobs of glass are handled in one of two different methods by two totally different types of forming equipment to make glass containers. The two glass container–manufacturing processes are referred to as blow–blow and press–blow techniques when describing the blow molding operations.

Blow–Blow Molding of Glass

In a blow–blow container–manufacturing operation, a forming machine delivers the gobs of glass by gravity into a "blank mold" designed to form the finished container (Fig. 2) (5). This is first of the two molds used to make the container. The mold closes and is sealed off at the bottom, and the bottle-making process begins. The hot glass is forced by air pressure into the neck ring. This portion of the mold forms the neck of the bottle and the finish, the name given to the opening's size and shape combined with the type of threads or closing attachment flair on the neck of the bottle. At the same time, a plunger enters the gob of glass at the other end to begin to force the glass out against the sidewalls of the "blank" or parison mold. The design of the mold is critical for placing glass in the proper position needed to get a uniform distribution of material in the final bottle. Blowing air through the plunger produces a hollow cavity on the inside of the parison (gob) and helps the distribution of glass on the outside of the cavity

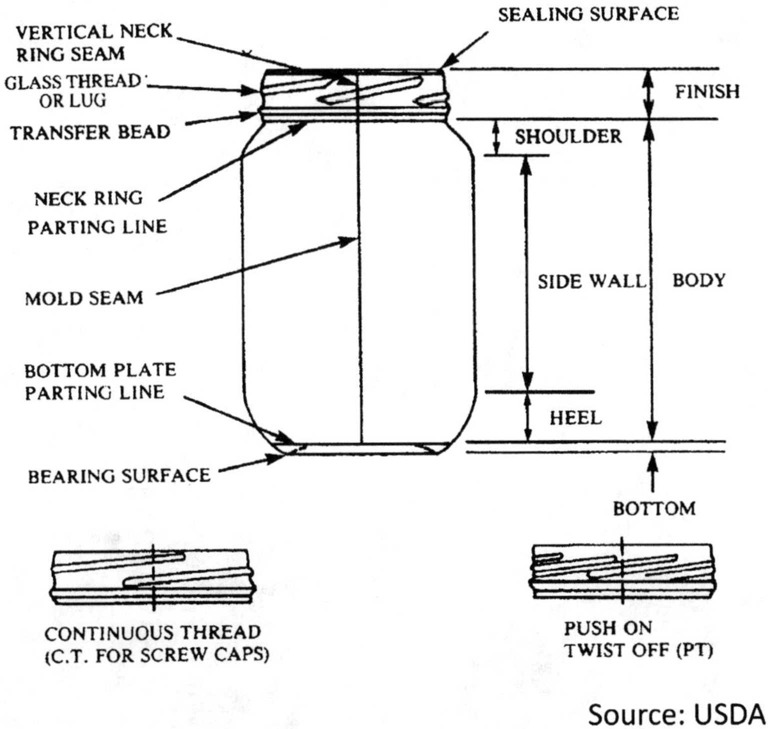

Figure 1 Glass bottle construction and terminology.

prior to the second phase of blow molding the container. The partially shaped container is referred to as parison. The parison is removed from the first mold and placed into a second or finishing mold. More heat is added to the parison, and air is blown for the second time, forcing the glass parison to expand against the walls of the second mold, completing the formation of the container.

Press–Blow Molding of Glass

The press–blow process is very similar to the blow–blow process for making a glass bottle. Originally, the blow–blow process was used to make narrow-necked containers, and the press–blow process was used to manufacture wide-necked bottles and jars (Fig. 3). The press–blow process has more control in the manufacture of the initial parison and produces bottles with container walls that are more evenly distributed than the blow–blow process. This uniformity translates into thinner sidewalls that reduce weight and cost of the finished bottle and speed in the manufacturing process. This advantage and improvement in the manufacturing process are what led engineers to refine the process and expand its use to narrow-necked containers. The steps for making a bottle with this process are

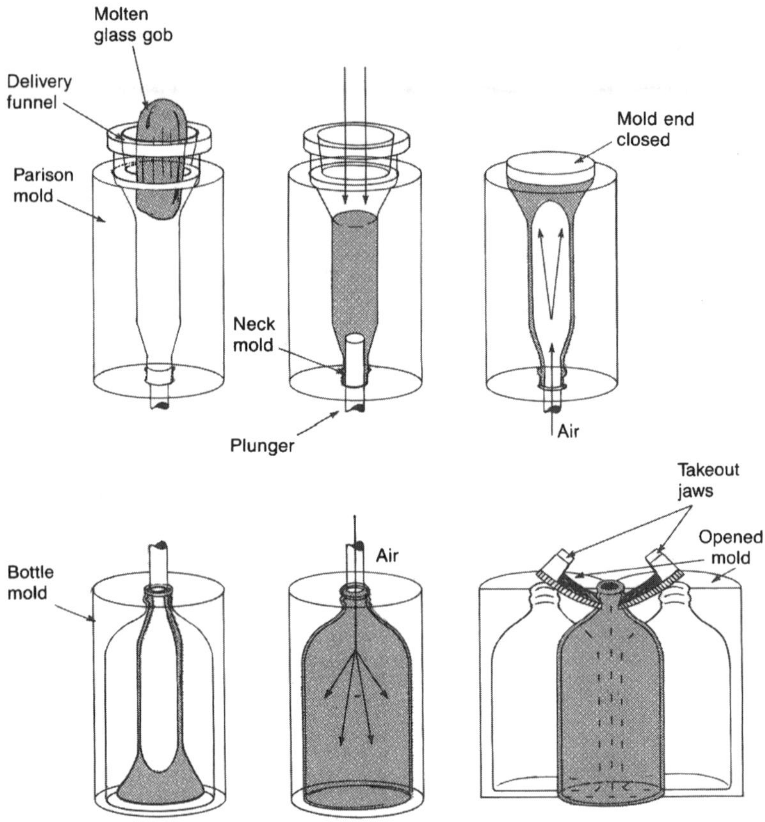

Figure 2 Blow–blow bottle process.

similar to the blow–blow process with the major difference being the plunger now pushes glass against the sidewalls of the parison mold with minimum use of air pressure to force the glass out to the cavity walls. The remaining steps in the process are the same. The completed parison is reheated and placed in a second mold cavity to produce the final shape of the finished bottle. Pressurized air is forced into the parison, expanding it to the cavity walls.

Annealing and Treating—Glass Finishing

Following molding, the hot glass bottle moves to an annealing furnace called a lehr. The bottles are reheated and slowly cooled to relieve any stresses or strains created in the glass during the molding process. Traditionally, glass was inspected at this point in the process, and the visual results were used by the operating crew to adjust the entire process. It was not unusual, prior to more

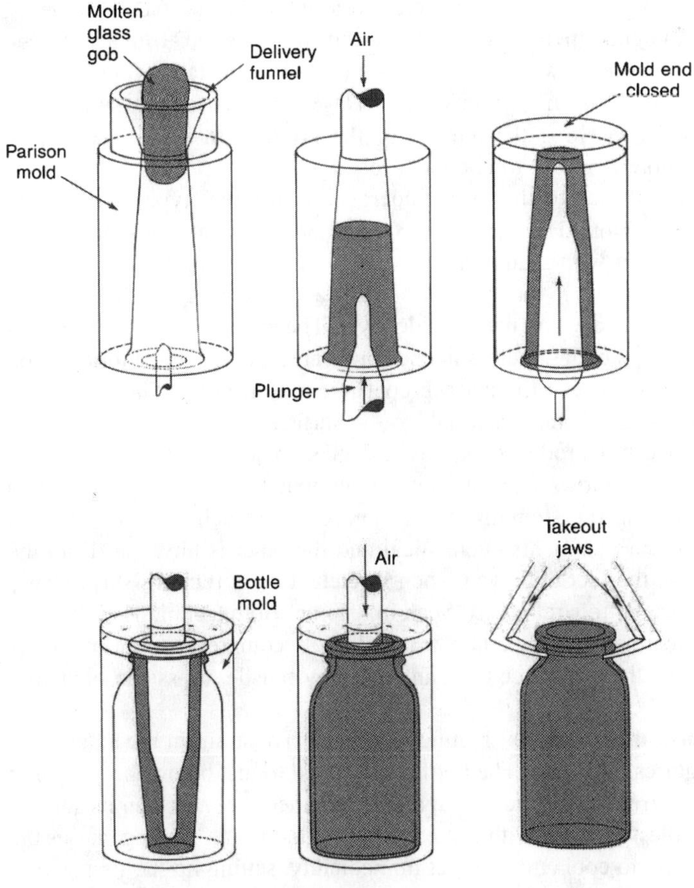

Figure 3 The press blow bottle process.

sophisticated process statistical modeling and feedback controls, that 80% of the visually inspected glass was rejected and returned to the furnace as cullet. "Cullet" is the term used for broken or waste glass that is recycled in the glassmaking process. The cullet is remelted and may pass through the process multiple times as the process is fine-tuned. Eventually the process conditions are refined or adjusted to the point that approximately 90% or more of the finished glass emerging from the lehr is of acceptable quality. This slow and skill-related process has slowly been replaced by statistical modeling and computers that quickly analyze conditions in the furnace and the blowing operation and make adjustments accordingly to produce a high percentage of quality glass almost immediately after startup. The addition of statistical modeling and computers moves the startup operation to high efficiency and high yield production.

After annealing, the glass is subjected to a number of surface treatments. The interior of the container is treated with sulfur dioxide (SO_2) that is forced into the container to react with sodium oxide (Na_2O), a material found on the surface of the glass, to form sodium sulfate (Na_2SO_4). The sodium sulfate residue is removed by washing the interior of the bottle with water, leaving an interior surface more resistant to chemical reaction.

The outside surface of the glass undergoes a different type of treatment than that in the interior glass surface. Exterior surface treatments permit the outside of the glass to withstand handling and to improve strength.

A common treatment for the exterior surface is coating with tin or tin chlorides; this forms a very thin metal oxide layer. The metal oxide layer provides adhesion for coating the glass with silicones or stearates that act as lubricants on the glass surface and reduce the surface coefficient of friction. These coatings also enhance adhesion of other materials to the surface of the glass.

Glass science has produced bottles and glass objects that can withstand large drops and even blows from a hammer without breaking. There are two methods of improving the strength of glass, making it much more resistant to breakage. One of the treatments is chemical and the other is physical. Both the techniques rely on the fact that even when scratched glass remains surprisingly strong in compression, it is relatively weak in tension. Both techniques introduce a prestressing treatment to the glass to produce a compressive strain in the exterior surface of the container to counteract any tensile stress the container may encounter.

The physical treatment for creating a compressive strain in the exterior of glass bottles requires reheating. The bottle is reheated to just below the softening point, the surface irregularities reflow and are smoothed out, and then its surface is chilled with a blast of air or with some type of oil bath. The chilling causes the exterior of the glass to cool and contract immediately, setting up the first step in the process to create the exterior strain (compression). The interior of the glass remains hot, and because glass is a poor conductor of heat, it remains viscous after the exterior surface has set. The interior still must contract as it cools, and as it does, it continues to pull or draw the exterior glass into compression. A compressive strain is built into the outer layers of the glass. When the glass encounters a tensile strain, the compressive strain built into the glass counteracts the force and prevents the glass from breaking. Only when the compressive strain is exceeded does the glass break.

This technique works only as long as the exterior surface to the glass is not scratched or gouged. If the exterior surface is broken, the stresses set up by the interior contraction create tension, which causes the glass to break. It will shatter into many small pieces as these stresses are relieved.

The second method used for exterior compression treatment of glass containers is a chemical reaction. The reaction uses ion exchange on the exterior surface of a bottle to put the material into compression. This exterior treatment process reacts the outside surface of the container with molten potassium salts that convert the sodium oxide (Na_2O) to potassium oxide (K_2O). The larger potassium ion puts the entire outside surface into compression and thus strengthens it. The first step of this process is to place the container into a molten bath containing the potassium ions. These ions then replace the sodium ions on the surface of the glass. The size of the potassium atom is larger than the sodium atoms it replaces, and this replacement on the surface crates requires slightly more space between the various atoms. The packing of the atoms induces the compressive strain in the glass surface. Glass produced this way is dramatically increased in strength, as much as a factor of 10 in kilonewton per square meter (kN/m^2), making it extremely shatter resistant and useful for containers.

Tubular Glass Fabrication—USP Type I Glass

Ampule and Vial Manufacture

Making pharmaceutical containers from borosilicate glass requires a different glassmaking process than the bottle-making processes just described. Again, the starting point is a furnace where raw materials are batched into a continuous melt and manufacturing operation. This is the same as that described for the soda lime glass bottle process. The difference in the process begins as the glass moves out of the furnace. It is converted into glass tubing using one of two processes, the Danner process or the drawdown process.

The Danner process consists of an angled rotating sleeve that permits the introduction of air to inflate the tubing. The glass flows past a rotating, hollow, water-cooled mandrel or sleeve that introduces inflating air into the tubing. The inflating air controls the process by controlling the outside diameter of the finished tubing.

The drawdown process is very similar to the Danner process. The glass is extruded and forced through an annular die that has an orifice for the introduction of air. The outside diameter of the tube is set as the tubing moves through this die. For both processes, controlling the rate of glass withdrawal, that is, the speed glass is pulled through the forming area, controls the weight and thickness of the tubing wall. The tubing produced by both the processes is uniform and exact in its tolerances.

Control of the tubing is essential to controlling the next steps in the process, which is conversion of tubing into ampules or vials. The tubing is formed in a continuous straight-line process. The tubing is supported after being pulled through the dies for cooling and hardening, and during this process step it is annealed to some degree. At the end of the line is a device that actually pulls the glass through the process from the furnace. A cutoff mechanism completes the tubing manufacture by cutting the tubing to proscribed lengths.

Ampules Following tube forming, the finished lengths of tubing move to separate machines for conversion into ampules or vials. The process for

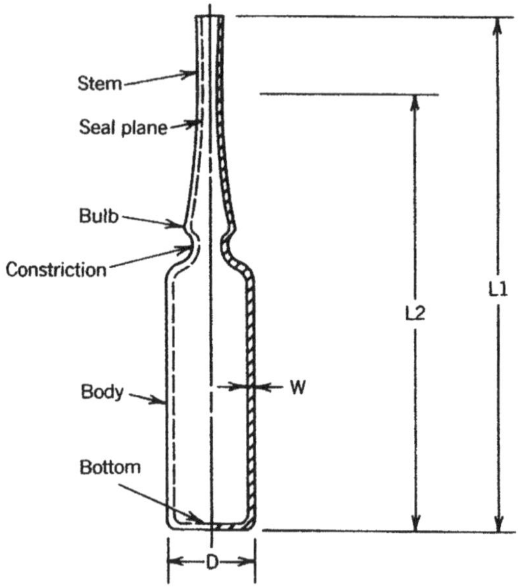

Figure 4 Diagram of a standard ampule.

producing both containers is very similar. The major difference is the introduction of a die in the formation of vials to create the opening and its lip.

Drug- and serum-containing ampules (ampoules) are formed on a rotary machine that moves the glass through a number of stations or steps to form the finished ampule (Fig. 4). In this process, tubing is first heated and then pulled to form a bulb, stem, and constriction. The tubing is continuously heated and constantly rotated in the equipment to produce a uniform shape in each of the multiple forming stations used in the process. All the contours and shapes are controlled by heating of the tubing in a measured way while the tubing is stretched and mechanically pulled into the shape of the finished ampule.

Equipment used in this process sometimes uses a die to assist in the formation of the ampule in the same way a die is used for formation of a vial. The process produces ampules with very accurate sizes and uniform shapes. Accurate size is critical to filling and storage of the drug in the ampule (Table 1). After the initial ampule formation for shape, it is transferred to another piece of equipment. In this step of the manufacturing process, the ampule is trimmed in length, glazed, color banded, or given an identification band. The opening properties of the ampule are created by scoring the constriction or by adding a ceramic paint at the constriction in the neck of the ampule. The ceramic paint introduces stress to the constriction, making it open reproducibly at the point of stress. Following completion of the forming and scoring or painting process, where additional

Table 1 Long-Stem Ampule Dimensions

Capacity (mL)	Diameter (mm)	Width (mm)	Length 1 (mm)	Length 2 (mm)
1	10.40–10.70	0.56–0.64	67	51
2	11.62–12.00	0.56–0.64	75	59
5	16.10–16.70	0.91–0.69	88	73
10	18.75–19.40	0.66–0.74	107	91
20	22.25–22.95	0.75–0.85	135	120

bands may be placed on the ampule for identification, the completed unit is annealed to increase strength in the glass and remove residual strains in the glass created by the manufacturing process. Following annealing, the finished ampules are moved into a clean room, a room with positive interior pressure created by high-efficiency particle filtration (HEPA). The highly filtered air maintains cleanliness and the positive pressure surrounding the container prevents airborne contaminants from entering the filled ampule. Here, ampules are inspected automatically, accumulated, and packaged for shipment to the pharmaceutical company. During filling, the glass tip is heated and sealed (Fig. 5).

Vials Vial formation is almost exactly the same as ampule manufacture, with the major difference being multiple heating and tooling applications to the tubing that form the flanged lip of what appears to be a miniature bottle (Fig. 6). The introduction of a final forming tool into the neck of the vial precisely sets the diameter of the opening. The complete finish on a vial, both inside and outside, must be precise to accept the elastomeric plug and the aluminum band that holds the plug in place (Table 2).

Vials, ampules, and glass bottles represent a major category of containers used for pharmaceutical packaging. Other glass products, such as test cells for spectrometric analysis, syringes, tubes, rods, and mixing implements, are a few of the wide variety of other glass products used in pharmaceutical packaging. Glass is a versatile, well-understood, and stable material that has always been the material of choice for pharmaceutical products. Plastics produced in form, fill, and seal operations as well as plastics produced in separate bottle-manufacturing operations have begun to replace glass, but remain a small portion of the vial and ampule market.

METAL CONTAINERS—CANS

Metal containers (cans) used for pharmaceutical packaging are made by the same fabrication processes used to manufacture food and beverage cans. Pharmaceutical and nutraceuticals products are packed in both two-piece and three-piece cans. There are three different processes in use that produce a majority of metal

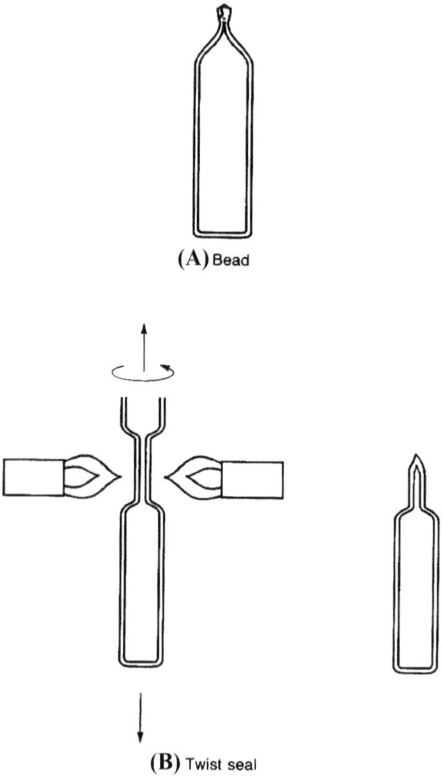

Figure 5 Ampule sealing.

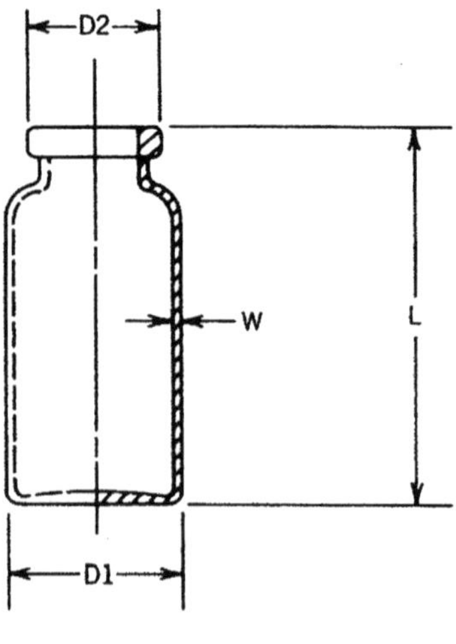

Figure 6 Diagram of a standard tubular vial.

Table 2 Tubular Serum Vial Dimensions

Capacity (mL)	Diameter 2 (mm)	Width (mm)	Length 1 (mm)	Diameter 2 (mm)
1	13.50–14.00	0.94–1.06	27	12.95–13.35
2	14.50–15.00	0.94–1.06	32	12.95–13.35
3	16.50–17.00	1.04–1.16	37	12.95–13.35
5	20.50–21.00	1.04–1.16	38	12.95–13.35
10	23.50–24.00	1.13–1.27	50	19.70–20.20
15	26.25–27.00	1.13–1.27	57	19.70–20.20

containers you see in pharmaceutical packaging and in food and beverage packaging. The three processes are described as draw and iron, draw–redraw, and welded. They account for well over 90% of all cans and make all cans for pharmaceutical packaging as well as those used for food and beverage products. Although there are other methods of making cans, these three dominate all commercial markets.

The term "three-piece can" describes the oldest and one time predominant method for making metal cans. The term three-piece describes the can body (cylinder) and the two ends that are attached to the cylinder to complete the can. Welding has replaced a mechanical soldered joint originally used to produce a three-piece can. The three-piece process originally used pure tin and tin/lead solder combinations to seal a folded mechanical seam on the side of the cylinder. The welding process replaces the mechanical overlapped and soldered seam with a butt-welded joint. It is interesting to note that the methods for making three-piece cans stayed the same as those found in the original automated processes developed during the early 20th century. Automation made cans available and affordable when it replaced hand soldering and hand assembly of cans. Welding is one of a number of significant evolutionary changes made to the elements of the process for improvement, but the steps for making a three-piece can have remained the same.

Draw and iron cans are made by a new process developed during the 1950s and 1960s. The process eliminated the need to weld or solder a side seam and produced a cylinder with one end already formed, reducing the number of steps needed for manufacturing. This process has replaced three-piece cans with two-piece cans in the majority of food, beverage, and specialty applications. It is the preferred type of can for pharmaceutical nutritional products.

Draw–redraw cans are an offshoot of metal stamping and forming. This type of can is almost as old as the three-piece can. Early on it was used for shallow draw applications such as tuna and sardine cans. It was simple and easy to make from a metal standpoint, but the coatings needed to withstand the forming operations did not become available until much later. This limited its early use to products that would not interact with bare metal.

Draw–Redraw Cans

Smaller size (8 oz or less) cans are made using the draw–redraw process, although the other two processes, welding and the draw and iron process (D&I), also supply small-size cans. Larger cans are made with this process, but it is not the most efficient use of metal. Both standard can-making materials, aluminum and steel, are used in this can-making process.

Lower capital costs for a complete can production line, the ability to buy thin-gauge organically coated steel sheets or coils as the starting material, and the ability of this process to produce multiple sizes and shapes with the same equipment make it attractive for a pharmaceutical or food company. Its versatility in size and shape make it adaptable to producing a wide variety of small volume cans. The starting precoated steel or aluminum, which is slightly more expensive to use than coating cans in the draw and iron process, permits companies to produce cans with a wide variety of performance properties. It also permits them to avoid an expensive and difficult to manage environmentally regulated coating process and instead rely on suppliers specializing in precoating materials for cans. The thermoset coatings used to coat the metal are made from acrylic, vinyl, polyester, urethane, and other polymers. All materials used to produce the can coating are regulated by the Food and Drug Administration (FDA) and meet FDA regulations for extraction and interaction with products. They provide the product contact layer needed to protect the metal from interaction with the product. All organic finishes used in can making undergo a curing or cross-linking step that heats the coating in an oven or subjects it to high-energy radiation (UV) to promote "curing," the term used to describe polymerization of the coatings components.

Many companies use a variety of can sizes depending on the product packaged, and this process produces a large number of different size containers with a minimum investment in tooling. It is well suited for operations producing small volumes of cans in a particular size. The different can sizes (diameter) produced by this process are matched to standard sizes of commercially available easy open or flat panel ends. Any variation in the volume of a draw–redraw can is produced by combining a standard can end of fixed diameter to a can of different heights or depths of draw.

A draw–redraw can-making process requires more metal to form a can than do the other two processes. The can size (depth of draw) is limited by the ductility of the metal and organic coating. The coating is used to insulate the metal from the product and also provides part of the lubrication necessary for the metal to flow over the forming dies. This ability to move or flow metal over forming dies is analogous to thermoforming a plastic container. The amount of metal that can be moved cost-effectively over the die governs the amount of thinning and weight reduction the process can introduce into the sidewall of container. This process leaves the majority of metal in the can at the bottom and at the top of the can sidewall. As with all metal cans, the finished container cost is determined almost exclusively by the amount and cost of material used to produce the container.

When a can size requires more efficient use of material (usually based on total volume of cans produced), the draw and iron process or the three-piece process is typically used to produce cans greater than 8 oz in volume capacity. This does not mean that the process cannot produce large-size cans, only that other more cost-effective processes are used when large sizes are needed. The draw–redraw process forms both round and square cans.

A can made using the draw–redraw process begins with a coil of metal precoated on both sides with organic thermoset coatings. The metal can be coated and maintained in coil form or it can be coated after being cut into individual sheets of metal. The coatings provide, along with an external lubricant such as high purity mineral oil, the lubricity needed to form the can by stretching and moving the metal over a die and are flexible enough to stretch and maintain a continuous film on the surface of the metal after forming. This film becomes the protective layer of material that prevents the interaction of the can's contents with the metal from which it is fabricated.

A punch press stamps cups from a flat sheet or coil of metal in the first step of this can-manufacturing process. For small shallow cans, this is the only step needed to make the can. For larger cans with deeper depths (draws), the cups are controlled through the press for orientation and handling into the second forming process. The standard manufacturing configuration for draw–redraw cans is a two-stage process where the blank is first formed into a shallow cup of large diameter. The cup is mechanically transferred to a second set of dies that form or punch the cup into its final diameter and height. A third operation, or in some cases, part of the second operation, trims the ragged metal edge formed in the process by the uneven stretching of the metal and then shapes the edge into a flange for attaching a metal end after filling.

Cans are fabricated from coated steel, tinplate, and aluminum using the draw–redraw process.

Draw and Iron Cans

Another way to produce metal cans is called a draw and iron process. The first steps in the process are the same as those used for a draw–redraw can. Both tinplate and aluminum are fabricated into cans with this process, but its most common application is in the manufacture of aluminum cans. Following the punching of a circular blank, the container is first formed into a shallow wide diameter cup in the same way a draw–redraw container is made. After this formed cup is stamped out of the incoming coil stock, it is forced axially through a die. The diameter of the die is much smaller than the diameter of the cup. The bottom of the can is shaped in this step of the process by a punch. A large amount of development work has gone into the design of can bottoms to optimize strength and minimize the amount of metal used to make the can. As the metal passes axially through the die its diameter is smaller than the wall thickness of the container. The result of this action on the aluminum in the sidewall extrudes or draws the metal up the sidewall of the can to produce a portion of the can height. This process is repeated two or three times with the punch passing through the dies in a single stroke. Lubricants are flooded on the container throughout this process to permit the metal to move through the punch and die sets. The bottom of the can is reduced in metal thickness in each operation. The gap between the punch and die is smaller than the thickness of the metal, and as the bottom of the container is shaped to the final configuration, sometimes called sizing, the metal flows out of the bottom into what becomes a sidewall of the container. The metal is further ironed into the final wall thickness and height of the finished container. The wall-ironing process leaves more metal at the top of the can than in the sidewall.

The closed cylinder is trimmed to a standard height and then moves through a series of steps to remove the drawing lubricant and pretreat the metal. Chemical pretreatment causes the reaction of the metal can surface with another material to improve coating adhesion.

Following pretreatment, the outside of the closed cylinders are printed ("decorated" is the industry term) using a multicolor offset press. Each cylinder is loaded on a mandrel and rotates over multiple blankets to produce the multicolor label. The decorated cylinders are then moved through an oven, usually a "pin" oven. The term is derived from the posts that extend into the cylinder and carry it through the oven to bake or cure the exterior finish.

After the outside of the cylinder is decorated, an inside spray of coating is applied to seal the inside surfaces. Both the bottom and the sidewalls of the decorated cylinder are coated in one step by airless spray. The cylinder again passes through an oven to cure (cross-link) the inside coating.

The metal at the top of the can, which is thicker than the sidewalls of the can, is further formed and reduced in diameter to a size smaller than the bottom of the can. This process, called "necking in," reduces the amount of metal used in the can end and creates an overall reduction in the total metal used to make the package. Reducing the diameter of the end saves as much as 15% of the aluminum that would be needed to make a straight sidewall can end. The savings in metal is a significant cost reduction of the finished package. The necking in of the container also increases the vertical strength of the can and permits the cans to fit together in a stack.

After necking in, the cylinder passes through additional metal-working machinery that creates a flange or "body hook" on the open end of the closed cylinder. The flange on the open end of the closed cylinder is "double seamed" (Fig. 7) or rolled with a mating flange on the can end to form the seal between the can body and the can end after the can is filled.

Part of the bottom design in any two-piece can is the use of a standard size that matches the interior diameter of the double-seamed end to permit positive stacking of containers. Stacking of all cans, not just those used for pharmaceuticals

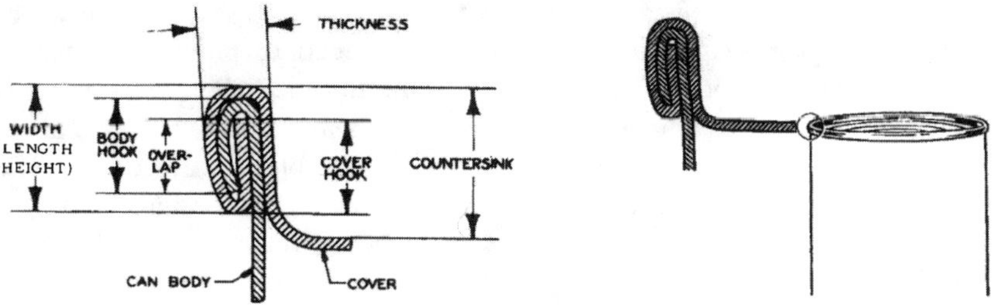

Source: USDA

Figure 7 Metal can double seam.

or medical nutritionals (nutraceuticals), on the shelf and in the pantry is a required consumer attribute.

Welded Cans—Three-Piece Cans

The three-piece soldered can was the standard of the industry until the 1970s. During that decade welded three-piece cans and two-piece draw and iron cans began to replace them. The soldered side seam contributes to lead levels in products, and during the 1970s, regulations regarding the amount of lead and tin in food spurred the movement to welded cans. Welded cans in pharmaceutical products are primarily aerosol containers.

Three-piece can manufacture is significantly different from two-piece manufacture. The starting material still begins as thin-gauge coil stock, but the processing steps are totally different. A number of different metalworking operations that together produce the finished can define the process.

The first steps in making welded cans begin with a coil of tinplate. The tinplate is produced using a process identified as double-cold-reduced plate or 2CR plate. This process produces a lighter gauge of steel with higher tensile strength prior to coating with tin. The 2CR process produces steel in three different tempers, DR-8, DR-9, and DR-10. The plate or steel designations for the substrate under the tin coating are in order of increasing strength and hardness. Older methods of rolling steel for tinplate are still in use, but the increased strength of the 2CR process has resulted in its substitution and use in most can making.

The steel, after completing the cold-reduced rolling process is electrolytically coated with tin. By controlling the deposition process and masking one side of the metal from the other, the tin is applied in the same or differing coating weights (differential tin coating) on both sides of the coil stock, depending on the end use of the container. Variations in the deposition process produce one of three finishes, designated as bright, matte, and satin on the surface of the tinplate.

The finished coil of tinplate then begins a number of steps to process it into finished cans. First, the coil is cut into large sheets in an operation called sheeting.

This operation has historic roots and a unique term, “base box,” which was developed to quantify the amount of tinplate needed to produce a specific number of cans. The term is still used from time to time today. A base box is defined as 112 sheets of tinplate, 14 in × 20 in in measurement or 31,360 in^2 of surface on each side of the sheets or 62,720 in^2 of total plated surface. A bundle of 112 sheets of this size produced 400 number 2 cans, also designated 307 × 409 cans, and most often called a 307 can (3 and 07/16 in × 4 and 09/16 in cylinder size). In the United States, all can size designations are abbreviations, with the final two digits in the three-piece can size referring to the size of the can in 16ths of an inch (e.g., a 211 × 400 can is 2 and 11/16 in in diameter and 4 in in height). At one time, the term base box was applied not only to the amount of metal in the can but also to the amount of other materials needed to make cans. For example, a gallon of organic coating would cover *x* base boxes of metal.

The stacked sheets move to the next step in the operation, which is organic coating. The sheets are first coated on what will be the outside of the finished can and then coated on what will become the inside of the can, with organic coatings necessary to protect the tinplate from interacting with the product.

Coatings are applied and “cured,” a term referring to the removal of solvent and the initiation and completion of the thermosetting reactions in the coating by heating in a large oven. The oven has a continuously moving rack system that stands each sheet on end following the application of coating. The coated portion of the sheet is held in the rack and does not touch another sheet or any part of the oven. After it passes through the heating zone of the oven, it passes through a cooling zone and then is restacked for the next operation. The sheets may be coated with multiple applications of organic coating (sometimes called lacquer) measured and recorded in milligrams per square inch (mg/in^2). This is where the idea of base boxes comes in, with the amount of metal coated by a volume unit of coating being used as one possible measure to determine the material costs of the coating or to compare the cost of different coatings and coating weights needed to make the finished can.

The last coating operation, which coats the sheet on what will be the inside of the can, is unique in that the coating is applied with gaps creating a grid pattern. The sheets are not coated completely in this operation. A thin strip of exposed metal is left where the sheet will be slit to produce body blanks. This exposed metal is required for welding the two sides of the metal cylinder together. Older soldered cans also require a coating-free area. Coating would act as an insulator and stop the welding or the adhesion of tin solder to the mating surfaces.

Following the coating process, the large completed sheets are moved to body making, the name used for the cylinder-making operation.

Preparation of tinplate for flat panel (no easy open feature) ends follows the same steps through coating, with the exception that the sheet is completely coated or is spot coated. The spot is circular and corresponds to the portion of the sheet punched to produce the end. Many manufactures spot coat the material for metal ends as a cost reduction step. The flat sheets are cut in a sawtooth pattern to maximize metal utilization. The pattern permits ends to be stamped with less metal waste between the ends. The sawtooth strips, only slightly larger than the round ends, are fed into a punch press that produces a round blank, the starting point for the end. This circular blank contains the complete profile of the end that includes the multiple circles one can see on a can end, called a flat panel end. The circles are called beads, and they are there to increase the strength of the end and reduce the amount of material used. The outside of the round blank is flanged in a roll-forming operation to complete can end manufacture.

There are two components to a body-making line. The first is a slitter that makes two cuts in the large metal sheets. This operation consists of two sequential cuts perpendicular to each other that reduce the large sheets into small rectangles of metal called body blanks.

One by one the body blanks are fed into a body maker that forms the cylinder over a mandrel. It then adds a mechanical hook if the can is soldered, or it butts the two ends of the cylinder together if a weld is made. For a soldered can, the mechanically hooked section of the cylinder is treated with solder that flows completely into the joint and completes the sealing of the joint. For a welded can, the two sides of the cylinder are butted together and held in place by a copper wire as the welding process mates the two sides of the cylinder at the "side seam" of the can. One of these two processes is used to produce the can cylinder.

The cylinder is then flanged, that is, it is worked on both ends to produce a curl of metal called a body hook. The body hook is required to mate the cylinder with an end in a process called double seaming (Fig. 7).

The can is beaded after this operation. Beading is the creation of ridges in the sidewalls of the cylinder to improve the containers' strength and reduce the amount of metal required for can performance through filling, processing, and, later, distribution. A can is a highly engineered product that has been refined over many years to use the minimum amount of material while maintaining or increasing its strength.

The last step in the operation is the double seaming of a metal end onto one end of the cylinder (Fig. 7). This is called the manufacturer's end. The other end of the can is added after the can is filled. The finished cylinder with an end attached is then ready for filling.

Refinements of the can-making process have pushed the speed of manufacture and automated many of the steps, particularly the multiple transfers of material from process to process. High-speed can making is standard industry practice and produces containers at rates greater than 500 cans per minute. Considering that cans are produced in the billions, this speed is essential to meet demand. What is even more remarkable is the fact that the billions of cans perform reliably in all types of applications. It is very unusual for cans to be recalled because they did not perform.

Metal Tubes

Metal tubes, much like metal cans, were the original standard package for all types of creams, ointments, and viscous materials. Metal tubes are manufactured by an extrusion process and use both tinplate and aluminum as their starting materials. A slug of metal in the shape of a ring is heated and then forced into and through a die that forms the tube shoulder end. This is the dispensing end of the tube that receives the screw-on cap. The sidewall of the tube is formed by forcing metal up and around the punch that initially forms the shoulder and end extruding it as it moves through a die. The extruded walls of the tube are cut to length, and a thread is added to the nozzle end of the tube.

The interior of the tube is coated with an organic protective coating made from vinyl-, acrylic-, and epoxy resin-based coatings, all of which are thermosetting. The coated tube is passed through an oven for curing. The exterior of the tube may be coated and printed as separate steps in the process with organic coatings and inks. The cap for the tube completes tube manufacture and is applied in a separate operation.

The bottom or crimped (folded) end of the tube is left open for shipment to the pharmaceutical manufacturer. This is the end of the tube that is filled with product. The filled tube is crimped to complete the final closure of the tube. The expiration date and lot number for the product is impressed into the crimped portion of the tube. For small tubes, only one of these two pieces of information is crimped into the bottom, usually the lot number. The expiration date is printed on the side or in some other location on the tube. This method of marking is used for both metal and plastic tubes.

PLASTIC CONTAINERS

Pharmaceutical plastic containers come in a number of forms, bottles, blisters, vials, pouches, tubes, cups, and plastic cans. Each of these packages is produced in a different way, and some may be produced by two or more standard plastic container–manufacturing processes. The fact that two containers, very similar in appearance and performance, can be produced by more than one manufacturing process requires the packaging engineer to understand the differences. Different manufacturing techniques have different strengths and weaknesses that must be well understood to choose the most suitable for the product. The engineer must also be familiar with the pharmaceutical product and how it is produced. One method of container manufacture may align better with the needs or volumes of the product being packaged, and the engineer is expected to determine the best combination for the product. Understanding of how packages are produced is also important in manufacturing validations. In order to develop suitable validation packages for a product, the engineer must understand the types and amounts of variably different manufacturing processes imposed on the finished product and plan for that variability to ensure that the packaged product is always safe and efficacious.

The ability of plastic fabrication to use multiple technologies is one of the reasons it has displaced metal and glass containers for many pharmaceutical products. The adaptability of plastic-manufacturing processes many times better match the small volume needs and requirements of a product. This adaptability extends to providing products that add convenience features useful in applying or dispensing product. The ability to develop a custom packaging approach for manufacturing and filling is a continuing trend in pharmaceutical manufacturing. It permits the production of pharmaceutical containers on a smaller and more customized scale that better meet the needs of the product, the patient, and the health care professional. Much of the plastic container–manufacturing equipment is extremely flexible in capabilities, permitting the same equipment to be used for a wide range of container shapes and sizes.

Bottles and Vials

The standard methods for manufacturing bottles and vials includes extrusion blow molding, coextrusion blow molding, injection blow molding, and reheat blow molding to produce bottles for food, beverage, and pharmaceutical products (4). The technologies have multiple variations within a given process, but the basics are straightforward to understand. Each of the technologies has characteristics that make it the most efficient or the most cost-effective method to produce a specific type of bottle or container. Some minimize material, some produce containers with improved protection properties, some produce containers on a small scale, and all are used as standard manufacturing processes for pharmaceutical bottles and vials (3).

Different blow molding processes are used to produce bottles. Bottles for tablets and solid dose forms are manufactured and handled differently than bottles that are liquid filled and undergo sterilization. The bottles may be produced by the same container-manufacturing process, but their characteristics, design, and handling place different needs and requirements on them.

Blow molding operations producing pharmaceutical containers for solid dose products (tablets, powders, capsules, etc.) are done in a clean room (the FDA provides specific definitions, e.g., class 100 or class 10,000) with HEPA-filtered laminar air flowing over the molding machines from ceiling to floor to prevent dust or other contaminants from entering the molding area or the finished packages during manufacture. The high heat of the extrusion process eliminates any pathogens in the container and on the surrounding tooling. The finished packages must be handled and maintained in a clean environment until they are sealed in a sterile outer package for shipment to the pharmaceutical manufacturer or repackager. The heat of the blow molding process kills and eliminates microorganisms making bioburden almost nonexistent. Most often any bioburden found when testing blow molded containers is introduced by handling after the actual molding of the bottle.

Extrusion blow molding is the most common technology used to produce bottles for pharmaceutical products. High-volume products, generally the over-the-counter products, use this technique when polyolefins and polyvinyl chloride (PVC) are the starting materials. The process changes when polyethylene terephthalate (PET) is the packaging material. Here the container-manufacturing process changes to injection blow molding or injection/reheat stretch blow molding.

Small bottles and vials are produced by injection molding (7) or injection blow molding. These manufacturing techniques are offshoots of other container-manufacturing methods. The differences in these two techniques produce bottles with very different performance characteristics.

Injection Blow Molding

Blow molding of plastic is very similar to blow molding used for making glass containers. It differs in a number of significant ways, which are discussed in this section, but the basic idea of forcing a material to conform to the shape of a bottle and the idea of using more than one step to shape the finished bottle are analogous for both.

Injection blow molding (Fig. 8) is most often used for making vials and small bottles less than 50 mL in volume. The process produces pharmaceutical containers that are very accurate and reproducible in size, neck finish, and material distribution. For small containers, accuracy in the neck finish is very

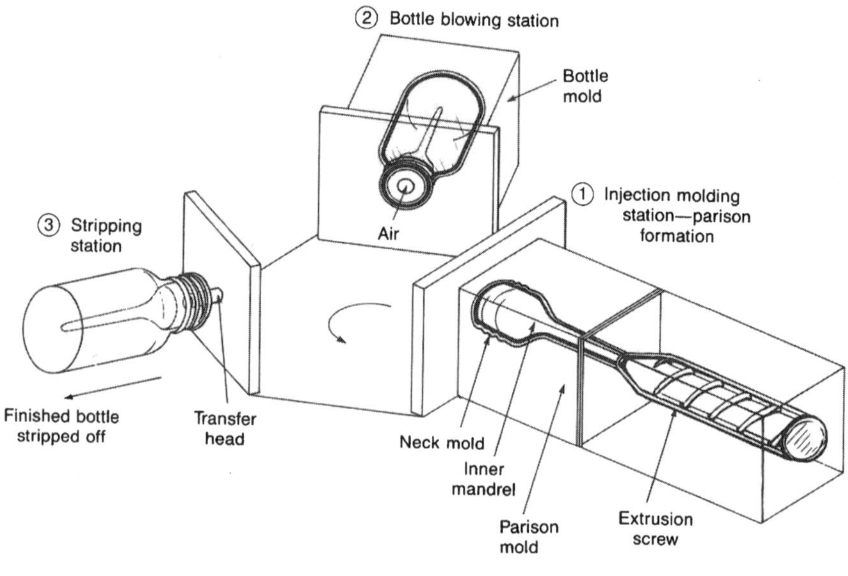

Figure 8 Three station injection blow molding.

important. It permits high speed and closing and sealing technologies to work reliably. This is difficult in small parenteral vials because a small variation percentage in tolerances is quite large dimensionally and could cause seal failures.

Injection blow molding produces a bottle finish that can accommodate unusual or unique closures for measured dispensing or for inclusion of tamper-evident and anticounterfeiting features. Sometimes the easiest and best anticounterfeiting feature is nothing more than molding the company or product name directly in the bottle or vial. The injection blow molding process produces bottles and vials from all common plastic materials, including polyolefins, the traditional pharmaceutical packaging material, and with other newer resins like (PET, polyethylene naphthalate (PEN), and polycarbonate (PC).

Injection blow molding has a minimum of three steps (Fig. 8). The process starts with an extruder; this piece of equipment receives the raw plastic resin normally supplied as small pellets and applies heat and mechanical energy to melt and mix the starting material into a homogeneous liquid under high pressure. The hot plastic is injected under high pressure into a die cavity and around a core rod to form a parison. The parison looks like a test tube or some other tubular shape with a neck finish on one end. The precise neck finish is molded in the first step of the process. The tubular section of the parison is customized to the finished container by varying the amount of plastic coming from the extruder to ensure that the proper amount of material needed for performance and protection is present at all points in the bottle. The injection molded parison, which is extremely soft and pliable, is transferred to a second station in the machine where a mold encloses the parison and core rod. In the second step, filtered air is forced through the core rod into the tube portion of the parison to expand it into the bottle shape contained in the second mold cavity. The finish section (top of the bottle where the closure is applied) of the bottle is held firmly in a die to ensure it does not move during the second part of the molding operation. The mold used to produce the final bottle shape is chilled with water circulating around the outside of the mold cavity. The cool surface of the mold freezes the thermoplastic resin into its final shape.

The finished bottle then moves to the last step in the process for removal from the mold. Although this is described as a process for one die, in actual operation, multiple molds are used to produce the package. This speeds the process and provides the time necessary for the plastic to fill the molds and then cool, fixing it in place. After removal from the injection blow molding process, the containers are maintained in an aseptic environment (clean room) for bulk packaging. The bulk package maintains the extremely clean/sterile condition of the bottles during shipment to the pharmaceutical customer.

Extrusion Blow Molding

Extrusion blow molding is very similar to injection blow molding. The major difference between the two manufacturing processes is how the parison is produced (Fig. 9). Molten resin from an extruder is pushed out from an extrusion die

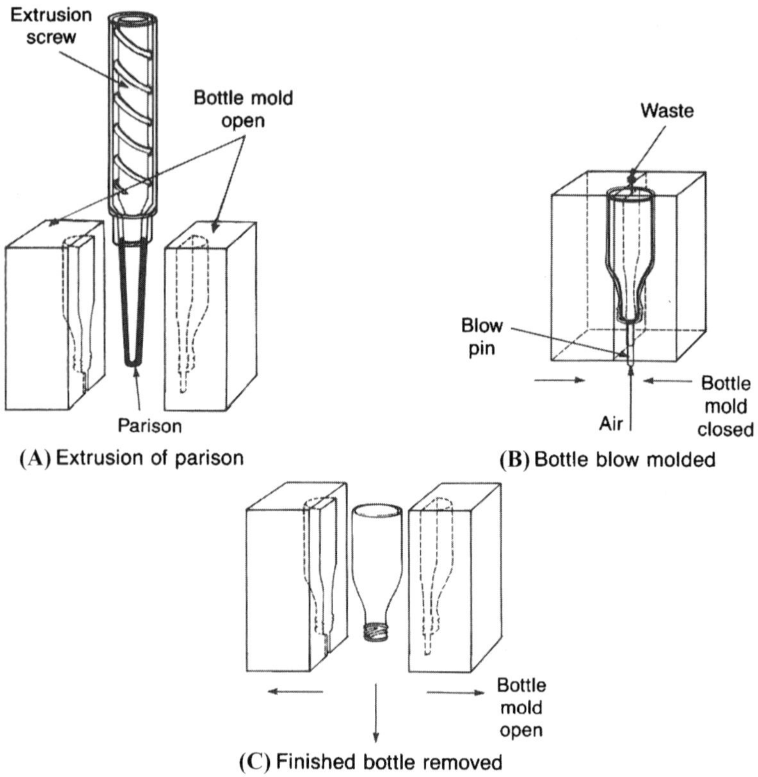

Figure 9 Extrusion blow molding.

or head into free space; no die is used to shape the parison. The die for extruding the parison resembles two cylinders, one inside the other. A gap between the cylinders is fed plastic by the extruder through the side of the outside cylinder. The plastic travels around the cavity between the two cylinders and can only escape at the open bottom of the die. Plastic in this extrusion step can be pictured as toothpaste squeezing into the space between the two cylinders and then out the bottom. The inner cylinder is hollow, permitting air to be introduced from the top to the bottom of the die. The size and shape of the parison that emerges at the bottom of the die is controlled by the temperature of the melted plastic, the speed plastic is fed through the extrusion die by the extruder, and the amount of air pressure used to inflate the parison. One must think of air inside the parison as a core rod, and in this process the plastic is forced around a hollow core rod. As the plastic exits the hollow cylinder, air is introduced to continue its expansion in space.

This expanded tube of hot plastic is captured, either by molds on a shuttle that pinch, cut, and move the trapped plastic tube to a second station or by molds arranged around a wheel that trap and move the plastic away from the bottom of the extrusion die. The capture of the plastic parison closes one end to permit the extruder to continue to produce the expanded "balloon" and cuts the plastic from the continuous tube formed from the flow of plastic at the bottom of the die.

The parison in the mold then moves to a second station on the machine, while the plastic from the extruder continues to move out of the parison die. At the top of the mold, the hot plastic tube is captured by a portion of the die that forms the neck finish of the completed bottle. A blow pin or blow rod is inserted through the top of the mold into the open balloon held in place by the closing of the top and bottom of the bottle mold. The blow pin injects air into the parison, expanding it to take the shape of the finished container. After cooling in the die, the finished bottle is trimmed on both ends, top and bottom, to remove flash (a flat piece of plastic left over from forming the bottom of the bottle, at the bottom parting line, and the waste plastic above the lip of the bottle finish). The parting line refers to the small line, visible in bottles produced by the process, where the two halves of the mold separate to release the bottle from the cavity. In smaller bottle sizes produced by extrusion blow molding, the amount of plastic waste becomes a significant cost issue. For this reason, other more "scrapless" techniques such as injection blow molding are used to produce small-size bottles.

Extrusion blow molding can produce bottles with built in or molded handles. This is important for large-volume products, which are heavy and hard to handle. Handles cannot be produced by the injection blow molding process. Extrusion blow molding produces majority of polyolefin bottles greater than 100 mL in volume used in pharmaceutical packaging. Bottles between 50 and 100 mL in volume may be produced by this process or by the injection blow molding process.

Reheat Blow Molding

Reheat blow molding is a process similar to injection blow molding of bottles. It is the process used to produce soft drink, water, and some beer bottles. The process was developed to increase the speed of PET bottle production and overcome limitations of melting and molding PET.

Crystalline and semicrystalline resins with well-defined melting points do not permit the formation of a parison in free space. For materials with this characteristic, the parison is first injection molded in a process separate from the blow molding process. The parison produced in the separate injection molding process is called a preform. A preform resembles a test tube with treads at the open end. The preform with the final neck finish accurately injection molded in place is transferred to a second operation.

The second operation can be at the filling site or at a separate manufacturing location. Shipping preforms is more efficient than shipping finished bottles.

In the blow molding operation, the preform is reheated, most often by a quartz high-intensity lamp, in the area below the neck finish, the test tube portion of the preform, and then blow molded into bottles using standard blow molding cavities as described for the other processes. PET is the material that utilizes this process for manufacture for the majority of larger-size bottles used in food beverage and pharmaceutical applications. Most over-the-counter cough syrups and liquid cold remedies use bottles made this way. This process is used to mold bottles from PEN and engineering plastics.

THERMOFORMING OF PHARMACEUTICAL CONTAINERS

Blister Packaging

Blister packaging is a familiar form of pharmaceutical packaging (8). Each tablet or capsule is incased in a small custom-formed cavity of plastic or aluminum and sealed in place (Fig. 10). It is the fastest growing segment of pharmaceutical packaging in the United States, and is well established in Europe and other parts of the world. Blister packaging provides a unit dose of product directly to consumers in a convenient easy-to-use form.

Blisters are produced from two different materials, plastic and aluminum. The materials are highly engineered multiple-layer structures made up of adhesive laminated and extrusion-laminated components (Figs. 11 and 12).

Metal blister packages use cold forming (stretching of foil) in combination with thermoforming to produce the blister. Plastic blisters are made by vacuum forming and plug-assisted thermoforming (Fig. 10). Metal packages for blisters and strips are made from foil contained in a plastic laminate, which is cold stretched into the shape of the cavity holding the tablet, with a second foil laminate web used to seal the package.

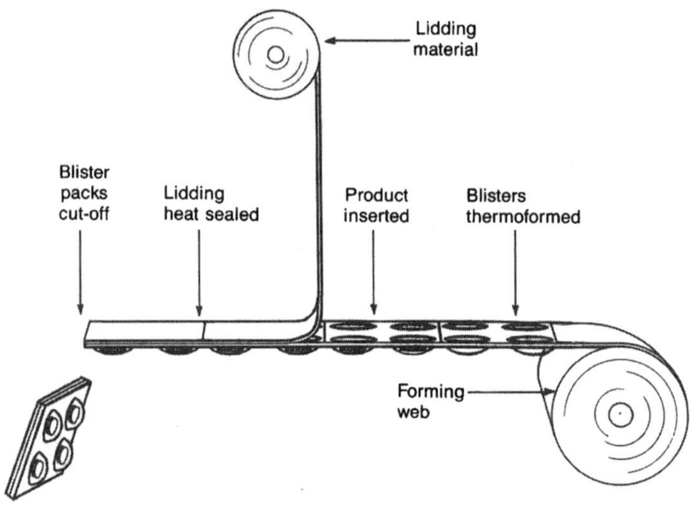

Figure 10 Thermoform blister machine schematic.

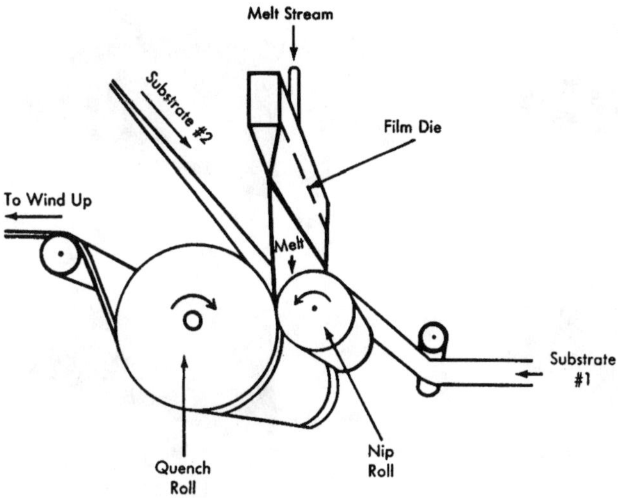

Figure 11 Extrusion lamination.

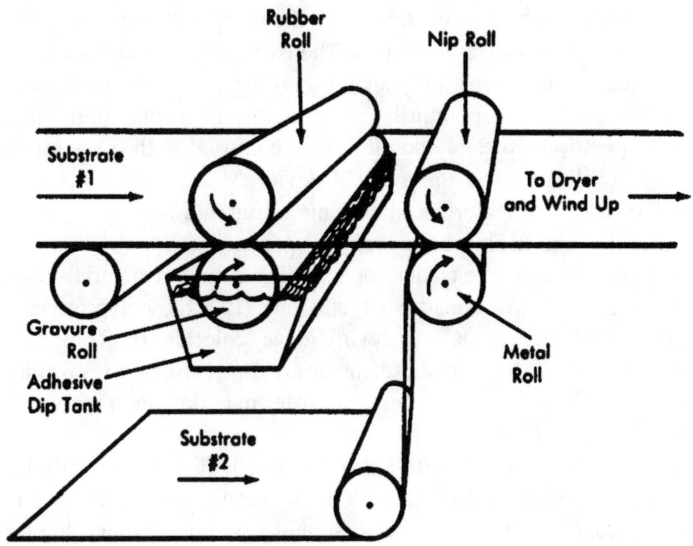

Figure 12 Adhesive lamination.

The distinction between blisters and strips is blurred depending on the pharmaceutical company employing the technology and the way the packaging manufacturer describes the package. A blister refers to the single cavity containing the drug. A strip refers to multiple blisters separated by perforations that deliver a unit quantity (day, week, etc.) of the drugs. Strips are multiple blisters

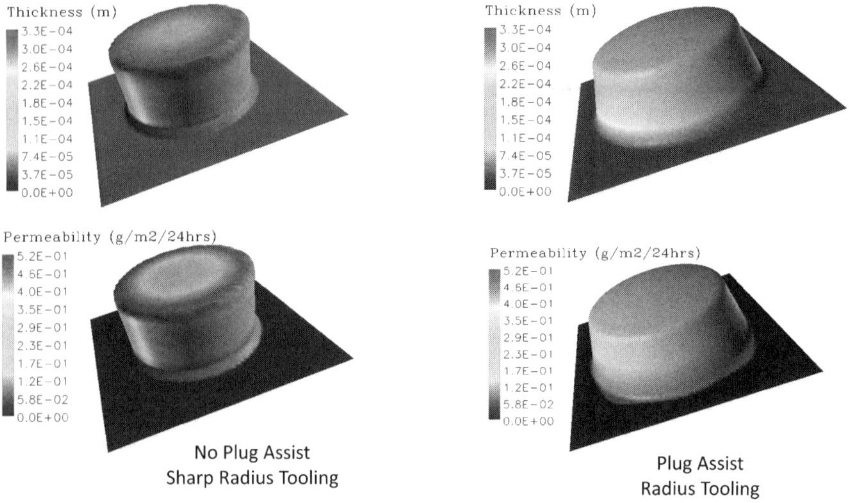

Figure 13 Comparison of material distribution and permeability.

formed, filled, and sealed at the same time and then cut and separated into prescription or over the counter-size quantities. The two terms can and are used interchangeably whenever this type of packaging is described or discussed. Blisters and strips are produced from multilayer or monolayer sheet material in a form, fill, and seal operation. Blisters should not be confused with other much larger thermoformed packages used for medical devices. Larger packages that contain medical instruments or materials, including a drug and its diluent, are a separate type of thermoforming for pharmaceutical packaging.

Blister or strip packages may be clear or opaque. The common materials used for these packages are PVC, low-density polyethylene (LDPE), polypropylene (PP), cyclic olefin copolymer (COC), polyvinylidene chloride (PVDC), and chlorotrifluoroethylene (ACLAR). All these materials are resistant to moisture transmission (Fig. 13). PVC is the most common blister material, and monolayer blisters are almost always PVC or PP.

The other polymers are laminated to a thin layer of PVC for compatibility with lidding (sealing) materials and to improve their thermoforming characteristics. Newer drug products and some super disintegrents require much higher barriers than those used in the 1990s and early 2000s. Laminated blister material is evolving into three-, four-, and five-layer structures, some with desiccants in one or more of the layers to meet the new high-performance demands (Figs. 14 and 15). The same is true for aluminum blister materials. Standard three-layer material is being replaced by four- and five-layer structures, and desiccant is available to further improve moisture resistance. The desiccant in this case counteracts moisture ingress between the foil-sealing material and the multilayer aluminum blister material.

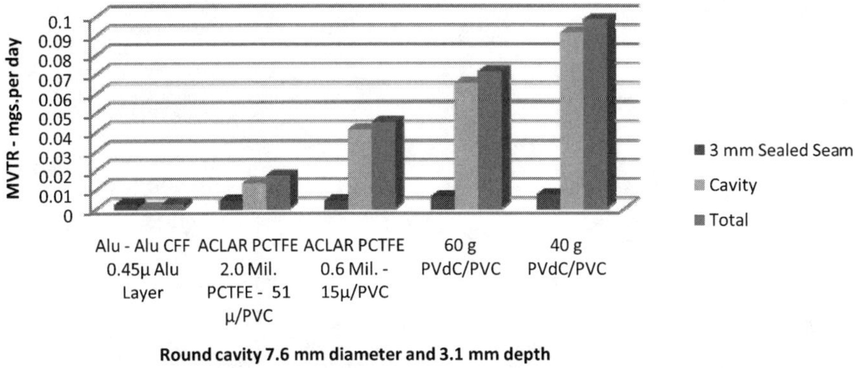

Figure 14 Comparison of MVTR on a formed blister cavity. *Abbreviation*: MVTR, moisture vapor transmission rate.

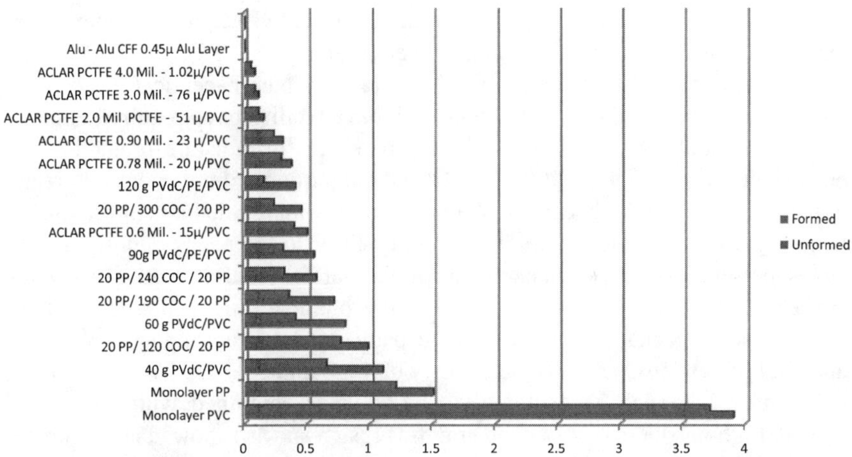

Figure 15 Comparison of formed and unformed blister material performance.

Plastic blisters are clear materials that use laminates of plastic, paper, and foil for lidding, the thin material on the back side of the blister used for sealing. These laminate combinations form the seal on a finished blister. Blister-sealing materials must provide a portion of the child-resistant attributes needed for blisters to meet U.S. Consumer Product Safety Commission requirements (15 USC 1471–1476). The laminated materials are supplied as peel-only and peel-push for child-resistant blister manufacture.

All materials for blisters are supplied in roll form. Blister materials start at thicknesses of 5 mils and are either monolayer or multilayer material. New high-barrier material uses polychlorotrifluoroethylene (PCTFE) in thicknesses from 4 to 8 mils laminated to 10 mils of PVC. The plastic structure can be made of more than one layer of material to impart different properties to the blister for protection of its

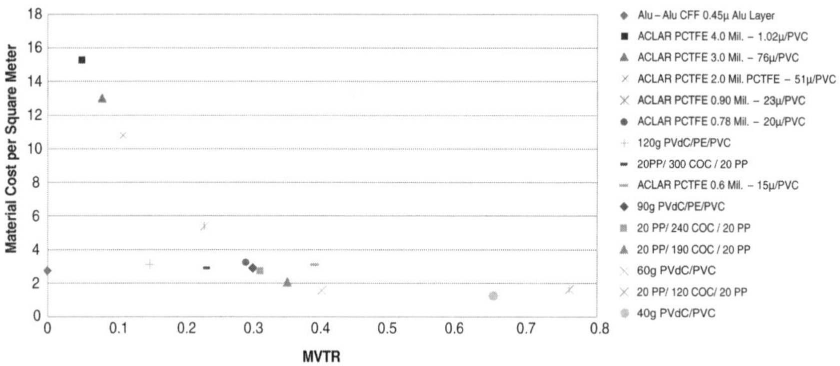

Figure 16 Cost and MVTR performance of blister materials.

contents. The fluorinated materials PVDC and ACLAR impart moisture resistance and oxygen barrier to blister materials when laminated to PVC. These materials are expensive, usually 5 to 15 times more expensive than plain PVC or PP (Fig. 16).

In the past decade, drug products [active pharmaceutical ingredients (APIs)] and new excipients that enhance bioavailability have required extreme barriers particularly to moisture. The improvement in moisture protection has been achieved by offering PCTFE (ACLAR) in particular at greater and greater thicknesses. PCTFE had been supplied up to a maximum thickness of 2 mils, but is now available in thicknesses up to 8 mils. Prior to the needs imposed by the new supersensitive drugs and excipients it was hard to justify the use of extreme-barrier materials. The overall caution of the pharmaceutical industry and the desire to keep packaging off the critical path for new drug application (NDA) approval of new products was the primary driver behind the use of these barriers. Other materials such as cyclic olefin copolymer combined with PVDC are another high-barrier choice. As the graphs (Figs 15 and 16) show, a large number of different materials used to achieve barrier in blisters are clustered in the same general area, making the choice and qualification a much more difficult decision for the packaging engineer.

Coating of plastic films to improve moisture transmission properties has become an accepted way of minimizing cost while producing films that approach the performance level of thicker, more expensive materials. Plastic extrusion coating (Fig. 17) extrudes a thin layer of plastic material onto a film or composite substrate. Plastic film coating (Fig. 18) is a dip process that then uses rollers and air to remove excess coating before it is dried or cured. In the first instance, the polymer material adheres to the substrate, and all that is needed is removal of the solvent. In the second case, the coating is exposed to UV light or heat to cross-link its components.

PVC has been a controversial packaging material for the past 15 years. Environmental concerns are one part of the problem, with the majority of the environmental questions centered around dioxin production when the material is

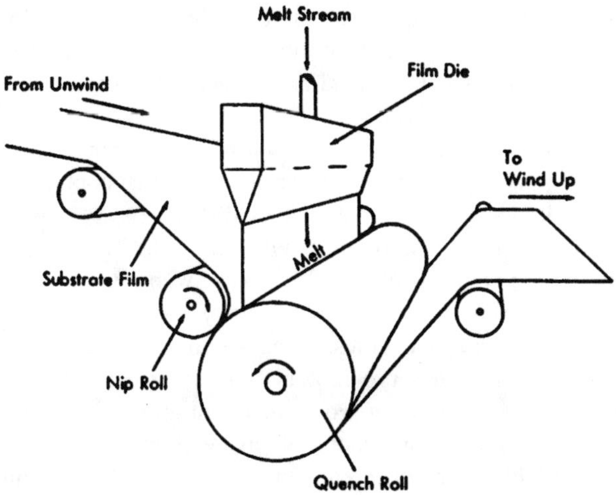

Figure 17 Schematic of an extrusion coating operation for flexible material.

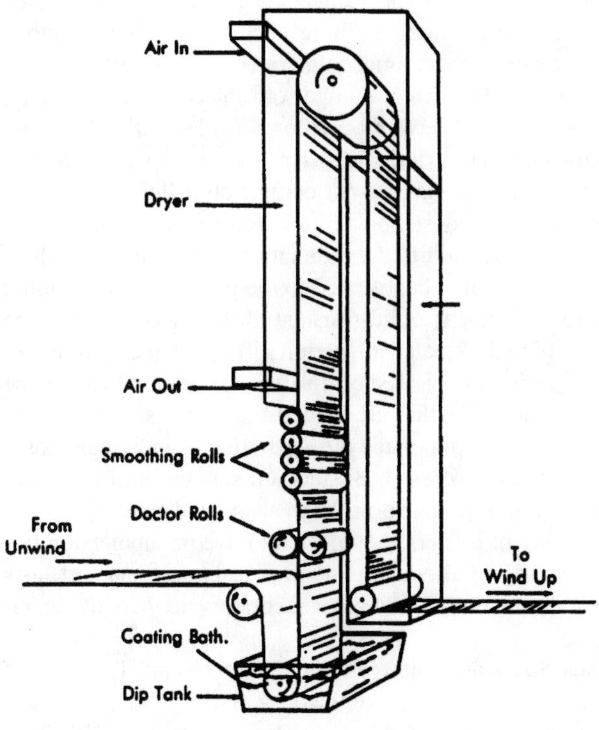

Figure 18 Schematic of dip coating of flexible materials with a drying tower. Multiple rolls smooth and adjust the thickness of the coating application and the drying tower removes solvent from thermoplastic coatings or cross-links (reacts) thermosetting coatings.

incinerated. Secondary questions regarding the possible leaching of vinyl chloride monomer have also been examined, and even though no deleterious effects have been documented, the questions have moved many package developers to use PP and LDPE.

LDPE and PP are alternatives to PVC and are used when superior sealing characteristics are needed for packaging oily products or products that contain an organic base. The materials do not possess the clarity of PVC.

Manufacture of the blisters from a web along with filling the blisters and sealing them are all done on the same machine. The plastic web material that forms the blister is heated and vacuum formed into a cavity. Newer machines form the cavity, even small and shallow cavities, using a plug assist. This is a male die that is mechanically (or with servo motors) pushed into the blister cavity prior to the application of vacuum to improve the distribution of material throughout the finished blister cavity. The graphs of blister thickness formed with and without plug assist highlight the difference in material distribution and permeation performance of the same package (Fig. 13).

It is not unusual for 20 to 30 blister cavities to be arranged in multiple rows on a blister machine. Following the formation of the cavity in the plastic, the tablet or tablets are placed into the cavities. Tablets are flood fed onto the web and fall into the cavities as one method of filling each cavity, or they move through a specially designed feeder that aligns guide rails or tubes with the rows of blister cavities and dispenses the correct number of tablets into each cavity. Sensors or an operator examine and detect any cavities not filled with tablets and automatically reject unfilled blisters. Modern blister operations will open and recycle tablets from strips of blisters that are not completely filled.

A specialized laminate film consisting of foil, foil/paper, coated paper, paper/plastic, foil/paper/plastic, or modified combinations of these materials is used to seal the blister. The material is heat sealed to the plastic web containing the blister cavity. To facilitate access to child-resistant blisters, a corner or some area of the blister is only partially sealed to permit lifting of the lidding for gripping and tearing. The entire web is sealed in this operation and then the blisters are cut from the web in strips that are loaded into cartons. Perforations between blisters are added at this step to make the separation of individual doses of product from a blister strip or card easy. Blisters are supplied in many configurations normally based on dosage or on package shape and size.

After sealing, a sampling of blisters is subjected to dye or submersion test in water to confirm the quality of the seal. The two tests are standard tests performed at the manufacturing line or within a laboratory as part of manufacturing quality assurance.

Blisters in the United States are required to be child resistant. The level of child resistance is determined by the toxicity of the drug in the package. Industry will refer to an "F" level usually represented as $F = 1$ or $F = 4$ as examples. This number refers to the maximum number of packages a child can access in testing specified by the "Consumer Product Safety Commission" (CPSC). The term does not appear in the regulations, but is commonly used by suppliers and pharmaceutical companies to describe pharmaceutical blister packaging.

Aluminum Blister Packages—Cold Form Blisters

Aluminum laminated to plastic in three-, four-, and five-layer structures is another common blister material. The material is formed on the same equipment as plastic blisters and follows the same heating and sealing processes. The difference between aluminum and plastic blisters is the use of a male die or plug to stretch the foil while molding the blister cavity. The amount of stretching and the draft angles (curves) in the blister shape are much softer and more rounded than those found with plastic packages. In general, an aluminum blister and a strip of aluminum blisters will be larger than plastic blisters because of the separation needed to form the material without creating pinholes or tears when stretching the metal. Foil tears or rips easily, and this is the reason it is always supported by a plastic material, typically, PVC or PET. Both PVC and PE serve as the heat seal layer in an aluminum blister laminate.

Blister packages are viewed as a method to enhance compliance. New FDA requirements specifying a bar code on each individual drug dose supplied in a hospital or nursing home make blisters a good packaging choice for automation and safety. Bar codes on each dose of product permits the introduction of scanning the patient identification bracelet and the drug before dispensing to reduce medical errors. Blister packaging is also gaining acceptance with large retailers who operate pharmacies. They eliminate the need for a pharmacist or an assistant to count tablets, saving labor and time.

Large Thermoformed Packages—Strip, Tray, and Clamshell Packages for Medical Devices

Larger packages, mainly used for medical devices, trays, or clamshells, are made by more involved methods of thermoforming. The manufacturing steps for forming these packages are similar to those described for blister packaging, but increase in complexity because of the size of the package. A plastic web is extruded (Fig. 19) to produce the starting material for thermoforming. This is much thicker than that used for tablet blisters, although the process is the same (Fig. 20).

The web is larger, and the thermoforming equipment used is much heavier. Heating and control of the hot web during forming is much more difficult. Thermoforming, either by vacuum only or by vacuum, pressure, and plug assist, is the plastic-manufacturing technique of choice for making these packages.

The key difference in molding large packages is the amount of design and development required to maximize material distribution without compromising package strength. The use of adjusting bolts or plastic feed variations are

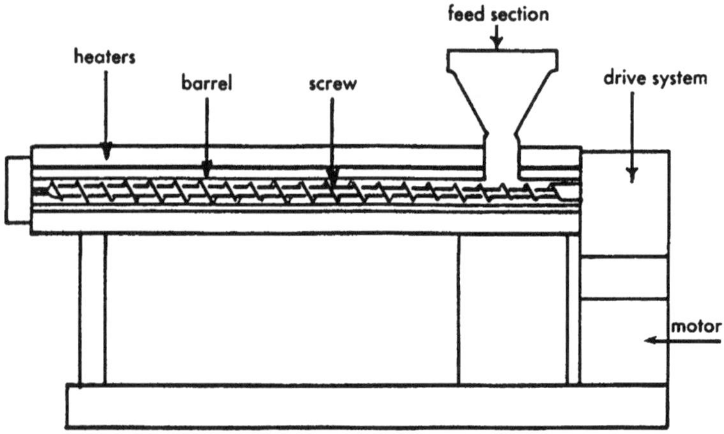

Figure 19 Film extruder schematic.

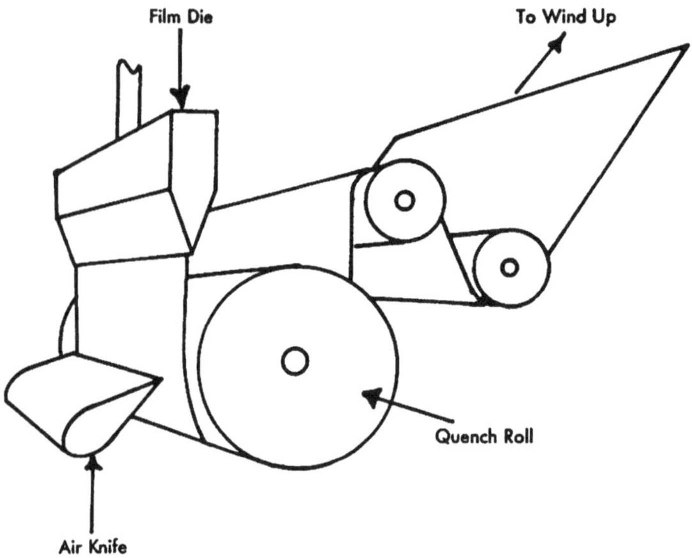

Figure 20 Flat die extrusion process.

methods used to produce starting materials that can enhance this material distribution (Fig. 21). Thermoforming, much like metal stamping described for draw–redraw cans, tends to leave a majority of the starting material at the bottom and top of a container, producing relatively thin and weak sidewalls. Plug design, cavity design, and programmed heating of the material being formed alleviates this problem to varying degrees, but it remains an issue with any thermoformed part.

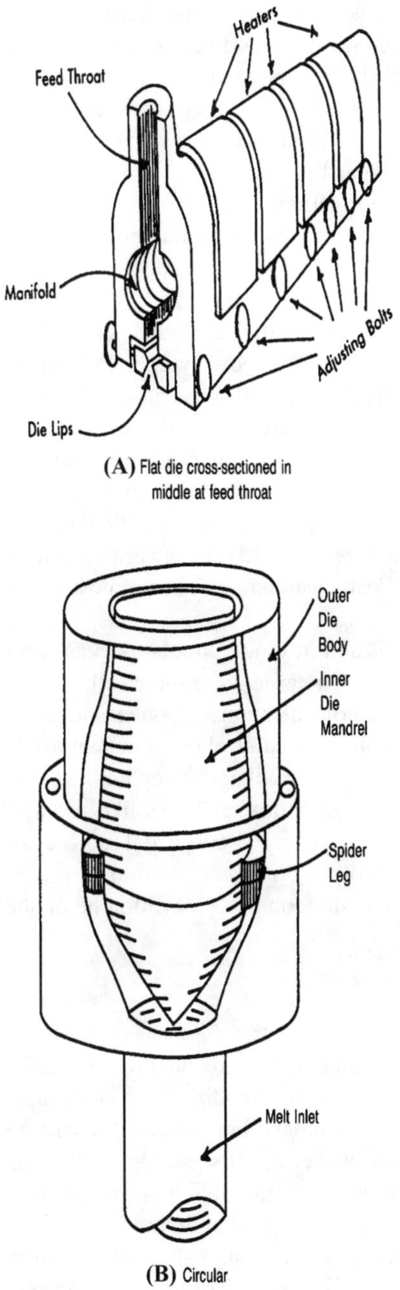

Figure 21 Flat and circular monolayer extrusion dies.

The hinged top of a clamshell package is made in a single thermoforming operation. The package is molded in the open position, and the design of the die induces thinning at the hinge. Interference fit blisters or snaps are part of the tooling design and molding to form a built-in closing device. Interference fit refers to the tight fit of a blister into a molded cavity in the package used for closing the package; the blister is slightly larger than the cavity into which it is pressed, creating friction that holds the parts together.

Injection molding (7) is sometimes used either as a stand-alone process or with a paper or plastic molded inserts to produce a small percentage of this style of large medical device or kit packaging.

The common materials used for large thermoformed packages include PVC, polyethylene terephthalate glycol (PETG), PE, polystyrene, PC, amorphous PET, and acrylonitrile butadiene styrene (ABS) materials. All these materials, with the exception of amorphous PET, have a good melt strength, making them easy to mold. Recycled plastic, both manufacturing waste and postconsumer in origin, may be used in thermoformed packages in noncritical applications. The recycled material may be extruded as the middle layer in multiple layers of a coextruded plastic structure (Fig. 22) if concern regarding contact with a part or product is present. Postconsumer waste is almost never recycled into pharmaceutical primary packaging.

Large thermoformed molded clamshells and trays are popular packages for medical devices. The shape of the interior of a large thermoformed tray or package can be molded to precisely fit and hold a medical device and other materials. They are strong and ideal for protecting multiple components held firmly in place in cavities created when molding the package. They are ideal for surgical or trauma kits, which contain both drugs and other items like needles, syringes, or dressings to treat a wound. In the case of surgical kits, the kit contains all the disposable materials needed to perform the operation or procedure. They are an effective way to guarantee a surgeon has everything he or she needs to perform a procedure.

Pouches

Pouches are a largely overlooked form of packaging that is excellent for delivering a unit dose of a drug product. They equal blisters in providing very convenient easy-to-transport and easy-to-use packages to consumers. For tablets, it is easy to see someone putting a pouched product into a pocket or purse and know that the tablet(s) will be protected until they are ready to use it, and that it is easy to open and access. Probably the best-known pouched product is Alka-Seltzer®.

Pouches can be fabricated at a packaging supplier and shipped to a drug manufacturer for filling and sealing, or they can be part of a form, fill, and seal operation within a pharmaceutical manufacturing operation. Pouches vary in size ranging from small flat sachets to large gusseted stand-up containers. They can hold solids or liquids. They can provide dispensing, measuring, and gripping

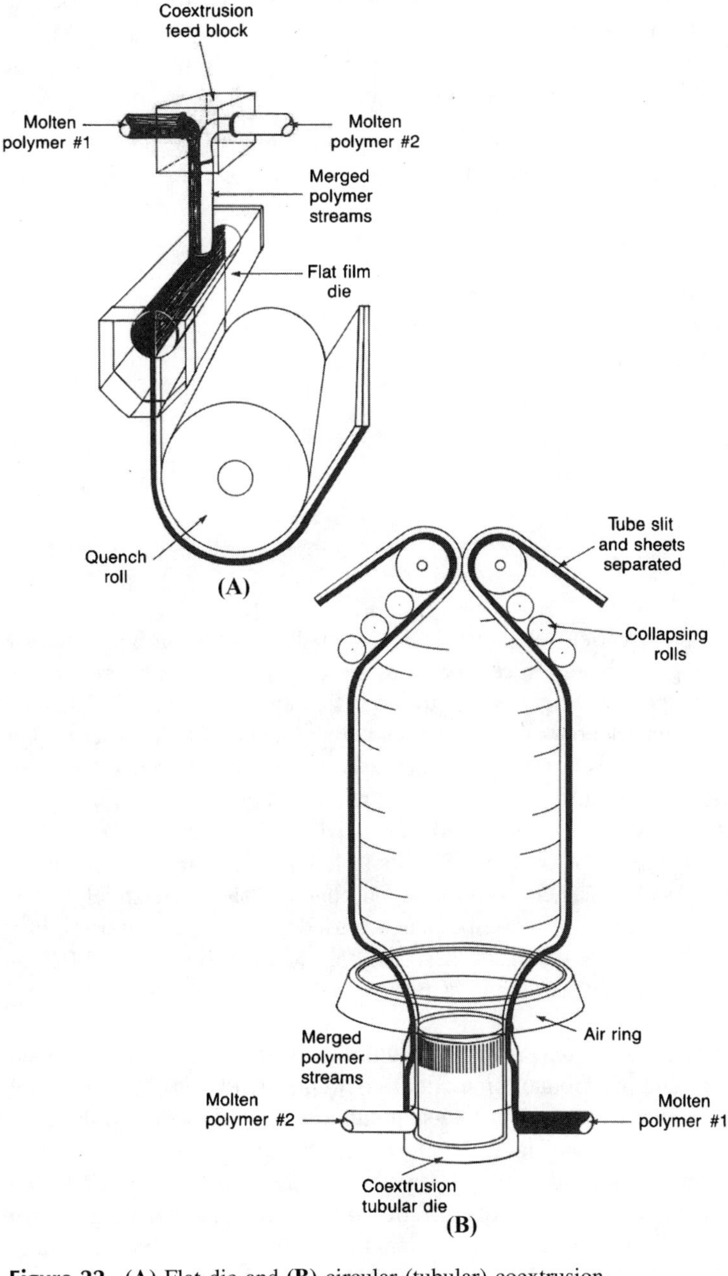

Figure 22 (**A**) Flat die and (**B**) circular (tubular) coextrusion.

features. Their versatility is just beginning to be recognized for both food and pharmaceutical applications.

Form, Fill, and Seal Pouch Operations

Form, fill, and seal pouch-making equipment is a common operation for most drug and pharmaceutical manufacturers. Pouch machinery is referred to as "horizontal or vertical." The term describes how the pouch is formed and filled. Pouches come in a variety of styles and use these designations as descriptors:

1. Three-sided fin seal
2. Four-sided fin seal
3. Four-sided no-fin seal
4. Single gusset
5. Double gusset
6. Pillow pouch
7. Sachet
8. Shaped seal
9. Other unusual configurations such as parallelograms, tetrahedrons, and Chubb

The complete variety of pouch styles is made on either a horizontal or vertical pouching machine. Fin seal pouches are packages with a seam running the length of the pouch at right angles to the sealed ends. It is the small flap or "fin," hence, the name. Four-sided no-fin seal are pouches that are either folded and sealed on three sides to form the pouch or are sealed on all four sides using two separate pieces of multilayer material. A pouch with one or two gussets is a package that has extra material folded into the bottom or sides to produce a package that can expand on opening. This looks like an accordion fold on one or two sides of a pouch. "Sachets" refer to small pouches that are relatively small and flat. This term is used to describe pouches used for desiccant insertion into bottles. Pouches are made with seals designed for breaking to mix two different components such as pouches for cold packs used to treat athletes at sporting events.

Pouches are made from single and multilayer materials specially designed for the package and the product protection requirements. Plastic film, produced with a circular die and cooled in a free-standing tower, produces single-layer or multiple-layer plastic films (Fig. 23).

Tyvek®, a product widely used for medical device pouches, is a nonwoven material made with polyolefin. Paper, plastic, foil, and composite materials form the wide range of material choices for making a pouch. Just about anything available as a thin flexible film can and is laminated together to produce the starting material for making a pouch. The various layers of material found in a multilayer pouch are produced by adhesive lamination, extrusion lamination, or a combination of these techniques. Individual layers may also be produced in a

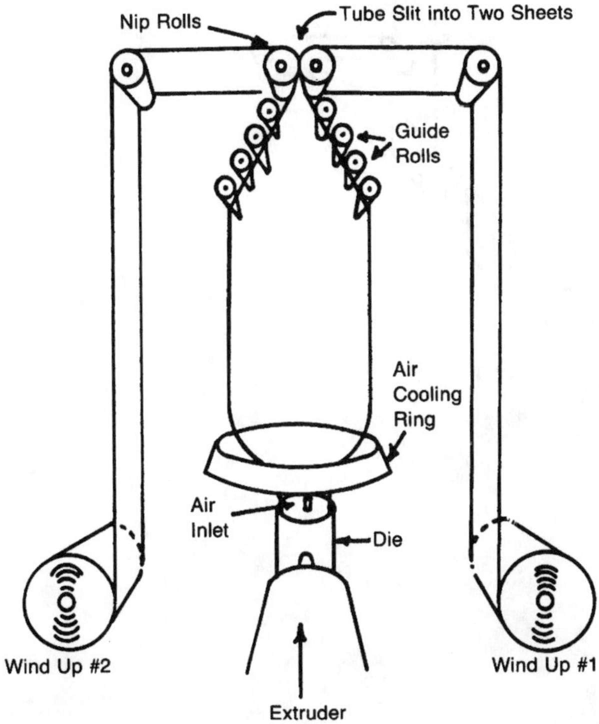

Figure 23 Circular die film extrusion process.

sandwich with only the outside materials fused together. Each layer contributes one or more attributes needed to either make the package or protect the product.

Horizontal Pouch Equipment Pouch machines, referred to as horizontal machines, are the type of equipment most often used for pouching operations (Fig. 24). These are intermittent motion machines that can handle pouches made in locations separate from the filling location or can be configured to make, fill, and seal the pouch in one operation. Premade pouches on a horizontal machine are separated and transferred one at a time to a conveyor that uses multiple grippers that open and hold the pouch open for filling and then move it to the sealing operation. If the pouches are produced at the point of filling and sealing, a pouch-forming operation is added. A horizontal machine normally produces a pouch with a fin seal, but can be configured to fold and seal the pouch-making material eliminating the side or fin seal. The intermittent motion used in horizontal pouch machines generally translates to a slower-speed (100–300 units/min) operation that maintains the pouches in a straight line through multiple filling stations and then to a final sealing station. A variation of the horizontal

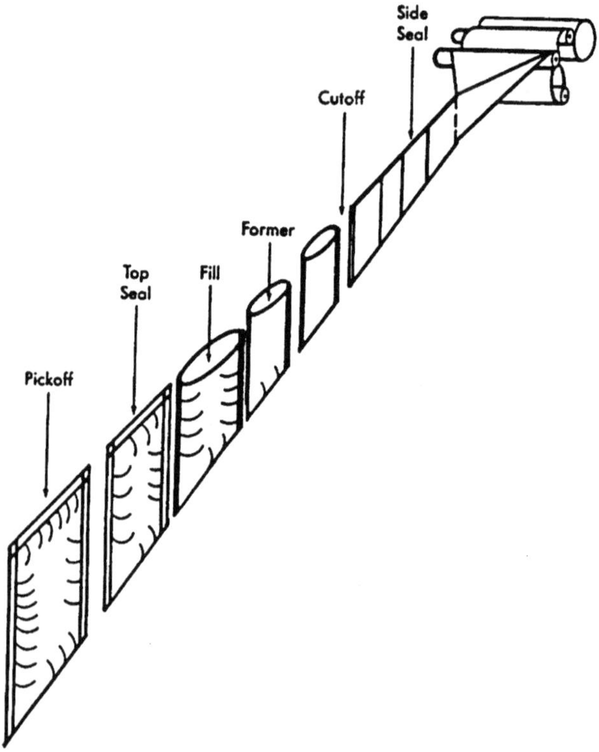

Figure 24 Schematic of a horizontal pouch machine.

machine employs a rotary arrangement to increase the speed of pouch manufacture and filling. A complete range of pouches, including gusseted pouches, can be made on a horizontal pouch form, fill, and seal line.

For a high-performance pouch operation with no oxygen or moisture ingress, a four-sided seal is required. Folding material to eliminate one seal creates the potential for a channel at the fold. All four seals on a high-performance package must be smooth and wrinkle free. Helium leak testing is used to determine the hermetic nature of high-performance pouches. Channels or minute leaks in the pouch will not be detected by dye or water immersion/vacuum testing but will be identified with helium leak test equipment.

Vertical Pouch Equipment The second type of pouch-making equipment is a vertical form, fill, and seal operation (Figs. 25 and 26). Again this equipment can produce pouches with and without a fin seal. A vertical pouch-making machine maintains the pouches in a vertical arrangement until they are cut and separated (Fig. 26). The material enters the machine and is pulled around a forming mandrel sometimes called a plow or horn.

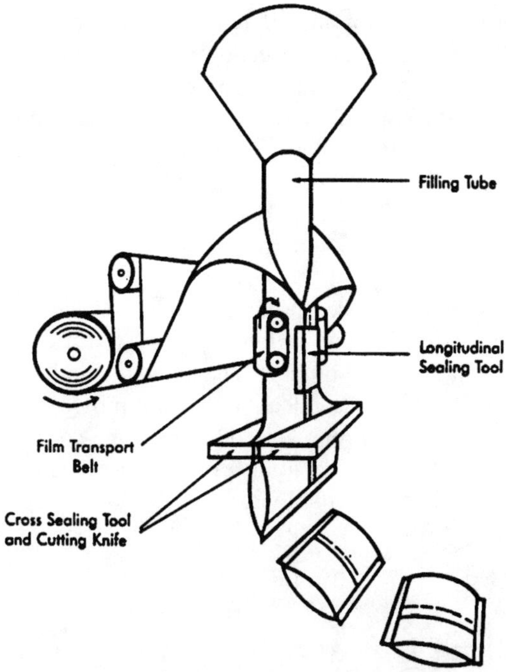

Figure 25 Vertical form, fill, and seal pouch machine.

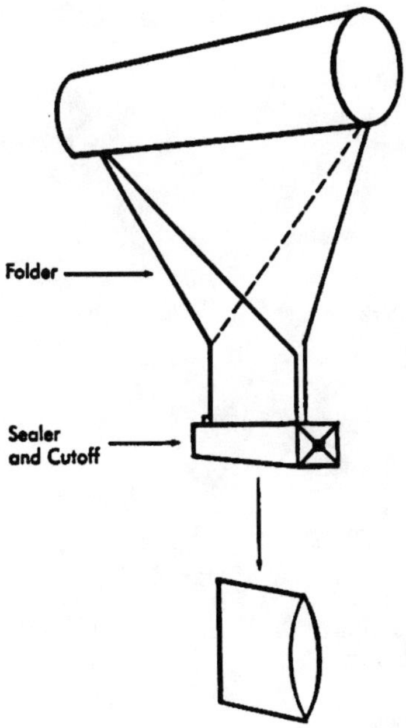

Figure 26 Vertical pouch-making machine.

It is sealed either with a fin or on two sides if two separate pieces of material are being made into the pouch. The second sealing operation uses the perpendicular seal as the closing seal and the bottom seal of a finished package and the next pouch to be filled. After the bottom seal is made, the product drops into the pouch, and as the material indexes down (vertically through the machine) the top seal is made.

Form, Fill, and Seal Bottles

The form, fill, and seal process for pouches, used primarily for solids and dry drug products, is also used for liquids in a slightly different form. An extrusion blow molded bottle becomes the package for the product. The blow, fill, and seal process (Fig. 27) is extensively used for pharmaceutical products. It provides an excellent unit dose option for eye drops, eardrops, and other over-the-counter products. The process can produce bottles ranging from 0.1 mL to as much as 10 L for irrigation products and some infusion solutions.

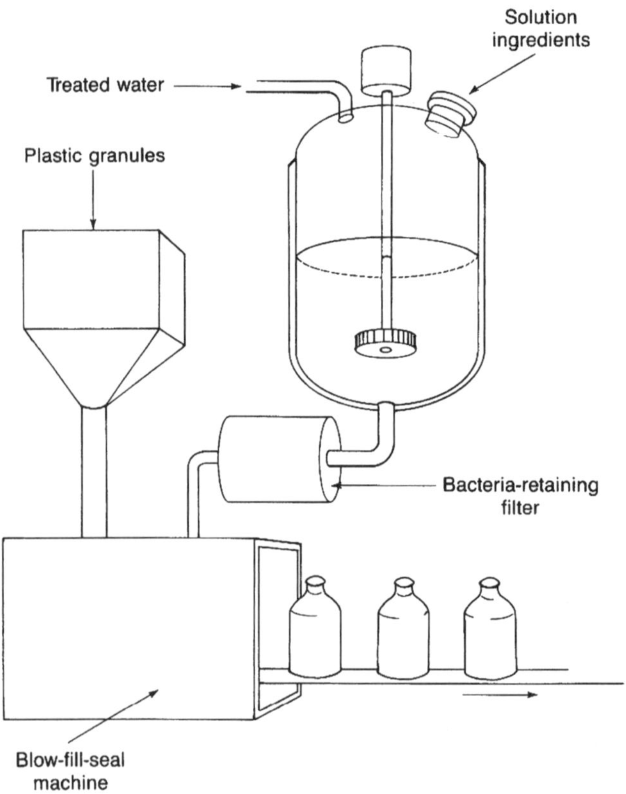

Figure 27 Overview of the blow, fill, and seal process.

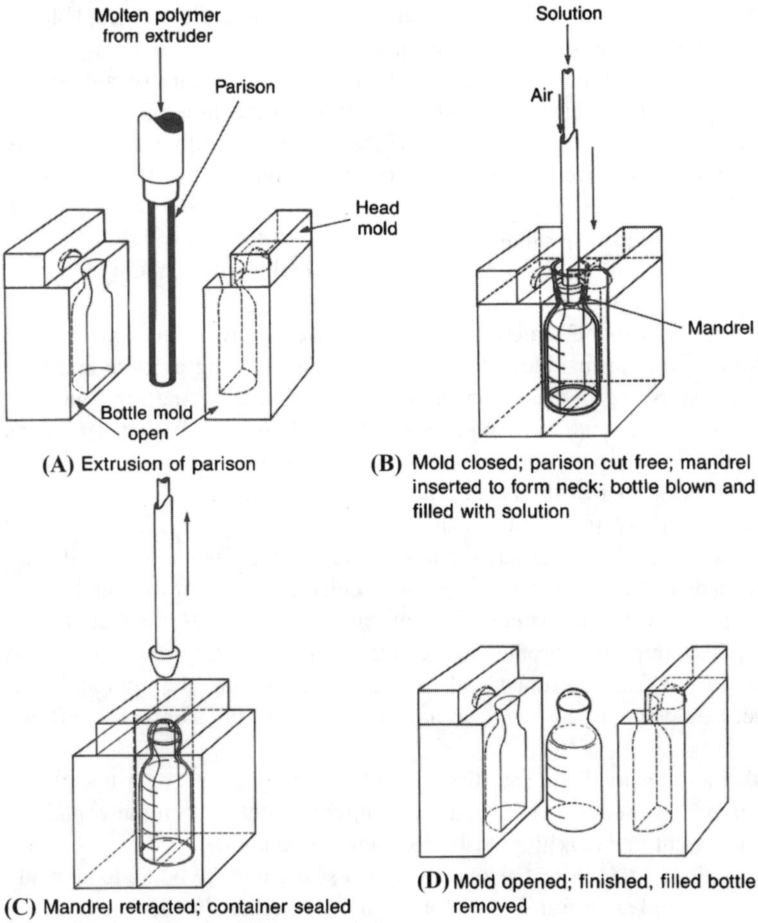

Figure 28 The blow, fill, and seal process inside an aseptic machine.

The process as practised for pharmaceuticals is aseptic (Fig. 28). The equipment takes advantage of the heat used to melt the plastic during extrusion and marries it with aseptic filling and sealing within the blow molding enclosure. The bottles are extrusion blow molded with HEPA-filtered sterile air. The sterile environment within the blow molding enclosure is maintained using sterile air, also produced by HEPA filtration, under positive pressure. The positive pressure within the enclosure prevents contamination during the blow, fill, and seal process.

The process eliminates the need for sterilization of bottles prior to filling, and it eliminates the bottle inventory required for operation. The bottle inventory is replaced by plastic pellets, the starting material for extrusion blow molding. The system uses LDPE, high-density polyethylene (HDPE), PP, and PE/PP copolymers as the primary bottle-making resins. Other plastics capable of extrusion blow molding may also be used.

Sterile product is prepared separately from the packaging operation and transferred by aseptic connections to the presterilized machine. The blow, fill, and seal units have a preprogrammed sterilization sequence that automatically cleans all interior surfaces. This would cover all internal surfaces, product passages, blow pins, valves, and molds contained within the sterile zone of the machine. A positive pressure laminar airflow across all the surfaces at the end of the sequence maintains the equipment in a sterile condition throughout the filling and sealing of product.

The process can be structured in two different ways. The conventional method follows the normal steps for extrusion blow molding a bottle and then internally transfers the bottle within the aseptic chamber of the machine to a second station where it is filled and sealed. The neck area of the bottle remains hot during the filling operation, and at the completion of filling, a second mold closes on the neck area of the bottle, forcing the hot plastic together and completing the molding of the bottle finish.

The second method inserts two tubes (which look like one unit) into the hot parison immediately after extrusion. One tube delivers sterile air to complete the blow molding of the bottle. After a delay of one or two seconds, the second tube delivers a preset amount of product to the bottle. The two tubes are retracted, and another mold encloses the open end of the bottle, sealing the contents with hot plastic. Bottles made this way can have twist-off openings or a twist-off cap (Fig. 28).

Both molding methods seal the contents in the package with hot plastic from the extrusion process. This creates a tamper-evident seal on the container. A person must cut or break the plastic to open the container.

A variation of the blow, fill, and seal process permits the bottle to be made in one location and then transferred for filling at another. In this method, the bottle is extrusion blow molded and resealed using the hot plastic from extrusion. The sealing area in this step is designed to be part of the bottle scrap created just before filling. This follows the steps described above without filling of the bottle. The bottle can then be stored and transported to a filling location. The interior of the bottle, which was sealed at the time of molding, remains sterile. No effort is made to maintain sterility of the exterior of the bottle. It is handled in clean conditions, following good handling practices to avoid gross biological contamination.

At the filling operation, the bottle is washed and resterilized on the outside. The top of the bottle is trimmed, and it is filled and resealed. An aseptic chamber that encloses only the neck finish of the bottle and an area large enough to introduce a presterilized closure also maintains the aseptic barrier between the outside and inside of the bottle. If a lined threaded closure is used to seal the bottle, the closure liner will contain a foil layer that is induction sealed to the lip of the finish.

Although this process has seen use outside the United States, many manufacturers are reluctant to use it because of one critical variable that cannot be tested. The seal made at the blow molder must be completely hermetic to maintain sterility inside the bottle. Without a method to test every bottle, the possibility of contamination during storage and transport is always present. The advantage of a system like this is the decoupling of the blow molder and the filler.

Plastic Tubes

Plastic tubes have slowly replaced metal tubes for ointments and other viscous pharmaceutical products. Tubes come in two different types, single layer and laminated (9). The single-layer tube may be coated, but in general does not provide true barrier properties. Laminate tubes are a composite structure that contains a barrier layer of foil in the laminate structure. The laminate tube is sometimes referred to as a glaminate tube.

Tube making is a two-step process. The first step is the manufacture of a tube, called a sleeve, and the second process is the attachment of the head or threaded end of the tube. The Downs process and the Strahm process are the two common manufacturing processes used to make and attach the head to a sleeve.

The first step in the process for making single-layer tubes is the manufacture of a uniform piece of tubing called a sleeve. The tubing ranges between 12 and 20 mils in wall thickness. An extruder forces a plastic material, usually LDPE, through an annular die to produce the uniform tube. As the hot plastic tube emerges from the die, some type of flame or electric discharge (corona) treatment is applied to the plastic to eliminate low-molecular-weight fragments from the surface and prepare it for coating. The tube is pulled across a chilled mandrel to cool and harden the hot plastic. At the end of the mandrel the plastic tube is cut into uniform lengths with a rotary knife.

Each sleeve is then coated and decorated. Because the tube is uniform and can be placed on a mandrel for rotation, the decorating and coating process is simplified. The surface treatment of the outside of the tube enhances the adhesion of the printing inks and the coatings used. The coating used in this step enhances the moisture barrier characteristics of the thin LDPE tube by as much as a factor of 10.

Following the manufacture and decorating of the tube is a second process called heading. Heading is the process that attaches the threaded conically shaped end to the tube. This is the dispensing part of the tube that receives a separate closure. There are two processes used to attach the head to the tube, the Strahm process and the Downs process.

The Strahm heading process (Fig. 29) is an extrusion insert injection-molding process for adhering the sleeve to the head. A sleeve is inserted into one end of a die, which is an injection mold. From the other end of the die, molten plastic is injected from a low-pressure extruder. The injection mold tooling holds

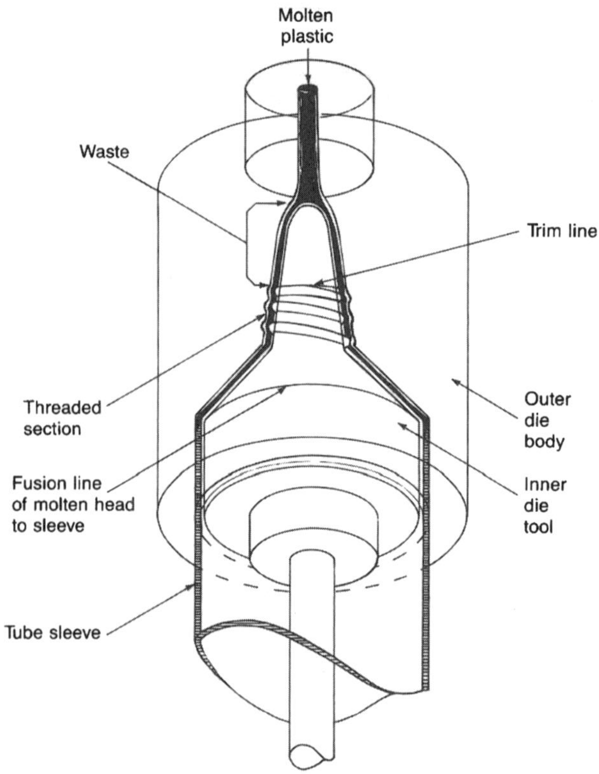

Figure 29 Tube heading using the Strahm process.

the sleeve in contact with the molten plastic until it cools, producing a fusion seal. After cooling, any excess plastic must be trimmed or removed from the fusion area. The injection mold may contain a method for the insertion of some type of pin at the opening end of the tube to mold the heading with a uniform orifice. The orifice can also be created later by cutting or puncturing the heading.

Following heading, the finished tube receives a closure and is packed for filling in some type of bulk shipping container. The open end of the tube is heat sealed after filling.

The Downs process (Fig. 30) is a completely different method for attaching the heading to the sleeve. This process uses a tool to carry the sleeve to a punch. The tool used in the Downs process is a male tool around which the sleeve is carried. As the male tool enters a punch, it encounters a web of hot plastic, the same plastic used to make the sleeve. As the sleeve and the punch clamp on the hot plastic web, the male tool moves forward and punches a hot plastic disk from the strip of plastic. The combination of the sleeve and the hot plastic disk is then moved to a second forming die that creates the shape of the end of the tube.

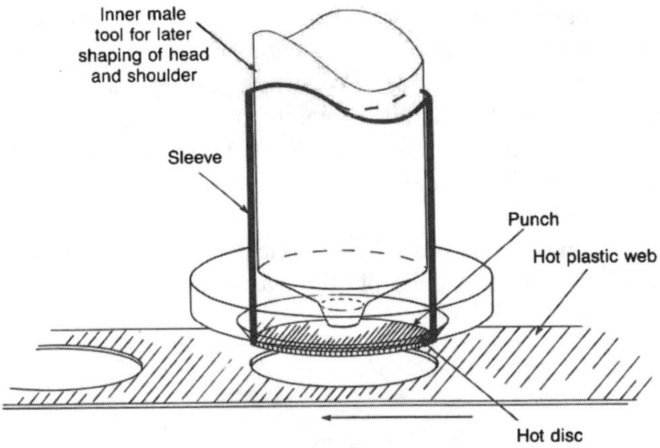

Figure 30 Tube heading using the Downs process.

The combination of pressure and hot plastic welds the sleeve to the heading, while the final shape of the heading is created by the mold. The Downs process eliminates the trim step of the Strahm process.

Laminated Tubes

Laminated tubes are very similar to single-layer tubes (5). The difference is in the construction of the sleeve. A material laminate made with between 6 and 10 layers of material, including 1 aluminum layer, is used to make the sleeves (Fig. 31).

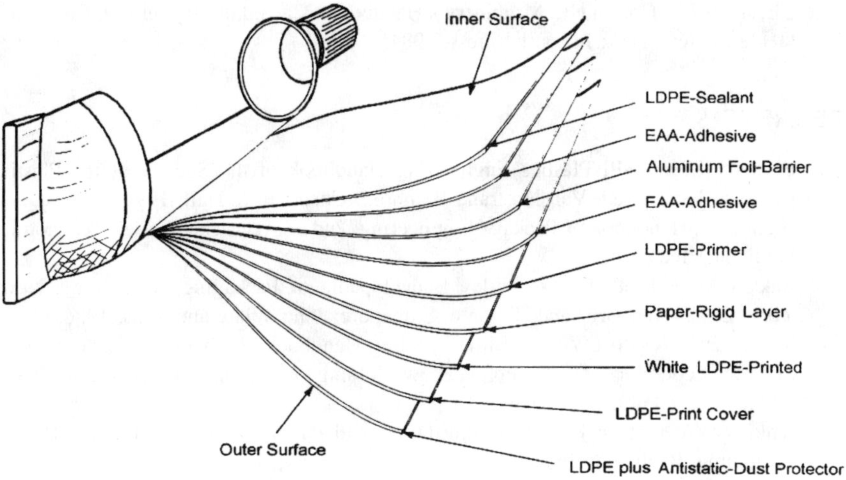

Figure 31 Structure of a laminate tube.

The laminate material is wound around a mandrel with a small overlap. Because aluminum is part of the laminate construction, induction heating is used at the overlap to create a seal on the side of the sleeve. Induction heating is a clean and easily controlled method for heating materials containing metal.

The sleeves are converted into tubes using the Strahm process. The molten plastic from the insert injection molding die locks the laminated sleeve into place with the head.

Printing and decoration of laminated tubes is done on one of the layers of the laminate before it is assembled. A clear plastic layer is used to protect the printing on the finished tube.

SUMMARY

Container fabrication is a very broad topic. Although this chapter highlights the methods used to produce packaging, it does not touch many of the nuances related to the technologies highlighted. There are many clever ways of making things, and pharmaceutical packaging has employed most of them to provide product to patients in convenient and safe ways. Improvements in all these technologies is ongoing, each generation of packaging fabrication equipment has expanded or improved on the last and has made the technology more competitive with other package-making technologies. New package machinery uses less material, less energy, and less mechanical parts to produce packages with amazing accuracy in dimensions and consistency.

FURTHER READING

Oberg E, Jones FD, Horton HL. Machinery's Handbook. 22nd ed. In: Ryffel HH, Geronimo JH, eds. New York: Industrial Press, 1984.

REFERENCES

1. Berins ML, ed. SPI Plastics Engineering Handbook of the Society of the Plastic Industry, Inc. 5th ed. Van Nostrand Reinhold: Chapman & Hall, 1991.
2. Hanlon JF. Handbook of Package Engineering. 2nd ed. Lancaster, Basel: Technomic Publishing Inc., 1992.
3. Bakker M, Eckroth E. The Wiley Encyclopedia of Packaging Technology. New York, Chichester, Brisbane, Toronto, Singapore: John Wiley and Sons, 1986.
4. Rosato DV, Rosato DV, eds. Blow Molding Handbook, Technology, Performance, Markets, Economics, The Complete Blow Molding Operation. Munich, Vienna, New York: Hanser Publishers, 1989.
5. Jenkins WA, Osborn KR. Packaging Drugs and Pharmaceuticals. Lancaster Basel: Technomic Publishing Inc., 1993.
6. Throne JL. Thermoforming. Munich, Vienna, New York: SPE Books, Hanser Publishers; Cincinnati: Hanser/Gardner Publications, Inc, 1987.

7. Pötsch G, Michaeli W. Injection Molding: An Introduction. Munich, Vienna, New York: Hanser Publishers; Cincinnati: Hanser/Gardner Publications, Inc, 1995.
8. The United States Pharmacopeia, USP 28 (28th revision). Prepared by the Council of Experts and Published by the Board of Trustees, 2005. Official from January 1, 2005, United States Pharmacopeial Convention Inc., 12601 Twinbrook Parkway, Rockville, MD 20852.
9. The Tube Council. How plastic and aluminum tubes are made, 2007. Available at: http://www.tube.org/i4a/pages/index.cfm?pageid=3282.

9

Sterilization Technology

INTRODUCTION

Product and package sterility are expected and assumed by anyone taking a drug. It is viewed as a requirement, a given, a normal expectation for all pharmaceutical products and all pharmaceutical packaging. Few people realize that producing a sterile product means sterilizing the product and the package in the same process. Sterilization places many unique demands and requirements on a package. Sterilization is a key process component for any product and package development program. It is here that the two entities, product and package, become inextricably linked. The packaging is considered part of the product by the FDA and must be capable of preventing any contamination or adulteration of the product. What few outside observers do not realize is that the packaging provides the means for sterilizing the product and that the primary package is the most important barrier for maintaining sterility. Many products are sterilized after packaging, particularly medical devices that require intricate assembly and contain many different materials and components. Many of the components and materials cannot be sterilized by themselves and must rely on the packaging to protect them while they are undergoing sterilization. Even simple items like infusion sets have multiple components that must be sterilized after final assembly. These devices are made of many diverse materials needed to make them work, which complicate their packaging and sterilization.

One must understand, at least on an introductory level, the microbiology of the harmful agents and the techniques that are effective in eliminating them. This is another dimension that a packaging engineer must consider in designing packaging for a product. The choice of sterilization method requires the packaging professional to understand how the materials, seals, and characteristics of the package will behave during sterilization. Tablets and solid dose forms are relatively easy to navigate through this requirement, but liquids, medical devices, ointments, gels, and other pharmaceutical products can be extremely problematic. Products that are heat or light sensitive, for example, pose difficult packaging challenges along with difficult sterilization challenges.

The range of sterilization methods is broad and encompasses everything from heat and steam to ionizing radiation. Sterilization of products takes many different forms and may require more than one sterilization technique to achieve complete sterilization of the product and package (1). Each sterilization technique has its strengths and weaknesses, and the one chosen, first and foremost, considers the chemical attributes, stability, and physical properties of the product along with the capabilities and requirements of the package (Table 1). Absolute sterility is the goal, but it is not always achievable for all products. Some sterilization techniques are limited in their capabilities because the process has built-in limitations, for example, sterile filtration. This process will not work with large molecules that approach or exceed the size of a virus. The idea of commercial sterility, discussed in chapter 4 on medical foods, and the sterilization requirements for nutraceuticals and food products that fall between true consumer food products and pharmaceutical products differ from the absolute sterility required of drugs, medical devices, and sundries. Because of these variations and diverse requirements, this overview of sterilization will discuss common techniques, their limitations, and what can be achieved with the technique.

Sterilization has become multifaceted with new infectious agents like prions (2) placing significantly greater demands on existing methods and systems, while new products, sensitive to many existing sterilization techniques, require new or modified methods to achieve sterility in the finished package. Many new products require innovative solutions for sterilization to solve the problem. This is particularly relevant when the targeted microbe is resistant to the sterilization techniques compatible with the molecule or drug. The needs and demands of sterilization are all part of the design and development of new packaging. Sterilization is a hidden and extremely important component of the manufacturing system needed to deliver a product. Understanding the different methods used for sterilization of products and how they impact both the product and the package are part of the project plan for new product development. Drug, package, and sterilization systems are the three components necessary for the delivery of any product. They are like a three-legged stool, where each leg or component is required, and no two components alone can make the product work.

OVERVIEW OF STERILIZATION REQUIREMENTS

The world is full of dangerous microscopic creatures and biologically active molecules. Microbiology, the microscopic world of these creatures, is a fascinating and strange world of things that are both beneficial and harmful. Fungi, mold, and spores are part of this microscopic world. Microbiology is the study of

Table 1 Comparisons of Sterilization Techniques

	Heat		Chemical			Radiation	
	High temperature/ pressure/steam (Autoclave)	Dry heat	EtO	Hydrogen peroxide Plasma	Glutaraldehyde	γ	Other E-beam
Temperature (°C)	121–148	160–190	50–60	45–50	20–45	30–40	20–30
Cycle time	10–90 min	Minimum 2 hr	6–20 hr	45–90 min	12–20 hr	10 hr	Varies by type of item
Sterilant	Steam	Heat	EtO gas	H_2O_2 and plasma	Glutaraldehyde	High-energy photons	α- particles (electrons)
Environmentally friendly	Yes	Yes	No gas residuals	Yes	No	Yes	Yes
Advantages	Safe Reliable Well understood Available Economical Fast	Can be used with powders and materials that are moisture sensitive Does not cause rusting of steel instruments	Reliable Low temp Well understood Handles a wide variety of materials Handles heat-sensitive equipments Handles moisture-sensitive equipments	Safe Low moisture No residue No toxic by-products No aeration or flushing required Low temp	Reliable Simple Low temp	Reliable Safe Simple Fast Cost Competitive	On/off Source Simple

Table 1 Comparisons of Sterilization Techniques (*Continued*)

	Heat		Chemical			Radiation	
	High temperature/ pressure/steam (Autoclave)	Dry heat	EtO	Hydrogen peroxide Plasma	Glutaraldehyde	γ	Other E-beam
Disadvantages	Wet heat Not suited for soluble materials Unsuitable for heat-sensitive materials	Long cycle Damage to heat-sensitive materials	Long cycle Potentially hazardous to operating personnel Costly Long turnaround time with verification	Expensive Small chamber size Damages nylon materials	Costly Short shelf life Long exposure times Hazardous to personnel	PVC PTFE and acetal are affected by γ-radiation	Limited penetration of solids

Abbreviations: EtO, ethylene oxide; PVC, polyvinyl chloride; PTFE, polytetrafluoroethylene (Teflon®).

small microorganisms and their effects (3). The organisms are unicellular or cell cluster microscopic. Viruses, although not strictly considered microorganisms, are part of this field of study. Some basic background in microbiology and the various organisms and proteins that cause disease is necessary to understand what sterilization is designed to eliminate (4).

Microbiology is in its infancy compared with other biological sciences like zoology or botany. It is a broad term that includes bacteriology, virology, parasitology, and mycology to name a few of its subdivision specialties. It is estimated that less than 1% of all microorganisms have been identified and studied. Although considered part of the microbiological world, even smaller entities, viruses and proteins, are every bit as effective in producing beneficial effects on animals, plants, and humans as well as carrying and transmitting disease. Viruses that cause influenza (flu) and extreme diseases like HIV and hemorrhagic fever are part of peoples' consciousness just as bacteria with disease names like staph and pneumonia have been for generations. The potential of a pandemic from a new virus is quite real, whether it is a new strain of influenza like the Spanish flu of 1918, or the worrisome bird flu or avian flu identified in Asia. The study of viruses and more recently prions has increased significantly and is part of the revolution in genomics that is part of the ongoing changes in biological sciences. The study and funding for this research now match and in some cases exceed that devoted to understanding bacterial disease. These entities need increased understanding to better combat the potential health threats they pose.

The larger microbiological world of eukaryotes, which refers to fungi and protists, and prokaryotes, which refer to bacteria and some forms of algae, are the more studied parts of microbiology. Abundant evidence from the time of Pasteur and Koch has proven that many bacteria, fungi, mold, and spores are transmitters and causation agents for disease. Pasteur and Koch are considered the founders of modern microbiology.

Initially viral diseases were not understood, although Edward Jenner in 1796 successfully used cowpox to vaccinate a boy against smallpox. Eugene A. Rolfjohns first identified a virus as the cause of disease in 1892. His work showed that the tobacco mosaic virus was transmitted to other plants from extracts that had passed through filters too small to permit the passage of bacteria. Martinus Beijernck in 1898 showed that the extract was not a toxin, but rather something that grew in the host cell. The question of whether a virus is alive or dead is still a question for debate today. A virus cannot grow or reproduce outside a host cell.

A virus consists of strands of DNA or RNA encapsulated in a protein shell called a capsid. Viruses infect cellular forms of life and are classified as animal, plant, or bacterial viruses. Viruses that attack bacteria are called bacteriophages.

Bacteria, viruses, spores, algae, and fungi are the common agents of disease, and a large amount of pharmaceutical research and a wide variety of products have been developed to counteract the conditions they cause. Remarkably, humans without understanding the reasons behind diseases have developed methods and procedures from the earliest times to kill or mitigate the effects of these unseen organisms in food, on surfaces, and in the treatment of wounds. The Romans actually flamed any medical instrument used for the treatment of a disease or wound, and wine was used as an antiseptic.

The ability to identify new and potentially dangerous proteins is an example of our expanding understanding of disease. Our life spans have increased directly from our understanding of microbiology and how disease is transmitted. Public health measures that cleaned our water and air to create a better living environment are good examples of widespread sterilization techniques. As society gained understanding about the common causes of disease and acted to mitigate some of the most damaging ones, we began to enjoy the benefits of living longer. Long life has revealed many new diseases that were previously unknown and are only found as life spans increased. From study of the diseases of old age, the prion was identified as the causation agent for Alzheimer's disease.

Sterilization is the process that eliminates, kills, or inactivates any organism capable of compromising a product or causing disease. Microbiological agents are present everywhere. They can be infectious agents clinging to hard or soft surfaces. They can be a component of a raw material. They may be airborne in an aerosol droplet from a cough or sneeze. They may be growing within our homes (mold and mildew). They can be present in processed foods and not cause harm (thermophiles—bacteria inactivated to the point they can only grow in very unusual conditions), or they can be present in food and be beneficial like the bacteria found in yogurt.

Pharmaceutical products require complete sterility. Surgical instruments, blood, fluids, parenteral drugs and syringes, and anything that is introduced into an opening in the body has the ability to cause severe disease.

Heat, chemicals, ionizing radiation, and combinations of these agents all are used to sterilize drugs, implants, medical devices, analytical testing materials, and the manufacturing environment used to produce a product (Table 2). Each has its strengths and weaknesses regarding application to a particular circumstance or need. Each requires understanding and testing to prove that they produce complete sterilization of a medical product.

Heat is typically the first choice for sterilizing drugs. Medical foods, discussed in chapter 4, provide some insight into food sterilization with heat. Heat as a sterilization method is not the only option available. Other methods of sterilization including ionizing radiation, ultraviolet (UV) light, filtration, and chemicals are all possible alternatives to heat sterilization. Many of the agents that attack the body have developed defenses to many of the sterilization methods described. Many of the drugs and the device products are equally sensitive to different sterilization methods, and understanding how to achieve the desired elimination of infectious agents without damaging the product is an area of study broad and unique in scope (Table 1).

Table 2 Overview of Sterilization Techniques

Description	Common methods	Other methods
Heat/temperature/pressure	Steam autoclave	Dry heat
Chemical	EtO gas Hydrogen peroxide vapor	Hydrogen peroxide vapor and plasma Chlorine solution Ozone Glutaraldehyde Formaldehyde
Radiation	γ-Rays	Electron beam X-rays UV light (Limited applications)

Abbreviations: EtO, ethylene oxide; UV, ultraviolet.

In the last few years, a new disease-causing agent, called a prion, has joined the list of disease-causing agents, and it is part of this discussion on sterilization. A prion attacks the body in a completely new way and is the attention of a large research effort into neurological disease. Proteins are prions, and a prion is a new entity added to the list of dangerous substances that transmit disease. The name, prion, is short for proteinaceous infectious particle and refers to an infectious agent comprising only protein. A prion does not contain nucleic acids, as do all other infectious agents, and are only now beginning to be investigated and understood. A prion is a protein agent that can transmit disease.

Stanley Prusiner discovered prions in 1982 and showed that they are the cause of scrapie, a neurological disease found in sheep. His pioneering work was recognized with the Nobel Prize in Medicine in 1997. Bovine spongiform encephalopathy (BSE), referred to as "mad cow disease," is caused by prions infecting cattle. Creutzfeldt–Jakob disease is an analogous version of this neural disease in humans. It should be noted that a number of scientists dispute the idea of a prion as a causation agent for disease and point to an unidentified slow-acting virus as the real disease carrier. This objection is slowly fading but remains a point for debate.

All identified prion diseases affect the brain or neural tissue, and all are fatal. There is no medical treatment available at this time. Prions infect and propagate by refolding a normal protein into an abnormal configuration. A prion, once formed, has the ability to cause abnormal folding in other normal proteins. This is somewhat analogous to a free radical in chemistry that once formed can cause a large number of molecules to change. Prions create a fold in a protein referred to as an amyloid structure, and as we discuss sterilization methods, it is necessary to note that eliminating this extremely stable form of a protein with common sterilization methods is extremely difficult.

This overview briefly outlines the microscopic entities we must understand and eliminate. The world of the microscopic entities is a broad and diverse world that has only been partially understood. Tremendous amounts of information are available on it and on many of the disease causation agents. People and industry have successfully overcome most of the difficulties in sterilizing our food and drugs. Think how big the news story would be when a food or drug has to be recalled because of inadequate sterilization. This does not happen by accident. The application of the techniques discussed in this chapter must be almost foolproof. They require long-term development and constant surveillance once employed to reproducibly deliver safe and efficacious products.

HEAT STERILIZATION TECHNIQUES

Sterilization Using Steam and Pressure (Autoclave)

The use of heat, in many different forms, is the first choice for sterilization. It is well understood, and it is familiar to regulatory agencies charged with approving the sterilization of pharmaceutical products and procedures. Most large companies employ an individual or group of individuals to oversee development and review of sterilization procedures used in manufacturing. Smaller companies will hire an outside consultant, recognized as such by the Food and Drug Administration (FDA), to develop and put in place their procedures for sterilization. The "process authority," the individual or group charged with developing safe sterilization techniques for products, almost always starts with heat as the first choice for product sterilization.

Heat can be applied to a product from a variety of sources: infrared, wet and dry steam, and heated air (sometimes called dry heat). The use of heat as a method for sterilization is both well known and well understood. It is used extensively for food and other products requiring sterilization, and not just for pharmaceutical products. Flaming, the application of a flame to a surface requiring sterilization, is another way of using heat to sterilize an object.

Boiling water, or heat applied at 100°C (212°F), will kill most bacteria and disease-causing microbes, including the spores of some bacteria. This treatment does not render a product or a food completely sterile, it only kills the majority of microbes and organisms in the product making it safe for a person with a healthy digestive and immune system.

Pharmaceutical products require a much more stringent treatment to render items sterile. The most common way to achieve this is in an autoclave. An autoclave is a pressure vessel that reaches temperatures above 100°C by increasing the pressure inside the chamber. Steam is introduced into an autoclave and held at 103 kPa (15 psi) permitting the temperature inside the vessel to reach a range of temperatures between 121°C and 134°C. Holding a medical instrument or liquid at this temperature will effectively sterilize the item. The difference in the two temperatures equates to a time difference for the exposure cycle, normally a range between 15 and 3 minutes, respectively, for an item undergoing sterilization. The higher temperature effectively reduces the time needed to sterilize the item. This time frame is conditional and is based on the proof that the high-temperature-saturated steam penetrates and heats every item being sterilized and that the items are held for a sufficient period of time to reach and remain at the maximum temperature of the sterilization cycle. This requires that preparation of the load of items being sterilized, and their arrangement in the autoclave permits steam and heat to permeate through the items and to heat all surfaces. Testing and evaluation of the loading of an autoclave to insure that the steam and heat penetrate to all parts of the load is important. Testing can determine if items are shielded from steam and heat penetration, or if items do not reach lethal temperatures because too much mass (e.g., too many instruments) has been loaded in the autoclave. If empty jars or containers or laboratory items like Erlenmeyer beakers are being sterilized, the closures must be removed and placed in the autoclave in separate locations. Again, this is to permit the high-temperature steam to touch all surfaces and heat them to sterilization temperature. Thermocouples, temperature indicators, and biological indicators are a few of the methods used to test the effectiveness of an autoclave sterilization cycle and the amount of loading in the autoclave.

Autoclaves must vent any air within the sterilizing chamber as part of the process. By venting the air, the atmosphere within the chamber is completely saturated by the high-temperature steam. Residual air within the chamber prevents the complete introduction of saturated steam.

Food is sterilized in the same manner. In the home, a pressure cooker is analogous to an autoclave. The high pressure within the vessel allows steam generated within the cooker to reach temperatures similar to those found in an autoclave, thereby cooking meat and vegetables faster than normal heating, boiling, or roasting methods. Many foods, just like drugs, are very difficult to cook or sterilize, and care must be taken to insure that the temperature inside a food or drug reaches the temperature required to render any microbe harmless.

Liquids and their containers, when sterilized in a pressure vessel, must be heated and cooled under controlled pressure and temperature conditions to avoid expansion and boiling when pressure is removed. Both food and pharmaceuticals share similar equipment, designed to ramp temperature and pressure up and down in a controlled manner to keep a liquid from instantly boiling when the pressure is released and to keep the container from deforming or losing integrity. Extremes of temperature on an enclosed liquid produce pressures and vacuums inside containers. By matching the outside pressure in the sterilization chamber with the inside pressure of the container, the stress and strains on the package are minimized.

The autoclave, found in most doctors' and dentists' offices, is used to sterilize medical instruments. The proper use of an autoclave will eliminate bacteria, fungi, spores, and viruses on surfaces and within the hard-to-reach parts (nooks and crannies) of medical instruments. The time it takes to sterilize instruments is determined by the quantity of instruments loaded into the autoclave and the time it takes for heat to penetrate trays or stacks of items undergoing sterilization.

It is extremely important to remove dirt or extraneous materials from an item undergoing sterilization. Dirt, grime, oils, or biological material (blood, tissue, etc.) should all be removed from a surface before autoclaving. These contaminants can act as insulating blankets that may protect the microorganism from the heat of the autoclave, permitting it to survive. Prior cleaning removes a large amount of the visible microorganisms. Physical scrubbing mechanically removes organisms from smooth and accessible surfaces. Hot water with detergent will aid and improve the process. Some organic materials may coagulate when subjected to hot water, and care must be taken to insure that materials that simply fall out of solution must be rinsed and removed from the surface of the item prior to placing them in an autoclave. This is important because the time/temperature cycle of sterilization programmed into the autoclave or heat treatment is based on a likely bacterial load carried into the process by the items being sterilized. The sterilization cycle of the autoclave and the worst-case biological load of organisms to be killed are determined by testing with biological indicators. If the amount of bacterial contamination carried into the autoclave is greater than what was used to develop a worst-case scenario, incomplete sterilization may result.

Normal bioburden testing is done and designed to determine an expected range of viable organisms on a medical device, container, or component that must be eliminated in the sterilization process. It is conducted on medical devices and manufactured items after all manufacturing and assembly steps are complete. It may also be done as a spot check to determine if the amount of biologic materials carried into an autoclave will render the cycle ineffective. This step may be done after simple washing or cleaning of items before they are placed in the autoclave, and a procedure detailing the cleanliness level of the materials to be autoclaved is developed.

Chemical cleaning with alcohol or another astringent, as well as washing of items with soap or detergent, and hot water prior to the autoclave cycle is effective in eliminating microbes, fungi, and viruses. Because conditions cannot be closely controlled, the cleanliness achieved is variable. By adding an autoclave cycle to the cleaning and astringent procedure, a controlled and proven level of sterilization is achieved.

Packaging is a part of the sterilization process. Items may be packaged and then sterilized, or the packaging process may take place immediately after sterilization. Packaging must provide a barrier throughout the sterilization process, and be robust enough to maintain package integrity throughout the heat and pressure cycle. It must continue to maintain this integrity until the product is used, meaning sterilization cannot degrade the package to later compromise product protection.

Packaging must be bioburden tested to determine if it introduces a significant amount of bioburden, or if part of the organisms targeted for sterilization (bacterial load) is present and to what amount as part of the packaging. Package contamination may come from the manufacturing process (oils, dirt, etc.), or may be introduced when a packaging component is packaged and shipped from the manufacturer to the customer for filling and closing. Examples would be cans or jars bright stacked or palletized in bulk for shipping to a filling operation, or plastic bottles palletized or bulk packaged for shipment in corrugated containers. Packaging can also become contaminated through human handling of packaging components. Plastic packaging, for example, is typically sterile when it exits an injection molder. The heat needed to melt the plastic destroys any organism. After a component exits the molder, it can be maintained in almost sterile condition and most likely will receive cautious handling to preserve the sterile condition of the part surfaces. When the part is packaged for shipment to the end user, dust, airborne spores, and other potential disease-causing materials may deposit on the surface of the part. Plastic can attract the contaminants if an electrostatic charge is present on its surface. By testing and determining a maximum bioload for packaging along with the maximum load of contaminants in the product, a safe and effective sterilization protocol can be devised.

Sterilization by Boiling

Boiling is a common and well-known method of sterilization that works for a wide variety of items. It is capable of killing waterborne diseases and many bacterial and viral disease-causing agents. It is not completely effective in killing many bacteria and fungi in their inactive or spore state. This is when the organism is most resistant to environmental conditions, including heat, and can be considered the way an organism hibernates when conditions are not conducive to growth. Boiling will not kill thermophiles, bacteria that thrive at extremely high temperatures. Boiling is also ineffective against many varieties of fungi and prions. Because boiling cannot provide a wide spectrum of absolute sterilization, it cannot be considered as a method of sterilization for pharmaceutical products. The value of boiling is its ability to reduce the number of organisms to a level low enough to minimize risk in an individual with a working immune system. The technique is well known and only requires a vessel to hold water and a source of heat. It is still the most common method used in field situations when an autoclave or chemicals are not available.

A variant of boiling that can be used is a process called tyndallization. This process, which only works in a medium that permits bacterial growth, involves multiple boiling and cooling cycles. The item being sterilized is boiled, allowed to sit for a day, and then boiled again. The process is repeated three times with the idea that the heating will shock bacterial spores into growing and changing to their vegetative state. While in their vegetative state, they are vulnerable to boiling and can be killed. This process will not work with water and prions.

Dry Heat

Dry heat is another method to sterilize drugs, instruments, and packaging. Dry heat is used to sterilize anhydrous oils, petroleum products, and bulk powders. Death of microbial products by dry heat is a slow oxidation process of protein coagulation taking place within the cells. The cells are destroyed by the slow take-up of heat by conduction. The reason dry heat is not a method of choice for sterilization involves the amount of time it takes to transfer heat in a chamber to an object. Wet steam in an autoclave carries a large amount of heat to an instrument or medical device and deposits it quickly. With dry heat, the transfer relies on air and convection, which is much slower and less efficient. Because the process is much slower, the item or product being sterilized by this method must be able to withstand relatively high heat for extended periods of time. A normal cycle for dry heat sterilization requires items in the process to reach and be held at temperatures between 160°C and 200°C for a period of time ranging from 6 to 15 minutes. It may take an object or a powder three or four times longer to reach these temperatures compared with autoclave methods. Dry heat has the advantage of removing water, which can react with mild steel and other objects causing rust or corrosion. A drug product that is soluble in water is another example of a product sterilized with dry heat.

Packaging for objects sterilized with this method must be heat resistant and must permit rapid heat transfer to the item it contains. The package may be left open to speed heat transfer by convection by removing the closure and permitting the heat to penetrate the package.

Other Heat Sterilization Methods

Flaming and Incineration

Two other methods based on heat are flaming and incineration. Flaming is used in laboratories on wire loops and small metal or ceramic items used to transfer or culture microorganisms. Incineration is used to eliminate hazardous biological waste.

Flaming is an old technique, dating back to Roman times or before, and consists of nothing more than placing the object to be sterilized in a Bunsen burner or alcohol burner flame and holding it there until it glows red from the heat. This kills and removes any biological materials on the surface of the object. It has its drawbacks; during the heating process pieces of material may be sprayed into the area around the flame as water heats and boils. If the liquid is contained in a tissue, the flash boiling can propel the tissue or material into the air around the burner.

Incineration is used to destroy biological waste and reduce it to ash. It was the common way to eliminate hospital-contaminated waste through the 1970s but lost favor as strict environmental laws and their limits on emissions closed many small incineration facilities at hospitals and health care facilities. Incineration is still used but it competes with autoclaving as a process to render harmless all biologically contaminated items from hospitals and nursing homes. The ash produced by incineration, if free of heavy metals, can be considered safe and buried in a landfill.

CHEMICAL STERILIZATION

Chemical sterilization is used for medical devices and other implements and is the method of choice for sterilizing medical devices. A wide variety of chemicals are used for sterilization, but the predominant method and sterilant of choice is ethylene oxide (EO or EtO), a gas used to sterilize a wide variety of instruments and medical devices. The advantages of using chemical sterilants are the low temperatures required during the process and the ability to tailor the sterilizing material to the material being sterilized. This makes the method compatible with plastics, fiber optics, sensitive electronic components, and other materials sensitive to heat. Many biologically derived materials are sensitive to heat and must use another method of sterilization.

Chemical sterilization is not without its limitations. The handling of hazardous materials and environmentally sensitive materials is the biggest problem. Safety and training of workers are required for the workforce to build understanding to use chemicals safely.

EtO Sterilization

The most common method for sterilizing medical devices is exposing them to EtO gas (Fig. 1). It is estimated that more than 50% of all medical devices use this sterilization procedure. Because EtO sterilization conditions range between 120°F and 140°F (50–60°C), it is extremely compatible with most devices and materials, particularly plastics. This low temperature does not cause changes in the finished product. The gas will penetrate almost all packaging with the exception of foil or barrier pouches, or materials sealed within barrier plastics. It cannot be used for liquids.

EtO is a chemical sterilant that kills microorganisms including bacterial spores by disrupting normal metabolism of protein and reproductive processes that result in cell death.

The highly successful introduction of kits, which is the bundling of multiple products needed for a specific procedure or emergency situation such as a

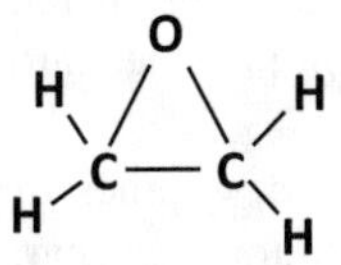

Ethylene Oxide
C_2H_4O
(epoxyethane)

Figure 1 Chemical Structure of EtO. *Abbreviation*: EtO, ethylene oxide.

heart attack, has become widely accepted as an effective method to manage inventories. The kits are used for a wide variety of surgical procedures and by paramedics and emergency responders in the field, for example, at the scene of a heart attack; a kit contains everything necessary including drugs to stabilize and treat the patient, such as preparing them for transport to a health care facility. Kits consist of multiple instruments and sundries, which are all the drugs, sponges, needles, and other materials needed to treat a problem in the field, or all the materials necessary to complete a surgical procedure. Items contained in a kit are for one-time use only, and often contain drug products including syringes and other materials for parenteral administration of the drug. The drugs are sterilized during their manufacture, and the EtO sterilization of the finished kit is for all the other items in the package. They have replaced expensive reusable materials like stainless steel with plastics to enable easy disposal of used instruments and materials. They protect everyone from highly contagious transmissible diseases like HIV.

EtO in gaseous form is the sterilant introduced into a closed chamber conditioned to receive the gas and to maintain a mixture of EtO and air below the combustible or explosive limits of the material. The process normally requires four steps to treat items being sterilized. The treatment chambers are large and typically multiple pallet loads of material are sterilized at one time, a fact that surprises people when the process is first explained. The process requires multiple steps to insure the gas reaches all parts of a pallet, including inside the individual packages the corrugated cases contain. It also requires extensive testing for confirming sterility and confirming the elimination of residual gas within the packaging that might be harmful to the environment or people operating the equipment.

Packaging used for devices undergoing EtO sterilization must be permeable and permit the gas to contact all parts of the products they contain. Tyvek®, paper, and other gas-permeable materials are used. Concurrent development of packaging with the device is a must to optimize performance of EtO sterilization. Packaging development includes managing the density of the product packaging, both in a corrugated case and in the pallet load to permit an acceptable sterilizing cycle to be designed. Originally the EtO process typically took 9 to 14 days to complete. The actual sterilization process ranges between 6 and 20 hours. The additional time to completion increased because of biological indicators. Biological indicators placed in the load for later culture testing (i.e., attempting to grow the organism in media) may take as many as 10 days to produce results, that is, the paperwork needed to release the product lots for commercial sale and the time needed to ship and move the product from a sterilization provider into the distribution chain.

EtO gas sterilization is based on the handling of very large volumes of packaged product, normally in the millions of cubic feet over the course of a year. When a manufacturer produces less than a million cubic feet of total package volume, they often use a third-party contract sterilizer. This arrangement may be both an advantage and a disadvantage because a small volume of product creates high operating costs when maintaining a sterilization facility, an unsuitable burden to small device producers. On the other hand, running a sterilization facility provides increased control and decreased turnaround time, advantages to manufacturers with high-cost devices. Many smaller units are found at individual manufacturing locations for this reason.

There are four steps to an EtO sterilization cycle.

1. Air removal
2. Steam injection, temperature conditioning, and dwell
3. EtO injection and dwell
4. Gas purge and replacement with air

The critical parameters used to identify what happens in the process are temperature, humidity, pressure, EtO concentration, and dwell time. Development of the process by a process authority, as described by the FDA, establishes the interaction of the process with the products being sterilized. The process is designed to deliver a six-log reduction (10^{-6}) of critical organisms, the same standard used for liquids and food products without causing harm to the product. It must also be designed to protect the people operating and monitoring the process and the equipment being used, and it must protect the end user from harm that could be caused by residual EtO gas.

The sterilization development process must include many different considerations. These include heat tolerances for all the components within the packages, sensitivity to both positive and negative pressures on the package seal, raw material composition, and sensitivity or reaction to water vapor/EtO concentrations used in the cycle.

Products containing exposed chemical salts may react with EtO to produce ethylene chlorohydrins or ethylene glycol; some products may bind with the gas, producing high levels of residual EtO gas or residual ethylene glycol within the product. Pouches and packages that are sealed prior to sterilization may be sensitive to extreme changes in pressure, both positive and negative, and may break or compromise pouch seals.

The design of an EtO process requires an understanding and identification of the issues sterilizing packaged product with a gas presents. The issues identified and the testing or validation challenges used to prove successful product sterilization, sometimes called the sterility assurance level, is met in the worst-case (for sterilization) situation. Some of the problems include the use of natural fibers that restrict easy access of the gas to all parts of the fibrous material or physical configurations based on the package design that may slow or impede the gas sterilant from reaching all parts of a product.

Every level of packaging must be designed and evaluated to confirm whether it would permit permeation of the sterilant gas into cases of product and inside the unit package based on heat, moisture, and gas concentration used in the sterilization process development. In addition to testing to insure the gas reaches all parts of the product within the packages, it must also be evaluated to determine if it releases the gas in an acceptable manner to eliminate any risk of residual gas (outgassing) remaining within the packaging.

The objective behind all the evaluation is to determine the fastest and most cost-effective sterilization cycle that is safe for all possible conditions and all personnel coming in contact with the product or the process at any time during and after sterilization. The process has the option to use a high concentration of EtO gas for a short time or a lower concentration for a longer time to achieve the final result. Factored into the gas exposure is the time it will take to replace the air in the chamber and the packaging with an inert atmosphere to prevent possible explosion or fire when the gas is introduced. Because large amounts of packaging are being moved, the potential for static electric buildup and a spark from the packaging igniting the gas is always a concern.

The four steps outlined above highlight conditioning of the environment within the sterilization equipment and the packages themselves for sterilization. Both product manufacturers and contract sterilizers may employ multiple chambers to speed the process. The additional chambers permit conditioning of the product and the packaging to be receptive to the gas sterilant, while other materials undergo actual sterilization or removal of the EtO after sterilization. Water vapor, in the form of steam, is part of the conditioning process. This insures that the gas permeates and penetrates throughout the packaging and is not absorbed to replace moisture lost when the product is heated in the initial conditioning of the load. It also may eliminate the possibility of the corrugate or other package materials failing as water is removed along with the oxygen-containing air as indicated in the first process step.

Step 1

The first process step, air evacuation, consists of placing the pallet load(s) in the sterilizing chamber and drawing a vacuum inside the chamber to evacuate the air and the oxygen it contains. Depending on the sensitivity of the product and package to partial pressure, the vacuum used is designed to remove the maximum amount of air each time it is repeated. As the air is evacuated, the product and packaging are held for a period of time to insure whether all the air removal required by the cycle is complete before the next step or a repeat of the evacuation cycle restarts. At the end of the air evacuation cycle, the pressure is returned to the atmosphere by the introduction of nitrogen or carbon dioxide. Multiple sequences are used to eliminate the atmospheric oxygen to a safe level, that is, one that eliminates any danger of fire or explosion.

To further illustrate this point, when the sterilization equipment is loaded, the chamber is completely filled with air. If the pressure within the chamber is reduced to 50% of the atmospheric pressure, half the air has been removed. This is a simple Boyle's law of application. If the chamber is then returned to atmospheric pressure using nitrogen or carbon dioxide as the backfill gas, half or 50% of the oxygen contained in the air has been removed. If this cycle is then repeated as described, the next cycle has reduced the oxygen within the chamber to 25% of the original concentration. This is repeated until the concentration of oxygen is below the flammable limits or explosive limits of EtO gas when it is introduced.

Step 2

The second phase of the sterilization process may begin prior to the load actually entering the sterilization chamber or after the air removal step if the sterilizing unit has only one chamber. Pallet loads are preconditioned to a specified temperature and humidity. This is important in two ways to the sterilization cycle. The conditioning for temperature is most important to prevent localized hot or cold spots forming during the third phase of the cycle when the EtO gas is injected into the chamber. These variations would interfere with the process, and if severe, it could prevent complete sterilization. Separate chambers are used to speed the actual sterilization cycle and maximize the revenue-producing time of the sterilizer. The second phase of the four-phase sterilization cycle consists of steam injection and maintenance of the chamber at a target of relative humidity (RH) level. The RH level for most products falls in the 40% to 80% RH range. The actual RH level is determined as part of the process validation. This process can be part of the evacuation process outlined above, or it can be a separate step in the process. The goal is to prevent the packaging from undergoing a significant change from the removal of moisture as part of the vacuum evacuation programming. The replacement of moisture is necessary before the introduction of the EtO to guarantee the correct concentration of gas reaching all parts of the product load.

Step 3

The third phase of the cycle is the introduction of EtO into the sterilization chamber. The liquid EtO is heated in a chamber, sometimes referred to as a volatilizer, to create the gas and is then distributed into the sterilization chamber. Manifolds, sometimes called recirculation headers, of varying design are used for the injection to evenly distribute the gas throughout the chamber and around the pallet load(s). The concentration of the EtO level is based on a number of factors including product and package permeation rates, flammability concerns, and the desired microbial lethality. The gas concentration and dwell time will directly affect lethality with higher concentrations of gas permitting the load to reach its sterility requirement, sometimes called the sterility assurance limit (SAL), in the shortest period of time. The ability of a particular product and its packaging to withstand rapid ramping during the various steps in the cycle is a major factor in determining a sterilization cycle.

The rate of EtO injection is important. Products that absorb EtO or chemically react with the gas may require a fast rate to minimize absorption or the amount of reaction that takes place. In case of a complex product, for example, an electromechanical instrument, a high concentration of EtO may be required but may need to be introduced slowly because the device or kit absorbs the gas slowly from the headspace of the chamber. During the process development cycle, an understanding of how the gas is absorbed can be developed, and this understanding permits the development of allowances for sterilant introduction. In a case like the one just described, a larger amount of total sterilant may be required in the cycle based on how it is absorbed over an extended period of time.

Design of gas sterilization cycles almost always follows a "one-charge method" of administration. This is for safety reasons and because it permits the steady introduction of the gas until the required concentration is achieved in the headspace of the chamber. The concentration amount can be based on a calculation or on a final partial pressure within the chamber. There is no additional gas added as the EtO concentration changes during the dwell time of the cycle when the gas is absorbed throughout the load. The biggest problem with a product that has a high absorbency of gas is seen in the decrease in pressure within the chamber as the gas is absorbed. To counter the absorption problem, multiple infusions of gas are used, but this creates a different problem because each time the partial pressure is adjusted, the concentration of EtO changes, making it impossible to calculate. This is a serious safety concern and can be alleviated if a gas chromatograph or spectrometer is used to constantly monitor the EtO concentration.

The temperature of the EtO entering the sterilizer is very important. It should always be at or above the temperature of the load within the chamber and the temperature of the process cycle. Gas introduced below the process temperature can cause localized cooling, which may interfere with the sterilization and microbial inactivation during the process in localized areas within a pallet load. Gas introduced at much higher temperatures also creates problems and can desiccate the outside layers of a load but not reach completely into the load of product. Localized heating or cooling can cause sterility failures, product damage, or inconsistent sterility results, all of which are major process problems.

When the EtO addition is complete, nitrogen or carbon dioxide is added to bring the pressure within the chamber to a process set point and is used to hold the pressure at that point during the dwell time the product spends under EtO exposure. Throughout the dwell period, the temperature within the chamber is kept constant.

It should be noted that design of a sterilization cycle does not require the gas to never achieve a flammable concentration. It is highly recommended, but on the basis of all the factors described, there are instances when the flammability limits are exceeded during introduction and absorption of the gas. This safety concern must always be addressed and should be evaluated on the basis of state and local regulations as well as insurance requirements for the facility. A decision to use a cycle with potential flammability problems is a major management decision.

Step 4

The final step in the EtO sterilization process is the removal of the gas back to safe limits for handling and exposure of operating personnel. EtO must be removed to a level of 3% by volume or 30,000 ppm to be safe. To achieve this level, four sources of EtO must be accounted for and measured as part of bringing the load of material back to ambient pressure and temperature.

The four sources of EtO in the evacuation process are as follows:

1. Gas in the headspace of the sterilizer chamber
2. Gas contained in the packaging of the product
3. Gas absorbed by the product and packaging
4. Gas adsorbed by the product and packaging

The first source is obvious and consists of the headspace around the product in the sterilizer. The second source is the areas within the packaging that can trap the gas. This would include the flutes in corrugated cases, all bundles, cartons, individual packages, and the device or product itself. The third source is gas that is absorbed by plastic or other semipermeable material that holds the gas in the matrix of the material. This would include plastics, fibrous materials, and other materials that actually absorb or take in the gas and in some cases weakly bind the gas to the material by weak hydrogen bonds. The humidity within the load and within some of the materials is a contributor to this source of moisture, which weakly retains the gas. The gas is bound by weak hydrogen bonds to the moisture contained in the product and package components. Finally, the last source of gas is the material that actually binds to materials in the product load by chemical reaction. This is dependent on the material composition and its ability to react with the gas. This source of gas presents hazards to personnel and is a source of continued outgassing, which occurs when temperature, humidity, or other environmental conditions reverse the chemical reaction of the gas with the material, and EtO is released.

The final step of the gas sterilization process is similar to the first step. A vacuum is pulled on the chamber after the sterilization cycle is complete, and then the vacuum is relieved by the addition of inert gas, usually carbon dioxide or nitrogen. The inert gas is still needed to insure that the EtO/air mixture in the chamber does not have the oxygen necessary for combustion. This is repeated until the concentration of EtO drops below the 3% level. At that point a final evacuation cycle is completed, and the chamber is backfilled with ambient air.

The amount of gas trapped in the packaging is removed by the amount of vacuum pulled on the chamber in each cycle after sterilization is complete. Time is also a factor in these steps because the gas must still migrate out of the areas it is trapped in and through the packaging before it is removed by the reduced pressure in the chamber.

The remaining two sources for EtO are the most difficult to address. The most common method is to raise the temperature of the sterilizer and its contents to the maximum level the product will tolerate and then draw as deep a vacuum as possible. These conditions are normally the most conducive to remove both absorbed and adsorbed EtO from the product load being sterilized. The amount of gas released will reach equilibrium in the headspace of the sterilization chamber and continue to be diluted by multiple injections of nitrogen or carbon dioxide. Each time the sterilization chamber is backfilled with the inert gas to reduce the concentration of EtO, a vacuum is reestablished in the chamber, and the load is held under these conditions to provide the time necessary for the gas to release from its weak bonds and migrate to the headspace of the chamber where it can be removed.

All the gases used in the sterilization process and removed during flushing of the sterilizer must be mitigated by some type of an environmental device that oxidizes the gas or removed by some type of wet scrubber or incinerator. Each of these methods addresses any air pollution concerns and local criteria for emissions. The amount of EtO gas, nitrogen, carbon dioxide, and ambient air involved is substantial, and the system must be able to completely clean the gases coming from the chamber during each step of the process and particularly after the EtO is introduced and then removed. Wet scrubbers make the EtO react with water in an acidified solution to convert the gas to ethylene glycol. Dispersion tubes or packed columns through which the gas and acidified water pass are very dependent on the amount of gas introduced. The speed at which the gas can be converted to ethylene glycol is dependent on the capacity of the system or the medium being used to scrub the EtO from the gas mixture emerging from the sterilizer.

Catalytic oxidizers convert the gas to carbon dioxide and water. This process first mixes the gas emerging from the sterilizer with air to dilute it below the lower explosive limit. This diluted mixture is then heated to an optimum reaction temperature and passed over catalyst beds that convert the gas to carbon dioxide and water.

The final method of gas mitigation to the atmosphere is the use of an incinerator. This could be considered dangerous if the mixture of gas and ambient air is not strictly regulated before being introduced to the combustion chamber where it is converted to carbon dioxide and water vapor. This method has the advantage of recovering heat from the gas and heat from the combustion process, usually fired with natural gas, and then reusing that heat within the facility to heat and condition product being prepared for sterilization.

At the end of the repeated cycling of the atmosphere in the sterilizer, some residual EtO will remain in the packages, and internal conditions will not contain an atmosphere of ambient air. Prior to opening the chamber, or in some cases as the chamber is opened, a continuous pass through of ambient air through the manifold or through the door leading into the chamber is initiated. This is a simple flushing operation that is carried out without the vacuum cycles at atmospheric pressure. There is no need for vacuum in this air exchange step. It is designed to completely flush the gas within the chamber and insure that it matches the ambient air conditions found outside the chamber. This step is important to protect personnel and to remove any remaining residual gas as well as to bring the atmosphere inside the chamber back to ambient air conditions found outside the sterilizer. Proper flushing of the sterilizer with ambient air during this step protects plant personnel from potential harm. It eliminates small quantities of EtO gas that may be dangerous to health, and the possibility of a sterilization chamber with a low concentration of oxygen, a potential danger to operating personnel. Following the ambient air flush within the sterilization chamber, the air within the chamber must be safe to breathe.

The first and last steps in the sterilization cycle present the greatest hazards to employees operating the equipment. Air removal and the opportunity for problems and the potential for harm from differential pressures require extensive training, monitoring, and testing. The last step, removing the EtO gas, is equally dangerous to operating personnel. As described, the process must be monitored carefully and a true ambient atmosphere confirmed before opening the chamber for removal of product.

All of these steps should follow a controlled linear progression. Each process step (these are sometimes called ramps) should be designed to integrate into the system control features to easily follow calibrated and controlled pre-programmed rates that are validated and reproducible.

Process controls of all parts of the EtO sterilization cycle are critical to companies practicing parametric release. This method of release relies on verifying whether the process has performed as predicted during validation and permits release of product without post-sterilization biological indicator testing. The use of parametric release reduces the time during which the product is held in quarantine for determining sterility from weeks to days and in some cases hours.

EtO sterilization is governed by ISO 11135 and other American National Standards Institute (ANSI) and the Association for the Advancement of Medical Instrumentation (AAMI) regulations. An overview of the regulations, both for the United States and Europe, is listed later in this chapter.

Other Chemical Sterilants

A number of other chemicals are used as sterilants, but because of their unstable nature or individual hazards, they are found in very limited use.

Chlorine and Chlorine Bleach

Chlorine is the most common of the chemical sterilants that finds widespread use. Chlorine is the active component of household bleach, a common material available everywhere. Normal household bleach is a solution of 25% sodium hypochlorite. For sterilization the bleach solution is diluted before use, and the dilution factor is very important to the degree of sterilization achieved for many organisms are resistant to overly dilute solutions. Chlorine solutions kill many bacteria and fungi immediately on contact, but some organisms are resistant to bleach treatment, and a dwell time is necessary to achieve a complete kill of all organisms. Dwell times for sterilization run 20 minutes or more. Even with this long exposure time, some bacterial spores are resistant to bleach as a sterilization agent, notably some strains of tuberculosis. Chlorine is highly reactive and corrosive during this long exposure time, making it difficult to be used with many materials, particularly metals.

Hydrogen Peroxide

Hydrogen peroxide is a highly effective sterilant. It was originally developed for sterilizing containers but has been found effective for medical sterilization. Typically, peroxide is produced as a vapor and allowed to condense on the surface of items in sterilization. An exposure of six or more seconds can produce six-log reductions (10^{-6}). Hydrogen peroxide breaks down into water. It leaves no residue on the surface of the item being sterilized. A complete commentary on this method of sterilization is found in chapter 4 on medical foods. One problem with hydrogen peroxide is that it becomes an extremely strong oxidizer at high concentrations and will react dangerously under these conditions.

Hydrogen Peroxide—Sterrad® Process

A second method of using hydrogen peroxide is found in the Sterrad® process. This is a low-temperature combination of both hydrogen peroxide as a vapor and hydrogen peroxide generated as plasma. Plasma is a separate state of matter distinguishable from a solid, liquid, or gas. The system operates in a temperature range of 45°C to 50°C with dwell times of between 45 and 70 minutes.

The system uses a two-phase process. In the first phase, hydrogen peroxide in injected into the sterilization chamber and vaporized to begin the inactivation of organisms. This phase is followed by a second phase of the process that consists of reducing the pressure within the chamber and applying radio frequency (microwave energy) to the chamber. The reduced pressure permits the formation of low-temperature plasma. The plasma consists of ions, electrons, and neutral atomic particles that glow much like a florescent light. Free radicals are generated in this process step by breakdown of the hydrogen peroxide vapor. After interaction and killing of the organisms and bacterial spores, the plasma and vapor break down further into water and oxygen. This two-step or two-phase process is normally repeated at least once to complete the sterilization.

The system is generally designed for use on a small scale for a specific instrument, such as an endoscope. The chambers are not large, and the system is not practical for sterilizing packaged manufactured products.

Peracetic Acid

Peracetic acid is another chemical used in the same way as hydrogen peroxide. It has recently been granted FDA approval for sterilization of food containers. The material is more compatible with higher-volume container sterilization than hydrogen peroxide. It is viewed as a material more compatible with polyethylene terephthalate (PET) containers than hydrogen peroxide and is expected to gain acceptance in the sterilization of food and beverage containers. It can also be used to sterilize medical instruments.

Peracetic acid is an oxidant and sterilizing agent that requires the item being sterilized to be immersed in a diluted solution. The solution is usually heated approximately to the 40°C range. Chemically the material resembles acetic acid with an additional oxygen molecule added. It attacks multiple parts of the bacterial cell or spore. Peracetic acid is effective when large amounts of organic contaminants are present. As mentioned, the item being sterilized must be immersed in the liquid for sterilization. The item cannot be prepackaged, and this method would not be feasible with most manufactured medical device products.

Steris, a commercial system, relies on peracetic acid mixed with a proprietary anticorrosive agent and sterile water. The system and the equipment for preparing and handling the chemical are supplied by Steris Corporation. A normal cycle for the Steris system is approximately 12 to 15 minutes at 55° to 60°C.

Ozone

Ozone is a gas that is used to sterilize hard surfaces, water, and air. Ozone is formed by passing an electrical current through a concentration of oxygen gas. Some of the gas molecules split into individual atoms and reattach to other molecules forming an O_3 molecule. This form of oxygen is unstable. The ozone molecule destroys organic material by oxidation. For bacteria and fungi, it penetrates the cell and causes them to explode. It is difficult to handle and must be produced at the point of use. It is highly corrosive and damaging to skin and is able to oxidize most organic matter. The concentration of ozone used in sterilization ranges from 6% to 12%. Ozone works well for disinfecting hard surfaces and bulk items. It requires an apparatus for generation and is normally confined to nonmedical manufacturing settings.

Formaldehyde and Glutaraldehyde

These chemicals are accepted liquid-sterilizing agents provided the amount of dwell time of an article being sterilized within the liquid is sufficient to complete the bacterial elimination. Formaldehyde kills microorganisms by coagulation of protein within the cell. Dwell times of 12 hours with glutaraldehyde and even longer with formaldehyde are necessary to achieve sterilization. These materials are used to sterilize tissue, and again the problem of sufficient dwell time to completely saturate the tissue and render all organisms it contains harmless is the main problem with using the materials. Both of these materials are toxic to skin on contact and both are dangerous when inhaled. Glutaraldehyde has a very short shelf life, normally less than two weeks. It is also expensive. These make formaldehyde the more common of the two materials used as chemical sterilants. Formaldehyde will polymerize to paraformaldehyde unless treated with methanol as a stabilizing agent. Paraformaldehyde is found in some contraceptive creams as a fungicide and disinfectant.

Prions and Chemical Sterilization

Prions are extremely resistant to chemical sterilization. Most of the standard treatments are ineffective and do not produce acceptable log reductions of the prion level. The list of chemicals that produce less than a three-log (10^{-3}) reduction in prions after one hour of exposure includes hydrogen peroxide, iodine, formaldehyde, glutaraldehyde, and peracetic acid. Chlorine and sodium hydroxide will reduce prion levels by more than a four-log (10^{-4}) reduction after one hour of exposure. The resistance of the proteinaceous material to chemicals calls into question the effectiveness of most conventional wisdom about sterilization destroying dangerous organisms. Fortunately, prions are not common to the environment like bacteria, viruses, fungus, and mold. They have a different type of transmission mechanism.

RADIATION STERILIZATION

Radiation sterilization is a common technique for sterilizing items (5). Public perceptions of the technique are generally negative where any reference to radiation is viewed as extremely dangerous.

Ionizing radiation works by knocking electrons out of atoms or molecules creating ions. This has a cascading effect that is violent enough to cause the production of secondary collisions that knock electrons out of adjacent atoms as well. This process produces both thermal and chemical energy changes within a cell, and this energy disrupts the DNA within the cell, preventing cell division and the propagation of life.

The three common sources of radiation for sterilization are X-rays, γ-rays, and subatomic particles. UV light is sometimes considered part of this type of sterilization but has a number of shortcomings that must be well understood and controlled before it can be put to limited use.

Radiation has advantages over EtO and other sterilization methods. In the case of γ-irradiation the source is always "on," and it requires extensive shielding and other precautions to protect operating personnel. The high-energy radiation or particles have the ability to destroy bacteria and spores as well as make DNA and RNA inert.

Irradiation has appeared in the food industry to treat ground meat for *Escherichia coli* bacteria, and it is used to sterilize natural spices that contain natural contamination from the environment, which cannot be treated in any other way. Contrary to popular conception, irradiation of a product does not make it radioactive. Detractors of the process claim that all the potential mutations and changes in a product or food have not been proven to be safe; however, the FDA and other global agencies have found no evidence of problems and accept this method of sterilization.

γ-Ray Sterilization

γ-Rays are the most common method of sterilizing medical devices including medical products other than EtO sterilization. γ-Rays are high-energy photons with a very short wavelength (very high frequency). They are one of the three types of radiation created when the nuclei of a radioactive material break down into smaller elements. The releases of energy and/or particles are part of any natural radioactive decay process. γ-Rays have wave or cycle frequencies of 10^{-18} Hz and wavelengths of less than 10^{-10} m. These rays also bombard the earth from the cosmos. γ-Rays are similar to X-rays, and the distinction between the two comes from how they are produced. γ-Rays are produced or emitted by interactions in the nucleus of atoms, while X-rays are produced by processes involving high-energy electrons found outside the nucleus of atoms. Breakdown of the nucleus may also produce other types of radiation in addition to γ-rays, such as α- or β-particles.

The most common method for producing γ-rays for industrial sterilization of medical products is through the use of cobalt 60 or cesium 137 (chemical notation ^{60}Co and ^{137}Cs, respectively). Another spelling of cesium is caesium. Cesium 137 is produced by nuclear fission.

The radioactive isotope of cobalt is produced by the exposure of cobalt 59 to free neutrons in commercial nuclear reactors. The cobalt is left in the reactor for 18 to 24 months depending on the neutron flux of the unit. Cobalt 59 is converted to cobalt 60 by the absorption of a neutron during this period of exposure. Cobalt 60 has a half-life of 5.27 years, and cesium has a half-life of 30.17 years. Cesium 137 is limited to small self-contained dry storage reactors used for irradiation of blood and insect sterilization. Cobalt 60 is the only material used in large-scale wet storage commercial processing units that produce γ-radiation for product sterilization. Cobalt 60 decays into a stable and non-radioactive nickel isotope ($^{60}Ni_{28}$) by the emission of a β-particle. Nickel 60 is in an excited state when formed and immediately releases two photons with energies of 1.17 and 1.33 MeV in succession to reach a stable state. These two photons are the γ-rays responsible for sterilization in all cobalt 60 commercial processing units.

γ-Ray sterilization gained market share during the 1990s because of its fast turnaround time, but subsequent improvements in EtO sterilization have slowed this trend, and EtO remains the method of choice for most medical devices. γ-Ray treatment has the advantage of needing no preconditioning of the product load before exposure. Following exposure to the radiation there are no residual materials that must be mitigated. These advantages translate into turnaround times for the process in the 2- to 5-day range compared with 9 to 14 days with EtO sterilization when using older biologic indicators for proof of sterility.

The major drawback to using γ-irradiation is potential interaction of the high-energy rays with plastics and other materials. The interaction may cause odors, yellowing, cracking, or embrittlement of plastics. Teflon, polyolefins, and other common plastics used in a wide variety of medical devices, equipment, sundries, and packaging require extensive testing and monitoring to prove that the treatment does not cause the item to deteriorate over time. Testing of package seals for product integrity over time is another development requirement.

The problems of γ-interaction with plastics have been addressed by the suppliers with the development and introduction of numerous materials with proven resistance to γ-rays and other high-energy radiation (6). This compatibility of materials with irradiation is detailed by manufacturers in the data overview that they supply with their products. It is common to see this information as part of the product literature.

The cobalt-60 source is encapsulated in a skin of stainless steel. It is in the form of a thin rod sometimes referred to as a pencil. The rods are loaded into the source rack, which is raised and lowered from the storage chamber to the sterilization chamber. With a half-life of a little more than five years, the source loses approximately 12% of its strength each year. It would decrease to 50% of its strength in 5.27 years, the half-life of cobalt 60. All of the cobalt 60 in a facility is changed out in 20 years. All radioactive cobalt 60 continues to decay, and after 50 years is 99.9% converted by this natural process to nonradioactive nickel.

γ-Irradiation consists of exposing an item to a γ-emitting source, cobalt 60, in a specially designed enclosure. There are two types of basic designs used for γ-irradiation facilities, self-contained irradiators and panoramic radiators. With both types, the goal for sterilization is the same; the product being exposed must absorb the maximum amount of radiation possible in the shortest effective sterilization cycle possible while maintaining a uniform dose throughout the product.

Self-contained irradiators are small units found in research laboratories or pilot operations. They are designed for applications with very small throughputs and generally small-dose applications. Some of the most common uses are to sterilize blood to avoid TA-GVHD (transfusion-associated graft versus host disease) and other possible viral contaminants. It is also used to sterilize insects for development of potential pest management programs based on sterile insects mating with non-sterile insects and disrupting the reproductive cycle. These irradiators consist of a lead shield and contain either cobalt 60 or cesium 137. The units are small enough to fit in most laboratories. The radiation source rods (pencils) are arranged to create a well or cavity with a volume of about 1 to 5 L ($\sim$2 U.S. gal) inside the shielded chamber. Samples are lowered into this area for a predetermined dwell time to achieve the radiation dose. This type of irradiator produces a high-dose rate that is very uniform, and it is very easy to operate (7). The IAEA (International Atomic Energy Agency) classifies these irradiators as class I and class III (8,9).

Panoramic irradiators designated by IAEA as class II and class IV are large units used for pilot operation or full commercial operation (Fig. 2) (9,10). The units can be batch or continuous in operation. Continuous units use some type of conveyor system to move product past the radiation source. The source is raised from its storage chamber and positioned alongside the conveyors for exposure. Batch units consist of a shielded area that permits arranging the product either on pallets or by stacking around the radiation source. The source is raised and lowered into position from a shielded storage area to enable safe placement of product in the sterilization area. The product is left in the unit with

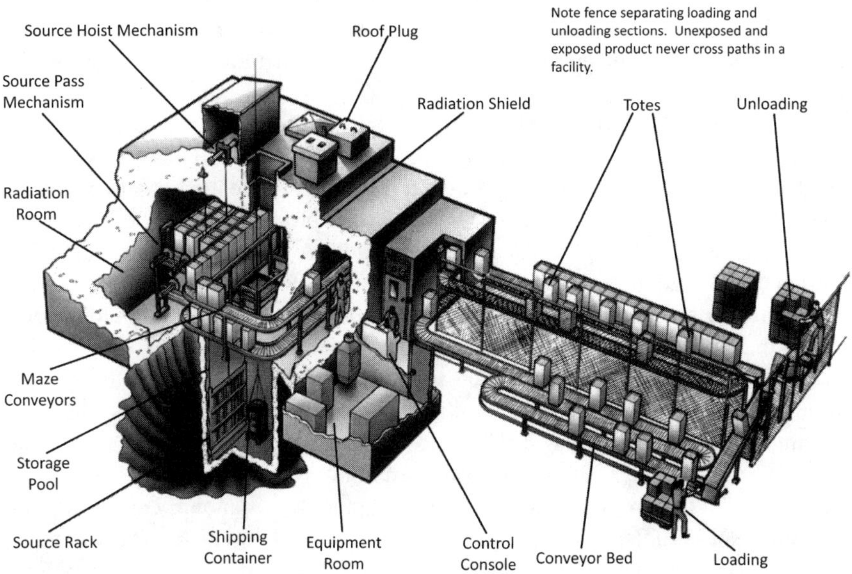

Figure 2 Schematic diagram of an industrial irradiator using γ-sterilization technology. *Source*: Courtesy of MDS Nordion.

the radiation source for a predetermined dwell time to achieve sterilization. At the end of the dwell, the source is returned to its shielded storage area, and the sterilized product is removed and replaced by the next lot of product to undergo sterilization.

Commercial enclosures consist of two chambers, an area where the product being exposed moves through the complex and a second room or area for storage of the radioactive source. The radioactive source is stored in a shielded area, which can be wet or dry. This chamber is usually below the area that the product transits, and the source is raised when in use and lowered into the safe area when not in use. This permits operating personnel to enter and perform maintenance inside the exposure area when the source is safely stored. Dry storage areas are lined with lead to provide shielding from the radiation. The wet storage method for a cobalt-60 source is used in all commercial applications. Water is an easily available material with excellent shielding properties that will not become radioactive. It can be circulated to maintain a constant temperature and remove any residual heat from radioactive decay. In wet storage facilities all materials used to house the radioactive source rack, the track or conveyor system used for moving the source from storage to the operating position and back, and any other equipment exposed to the source and water are made of stainless steel to prevent corrosion. The source is raised into position when in use and the lowered back into the pool of water when not in use.

The exposure area is heavily shielded with lead or concrete walls to prevent any escape of dangerous β-particles or γ-rays when the source is out of the storage area.

The unsterilized product is shipped to the facility and kept segregated from the product that has been sterilized by a physical barrier, usually a fence or a wall to make sure that the unsterilized product does not inadvertently get moved into or confused with sterilized product. The sterilized product is kept in a separate quarantined area. This physical barrier extends to the loading and unloading section of the irradiation facility. The product can only pass between the two sides of the facility by moving through the sterilization chamber. In continuous operations, the product is loaded onto a continuous conveyor on one side of the barrier, and the only way it can transit to the sterilized area in the facility is by passing through the irradiation enclosure. Continuous facilities often load the product into aluminum bins or totes that can handle corrugated cases, bulk bags, or other nonuniform packaging. By using a standard conveying unit, sterilization can be customized to the product placed in the container. This standard conveying unit also eliminates problems in conveyor design and operation. The totes slowly move on conveyor through the sterilization enclosure, with the speed of the conveyor set to achieve the required dose of radiation to achieve the level of sterility.

In most commercial facilities, a system called product overlap is used to maximize the dose and minimize the variation of the exposed products (Fig. 2). In operations of this type, the source is smaller in height than the product or tote. The tote must move past the source twice to achieve complete sterilization. To do this the conveying system is arranged to expose the top half or more of the product during the products' first transit past the source, and then the bottom half or second half of the product is exposed during the second transit. The conveyor system in operations like this is more complicated than in operations where the source is larger (source overlap) than the items moving past it. The conveyor system for both product overlap and source overlap operations typically turns the product on its axis each time it passes by the source to improve the uniformity of dose throughout the product.

Commercial batch sterilization facilities are designed to handle the product on pallets. By placing the pallet in the sterilizer, the operator eliminates any handling problems that may damage the packaging or the product. The pallets are arranged around the source location, and may be placed on a mechanism that rotates the pallet on its axis to minimize dose variations. Again the source is raised into position for a predetermined dwell time to sterilize the pallets of product. It is then lowered into its safe storage area, and the product is removed from the sterilization enclosure. Again the product flow through a facility is only one way, and the product is never allowed to cross paths to insure that the unsterilized product never becomes confused with the product that has undergone sterilization.

Some of the most common articles sterilized by γ-radiation are syringes, cannulas, needles, and IV sets.

X-Rays and Electron Beam (E-Beam) Sterilization

Two other types of ionizing radiation are used for sterilization, electron beam and X-rays. Both are used in small specialized areas for sterilization but have not reached the market penetration of γ-irradiation.

X-Rays

X-rays are another way to sterilize a product with ionizing radiation (11). X-rays are not as energetic as γ-rays but have sufficient ionizing energy to effectively sterilize a product (12). They have one big advantage over a γ-ray source; they can be turned on and off.

X-rays were discovered by W.C. Roentgen in the late 1800 while he was experimenting with cathode ray tubes. X-rays are produced when a source of electrons is accelerated to very high speeds and directed at a heavy metal target. A cathode ray tube used to produce X-rays is similar to the picture tube in a television with the picture screen replaced by a heavy metal target. Free electrons produced at one end of the tube are accelerated by an electromagnetic field and directed at a target of metal at the other end of the tube. In a doctor's office the electrons strike a tungsten target, and in industrial applications the target is usually tantalum. There are two different atomic processes that produce X-rays, one is called Bremsstrahlung and the other is K-shell emission. Bremsstrahlung is most useful for medical and industrial applications (13). Bremsstrahlung is a German word that means "braking radiation" or "deceleration radiation." High-energy X-rays are high-frequency short-wavelength electromagnetic photons. Bremsstrahlung X-rays are generated by the deceleration of a charged particle, in this case an electron, by another charged particle such as an atomic nucleus. The high-speed electrons give up some of their energy when they interact with the nucleus of the target atoms.

The second way for X-rays to form is for these same high-speed electrons to interact with electrons in the K-shell of an atom producing a similar effect.

The portion of the electromagnetic spectrum designated as X-ray wavelengths is very broad. The X-ray photon energy produced by an X-ray tube is roughly equivalent to the kinetic energy of the electrons striking the target. X-ray photon energy can reach 5 to 7 MeV, which is sufficient to penetrate pallet loads of a low-density product and provide sufficient energy for sterilization (14). Because X-ray formation in this type of apparatus is produced and concentrated in the same direction as the incident beam, the angular dispersion decreases as the kinetic energy of the electrons increases. This effect produces a high-energy beam that is concentrated in one direction in contrast to γ-rays, which are emitted in all directions for a radioactive source.

This characteristic allows X-ray of sterilization, with its energy concentrated in one direction, to use a smaller treatment room or area when compared with γ-radiation. It also makes it easy to vary the energy and time of exposure for small batches of product that comprise different densities and have variations in dose requirements.

X-ray sterilization treatment of a product still requires that two sides of the target, case or pallet, be irradiated from opposite sides to obtain optimum dosing and uniformity of dosing within the packaging.

X-ray photons can be generated in the 5- and 7-MeV range, high enough for sterilization but not as energetic as the γ-photons from a cobalt-60 source. This photon energy is capable of handling low-density pallets and some medical products of higher density (15). A commercial unit called a Palletron® produced by MDS Nordion of Ottawa, Canada, has demonstrated the ability to sterilize as many as four pallets at one time provided the packaged product meets the density criteria for penetration necessary for sterilization. X-ray sterilization remains an alternative to γ-irradiation but has not achieved the same acceptance level or market share.

Electron Beam Sterilization

Another method of producing ionizing radiation for sterilization is an electron beam. This method differs markedly from the other two methods of ionizing radiation because it uses electron, a true atomic particle, not photon, an electromagnetic wave form.

An electron beam apparatus is most akin to a television tube. It is a cathode ray tube, which permits the electrons generated at the filament and accelerated by an electrical potential between the two ends of the tube, which is from the cathode to the anode. A filament, usually tungsten, is heated to boil off electrons, and as the electrons escape from the surface, a electrical field created between the anode and cathode within the tube cause the electrons to accelerate from the cathode to the anode. The electrical potential between the anode and cathode determines the amount of energy imparted to the electrons. As the electrons arrive at the anode, they encounter a thin metal window made from aluminum or titanium that permits a percentage of them to exit the tube. The thin metal window is very important, for inside any cathode ray tube a vacuum must be maintained for the filament and the tube to function. It is like a light bulb, in a vacuum the electrons boil off the filament; if the vacuum is broken, the filament would immediately react with oxygen and burn out, just like a light bulb.

Electron beams incorporate a second electrical field to direct the beam in much the same way the picture tube in a television works. The second field directs or causes the electrons to be scanned across an area; just as the electrons in a television set scan the end of the tube we call the screen to produce the picture.

The high-energy electrons emerging from the metal "window" in the electron beam apparatus are the ionizing radiations that destroys microorganisms. Electron beam sterilization is approved by the FDA for use with foods. It was chosen by the U.S. Postal Service to sterilize mails going to sensitive locations in Washington D.C. to kill anthrax spores after powder containing the bacteria was discovered in letters. It works in the same way for the sterilization of medical products and medical devices.

Electrons produced by an electron beam–sterilizing unit are not as energetic and will not penetrate products to the same degree as X-ray or γ-ray photons. This fact and the relatively slow treatment rate of an electron beam, which must scan across a product, have limited its use with medical device products. It is a technology that has application but only to a very limited, low density, low thickness group of packaged products.

UV Light

The use of UV light is not a normal sterilization option, although it is viewed as such by the public at large. It can be used in combination with filtration to treat and sterilize water. Its main drawback is the ability of organisms to hide in shadows, including the shadows of dust and dirt. This makes the technique somewhat suspect, and not a method of sterilization of medical devices or other items.

UV irradiation under correct conditions will destroy microbes in the same way as other ionizing radiation. It will break bonds in an organism's DNA and create thymine dimers, which stop their ability to reproduce and render them harmless. In some cases it can actually destroy the organism.

UV light is found in another area of the electromagnetic spectrum just beyond the blue or violet portion of the visible light spectrum. It is more energetic than visible light, and a UV light source can cause sunburn or cancer. The wavelengths necessary for microbial elimination are approximately 244 nm or 2537 Å. At this wavelength the radiation is energetic enough to break the molecular bonds in DNA. These wavelengths of UV light are not present in normal sunlight because the atmosphere blocks them.

Germicidal UV is generated by mercury vapor lamps that emit light at the 254-nm wavelength, effective for sterilization. The lamps are monitored either with customized transformers or other instrumentation that measures current flow through the lamp, ensuring that the output produces the dose necessary for germicidal action. Power variations and other physical changes can affect the output of the UV source and render it ineffective in killing microorganisms. The bandwidth of lamps for germicidal action is very narrow, and everything possible is done to insure that the lamp output is providing the watt/density necessary for sterilization.

UV light is most effective when used for long-term exposure inside an enclosed space. Water treatment inside a tank and the treatment of the inside of ductwork to eliminate airborne bacteria are the most common and effective ways to deploy UV. Again the problem with this technique is that it is strictly line of sight and time dependent. One way to overcome this problem is to pass materials such as water or air past the lamp or lamps multiple times to maximize exposure. This will maximize the exposure of microorganisms to the light and will produce multiple opportunities for bacteria resistant to UV to be exposed and rendered harmless or killed. Bacteria can hide in shadows including those created by particles within the enclosed area. Dirt or oils can be very problematic because the energetic light can in some cases cause polymerization of the oils making them extremely difficult to clean and creating a place where bacteria can grow underneath its shadow. For these reasons UV radiation is at best a method to treat small amounts of water and hard-exposed surfaces.

Sterile Filtration

Sterile filtration is a technique for sterilizing pharmaceutical and protein solutions that are sensitive to heat, radiation, or chemical sterilization. The process consists of forcing a clear liquid through a membrane. For bacterial removal by filtration, a pore or opening of 0.2-μm size in the membrane will remove bacteria and spores. If viruses are to be removed, the pore size in the filter must be reduced to 20 nm.

These extremely small openings in the filter make the process very slow. Filtration will not remove prions.

The equipment for mounting and using these membranes is very specialized. The membranes are difficult to produce and require constant monitoring. Even a very small hole or tear that increases the size of the opening in one part of the filter will render it useless as a method for sterilization. Because the membranes are sensitive, they are typically purchased or supplied presterilized and most often cannot be reused or resterilized for multiple uses.

Constant monitoring of filtration equipment with biologic indicators is one method of confirming if the process worked properly.

Sterile filtration, much like aseptic processing, is carried out in very clean areas, usually class 100 clean rooms. These rooms, which maintain a positive pressure to keep out external dust and contaminants, rely on HEPA filtered air. In some cases the air moving through the room is designed to be laminar in flow. This type of flow constantly pushes any contaminant in one direction and out of the protected area. Eliminating air turbulence with this type of design is another way to eliminate possible contamination of a clean room.

REGULATORY OVERVIEW

Sterilization regulations are undergoing major revision and change for each of the methods discussed and for the development of standards that fit a global supply chain and economy. The primary effort is to establish standardized regulations between the United States, Europe, and other parts of the world (16). The sterilization process is unique because its effectiveness cannot be verified by testing and/or inspection. Testing and validation can determine if an individual article is sterile, but the testing requires breaking the sterile barriers. Biological indicators are in their own separate packages, which only prove that the process cycle was carried out correctly. Testing and validation are used to measure and determine if the process has been carried out correctly and if it is indeed effective. The use of testing and validation procedures has pluses and minuses depending on how the process is applied and how regulatory bodies in different parts of the world approach the question.

The differences are pronounced when understanding how regulatory bodies approach sterilization requirements for manufacturers versus health care facilities. Each of these users of sterilization equipment and technology has different needs. The variety of items sterilized in a hospital or health care facility is very different from products produced by a manufacturer. As emphasized these processes must be validated and tightly controlled to achieve reproducible results. Manufacturers approach the question by product line or product type, and many times pick the most difficult product within the product line for sterilization validation and use the results to represent performance for other similar products. They also use this approach with line extensions when sterilization testing and validation are carried out to determine the effect of minor changes to the product, not to the complete sterilization protocol.

Health care facilities on the other hand must sterilize a wide variety of products, many contaminated by blood and fluids, and all too expensive to discard. In this case the need to determine bioburden and cleanliness of articles entering the sterilization system becomes paramount. In both cases monitoring and verifying that the sterilization procedure, cycle, or protocol met all the required set points are the ways most facilities attack the problem. This coupled with biologic indicators, chemical indicators, and all the electrical and mechanical instrumentation attached to the process produce either a batch record for a product during manufacture or a cycle record for items undergoing sterilization in a health care facility.

The best way to represent the differences is to review the types of items sterilized in each setting. In a health care facility, sterilization is directed at multiuse products. These are items too expensive to replace with a one-time-use item. Examples would be surgical instruments or examination instruments such as a fiber optic examination devices used inside the body. The health care system is considered circular because the same items undergo repeated sterilization and reuse cycles.

Industry and manufacturers use sterilization procedures differently. In industry the sterilization process is directed at single-use items. It can be considered one way or linear because the item is sterilized and supplied to a health care provider and does not return to the manufacturer for additional sterile processing.

The most important thing to remember about both settings is that the final outcome is a sterile product that is safe to use. If nothing left its country of origin, this criteria would not be a problem once local laws and regulations were met. Today, we have a global economy that changes this expectation, and different parts of the world address the problem differently. Regulations and guidelines used in one region of the world may be different for health care facilities and manufacturers (U.S. position), while in other parts of the world the regulations for both types of facilities are the same (European position).

The question and challenge now become how and who regulates and monitors all the facilities that sterilize medical equipment and devices and how these regulations are presented to insure that international standards are consistent with local laws and regulations. Consensus standards are the answer, and they are replacing or have replaced local requirements and guidelines. The harmonization of standards is a constant process with the ultimate goal of achieving one global standard for every country. This includes any guidelines or technical requirements for developing and conducting validation protocols. This development and consensus is extremely important for manufacturers because products are shipped all over the world. They want to be able to meet one set of standards and have reasonable assurance that their validation and testing will be accepted wherever a medical device is shipped. The same holds true for health care facilities. Everyone everywhere wants complete assurance that any instrument or device undergoing sterilization will be completely sterile at the end of the process. Standardization requirements are important in guaranteeing that effective application of sterilization techniques takes place within any health care facility.

There are a large number of standard organizations, agencies, and government bureaus directly involved in the development of sterilization regulations and their enforcement. In the United States the AAMI develops consensus standards, technical information reports (TIRs) on those standards and practices, and recommended practices for each type of sterilization technology (17). The ANSI reviews and approves these standards for the health care industry. AAMI has the support of the U.S. FDA, and FDA has recognized and adopted many of the AAMI consensus standards (18). The reason behind this is twofold. Many FDA staffers actively participate in AAMI committees developing standards or TIRs. Because FDA staff members have active input in the development of documents and input into the acceptance of standards, the AAMI standards are listed in FDA's list of recognized consensus standards. Medical device manufacturers in the United States and abroad who conform to consensus AAMI standards on the list can state that they conform to FDA requirements.

AAMI must work through ANSI to be recognized as the official U.S. participant in the International Standards Organization (ISO). The ISO is the worldwide federation of National Standards Organizations and is responsible for the development of consensus standards, which are voluntary, but are usually adopted in some form, either as a whole or in part, by countries all over the world. The ISO has designated Technical Committee 198 to oversee sterilization of health care products and has broken this committee into a number of working groups, which deal with the different technologies in use. ISO's standards may be adopted as national standards, they may coexist with national standards, or they may be modified or changed to reflect local variations (sometimes called national deviations from standard).

In Europe the European Committee for Standardization (CEN—Comité Européen de Normalisation) is the counterpart for AAMI and is responsible for the development of sterilization standards in Europe that are part of the European Medical Device Directive (MDD). CEN-TC (Committee for Standardization-Technical Committee) 204 is responsible for the sterilization of medical devices. This technical committee comprises 10 working groups (WGs) that focus on different types and aspects of device sterilization.

In the European Union CEN standards must be adopted and used as national standards by each member state. They must be adopted unchanged, and all conflicting local standards or requirements must be withdrawn.

Sterilization in Europe is governed by the MDD. Article 5 of the directive requires that all devices carry the CE mark, which means that the MDD standards must be met. A product conforming to EN-harmonized standards is presumed to have met the requirements of the MDD as contained in the annex ZA. This is important because it provides a clear path for manufacturers to obtain CE marking for their product.

ISO and CEN have an agreement, called the Vienna Agreement, to ensure harmonization of standards issued by the two organizations. Under this agreement, when standards are considered for revision, one of the organizations takes the lead to develop the new standard so that both sets of standards are revised by a single committee. When they complete their work, the revised standard goes out as both an ISO and EN standard, and the member organizations in each country vote on the documents simultaneously. When a standard is approved, it is adopted by the member countries as an EN document.

All these three bodies, ISO, CEN, and AAMI, have produced sterilization standards that are very similar to that of each other's; however, as with other standards there are subtle differences that create problems for auditors and reviewers of products manufactured in one part of the world and shipped to other parts of the world. In the end, medical device manufacturers want to be able to manufacture products to one set of standards and be reasonably certain that those standards will be accepted as proof of regulatory conformance in all parts of the world.

During 1999 ISO and CEN technical committees agreed to a joint revision and harmonization of the sterilization standards under the terms of the Vienna Agreement. ISO was designated as the lead organization in the revision of sterilization standards. The standards proposed for harmonization and change were as follows:

1. Sterilization using moist heat: ISO 11134, ISO 13683, and EN 554.
2. Sterilization using EtO: ISO 11135 and EN 550.
3. Sterilization using radiation: ISO 11137 and EN 552 (19–21).
4. Biological indicators (BIs): ISO 11138 series and EN 866 series.
5. Chemical indicators (CIs): ISO 11140 series and EN 867 series.

Note: *ISO and EN references must be consulted to determine the current standard in use. This is often denoted by the year following the standard.*

The protocol for the revised standards was to follow ISO Standard 14937 formats with consistent definitions as defined by ISO TS 11139 and consistent quality control measures as defined in ISO 13485. The umbrella ISO 14937 Standard is titled:

Sterilization of Healthcare Products-General Requirements for Characterization of a Sterilizing Agent, and the Development, Validation, and Routine Control of a Sterilization Process for Medical Devices (ISO 14937:2000).

By 2005 all the standards had proceeded to draft revision status and were being finalized by the various working groups and voted on by the ISO members in various countries. AAMI was part of the ISO development of the new standards and created documents for the United States, which members of that organization reviewed.

The FDA had a number of objections to the new standards. These objections were due to different ideas on how to manage the standards, not unlike some of the objections detailed in chapter 7 regarding ISO standard 11607 parts 1 and 2. FDA's objections are centered on the differences in how standards were developed and applied to health care facilities and manufacturers. The FDA does not regulate health care facilities. AAMI and the FDA have issued separate standards for industry and health care facilities to recognize their differences in operation and function. In Europe, the European MDD treated the two types of facilities in the same way. The joint panel charged by the various participating groups with the development of consensus standards defined this standard as the one that will contain quality assurance standards that are applicable to moist heat, EtO, and radiation sterilization. As with 11607, many of the quality requirements listed in the broad ISO document do not meet FDA requirements, making this one of many major issues that will require substantial development, review, and acceptance before any new standard is accepted.

Another difference in views between the United States and Europe stems from the two different needs of the facilities conducting sterilization operations and the fact that the facilities are addressed in two separate ways.

In the United States sterilization equipment in health care facilities is installed and validated by the sterilizer supplier or the medical device supplier to provide the necessary information and background rationale for safe, proven, and repeatable sterilization of reusable equipment. Validation in this case is proof that various cycles within a sterilizer are repeatable, not validation for each medical device or each instrument undergoing sterilization. Regulation of the facilities is primarily a state oversight, not federal.

Industrial sterilization is the major regulatory purview of the FDA. Here the agency is responsible for monitoring and certifying whether a facility is operating according to statutory requirements. Its regulatory scope involves comprehensive validation and control of the process consistent with statutory regulations and guidance directives issued by the FDA. Each product or product family is expected to undergo validation and testing to confirm whether the process meets or exceeds standards for sterile product. It requires comprehensive understanding of the products and validation protocols that challenge both the sterilization equipment, the process, and the results based on biologic cultures, indicators, or other methods taken directly from items that have completed the process.

Validation is the big change that has taken place in how the FDA approaches risk management within pharmaceutical and medical device manufacturing. With the advent of microprocessors and programmable logic controllers, which control algorithms that monitor multiple operating needs and real-time instrumentation and monitoring of key parameters, the process is controlled in a way impossible only a few years ago. This change has ushered in a major increase in the development and complexity of validation protocols, and the need to develop and maintain documentation proving the process performs in the same way it was challenged (validated) to prove it produced sterile product. Record keeping and retention is one primary requirement for all systems regardless of the technology employed.

This increased control and monitoring capability has permitted many facilities to achieve parametric release of products because all the operating parameters met those specified for the process. This has replaced biologic indicators and other methods that increased the quarantine time of product after sterilization while they are incubated and evaluated for confirmation that the sterilization process was successful.

The problems this creates in developing new consensus standards or harmonizing existing standards stem from the two different approaches used in Europe and America by health care facilities and manufacturers (including contract sterilizers) and the general approach taken by the technical working groups to include FDA requirements. Moist heat sterilization and EtO sterilization are the two most prominent areas that have been particularly difficult to resolve in achieving consensus standards for the sterilization of medical devices.

In Europe, harmonization of many parts of the ISO and CEN standards was approached by moving many mandatory aspects of the standard from the standard to guidance documents that describe the intent of the standard. This creates a problem for FDA because many mandatory parts of the standard when moved to a guidance document are no longer mandatory. Many manufacturers, particularly those outside the United States, view many parts of a guidance as elective or window dressing for the standard, not procedures or requirements, that must be met and documented. ISO, the lead organization, and AAMI are attempting to address the problems by issuing TIRs and technical specifications (TSs) to directly address the FDA concerns. The TIRs and TSs are expected to directly address FDA concerns and provide a manufacturer with unequivocal expectations on what is required to meet FDA requirements. AAMI, with direct FDA participation on the committees developing consensus standards, will use this as their framework for developing and issuing a consensus standard that is acceptable. Much of the text and guidance documentation is being developed to indicate that the standard can be used for a health care facility to make them consistent with the European MDD. Unfortunately, this difference in approach and interpretation slows the process of developing consensus standards. It is extremely important for anyone in the United States to review the current list of FDA-approved international standards and not assume that consensus has been reached. The radiation standards listed above have achieved consensus status for the most part, but the moist heat and EtO standards are still in review and development as this is written in 2008.

MONITORING STERILIZATION PROCESSES

Mechanical, Chemical, and Biologic Indicators

All sterilization processes in both health care and industrial settings must be monitored and challenged regularly to document if the process is operating on a proven repeatable cycle validated to ensure sterility of the items completing the process. All sterilization processes rely on a combination of processes, mechanical, electrical, and radiation to achieve sterility in product. They are all based on the bioburden of the instrument or device being sterilized and the amount of exposure to the different types of environments required to destroy them.

Mechanical Indicators

Mechanical indicators include gauges, thermometers, timers, recorders, vacuum pumps, valves, seals, and locking devices that must be verified as working properly for the equipment to function as designed and validated. As equipment has become more automated, more and more mechanical functions are linked to programmable microprocessors that control the function of the sterilizer. Real-time readouts may be connected to alarms to constantly verify or warn operating personnel immediately if something is right or if something has failed. With algorithms built into the microprocessors, the entire sterilization process is monitored as never before.

All of these mechanical systems must be tested and calibrated at regular intervals to insure that the equipment functions are programmed. In some cases different monitors can detect if one parameter has not been met and highlight the other parameters that are affected. Examples would be pressure and temperature, which are interrelated in a sterilizer. If the pressure inside the vessel does not reach the set point, the operating temperature inside the unit will not be hot enough to achieve sterilization.

Chemical Indicators

Chemical indicators are used as a process check for sterilization. They consist of a dye or chemical that physically changes appearance, color, or shape if the proper process parameters within a sterilization cycle are met. Approved devices may measure one process parameter or multiple parameters. They may form a combination device that requires multiple conditions be met within the sterilizer before they change color or appearance.

A good example of a device of this type is a Bowie–Dick test pack. These packs require that a preset level of vacuum be achieved inside a sterilizer that uses vacuum to remove air before the introduction of steam. By removing the air, the steam is able to penetrate to all parts of the sterilizer and effectively transfer heat for sterilization. The Bowie–Dick test combines a series of air-filled chambers with a chemical indicator that changes color. The indicator will function only if the air within the test pack is removed, permitting the steam to reach the indicator. These indicators can be stored as verification that the vacuum level necessary for sterilization was reached. The AAMI publishes standards for construction and use of this test apparatus. The Bowie–Dick test apparatus is placed within the sterilizer at a point determined to have the greatest chance of not reaching process conditions. Typically the bottom shelf of a sterilizer in a health care facility or a doctor's office is the most difficult location to sterilize within the chamber itself and the coldest location within the sterilizer.

In addition to stand-alone chemical indicators for temperature, pressure, or moisture, packaging may be supplied preprinted with inks or other indicators that provide a positive indication of sterilization on each unit.

Biological Indicators

Biological indicators are devices designed for inclusion in the sterilizer or autoclave to independently verify that the monitored process achieved sterilization (22). Bioburden testing of the completed but unsterilized device must be done to determine the types and amount of biologically viable organisms that must be eliminated. The biological indicator device, normally purchased commercially, contains a known quantity of a microbe resistant to the sterilization process (Table 3). An example is *Bacillus stearothermophilus* (new designation *Geobacillus stearothermophilus*), an organism very resistant to heat, which is used as a standard independent test method for autoclaves. The microbe is supplied as bacterial spores, the most resistant form of the microbe. The indicator may be supplied as dry spore strips or disks in envelopes or sealed vials or

Table 3 Common Microorganisms Used as Biologic Indicators

Sterilization process type	Microorganism (spore) indicator
Steam	*Geobacillus stearothermophilus* (formerly *Bacillus stearothermophilus*
Dry Heat	*Bacillus atrophaeus* (formerly *Bacillus subtilis* var. *niger*)
EtO	*B. atrophaeus* (formerly *B. subtilis* var. *niger*)
Hydrogen peroxide	*G. stearothermophilus* (formerly *B. stearothermophilus*)

Abbreviation: EtO, ethylene oxide.

ampoules containing a standardized quantity of the organism. The indicator is always accompanied by a control sample that is not sterilized. Both the sterilized device and the unsterilized control are incubated to determine if the sterilization process was successful. The unsterilized control is needed to prove that the organisms used in the test were viable and would grow when placed in nutrient and stored under favorable conditions.

Many biologic indicators are supplied as completely self-contained devices. The devices contain a growth medium and an indicator prepackaged separately from the spores. After sterilization an internal seal is broken, releasing the spores into the nutrient solution. The device is stored in the case of *B. stearothermophilus* at 56°C (132°F) for 48 hours to determine if the sterilized microbes are still viable and can reproduce. Most nutrient solutions contain some types of indicator or dye that reacts with a metabolism product of the microbe causing a color change. A color change in the unsterilized control and no change in the sterilized unit are proof that the sterilization process was successful.

Another common organism used as a biologic indicator is *Bacillus subtilis* var. *niger* (new designation *Bacillus atrophaeus*). This is an organism very resistant to chemicals, and it is used as an indicator for EtO sterilizers. The microorganism is incubated at 37°C (98°F) for the relatively short time of four hours to produce a fluorescent change or luminescent change when exposed to UV light. The absorption of a photon results in the emission of a longer-wavelength photon with the lost energy becoming part of molecular vibration or heat. The growth of the organisms over a longer period of time (days) will slowly result in a color change. The florescence comes from an EtO-resistant enzyme present in the growing bacteria. The bacteria must grow and multiply to produce the enzyme creating a measurable change that confirms that the bacteria is growing.

Hydrogen peroxide plasma sterilization uses *B. subtilis* as a biological indicator. Peracetic acid chemical sterilization uses *B. stearothermophilus* as a biological indicator.

All biologic indicators are regulated by the FDA, and all must conform to United States Pharmacopeia (USP) testing standards. Hospitals and doctors' offices routinely use indicators as a daily or a batch check on the effectiveness of each sterilization operation. They are also used when installing a new unit or after repairs to a unit have been made. They are a fast and convenient way to test for changes in sterilization performance when making a packaging change either to the packaging materials or to the package design. All test results must be retained and filed for subsequent review in both an industrial and a health care application.

SUMMARY

Sterilization is an extremely complex topic. It can be done in many different ways, but the goal remains the same, make products sterile and safe for medical use. As new products and packages enter the market, the need to refine and speed sterilization will be viewed as a major process improvement goal. Investment in new techniques along with new test methods that provide sterility assurance much faster than cultured biologic indicators are needed now to improve the modern supply chain. Sterilization technology and its related test methodologies are poised for major change in the coming decade.

REFERENCES

1. United States Pharmacopeia (USP). Sterilization and Sterility Assurance. USP Vol. 28, Section 1211, NF 23, Rockville, Maryland: United States Pharmacopeial Convention Inc., 2005.
2. Prusiner SB. Novel proteinaceous infectious particles cause scrapie. Science 1982; 216(4542):136–144.
3. Madigan MT, Martinko JM, eds. Brock Biology of Microorganisms. 11th ed. Upper Saddle River, NJ: Prentice Hall, 2006.
4. Ryan KJ, Ray CG, eds. Sherris Medical Microbiology. 4th ed. New York: McGraw Hill, 2004.
5. Morrissey RF, Herring CM. Radiation sterilization: past, present, and future. Radiat Phys Chem 2002; 63: 217–221.
6. Hemmerich KJ. Polymer Materials Selection for Radiation Sterilized Products. MDDI, February 2000. Available at: www.devicelink.com/MDDi/archive/00/02/006.html.
7. International Organization for Standardization. Radiation Protection—Sealed Radioactive Sources—General Requirements and classification, ISO 2919, Geneva: ISO, 1998.
8. International Atomic Energy Agency. Emerging Applications for Radiation Processing, IAEA-TECDOC-1386. Vienna, Austria: IAEA, 2004.
9. American National Standards Institute. Safe Design and Use of Panoramic, Wet source Irradiators (Category IV). ANSI-N43, 10-1984. New York: ANSI, 2001.
10. American Society for Testing and Materials. Standard Guide for Dosimetry in Radiation Research on Food and Agriculture Products, ISO/ASTM 51900, Annual Book of ASTM Standards. Vol. 12.02. Philadelphia: ASTM International, 2004.
11. Fairand BP. Radiation Sterilization for Healthcare Products—X-Ray, Gamma and Electron Beam. New York: CRC Press, 2002.
12. Cleland MR. X-ray processing: a review of the status and prospects. Radiat Phys Chem 1993; 42 (1–3):499–503.
13. Standard Practice for Dosimetry in an X-Ray (Bremsstrahlung) Facility for Radiation Processing, ISO/ASTM 51608:2002(E). West Conshohocken, Pennsylvania: ASTM International, 19428–2959.
14. Meissner J, Abs M, Cleland MR, et al. X-ray treatment at 5 MeV and above. Radiat Phys Chem 2000; 57(3–6):647–651.
15. Stichelbaut F, Bol JL, Cleland MR. The palletron: a high dose uniformity pallet irradiator with x-rays. In: AIP Conference Proceedings, American Institute of Physics, Denton, Texas. Vol. 680, 2003, pp. 891–894.
16. Reich R, Schneider PM, Kinsley C. Global Sterilzation, Making the Standards Standard, MDDI. March 2005. Available at: www.devicelink.com/mddi/archive/05/03/008.html.

17. National Standards for Recommended Practices for Sterilization. Association for the Advancement of Medical Instrumentation. Arlington, Virginia: AAMI, Vol. 1.1–1.2, 1995.
18. ANSI/AAMI. Sterilization of Health Care Products—Requirements for Products Labeled "Sterile." ST67:2003. Available at: www.ansi.org.
19. ANSI/AAMI/ISO. Sterilization of Health Care Products—Radiation. Part 1: Requirements for Development, Validation, and Routine Control of a Sterilization Process for Medical Devices. 11137-1, 2007. Available at: www.iso.org.
20. ANSI/AAMI/ISO. Radiation. Part 2: Establishing the Sterilization Dose. 11137-2, 2007. Available at: www.iso.org.
21. ANSI/AAMI/ISO. Radiation. Part 3: Guidance on Dosimetric Aspects. 11137-3, 2007. Available at: www.iso.org.
22. Guidance for Industry and FDA Staff—Biological Indicator (BI) Premarket Notification [510(k)] Submissions. U.S. Food and Drug Administration. October 4, 2007. Available at: www.fda.gov/cdrh/ode/guidance/1320.html.

10

Container Closure Systems: Completing All Types of Filled Pharmaceutical Containers

INTRODUCTION

The most difficult part of completing a package is closing the container after filling. The design and application of closures to a container is always difficult. These packaging components represent some of the most sophisticated pieces of design and engineering found throughout packaging. The closure must be able to serve many masters, by providing a secure seal, and in many instances child resistance, tamper evidence, and consumer communication. This chapter will discuss the many common closures used in pharmaceutical packaging.

As stated in the first sentence, closing a container after filling is difficult and exacting. This is particularly true when filling and sealing liquids and ointments into the container. The closure must be sterile or fit into the type of sterilization process used to sterilize the container and its contents, and once sealed it must maintain a barrier between the outside world and the container contents. It also must be able to tolerate and overcome problems of contamination of the sealing surfaces created by the products being filled and provide an inert contact surface to the product. Liquid and ointment products have the ability to contaminate the sealing surface and the closure preventing the formation of the seal necessary to provide a sterile barrier between the product and the outside environment. The products can create micro voids and other sealing problems that make leak detection difficult. They tend to be the products with the most consumer complaints.

This does not mean solid-dose pharmaceutical products are easy to package and close; quite the contrary they present challenges different from liquids, but substantial nonetheless.

Pharmaceutical filling and closing operations present problems not found in other types of packaging, starting with the physical environment the package must endure in manufacturing and afterward in its journey to the end user (1). Because a closure system has a high potential for harm to a patient if it fails and product becomes contaminated, the need to develop nondestructive tests to monitor the package performance is another challenge. Closure and sealing problems are some of the more difficult problems in packaging engineering.

The closure must be unaffected by any contamination on the mating surfaces used to create the seal, it must be fast and easy to apply, it must be extremely reliable, and must make the contents reasonably accessible by the consumer. This is true of all closures, but pharmaceutical requirements place a number of additional demands on the closure not found in everyday food and beverage packaging (2). Child-resistant (CR) and senior-friendly packaging is not a requirement for food and beverage packages.

Pharmaceutical packaging places severe demands on designers and engineers of closures because the field is heavily regulated and required to supply product to many people who have physical and mental limitations or infirmities. The pharmaceutical closure is designed to meet a number of very difficult criteria, safety, ease of use, and product protection codified in law and regulations by the government and by the expectations of the public at large and individual end user. From a governmental perspective a pharmaceutical closure must meet child-resistance needs, tamper evidence requirements, and the ability of the elderly, particularly the elderly with arthritis, to open the package easily (3,4). Closures for drugs and medical devices must provide a method to seal a container and withstand a variety of sterilization techniques including heat, steam, gas, and radiation (5). For surgical medical product and device the closures must be easily opened in an operating room, at the scene of an accident, and in every situation where medical care is given. For many drugs the closure becomes the measuring and the dispensing device for the prescribed product dosage. Closures on ophthalmic products not only provide a dispensing mechanism, they also are color coded to guide the medical professional and help eliminate medication errors. Newer closures, used in clinical testing of drugs and biologics, have electronic and mechanical devices built into the closure to monitor patient compliance or to remind the patient when to take the next dose of medication.

From the consumer–marketplace perspective, the closure on over-the-counter (OTC) products must be able to provide complete protection of the product, be extremely resistant to failure even when mishandled, damaged, or dented and still be easy to open. The closure must communicate to the customer, and in many cases must supply the dispensing device for the product along with providing the seal and product protection.

Closures for OTC products underwent an extreme makeover in the early 1980s when Tylenol® from Johnson and Johnson (New Brunswick, New Jersey, U.S.) was deliberately contaminated with cyanide, killing a number of people and causing panic in the consuming public. Overnight, or at least it seemed that way, all closures on OTC products incorporated a temper-evident device in their construction or the application of a tamper-evident device to visibly show and communicate to the consumer that the product had not been opened before purchase or prior to their use.

Closures and all the methods used to close a package are really separate components, operations, or devices that are a highly engineered and specialized area within packaging. They range from bottle caps to metal can ends with easy-open features, to heat-sealed laminate structures, to uniquely filled and sealed containers whose design, filling and closing operation are all part of the packaging equipment. Many of the closures are highly specialized in their sealing and opening features that provide the user with a means of opening the package and using the product. Closures have evolved in all forms of packaging from items that required tools to open or were difficult to handle to the sophisticated and nuanced items that make accessing and using the product essential to successful packaging. This chapter will touch on closures for cans, bottles, jars, vials, tubes, and on sealing methods used to close a container without a separate closing device.

CLOSURE FUNCTIONS

To begin, you must have a clear understanding of what a closure is expected to provide. The features are different than those needed for the container but still emphasize the ideas of protection and containment. It's when you begin to go beyond these basic needs and understand that easy access to a product along with making that product convenient to use is not a set of attributes that extend from the container or the material used to make the container. Individual closures and all methods used to close and seal a package must provide a number of different functions. A list of these functions is as follows:

1. Protection
2. Containment
3. Complete and positive sealing
4. Access
5. Communication
6. Display
7. Metering and measuring

This list represents a number of requirements for closures that are easy to state and difficult to execute. Closing a filled container is difficult to execute, particularly when a product is produced in the millions and each and every package must provide and maintain its safe sealing performance throughout the containers usable life.

Protection

The closure completes the package after filling and provides consumer with the same level or in some cases a better level of protection than the container provides to the product. The closure must keep the product from escaping, and it must prevent the product from becoming contaminated by the external environment the package will endure in distribution, dispensing, and in its final place of use (1). The closure or the closure system must quickly and positively tell the consumer if a product has been opened (tamper evidence). The closure must also protect the product from misuse or poisoning of curious children.

Containment

Containment is an attribute that is dual functional. The closure must keep the product in the package and prevent anything extraneous from entering the packaging. The closure in providing containment must not interact with the product in a way that would affect its potency or purity. Containment and protection can be considered different facets of the same attribute. Containment focuses on the problems that can develop inside the package, while protection addresses problems coming from the outside of the package.

Complete and Positive Sealing

This function may be confused with protection, but it goes beyond protection and describes the need to design a closure with the ability to mold to or conform to the mating surface of the container to effectively complete the seal every time. Mating two parts of a container, such as putting together a bottle and a bottle cap (closure) will not always provide a complete seal to a package unless all the elements making up the seal are engineered into the design. The closure must provide a material interface at the sealing surface that is flexible and usually compressible, e.g., in bottles and metal cans, to fill in all the voids and irregularities in two mating surfaces and permit the creation of a reliable and reproducible hermetic seal. For heat sealing or other closing methods that do not use a pre-manufactured closure like a cap or a metal end, its design, the selection of materials, and the design of the apparatus or sealing equipment used to close the container is critical to achieve a melted or changed material surface that will mate, hold while hot (hot tack), and provide the same positive protection that other types of closures provide.

Access (The Ability to Open and Close a Package Repeatedly and Safely)

Access has a number of dimensions. One of the simplest examples is a core-pin closure that aids opening the product the first time (Fig. 1). They also include the ability to provide a visible and positive indication of tampering before the product is opened the first time; the ability of the closure to reclose and reseal

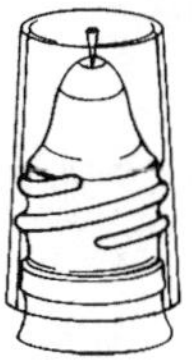

Figure 1 Core-pin resealable closure.

the package with the same level of protection and containment found when new; and the ability to pass CR closure regulations issued by the Consumer Product Safety Commission (CPSC) while remaining easily accessible to the elderly and infirmed.

One of the most difficult and unique sets of requirements is found in parenteral containers for injectable products. This is their ability to seal and protect the product while being penetrable with a needle. These closures must withstand multiple penetrations by a needle without exposing the contents to contamination, and must be strong enough to withstand damage from the needle like tearing, creating particles that could be transferred to a patient, or "coring," i.e., creating a plug in the needle. Highly formulated elastomers are the materials that deliver safe and repeated access to the contents of a vial. Parenteral containers are not required to meet tamper-evidence requirements; however, tamper evidence may be incorporated into the secondary packaging of these products as additional protection for the physician or the end user.

Consumer Communication

Communication with the consumer regarding the closure is another challenge found in pharmaceutical packaging. Communications with the consumer are labeling elements that describe how to open the package properly and include drawings, pictographs, and text. The graphics and wording are all used to communicate with consumers how to manipulate a closure, particularly a CR closure, making it easy to open. Words or pictures communicate to the consumer to press down and turn, align arrows, squeeze closure sides, and perform other manipulations required to open the container. Information about the product or its manufacturer may also be part of the labeling.

Display

In addition to the instructions for opening a closure may also incorporate the manufacture's name and logo and some type of security or tamper-evident device. This is a lot of information to compress and arrange in a straightforward, logical way on the closure. A closure may use a tamper-evident breakaway ring, some type of tape or adhesive seal, or other device to also communicate potential tampering information to the consumer. At times this information may also be part of the package label as well.

Other communication features on closures include color variations, spot labels, and foldable labels and leaflets that provide additional information about the product.

Color can and does play a significant role in communication by closures. Ophthalmic products are color coded to indicate the class or type of drug they contain. Colors are sometimes used to differentiate different strengths of the same product. Color many times plays an important role in enhancing the display of a product, or providing a color code between the label and closure that is easily recognized by consumers of OTC products.

Metering and Measuring

Many closures have features built in that permit the user to meter or measure the dose of product prescribed in the directions. A good example of this feature is on ophthalmic products where eyedroppers and other dispensing devices have long been incorporated into the closure to make it easy for patients to place a drop or multiple drops of solution in their eyes. Many closures are now measuring devices that the consumer is directed to fill for correct dosing. The use of the closure for measurement of a prescribed dose minimizes the chance for contamination of the product. The inside of the closure may have graduations indicating the volume or weight amount of liquid or powder for each dose.

Closures also provide features that permit dispensing products directly to specific areas of the body. Topical ointments with an applicator built into the closure or a pump and atomizer in the closure wipe or spray product directly on the area being treated. Pumps with sprays are effective for delivering product to the nose or throat. Other specialized closures designed for enemas and irrigation solutions highlight how diverse delivery devices built into a closure may be and demonstrate the ability to design closures for unusual needs or applications.

Closures with built-in pumps to atomize product, or aerosol closures with dose-specific and particle size–specific engineering are examples of highly sophisticated types of measuring and metering closures. These closures hold and dispense a metered dose of product, usually inside the nose while atomizing the liquid to make the drug ingredients available to the mucosa. Pump-style closures are also used to atomize medications for the lungs, although the depth of penetration of aerosols produced this way is limited.

Aerosol closures round out this general list of dispensing devices. Aerosol closures can be designed to produce a metered dose of product or a continuous stream of product. The aerosol closure that produces a metered dose is found on inhalers for asthma medications. These closures withstand the internal pressure in the container, act as atomizers to break up the stream of liquid product into a particle that can be inhaled, provide the device to concentrate and direct the mist of product into the body, and are reclosable.

TYPES OF CLOSURES

Closures for Metal Cans

The ends used to close metal cans come in many different sizes and shapes. They include standard flat-panel double-seam ends, double-seam easy open ends, full-panel easy-open ends, and a number of membrane or diaphragms bonded to a double-seamed metal shell. This last type of construction produces an easy-open feature on the closure that is separate from the method used to seal the container. In all of these examples the metal can end is attached to the body of the can with a technique called double seaming (Fig. 2). This technique is used almost universally on cans (6).

Closures for metal cans may also be applied by friction (spin) welding or induction welding to the container. Slip lids that rely on friction (often referred to as overcaps) or interference for sealing, shrink bands, or pressure-sensitive tape to hold the overcaps in place are a few of the other methods used to close a metal can. Friction closures, a good example is the closure for paint, are another method of sealing a can. This closure example makes a plug-type seal on the can and provides a good moisture and gas barrier.

Some highly specialized can ends are produced with screw threads that accept a plastic- or metal-threaded closure. These types of threaded closures are found on cans holding solvents and oils. They come in a variety of sizes and shapes.

Double seaming is the standard technique for affixing a metal end onto a can. The metal end is produced from tin-free steel or tinplate and is almost always coated with an organic coating to seal or insulate the metal from contacting the product in the container. After fabrication into a can end, the portion of the closure that becomes one part of the double seam is lined or filled with a metered amount of a rubbery material, typically a highly formulated product consisting of synthetic and natural rubber components, which, when dried, produces a gasket inside the mechanical double seam. The metal end is placed on the metal body of the can, and two sets of "seaming rolls," the name used for

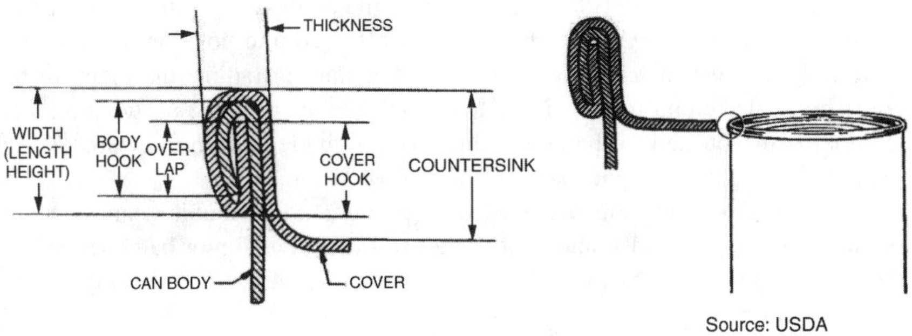

Figure 2 Diagram of a double-seam operation.

forming rolls with different angular profiles, contact the outside of the metal end at the cover hook. The cover hook is the small radius found on the outside of a formed end at the end of the seaming panel. The roll drives or pushes the metal inward and because of its rounded profile begins to curl first the metal at the outside of the end and then the metal around the rim of the can body called a "body hook," which is the portion of the metal cylinder designed to engage the can end. The two pieces of metal begin to fold over each other while trapping the end-seaming compound inside the sandwiched pieces of metal that form the mechanical seam. After the first set of seaming rolls creates the curves and partial mating of the two metal surfaces, a second set of seaming rolls engages the partially formed seam and finishes the operation by completing the curling or seaming process and then ironing or flattening the four layers of metal together tightly. This set of rolls has a different curvature or in can making terminology a different "profile" than the first operation rolls. The next time you use a metal can, look at one or both ends of the can where a metal end has been applied. The double seam is smooth and uniform around the circumference of the can.

Double seaming must be done right. The potential for contaminating a product inside a can is very real. The Food and Drug Administration (FDA) mandates that operators, mechanics, and maintenance personnel involved in heat retorting of products produced in metal cans attend a school to educate them on the proper alignment, shape, and force used to make the double seam (6). The size and shape of the double seam is very important to performance. Overlap, the amount of metal from the metal end (cover hook), which overlaps and engages the flange on the metal can body (body hook) is a primary determinant in the mechanical formation of the seam, and the specification of minimum and maximum overlap in a seam is a critical dimension used to determine seam quality. The seam height and the seam thickness are other measures of the consistency and quality of a metal seam. The FDA and the United States Department of Agriculture (USDA) both require a periodic destructive examination of the double seams on a metal can. They require the mechanic or line operator to physically tear the seam apart and measure the overlap, and "tightness" of the seam. The "tightness" refers to the amount of force used to force the metal in the seam together. It is measured by the amount of wrinkling observed in the cover hook of the seam after it has been forced into position. Because the radius of the finished seam is smaller than the starting radius, the metal in the cover hook of the end must wrinkle to a small degree to make up for too much material being pushed to the new radius. This wrinkle pattern is evaluated and read as a measure of "tightness" for the double seam.

All of the measurements needed to prove a good double seam is being produced and are recorded and kept as part of the production or batch record for the product. The cover hooks recovered from the destructive testing of the double seam are also kept.

A small leak in a double seam during the retort processing of a can will permit process water or external contaminants to enter the can. When the can cools and a vacuum forms in the can, the anaerobic atmosphere creates conditions that permit the growth of *Clostridium botulium*, the bacteria that produces one of the most serious spoilage conditions for food products, botulism. Medical foods produced in metal cans undergo the same heat retorting sterilization processes as food products do.

Bottles and Jars

Bottles and jars, both plastic and glass rely of two types of closures for sealing (7). These closures are either threaded or friction fit. There are a few variants to these two types of closures but their use is very specialized. The variants may provide opening features, such as hinged lids that for many are easier to open than turning a screw closure or a heat seal membrane on glass covered by an overcap. Some of the other highly specialized closures are ones that use the standard sealing techniques but incorporate multiple manipulations in their opening design to make them child resistant.

Threaded Closures

Threaded closures are the most used type of closure on prescription pharmaceutical bottles and bottles used for OTC products (8). There are three types of threaded closures used for bottles and jars, continuous thread, lug caps, and metal roll on closures.

Continuous Thread Closures

Continuous thread (CT) closures are the most common type used for drugs. Screw threads on the bottle and on the closure mesh to form a mechanical bond that when tightened generates torque (Fig. 3). Torque is the measured force that produces compression between the lip of the bottle or jar and the closure.

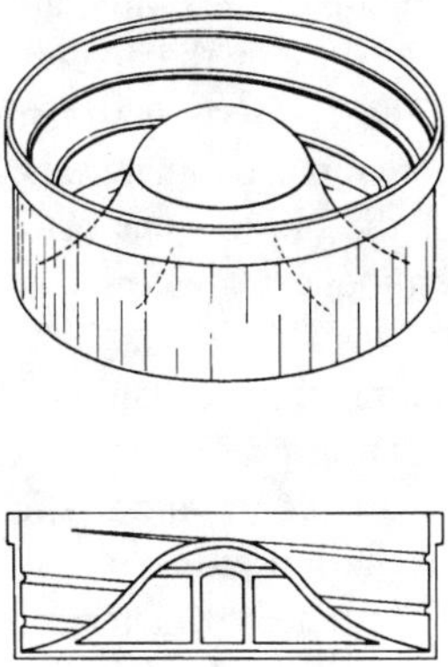

Figure 3 CT plug combination closure.

The amount of torque used to apply the closure is carefully controlled to produce a hermetic seal that withstands sterilization, handling, and distribution, but is still easy to open by the consumer. Closures may be lined with a gasket material that deforms when compressed to produce the seal, or they may incorporate a design in their construction where the cap and bottle meet that produces a gasket-like effect (9).

This second type of closure is called a "linerless" closure. The major complaint about bottles with threaded closures is the difficulty many people have with opening a container because the cap is too hard to turn.

It is interesting to note that the application torque used to tighten the closure in place is often considerably different than the removal torque or the amount of torque that must be applied to the closure to open the container. This is particularly true with thermoplastic closures used on glass bottles. The plastic has the tendency to creep or slowly flow because of the mechanical force. This flow or creep relieves the mechanical stress on the plastic, resulting in a decrease in torque over time. The creep or movement of the plastic is sometimes referred to as "cold flow." This phenomenon may also happen when thermosetting plastic caps are used on polyethylene (PE) bottles. In this case it is the bottle threads move (flow) to relieve some of the mechanical stress. When thermoplastics with high yield strength and tensile are used, such as PET, PP, and PVC, this problem does not occur.

Capping equipment for CT closures uses magnets, load cells, and in some cases a mechanical action to achieve the proper closing torque during production. The multiple heads on a capping machine must be able to produce bottles and jars at high speeds with reproducible closing torques from head to head in the capper. Modern cappers are rotary units that turn while applying the cap, and also turn the package as part of the process. These machines have multiple application stations or "heads" to apply the closures.

Originally CT or CT closures relied on a liner, plug, a paper, or plastic insert coated with an inert material or a liner made from a material that would not react with the product. The liner contacts the cap and the lip of the bottle and provides a gasket-like sealing action to hide any irregularities in the mating surfaces and produce an airtight and product-tight seal. These inserts or liners have been slowly replaced by linerless designs or foamed plastic materials. When no liner is needed, the design of the inside lip of the closure has been modified to take advantage of the elasticity of the thermoplastic material to produce a linerless closure.

The materials used to produce CT closures include aluminum, tinplate, tin-free steel, and plastics, both thermoset and thermoplastic. Drug packaging uses plastic closures made from almost all the commercially available thermoplastic polymers. The material most often used is PE. Metal and thermosetting plastic closures almost always require a "liner" or gasket-forming material. Thermoplastic closures may or may not require a liner. If the closure is molded with a unique mechanical design that engages with the mating surface on the bottle, the closure that produces two mating surfaces that form a robust seal that does not need a liner. Closures produced with thermoset materials may or may not require a liner depending on the end use. Linerless closures produce the same high-quality seals as lined closures and seals and can be reclosed with the same reliability.

CT closures with or without a liner are designed to produce a seal that meets the USP designations of "well closed" or "tightly" closed.

There are a number of combination designs that utilize both thermoplastic and thermoset components to produce a closure. An example of multiple material closures is one that incorporates a plug in the closure design. A plug made from one of the materials fits on the inside of the bottle finish, while the closure shell made from another material engages the outside of the bottle finish. Typically, the shell or the outside of the closure is made with the thermoset material and the plug, which also acts like a liner is on the inside. Many ophthalmic closures on squeeze bottles use this type of closure for producing a dispenser. One advantage of this combination is that it eliminates the slow creep or "cold flow" characteristics of an all-thermoplastic closure and maintains its application torque over an extended period of time.

Lug-Style Closures

Lug closures take advantage of the CT design on the bottle or jar for mating the two packaging components (Fig. 4). The difference with this closure is that it only contacts the threads in two, three, four, or six-point engagement locations around the circumference of the jar or bottle.

Lug closures were developed to improve the speed and efficiency of food packaging operations. The idea behind the lug closure was the quick application of the closure to the container and the equally quick engagement of the threads by the lugs to produce the torque needed to compress the gasket and seal the container.

A lug closure is pressed down on the finish of the jar or bottle and by using a very short quarter or half turn the closure fully engages the threads and achieves the torque necessary to seal the container. Compare this quarter or half turn to multiple revolutions of the capping head needed to apply a CT closure, and the

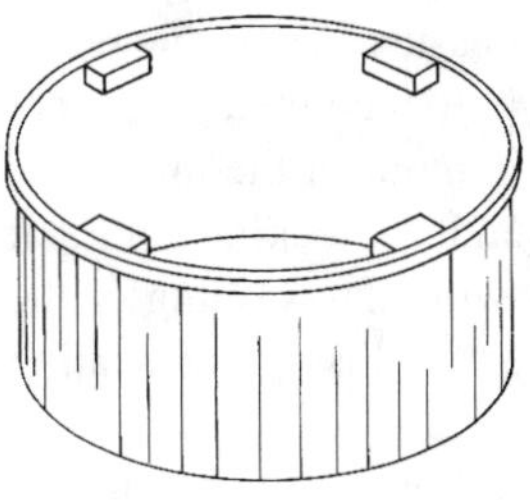

Figure 4 Lug-style closure.

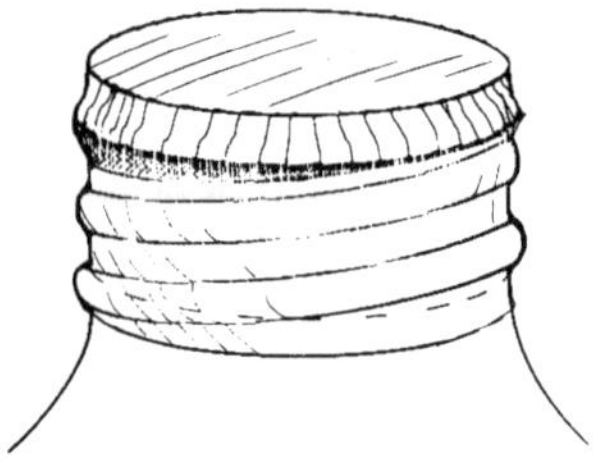

Figure 5 Picture or drawing of a roll-on closure.

speed advantage is evident. This short engagement also simplifies the design of the capper, eliminating the need for multiple complete turns of the capping head.

Because the closure only travels a very limited distance to achieve its seal, a liner or an elastic gasket material is used at the mating surface of lug closures to compensate for minor variations in application torque and adjust to the minor imperfections in the mating surfaces needed to create a tight seal.

Metal Roll-On Closures

This type of closure is a complete departure from the idea of mating two packaging components with premolded threads (Fig. 5). A roll-on closure takes advantage of a hard glass bottle finish (the threads at the top of the bottle) and the ductility of aluminum to produce a treaded closure on a container that does not require torque for application.

A roll-on closure starts as a flat sheet of metal. The metal often is spot decorated (printed) where each of the closures is to be punched out. The metal sheet then undergoes a metal stamping and pressing operation that punches out a circular disk and then draws or forms the flat circular piece of metal into a smooth cylinder closed at one end. This is analogous to the draw–redraw process for cans described in chapter 8 (Container Fabrication). If the closure has a tamper-evident feature, a secondary operation perforates the bottom of the skirt or bottom edge of the cylinder. A roll on closure shell with a tamper-evident feature is longer than one that does not incorporate tamper evidence into its design.

Application of a roll-on closure consists of placing the smooth closure cylinder or shell over the top of a bottle and then using heavy downward pressure (top load) to mate and seal the two surfaces. While the pressure is maintained on the top of the package, the entire unit is rotated against hardened metal dies that force the aluminum of the shell sides to conform to the contour of the threads in the container finish. The tamper-evident ring is forced under a locking ring that is molded into the bottle finish below the threads. Following the forming of the threads in the closure shell, the top load is removed and the closing and sealing of the container is complete.

The perforations in the tamper-evident band at the bottom of the closure are below the locking ring. When the consumer removes the closure, the metal breaks at the attachment points around the band and stays below the locking ring as the rest of the closure is removed.

Friction-Fit Closures

Friction-fit closures rely on a number of different methods to create friction or some type of interference fit between the closure and the bottle. They are most often used on glass containers but can be used on specialized plastic containers if the strain produced by the friction fit does not cause undue creep or cold flow of the plastic. There are four different types of friction closures widely used:

1. Bottle crowns
2. Snap-fit closures
3. Press-on closures
4. Elastomeric stoppers for parenteral products.

Crown Closures

Crown closures are best known as the bottle caps you pop off the top of a beer or soda bottle (Fig. 6). They are most often used in commercial beverage packaging on glass or the newer aluminum bottles for beer and beverages. They are used in pharmaceutical packaging on bottles for laxatives and other liquid products designed for complete consumption after opening.

Crown closures rely on the ability of steel or tinplate to bend without changing shape. The crown closure was invented by William Painter and patented in 1892. The Crown Cork and Seal Company was formed to make this closure.

The steel or tinplate material is crimped or folded into a number of indentations around the circumference of the closure. The closure is always lined, and it is applied over a ridge or formed area at the top of a bottle. The crimped circumference is slightly smaller than the maximum diameter of the lip of the container, and the indentations expand to permit the crown to fit over this area of the container and seat the liner on the lip. The smaller diameter of the lip just below its maximum diameter allows the indentations to return partially to their pre-application size, creating the friction to hold the closure on the bottle.

Figure 6 Crown-type closure.

Snap-Fit Closures

Snap-fit closures are most often found on bottles dispensed by pharmacies containing tablets. The closure is designed for use in two directions; in the first, the closure must be pressed down and then turned to engage the treads or the lip of the container. This provides a CR feature. In the second instance, the closure may also have regular screw threads on the outside of the closure that permit the cap to be turned over and applied to the bottle in the same way a CT closure is used. The second method of application is designed to make the closure more senior friendly and permit consumers to make the decision of whether they need child resistance or not.

The snap-fit style of the closure may be designed as a straight press-on application that forces the engagement mechanism of the closure over a ridge, lug, or series of partial threads on the top of the bottle, or as a press and twist configuration that engages and compresses the closure liner, creating both a seal and a friction fit. This type of feature is also used in many CT CR closures that are discussed later in this chapter. A variation of this design is a closure that snaps over a ring on the bottle finish. The ring is not continuous and the ring molded inside the closure also is not continuous around the complete circumference. By aligning to two components, bottle and closure, the closure can bend or move, permitting the closure to be popped off. An arrow or other alignment feature is present on both the bottle and cap to show how to align the closure for removal.

Snap-fit closures can be manufactured with a tamper-evident band at the bottom of the skirt of the cap. The band is a series of lugs or teeth that have been folded inside the closure. The closure is cut or perforated above this area on the skirt, creating a number of breakaway plastic bridges holding the ring to the remainder of the cap. When the closure is pressed on the bottle, the lugs or teeth at the fold slip over a retaining ring and the tops of the lugs or teeth engage the bottom of the ring. When the closure is removed, the plastic bridges break, leaving the ring on the bottle and providing a clear tamper-evident element to the closure.

Press-on Vacuum Caps

Vacuum caps are a unique style of closure used to protect oxygen-sensitive products. In pharmaceutical applications they are found on metal containers for powdered infant formula.

A vacuum is applied to a container, creating a partial vacuum in the container headspace below the closure. The closure liner seats on the container and is held in place by the vacuum until a permanent mechanical seal (double seam) can be applied to complete the closing operation.

When used on infant formula cans, the metal end is placed on the can with a friction fit usually created by partial crimping of the metal-seaming panel of the end to the top of the metal cylinder. The can is placed in a vacuum chamber (the number of cans processed at one time is based on the size of the vacuum chamber) and all air is evacuated. The metal end on the can acts as a one-way valve permitting the gas in the headspace of the can and in the product to be removed. When the external vacuum is removed, the internal vacuum in the can pulls the metal end down onto the top of the container, engaging the end-seaming compound in the seaming panel that embeds over the body hook of the can to hold the partial vacuum in the headspace. The vacuum inside the container holds the lid in place until the can passes through a double-seaming operation to complete the seal on the container.

VIAL STOPPERS

Vial stoppers are elastomeric friction seals for injectable drug containers (10). Elastomeric stoppers for vials are the only suitable closure for injectable products. The stopper acts as the seal in the mouth of the bottle and as a permeable self-sealing membrane that allows the needle on a syringe to be inserted for withdrawing the drug. They were originally made from cork before a variety of elastomeric materials replaced it. The elastomeric materials now in common use are described in chapter 6 (Pharmaceutical Packaging Materials).

When an elastomer is used to close a vial containing an injectable drug or vaccine, it creates a friction-fit closure that is unique in the number of properties it requires to produce a strong permanent seal of the parenteral vial. The elastomer must maintain a sterile environment inside the container, it must not interact with the contents of the vial, it must withstand sterilization and autoclaving, the material must not break or create particles when penetrated with a needle, and conversely the material must not core or be cut into a fine plug that would block the needle, and finally the closure must be easily and reliably inserted into filled containers on high-speed lines automatically.

There are four common designs for stoppers used in pharmaceutical packaging. These are

1. A flanged plug mechanically sealed with an aluminum band or overseal.
2. A flanged hollow plug with cutouts used for lyophilized products.
3. Plugs described in 1 and 2 above but sealed to the container with a plastic overcap.
4. A metal closure with a very small elastomeric disk attached to the closure.

Flanged Plug Elastomeric Stoppers

The first type of stopper, a flanged plug mechanically sealed with an aluminum band is the most common vial closure for injectable drugs (Fig. 7). The aluminum band has a hole in the center, permitting access to the elastomer seal by a needle (Fig. 8). After the plug is mechanically applied to the vial, the aluminum overcap is crimped over the elastomer and under a locking lip on the vial by

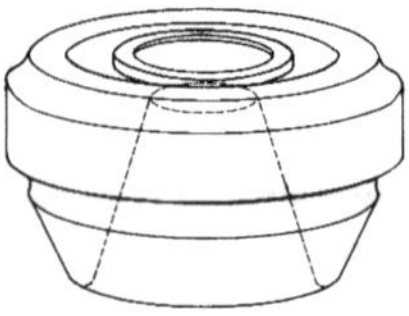

Figure 7 Flanged plug for vials.

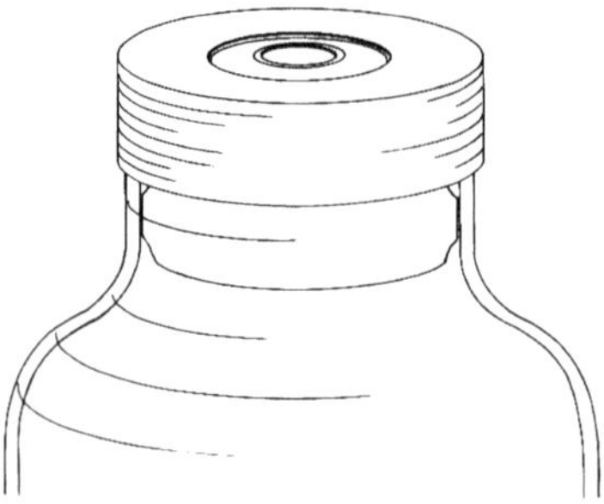

Figure 8 Aluminum overseal for vials.

rotating the vial against a stationary rail that pushes the metal into the required position. A second method of application for the metal band is the use of spinning rollers that contact the top of the container and force the metal against the glass of the vial under the locking lip. This is a variation of the roll-on thread application described earlier.

The flange design of the elastomer plug maintains a tight fit between the plug and the vial neck and the flange and the end of the vial. The aluminum overseal on the end of the vial locks this contact of the flanged plug in place.

Flanged Hollow Plug with Cutouts for Lyophilized Products

The second style of elastomer plug works much the same way as a crimped metal end on vacuum-sealed containers for infant formula. This type of plug is used on lyophilized drug containers (Fig. 9). The plug is partially inserted in the vial and the cutouts along the plug sidewall permit water vapor to exit the glass vial during the lyophilization (freeze drying) process. The water exits the container and leaves a dried cake or powder in the bottom of the vial ready for reconstitution and use. After completion of lyophilization, the plug is pushed the remainder of the way into the vial, creating the same type seal found in the first example with the flanged plug. An aluminum overcap is applied to the completed vial in the same way to lock the seal in place.

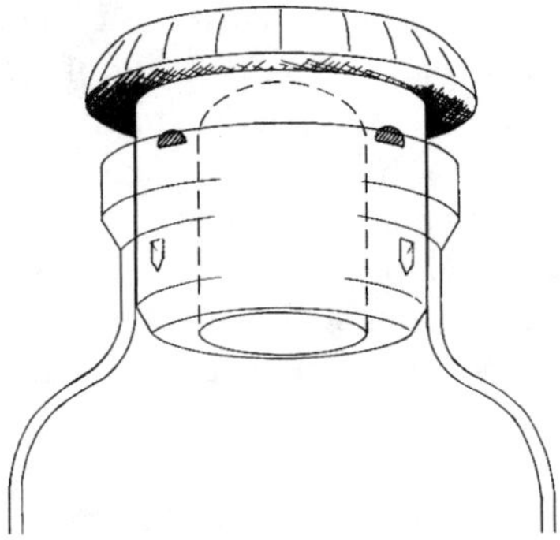

Figure 9 Flanged elastomer cutout plug.

Flanged Elastomeric Plug with Plastic Overseal

The third type of an elastomeric closure seal adds another feature to the closure process. A plastic overcap is fitted to the elastomer plug that has been put in place as described in the last two examples. The plastic overcap maintains the seal and the sterile conditions under the overcap where it contacts the plug. When the needle is inserted for the first use of product, the plastic overcap is removed and discarded.

Metal Closure with an Elastomeric Disk

The final method of capping a vial is very different. An aluminum cap, similar to that described in the first two examples of plugged vials is applied. The difference is the elastomer is not a plug placed in the neck of the bottle, but a thin gasket applied to the bottom of the overcap. This membrane acts as a gasket for sealing and a plug for inserting the needle. The top of the closure is open exposing the membrane. This type of cap is only used on very small vials and is not in common use. The closure design does not produce the same level of seal described in the first three examples.

Elastomeric Closure Performance

All closures for vials have a number of concerns attached to their use. The possibility of contamination or adulteration of a drug product by an elastomeric stopper requires extensive testing (8). The closure can interact with the product in a number of ways including leaching material from the elastomer closure, possible chemical interactions with the plug, coring or the cutting of rubber by the needle producing particulates, and selective sorption by the elastomer head the list of possible problems.

Sorption is almost always the major concern with these closures. Liquid injectables contain diluents or excipients to dilute and improve stability of the active pharmaceutical ingredient in solution. Included with the excipients are materials that maintain sterility in the solution by killing bacteria during storage and during use when the vial is penetrated multiple times by a needle used to administer its contents. The penetration of the package by a contaminated needle and the possibility of pushing bacterial contamination found on the surface of the stopper into the vial contents are some of the ways the liquid in the vial may become contaminated. Materials used for bacteriostats include phenol, benzyl alcohol, chloroxylenol, and a number of halogen compounds. All of these compounds have the potential to interact or be absorbed by the elastomer.

An older solution to this problem was to soak the elastomeric stopper in the material used as the bacteriostat prior to the stoppers insertion and use. This method has a number of drawbacks particularly controlling the amount of bacteriostat introduced to the drug from the stopper.

The standard method used to solve questions about sorption in the packaging system is chemical review of the elastomer followed by stability testing of the material options chosen to hold the drug. The difficulty with this approach is the number of elastomer components, sometimes as many as 15 to 20 combined with some non-elastomer components, and the time it takes to test for interactions. These materials may be leached from the elastomer into the drug solution, or they may sorb some of the drug or some of the bacteriostat or other component formulated in the drug solution. When stability testing reveals possible problems, there are a number of solutions to the problem. The two most common solutions are reformulation of the product with a different set of diluents or bacteriostats if earlier development work has indicated broad compatibility of the active ingredient with a number of approved diluents, excipients, and bacteriostats.

Another way to solve the problem is to coat the elastomeric stopper with Teflon® or other inert material. The drawback to this approach is the potential that the coating may lose adhesion to the elastomeric substrate and end up in the solution as particulate matter.

Elastomeric surfaces that are not adequately cured during the manufacturing process are another potential problem. Elastomers are crosslinked polymers that can retain small amounts of unreacted monomer. Particulates from the undercured elastomer surfaces can be broken from the elastomer by the needle. Until recently, the use of natural rubber was the common compounding change made to address particulate and coring problems. Unfortunately many health care professionals and patients have displayed allergic reactions to particular natural rubber components, and this compounding modification has become restricted or eliminated entirely.

Another source of particulates may come from washing or depyrogenation of the stoppers prior to use.

One major advantage of lyophilized products that are reconstituted immediately prior to use is the fact that the solid drug product is much less likely to interact with the packaging components. Many lyophilized products have strict schedules for use after reconstitution to minimize problems with hydrolysis or other chemical breakdown of the product in the liquid state.

Tube Closures

Tubes actually have two different closures as part of the package. A CT cap or snap-on pressure cap is used on one end, and a mechanical seal, heat seal, induction seal, or ultrasonic seal is used on the open end of the tube after filling.

The nozzle end of the tube usually has the reclosable cap or closure, permitting the product to be reclosed and used over an extended period of time. Tubes are most often used for ointments or gels that are designed for topical application. The nozzle end of a tube is made of metal or plastic and typically molded with screw threads just like a bottle or jar. It then is closed with a CT or snap-on closure applied over the threads. The tip of the tube is sealed as a tamper-evident indicator. Most CT and snap-on closures used on tubes have a recessed spike or puncture device molded into the outside of the closure (Fig. 1). When the closure is turned upside down, the puncture feature on the outside of the cap is pushed into the end of the tube. It is designed to break the seal in the nozzle of the tube for easy access to the product by the user.

Snap-on closures for tubes are plastic and rely on an interference fit of two different diameter sections of the closure and the tube nozzle or they rely on a molded ring on one or the other parts of the closure engaging a molded slot on the opposite piece.

Tubes are supplied with the end opposite from the tapered tip in an open position for fast filling. After filling the tube with product it may be crimped and then mechanically sealed if it is metal or it may be heat-sealed or induction-sealed if it is plastic. Both the mechanical seal on a metal tube and the heat-seal on a plastic tube are permanent and tamper evident.

Specialty Closures

There is a wide variety of specialty closures designed to make a product easier to use by the patient or consumer. OTC pharmaceuticals pushed this development and adoption along rapidly during the past decade.

A partial list of specialty closures includes

- Fixed-spout closures
- Movable-spout closures
- Flip-top closures
- Shaker closures
- Hinged-plug orifice closures
- Push–pull closures

Many of these designs began as closures for food or personal care products but were adapted to pharmaceutical products when their acceptance and benefits were proven. Flip-top closures do a good job of controlling the flow of low-viscosity liquids through small openings. Flip-top shaker containers do the same job with powders.

Pharmaceutical products that use these closures include rehydration solutions and closures that aid in the application of a topical product. Prescription dandruff products use these closures to dispense product in the same way as personal care shampoo products do.

One interesting adaptation of the flip-top closure is found on bottles containing antacid tablets. The solid tablets are made to conveniently dispense from the package with a flip-top closure that has a full mouth opening created by a hinged panel at the top of the closure. It permits easy access to the product after a tamper-evident induction seal liner or other tamper-evident feature has been removed. It is easily reclosable and can be produced in different colors, permitting a manufacturer, e.g., of antacids such as Tums® or Maalox® to highlight and distinguish different variations of the products on the shelf. It is hard to tell the difference between a true pharmaceutical package and consumer-oriented OTC drug package.

Dispensing Closures and Closures with Applicators

Closures provide many other features that are needed and required for pharmaceutical products. Closures that provide application and dispensing capabilities are good examples of a few of the additional features consumers expect in well-designed packages. The application features include closures that contain droppers, rods, and sponges that act as applicator pads. A short list of these closure designs includes

- Mechanical pump dispensers
- Closures with droppers
- Closures with brushes
- Sponge and cotton applicator closures

Closures for ophthalmic solutions originally contained eyedroppers. Droppers were also used in bottles of iodine. This was one of the first introductions of the public to dispensing closures. Later glass rods and more recently closures with plastic rods for dispensing and applying tinctures and other antiseptics replaced the droppers.

Sponge applicators, with the sponge made from a foamed or cellular plastic that meets FDA regulations, are found on containers that dispense topical gels, oils, and suspensions. These products are most often used to treat a skin conditions or other conditions requiring topical application of product.

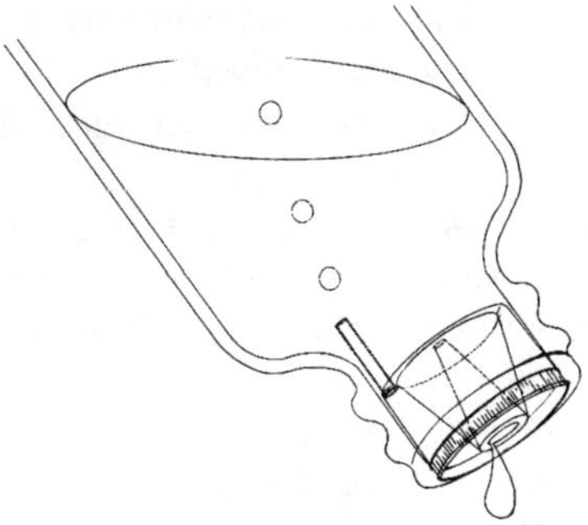

Figure 10 Precise dose-dispensing closure.

Fitment Closures

Fitment closures are another design of dispensing closures. These closures differ from the ones just discussed through the inclusion and use of a "fitment" or a portion of the closure that fits into the neck of the bottle and provides the dispensing mechanism (Fig. 10). The CT or snap cap fit over this fitment. Upon application, the fitment is driven down into the inside of the bottleneck, while the CT or snap closure engages the outside of the bottle finish. Most often the fitments are held in place by friction or the mechanical lock of a ring and groove.

Dropper closures with a fitment are designed to dispense product one drop at a time, with the size of the drop dispensed controlling the size and amount of the product dose. This type of closure is used with a squeeze bottle to give the user control over the speed the drop is formed and dispensed. Containers and closures for eye drops or other ophthalmic products are the most common example of this type of metered dispensing.

A large number of fitment closures are developed and patented. The variations in design solve a wide range of dispensing and application problems. It is prudent to review the patent literature before embarking on the development of a new dispensing-closure design. Many of the older designs have direct application to contemporary problems.

SPRAY AND PUMP DISPENSERS

Spray and pump dispensers have become very popular in the last 50 or so years to address other product application problems. These dispensers have slowly replaced aerosols in pharmaceutical applications and are found in a wide variety of forms. They are used to treat asthma, sore throats, and burns. Prescription drugs, a good example being steroids to treat allergy and nasal problems, rely on pump-actuated delivery of product rather than aerosol atomization and delivery. The mists delivered by these pumps and atomizing tips use the mucus membranes in the nose and throat to introduce a drug to the patient's circulatory system.

Spray and pump dispensers deliver liquids in spray patterns ranging from a steady stream to a coarse spray to a fine mist. They also deliver product in measured amounts for creams, lotions, and other high viscosity products that are spread on an affected area.

The different designations for delivery of product are a result of the spray or mist that they deliver. A pump delivering a coarse spray produces product with atomized particle sizes ranging from 100 to 500 μm in diameter. This type of spray is used when wetting an area or when covering an area with spray is desired. A pump delivering a fine mist, used for drug infusion across mucous membranes, produces a spray with a particle size less the 100 μm, with the majority of particles being under 30 μm in diameter.

The final type of pump sprayer, one that delivers a steady stream of liquid is not often used in pharmaceutical packaging, but is found when packaging bacteria-killing antiseptic cleaning products.

Viscosity of the material package determines the type of spray that is possible with a pump applicator. As viscosity increases to the range of 12 to 15 cP, the ability to produce a spray diminishes rapidly. Oily products with higher viscosities are very difficult to atomize, while aqueous solutions containing alcohol are very easy to atomize.

Bulk liquid pumps are used with creams and lotions. These closures work with any liquid that can be poured. If the product is extremely thick, it must be low enough in viscosity to flow into the dip tube of the pump and not cavitate or starve the pump. Pumps work best when the material delivered is homogenous and does not contain particles or materials that may separate during storage.

All pumps use common packaging plastics to fabricate the pump and closure components (8). Plastics may be combined with stainless steel or engineered plastic parts for springs and ball valves to improve the working action of the closure and its applicator. The plastics most often used to make pump closures include polyethylene (PE), polypropylene (PP), and polyvinyl chloride (PVC).

Single-Dose Closures

More and more closures are being developed to deliver a precise amount of product, a single measured dose from a bottle containing multiple doses of product (Fig. 10). Spray closures designed to deliver a measured dose of product are prescribed to the patient by the physician on the basis of the amount of product delivered in a single dose. This gives the doctor and the patient great flexibility in dosing options. It makes it easy to increase a product dose two or three times in strength without requiring a new closure or other measuring device. Applications where increased dosing is important are found with allergy patients when allergens are at higher than normal concentrations. These patients have the option of increasing the dose of product while the abnormal condition persists. This is common in asthma and allergy nasal sprays and inhalers.

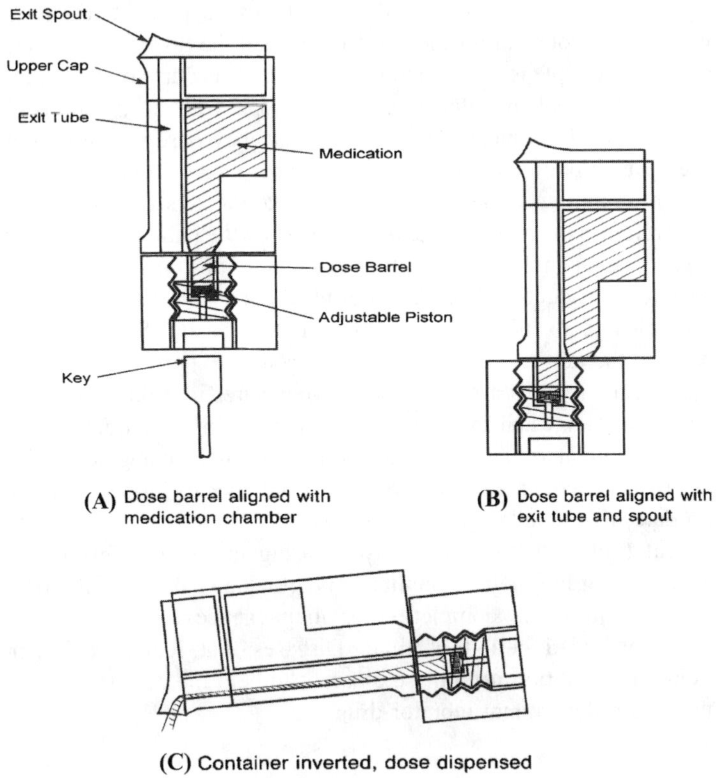

Figure 11 Preset single-dose container.

Other measured-dose closures contain visible chambers that are filled and then closed before pouring out the measured dose (Fig. 11). Some of these closures contain graduated chambers that permit the user to pump the correct amount of product into the chamber and then turn the container over to dispense the product.

Another type of measured-dose closure produces a very well-defined drop of product. The drops are designed to be multiples of the potential dosing regimens of the product. These closures may be linked to the graduated chamber described above to deliver a specific number of drops of product for a given amount of liquid.

COMPLIANCE (ADHERENCE)

Compliance is one of the most vexing problems facing health care today. A patient must complete the full regimen of prescribed doses of a product for it to be effective and for it to produce the complete therapeutic result expected by the doctor and the patient. Patients will stop taking a medicine when they feel better or will miss many doses of a medicine. This failure to use a product results in incomplete cures and the potential for the person to spread the disease while they are asymptomatic. An example is the need to take an extended course of antibiotics to treat tuberculosis. The problem with compliance in people with this disease is so severe that public health departments are requiring that they observe the patient taking each dose of antibiotic until the patient is declared tuberculosis free.

Examples of compliance design in closures are calendars and other tracking devices built into the closure. These can be set by the patient to establish when each dose is to be taken. Perhaps, the best example of this type of closure is birth control products. Products not produced in blisters rely on the closure and the ability of the closure to index to each dose of product and help alert the patient when to take the tablet.

More recently, microchips are fabricated into closures to produce audible reminders to the patient of when the next dose of medication in required. The closure becomes a dosing alarm clock. These audible closures are augmented by microchips that record opening and closing of the container to further monitor compliance. To date, this detailed monitoring has only been used in products undergoing clinical trials, but the possibility of using these mechanisms for critical products requiring long-term compliance is growing. Tuberculosis, HIV, and other diseases require the completion or maintenance of a prescribed treatment over an extended period of time. Diseases that require constant treatment of a chronic condition need these closures to help the doctor, patient, and possibly the health department monitor drug use.

CLOSURE LINERS

Closure liners are typically multiple material laminates called inserts that are added to a finished closure as part of the assembly process. Closure liners were originally developed to provide a gasket or seal to engage the lip or finish of the bottle and eliminate leaks by overcoming the minor imperfections in the bottle finish. They were originally developed for CT style closures. Originally closures were made of metal or a hard thermoplastic that could not be deformed to tightly contact the lip of the bottle or container. Without a liner compressing uniformly around the circumference of the container liquids would leak from containers that were turned upside down, and gasses and other contaminants could easily enter the container. Liners have become increasingly more complex and are used to not only provide a resilient gasket, but also act as a barrier material or to present a material that has no interaction with the product contained.

The gasket aspect of a liner is straightforward. The material is compressed by the action of the treads pulling down on the bottle or by the compression of the liner when a snap-on closure mates with the bottle.

Closure liners often are fabricated with an innerseal that acts as a tamper-evident device on the package. The closure is applied to the bottle, bringing the tamper-evident innerseal in contact with the lip of the bottle. The closure then passes through an induction-sealing device that heats the innerseal and melts the plastic or the coating on the innerseal to the lip of the bottle, creating a bond. The seal may be permanent, requiring actual destruction of the innerseal to remove it from the bottle lip, or it may be peelable and after one use it cannot be reused. A liner may be inserted in the closure before the removable tamper-evident seal. The liner remains in the closure to engage the bottle finish and provide a gasket-type seal after the tamper-evident seal has been removed.

The closure liner must possess a number of different attributes. These include resistance to attack and/or interaction with the container contents, compatibility with the sterilization method used for the product, and the ability to be removed cleanly from the bottle lip each time the closure is opened and reclosed.

COMPOSITION OF CLOSURE LINERS

Closure liners are made of a number of different materials that are combined in some type of lamination or coating. The terms used for the two parts of a closure liner are the "backing" and the "facing." The backing is a material thick enough and compressible enough to produce a gasket effect on the lip of the container, creating a seal. The facing material provides functional performance such as gas barrier, fat barrier, or moisture barrier, and isolates the closure from the contents of the container. The facing must also be inert to interaction with the product.

There are a large number of materials that are used as backing for closure liners. A partial list of materials includes

- Paperboard
- Pulpboard
- Chipboard
- Cork
- Foamed plastic
- Cork agglomerate
- Rubber

The backing material may be coated with an adhesive to hold the facing material to it, or the facing material may be extrusion or adhesive laminated to the backing. Extrusion lamination permits the molten material to form a bond with the surface of the backing.

The materials used for facings are a relatively broad category of materials that includes

- Plastic film
- Foil
- Foil/film laminations
- Coated foil
- Unsupported foil

- Paper laminations
- Coated paper

These materials have a very broad range of subcategories. These may include paper/foil, paper/PVDC, foil/PE (extrusion or adhesive laminated), foil/PVC, foil/PP, foil/ionomer, or foil/EAA, to highlight just a few.

When a tamper-evident seal is present, the seal is manufactured as an additional component and then attached to the finished liner with a peelable adhesive. If a tamper-evident seal is used, a foil material is not part of the closure liner, but is part of the tamper-evident seal. The traditional closure liner has been giving way to linerless closures.

Linerless Closures

Linerless closures eliminate the need for a liner by using all plastic construction combined with a number of molding techniques to produce the same sealing effect found in lined closures. A packaging plastic, usually PE or PP, is molded with a "seat" or contoured area designed to engage the lip of the bottle and produce a seal. The molded inner portion of the closure forms diaphragms, plugs, valve seats, deflecting membranes and rings, all of which contact the lip or bead of the container finish to produce the linerless seal. The compressibility of the two plastics provides the mechanism for intimate contact of the closure with the lip of the bottle.

The linerless closure may also have a gasket material applied only to the contact the lip area of the bottle as another method of producing the closure with good sealing characteristics but without a full lining. These high-viscosity liquid materials used as gaskets are made from organisols, plastisols, or semi-soluble elastomeric materials. The material is applied at the shoulder of the closure and then dried or cured to remove the solvent that made it a high-viscosity liquid. These materials are all compressible in the same way a molded rubber gasket is compressible. The major drawback with any gasket material is the need to fully remove solvent and cure the material. Undercured materials interact with product and can contaminate or adulterate a drug product.

Child-Resistant Closures

CR closures have become a common part of most pharmaceutical packages (4). They are required on both prescription products and OTC products. The United States CPSC administers CR packaging requirements called for in the Poison Prevention Packaging Act of 1970 (11). Originally the act placed enforcement under the FDA but then moved it to the CPSC, a newly formed agency at the time, in 1973. The Poison Prevention Act authorizes special packaging for hazardous household substances. The act covers 33 different substances and categories including most human oral prescription drugs, aspirin, OTC products, and many specific drug substances and supplements (Table 1). Pesticides are not

Table 1 Poison Protection Act–Regulated Substances

Regulated substance	Background and notes
Aspirin	Both flavored and unflavored, including effervescent tablets and aspirin containing preparations. This includes unit doses of the product.
Furniture polish	Non-emulsion-type liquid furniture polishes containing ≥10% of mineral seal oil and/or other petroleum distillates.
Methyl salicylate	Liquid preparations containing >5% by weight of methyl salicylate.
Controlled drugs	Any preparation for human use that consists in whole or in part of any substance subject to control under the Comprehensive Drug Abuse Prevention and Control Act of 1970 (21 U.S.C. 801) and that is in a dosage form intended for oral administration shall be packaged in accordance with the provisions of Section 1700.15 (a), (b), and (c).
Sodium and/or potassium hydroxide	Household substances in dry forms such as granules, powder, and flakes containing ≥10% by weight of free or chemically unneutralized sodium and/or potassium hydroxide and all other household substances containing ≥2% by weight of free or chemically unneutralized sodium and/or potassium hydroxide.
Turpentine	Household substances in liquid form containing ≥10% by weight of turpentine.
Kindling and/or illuminating preparations	Prepackaged liquid kindling and/or illuminating preparations, such as cigarette lighter fuel, charcoal lighter fuel, camping equipment fuel, torch fuel, and fuel for decorative or functional lanterns, which contain ≥10% by weight of petroleum distillates.
Methyl alcohol (methanol)	Household substances in liquid form containing ≥4% by weight of methyl alcohol (methanol), other than those packaged in pressurized spray containers.
Sulfuric acid	Household substances containing ≥10% by weight of sulfuric acid, except such substances in wet-cell storage batteries.
Prescription drugs	Any drug for human use that is in a dosage form intended for oral administration and that is required by federal law to be dispensed only by or upon an oral or written prescription of a practitioner licensed by law to administer such drug. Note there are 21 exceptions for specific drugs and substances listed in this provision of the code.
Ethylene glycol	Household substances in liquid form containing ≥10% by weight of ethylene glycol.
Iron-containing drugs	With the exception of: Animal feeds used as vehicles for the administration of drugs, and those preparations in which iron is present solely as a colorant, non-injectable animal and human drugs providing iron for therapeutic or prophylactic purposes, and containing a total amount of elemental iron, from any source, in a single package, equivalent to ≥250 mg elemental iron in a concentration of ≥0.025% on a weight-to-volume basis for liquids

Table 1 Poison Protection Act–Regulated Substances (*Continued*)

Regulated substance	Background and notes
	and ≥0.025% on a weight-to-volume basis for liquids and ≥0.05% on a weight-to-weight basis for nonliquids (e.g., powders, granules, tablets, capsules, wafers, gels, viscous products, such as pastes and ointments).
Dietary supplements containing iron	Dietary supplements, as defined in Section 1700.1(a)(3), that contain an equivalent of ≥250 mg of elemental iron, from any source, in a single package in concentrations of ≥0.025% on a weight-to-volume basis for liquids and ≥0.05% on a weight-to-weight basis for nonliquids (e.g., powders, granules, tablets, capsules, wafers, gels, viscous products, such as pastes and ointments), there are two exceptions to this provision, iron used only as a colorant, and powdered preparations with ≤0.12% weight-to-weight elemental iron.
Solvents for paint or other similar surface-coating material	Prepackaged liquid solvents (such as removers, thinners, brush cleaners) for paints or other similar surface-coating materials (such as varnishes and lacquers) that contain ≥10% by weight of benzene (also known as benzol), toluene (also known as toluol), xylene (also known as xylol), petroleum distillates (such as gasoline, kerosene, mineral seal oil, mineral spirits, naphtha, and Stoddard solvent).
Acetaminophen	Preparations for human use in a dosage form intended for oral administration and containing in a single package a total of >1 g acetaminophen. There are 2 exceptions to this provision for effervescent tablets and unflavored preparations containing ≤13 grains of acetaminophen in powder form.
Diphenhydramine	Preparations for human use in a dosage form intended for oral administration and containing >66 mg diphenhydramine base in a single package.
Glue removers containing acetonitrile	Household glue removers in a liquid form containing >500 mg of acetonitrile in a single container.
Permanent wave neutralizers containing sodium bromate or potassium bromate	Home permanent wave neutralizers, in a liquid form, containing in single container >600 mg of sodium bromate or >50 mg of potassium bromate.
Ibuprofen	Ibuprofen preparations for human use in a dosage form intended for oral administration and containing ≥1 g (1000 mg) of ibuprofen in a single package.
Loperamide	Preparations for human use in a dosage form intended for oral administration and containing >0.045 mg of loperamide in a single package (i.e., retail unit).

Table 1 Poison Protection Act–Regulated Substances (*Continued*)

Regulated substance	Background and notes
Mouthwash	Except as provided in the following sentence, mouthwash preparations for human use and containing $\geq$3 g of ethanol in a single package and mouthwash products with non-removable pump dispensers that contain at least 7% on a weight-to-weight basis of mint or cinnamon flavoring oils that dispense $\leq$0.03 g of absolute ethanol per pump actuation, and that contain <15 g of ethanol in a single unit are exempt from this requirement. The term "mouthwash" includes liquid products that are variously called mouthwashes, mouthrinses, oral antiseptics, gargles, fluoride rinses, antiplaque rinses, and breath fresheners. It does not include throat sprays or aerosol breath fresheners.
Lidocaine	Products containing >5.0 mg of lidocaine in a single package (i.e., retail unit).
Dibucaine	Products containing >0.5 mg of dibucaine in a single package (i.e., retail unit).
Naproxen	Naproxen preparations for human use and containing $\geq$250 mg of naproxen in a single retail package.
Ketoprofen	Ketoprofen preparations for human use and containing >50 mg of ketoprofen in a single retail package.
Fluoride	Household substances containing >50 mg of elemental fluoride per package and >0.5% elemental fluoride on a weight-to-volume basis for liquids or a weight-to-weight basis for nonliquids.
Minoxidil	Minoxidil preparations for human use and containing >14 mg of minoxidil in a single retail package. Any applicator packaged with the minoxidil preparation and which it is reasonable to expect may be used to replace the original closure.
Methacrylic acid	Except as provided in the following sentence, liquid household products containing >5% methacrylic acid (weight-to-volume) in a single retail package shall be packaged in accordance with the provisions of Section 1700.15(a), (b), and (c). Methacrylic acid products applied by an absorbent material contained inside a dispenser (such as a pen-like marker) are exempt from this requirement provided that: (*i*) the methacrylic acid is contained by the absorbent material so that no free liquid is within the device and (*ii*) under any reasonably foreseeable conditions of use the methacrylic acid will emerge only through the tip of the device.
OTC drug products	Any OTC drug product in a dosage form intended for oral administration that contains any active ingredient that was previously available for oral administration only by prescription. This requirement applies whether or not the amount of that active ingredient in the OTC drug product is different from the amount of that active ingredient in the prescription drug product. This requirement does not apply if the OTC drug product contains only active ingredients of any oral drug product or products approved for OTC marketing on the basis of an application for

Table 1 Poison Protection Act–Regulated Substances (*Continued*)

Regulated substance	Background and notes
	OTC marketing submitted to the FDA by any entity before January 29, 2002. Notwithstanding the foregoing, any special packaging requirement under this Section 1700.14 otherwise applicable to an OTC drug product remains in effect.
	Active ingredient means any component that is intended to furnish pharmacological activity or other direct effect in the diagnosis, cure, mitigation, treatment, or prevention of disease or to affect the structure or any function of the body of humans; and *drug product* means a finished dosage form, e.g., tablet, capsule, or solution, that contains a drug substance (active ingredient), generally, but not necessarily, in association with one or more other ingredients. [These terms are intended to have the meanings assigned to them in the regulations of the FDA appearing at 21 CFR 201.66 and 21 CFR 314.3 (2000), respectively.]
Hazardous substances containing low-viscosity hydrocarbons	All prepackaged non-emulsion-type liquid household chemical products that are hazardous substances as defined in the FHSA [15 U.S.C. 1261 (f)], and that contain ≥10% hydrocarbons by weight. There are 3 exceptions to this portion of the code covering: (*i*) Products in packages in which the only non-CR access to the contents is by a spray device (e.g., aerosols, or pump- or trigger-actuated sprays where the pump or trigger mechanism has either a CR or permanent attachment to the package), (*ii*) writing markers and ballpoint pens, (*iii*) products from which the liquid cannot flow freely, including but not limited to paint markers and battery terminal cleaners. For purposes of this requirement, hydrocarbons are defined as substances that consist solely of carbon and hydrogen. For products that contain multiple hydrocarbons, the total percentage of hydrocarbons in the product is the sum of the percentages by weight of the individual hydrocarbon components.
Drugs and cosmetics containing low-viscosity hydrocarbons	All prepackaged non-emulsion-type liquid household chemical products that are drugs or cosmetics as defined in the FFDCA [21 U.S.C. 321(a)], and that contain ≥10% hydrocarbons by weight. There are two exceptions listed to this provision of the code.
Sample packaging	The manufacturer or packer of any of the listed substances requiring special packaging shall provide the Commission with a sample of each type of special packaging as well as the labeling for each size product that will be packaged in special packaging and the labeling for any non-complying package.
Applicability	Special packaging standards for drugs listed in the act shall be in addition to any packaging requirements of the FFDCA or regulations promulgated thereunder or of any official compendia recognized by that act.

Abbreviations: OTC, over-the-counter; FDA, Food and Drug Administration; CR, child-resistant; FHSA, Federal Hazardous Substances Act; FFDCA, Federal Food, Drug, and Cosmetics Act.
Source: From Ref. 11.

part of the act and are regulated by the Environmental Protection Agency (EPA). The act has specific requirements for prescription drugs and OTC drug products. It also covers solvents and a wide range of other materials that may be components of health care products. It also carries provisions for sample products, a common sales aid in the pharmaceutical industry, used for sampling physicians and patients with product (12). It also permits exemptions when a product cannot be packaged in a form that produces the desired delivery or therapeutic effect and still be child resistant. This exception and many of the others qualifiers found in the act are becoming harder and harder to use because the act requires a manufacturer or packager to demonstrate why their products are different from similar classes of product packaged with CR features (11) (3).

The act was designed to prevent the poisoning of children under the age of five years. Poisoning incidents related to aspirin were well known before 1970, and in the first years the act was implemented, resulting in the number of problem incidents with aspirin dropping significantly. This example, which clearly showed the benefits of the new closures, justified the inclusion of all the other substances and packages listed in Table 1.

The regulations are flexible with regard to seniors and people with infirmities. It permits the patient or the prescriber to request non-CR closures (12). The act prohibits a pharmacist from making this determination. An additional provision of the act permits OTC products, except those dispensed directly by physicians and dentists (sample packages), to be produced in one size that is not child resistant, provided all other sizes of the product are packaged in accordance with the regulations. The size chosen for sale without CR protection cannot be the most popular size of the product. The one non-complying package must bear a label stating "This package for households without young children."

When a pharmacist dispenses a drug, both he/she and the manufacturer are responsible to supply the product with a CR closure. When a pharmacist repackages a drug, he/she assumes complete responsibility for complying with the act. Many drugs are packaged in large sizes (e.g., $\geq$500 tablets) and the pharmacist repackages the product for the patient. The manufacturer is not required to supply product with CR closures when the product does not go to the ultimate consumer as in the example above.

Prescription drugs that are intended for topical application to the teeth or for administration by inhalation are not required to have CR packaging, only orally ingested prescription and OTC products as described in the act and the summary table must comply.

CHILD-RESISTANT TESTING OF CLOSURES—AN OVERVIEW

Panels of children and seniors carry out testing of CR packaging. The act is very specific in detailing how the testing is done, and the number of children by age group and the number of people 50+ years of age required to ascertain results (11). It is interesting to note that the test does not require an absolute result of no

Table 2 Sequential Child-Test[a] Requirements

		Package openings (10 min)		
Test panel[b]	Total number of children tested	Pass	Continue	Fail
1	50	0–5	6–14	15+
2	100	6–15	16–24	25+
3	150	16–25	26–34	35+
4	200	26–40	—	41+

[a]From 60 FR 37736.
[b]Each panel uses 50 children.

children accessing the contents of a package, just a specific percentage that cannot. The provisions of testing and the criteria for pass/fail results were modified in 1995. The modifications to the testing procedures for children were designed to make the testing procedure easier to perform and for the results to be more consistent. The modifications to the testing procedures for adults were designed to increase the use of CR packaging and make that packaging easier to open. Seniors often complain about the difficulty in opening CR packaging, and many times do not replace the closure or only partially replace the closure to defeat the CR design.

The child-test protocol calls for sequential testing of groups of 50 children (Table 2). When the test is successful using the methods outlined below, 80% of tested children aged 41 to 52 months must not be able to open or access the package.

A summary of the child-test protocol is as follows:

CHILD-TEST PROTOCOL

Sequential Test50 children in four groups. Total 200 children.
Three different age groups are specified in months.

a. 42 to 44 months
b. 45 to 48 months
c. 49 to 51 months

The percentage of each group listed above is specified

a. 20%
b. 40%
c. 30%

Note: there is a standardized age calculation in the act to properly place the children in the right age group.

Gender Requirements: 50% boys 50% girls (there is a 10% tolerance factor for each group of children in the sequential test, i.e., worst case is a 60/40 gender distribution in one arm of the sequential test).

Test Time Requirements: Five minutes demo five minutes.

a. Initial—five minutes of testing permitting the children to try to open the package.
b. A demonstration by the tester of how to open the package followed by
c. Another five-minute period for the children to attempt to open the package.

The children are told the use of teeth is permitted. The procedure includes standardized test instructions to guide the person administering the test.

Performance Criteria: 85% after 5 minutes.
200 Children Total Tested: 80% after 10 minutes.
Tester: No more than 30% of children tested.
Site of Testing: No more than 20% of children tested.
Note: Children selected must have no overt physical or mental handicap.

The CR test protocols used for adults are some of the more contentious issues between older patients that have difficulty with manipulating the closure for opening and the CPSC. One reason the original test protocol was changed was that it specified the adults in the test be between 18 and 45 years. This and the fact that the test period stretched to five minutes for opening were considered by seniors to be unrealistic of how long they would struggle to open a package. They rightly pointed out that people in this younger age group probably did not have problems with arthritis and other infirmities that made manipulation of the closure difficult for them. These complaints were addressed in the changes to test methodology introduced by the CPSC in 1995 (4).

A summary of the current adult test protocols is as follows:

ADULT TEST PROTOCOLS—SENIOR
Number Participating: 100 adults
Age Groups:
Overall Group: 50–70 years
50–54 25%
55–59 25%
60–70 50%
Gender Requirements: 70% Female.
Test Time Requirements: 5-minute/1-minute test period.
Screening tests for unsuccessful participants.
Standardized test instructions.
Performance Criteria: 90% adult-use effective.
Tester: No more than 35% of adults tested.
Site: No more than 25% of adults tested.

Note: Test covers all regulated products except those in metal cans and aerosols.

Blister packaging, widely used in Europe and other parts of the world, when made child resistant are difficult to open. This difficulty leads people to use scissors and knives to open CR blisters and is one of the reasons more unit-dose packaging has not been used in the United States. Europe is in the process of adopting CR standards for blisters. None of the current individual country standards now used meets the CPSC performance requirements for child resistance.

DESIGN OF CHILD-RESISTANT CLOSURES

There are definite strategies or ideas behind the design of a CR closure. Most packaging manufacturers begin with the following assumptions for designing CR packaging.

1. Children are very persistent and will use tools such as the hard edge of furniture or their teeth to open a package.
2. Children's teeth and fingernails are sharp and small enough to fit into any gap in the packaging.
3. Children's motor skills will not permit them to perform two or more motions at the same time.
4. Children can learn quickly from watching adults.
5. Children in these age groups cannot read instructions and cannot determine alignments of components.

On the basis of these ideas, CR closures with reclosable features rely on the press-turn, squeeze-turn, and the combination (alignment) lock to produce the results required by the Poison Prevention Act.

Press-turn closures require downward pressure by the user to engage an inner shell that holds the threads and actual portion of the closure contacting the bottle. The combination of downward pressure and a simultaneous turning of the closure opens the bottle. When the cap is reapplied, friction between the bottle and closure threads requires the user to repeat this action to maintain the CR features of the container.

A variant of the press-turn closure relies on a locking lug on the bottle and a lug on the closure. These closures have a longer skirt or area below the threads than standard closures. When the closure is screwed onto the bottle, the lug on the outside of the cap engages a locking mechanism on the side of the bottle. Removal can only be accomplished by simultaneously pressing the outside locking mechanism and turning the cap. Normally an audible click or sound is heard as the lug and closure disengage. A similar click or sound is produced when the closure is reattached.

Squeeze-turn closure is similar to the press turn. It relies on the user squeezing the sides of the closure, usually at a designated location to engage the outer and inner shells of the closure. This closure has a variant in new one-piece designs that utilize squeezing of the closure along with a locking lug or mechanism on the bottle finish. The cap skirt is squeezed and deformed to clear the locking lug on the bottle.

Combination Closures

Combination closures rely on alignment for opening. The closure snaps over the finish of the bottle and cannot be removed unless the arrow on the cap is aligned with a slot in the bottle finish that permits the cap to deform sufficiently to pull out and away from the locking ring on the bottle.

Many locking style CR closure designs rely on two hands to open the closure. Newer designs that can be opened one handed have appeared and are slowly finding acceptance by consumers and pharmaceutical companies.

Aerosol Closures

As noted in the regulations, aerosol products are exempted when this method of packaging is the only method available for correct delivery of a drug. There are overcaps for aerosols that require a coin, knife, or other device to pry the lid from the can. The design relies on a reclosable locking mechanism and will reengage if the cap is replaced on the aerosol properly.

Non-reclosable Packages

Non-reclosable packages, blister packs, and pouches can be made child resistant. The most common method used for blister packs is the addition of an adhesive-laminated paper/PET film applied over the outside of the foil-sealing material. The consumer must first separate the blisters at a perforation and then attempt to peel the paper/plastic laminate from the foil before pushing the tablet through the foil and out. This is rarely viewed as easy open by adults and is a constant source of complaints from consumers. Examples of this type of CR blister are found on OTC products for colds and products containing ephinephrine.

Pouches

Pouches are used for drugs and aspirin but are extremely difficult to make child resistant. A determined three- or four-year old with teeth is a formidable challenge for the package to overcome. When pouches are used, the secondary packaging many at times provides the CR feature by providing a hard-to-manipulate box or case that requires multiple manipulations to open. This fact is ironic considering the number of complaints food manufacturers receive about hard-to-open pouches. The use of a specialized box to make a pouch child resistant is viewed by many consumers as over-packaging.

Successful CR pouches rely on plastic laminate materials that are impossible to tear unless notched. The films must be strong enough to not tear or be notched by a child's teeth. The notch adults use to open the pouch is hidden in one corner of the pouch and the person opening the package must first fold the pouch at the perforation to expose the notch for tearing.

With the move to managed health care in the United States, and a drive by the FDA to reduce incidents of the wrong medication getting to a patient, the blister pack and other unit dose options are gaining favor. In Europe and elsewhere in the world, the majority of products are produced in blister packs. Unit-dose packaging eliminates the need for large pharmacies and reduces the chances for a drug error when dispensing medication in a hospital or nursing home. Extensive work needs to be done with secondary packaging or with blister packaging materials if these are to become a primary form of packaging. The CR-laminated backing materials for blisters were developed over 30 years ago and have changed little since their first introduction.

Tamper-Evident Packaging Closures

The pharmaceutical and OTC packaging world changed drastically in October of 1982. Seven people in Chicago died after ingesting cyanide-laced Tylenol. This action rocked the packaging world with the reality that people could easily open and tamper with products without the ultimate consumer ever knowing something was wrong. This tragedy began an effort by the packaging industry, government, drug companies, and consumer groups that continues today. New tamper-evident designs from manufacturers, along with strong law enforcement against people caught tampering with products and a public now aware and educated to look for packages that display potential damage from tampering has kept this from becoming a widespread problem.

The FDA reacted quickly to tampering. During November of 1982, they published a series of proposed new rules in the Federal Register. These rules are now part of 21 CFR 211.132, which cover tamper-evident requirements for OTC drugs. Part 211 of the CFR covers Current Good Manufacturing Practice (CGMP) for finished pharmaceuticals (5).

At the beginning of the effort to develop tamper-evident definitions, the term "tamper resistant" was used. This became a misnomer because it really meant that no one could easily open the package. The FDA and other regulatory bodies that cite tampering changed the term to "tamper evident" soon after this incident. The definition now typically used to describe tamper evidence is "(a package which has) an indicator or barrier to entry which, if breached or missing, can reasonably be expected to provide visible evidence to consumers that tampering has occurred." The definition is found in 21 CFR 211.132. (5).

The thrust of the FDA regulations was to address the problem with products sold directly to the public. Products not accessible to the public were exempted from the regulations. These included

1. Products sold directly to hospitals, nursing homes, other health care institutions, and physicians.
2. Products sold from vending machines.
3. Products sold door to door or through the mail.

Over the years as concern about this problem has grown, and as manufacturers work diligently to pare the number of items they offer for sale, tamper-evident features are typically found on products sold through all distribution channels. Consumers are rightfully wary of products that do not contain tamper-evident features, and their voice and the benefit they perceive from products with this level of security forced manufacturers to adopt their use across the board.

The descriptions of what the regulations expect are very straightforward. They are as follows:

- The tamper-evident indicator or the barrier to tampering must be part of the design of the package, such as the case of an aerosol container, which would be extremely difficult to alter, or it must be an indicator that is unique (e.g., seals printed with logos, brand or company names, or unique patterns) and difficult to duplicate.
- The package must be labeled with a statement highlighting the tamper-evident feature and alerting the consumer to look for it and to examine the condition it is in before purchasing the product. The labeling must be on a part of the package that cannot be removed. For example, the highlighting statement cannot be on the tamper-evident feature itself and thus lost if the feature is removed or defeated.
- Products that require tamper-evident packaging are
  - OTC drug products.
  - Oral, rectal, nasal, vaginal, otic, and ophthalmic products.
  - Oral cosmetic liquids, including mouthwashes, gargles, and breath fresheners.
  - Contact lens solutions and the cleaning supplies and tablets used for their maintenance.

Note: a number of products such as insulin were exempted from the regulations and additional exemptions have been extended to lozenges, medical oxygen, and aerosol products.

Tamper evidence is even more crucial with the development of large scale counterfeiting of drug products in many parts of the world. This feature along with other overt and covert features designed to identify the genuine product and make it hard to copy, all are ways to insure that the consumer is receiving the correct and unadulterated product.

As part of the regulations issued by the FDA, the Agency cited examples of many different types of packaging and commented on what would be acceptable and what would not (13). This aided designers and manufacturers at the beginning of the program and has helped drive many of the innovations seen in the past 25 years following the original incident. The original list contained comments and examples that covered 11 forms of packaging. This list is found in the Federal Register (47, 50442-50456, November 5, 1982); note that the term tamper resistant instead of tamper evident was used in this publication at the beginning of the crisis in 1982.

FDA Examples of Tamper-Resistant Package Forms (Nov. 1982):

1. Film Wrappers: A transparent film with distinctive design is wrapped securely around a product or product container. The film must be cut or torn to open the container and remove the product.
2. Blister or Strip Packs: Dosage units (e.g., capsules or tablets) are individually sealed in clear plastic or foil. The individual compartment must be torn or broken to obtain the product.
3. Bubble Packs: The product and container are sealed in plastic and mounted in or on a display card. The plastic must be torn or broken to remove the product.
4. Shrink Seals and Bands: Bands or wrappers with distinctive design are shrunk by heat or drying to seal the union of the cap and the container. The seal must be cut or torn to open the container and remove the product.
5. Foil, Paper, or Plastic Pouches: The product is enclosed in an individual pouch that must be torn or broken to obtain the product.
6. Bottle Seals: Paper or foil with distinctive design is sealed at the mouth of a container under the cap. The seal must be torn or broken to open the container and remove the product.
7. Tape Seals: Paper or foil with a distinctive design is sealed over all carton flaps or bottle cap. The seal must be torn or broken to open the container and remove the product.
8. Breakable Caps: The container is sealed by a plastic or metal cap that either breaks away completely when removed from the container or leaves part of the cap attached to the container. The cap must be broken to open the container and remove the product.
9. Sealed Tubes: The mouth of a tube is sealed and the seal must be punctured to obtain the product.
10. Sealed Carton: All flaps of a carton are securely sealed and the carton must be visibly damaged when opened to remove the product.
11. Aerosol Containers: Aerosol containers are inherently tamper resistant.

EASE OF OPENING

It must be pointed out that the FDA and the CPSC have from the start indicated tamper evidence, and CR closure regulations are not related to each other. Both sets of regulations must be met, and one set of regulations has no connection with the other set of regulations. Both agencies realize that the inclusion of tamper-evident features and CR packaging features make packages more difficult to open, particularly for the elderly. Both agencies have studied this problem and the FDA declined setting a standard for ease of tamper-evident access to a container. The CPSC modified its requirements in 1995 to determine easier access of a container by the elderly and changed their test panel to include consumers between 50 and 70 years. Their changes were the only accommodation made to permit people to gain easier access to a container.

The FDA on the other hand after studying the access issue with tamper-evident closures concluded they would not develop an opening standard and would let manufacturers and the marketplace determine what was most acceptable under the existing tamper-evident regulation found in 21 CFR 211.132 for pharmaceutical products.

CAPSULE PROBLEMS

Following the Tylenol incident in 1982, a second incident occurred in 1986 when three people died after ingesting cyanide-laced hard gelatin capsules of Extra Strength Excedrin®. Bristol-Myers-Squibb (New York, New York, U.S.) discontinued the product in this form, and many other pharmaceutical manufacturers began to limit or phase out the use of hard gelatin capsules for drug products.

These capsules, made of two interlocking halves of hard gelatin, could be separated and the contents replaced. Companies that continue to use hard gelatin capsules because of product need began to seal the two halves of the capsule at the joint of the capsule. Most often it is easy to identify a hard gelatin capsule because the two halves of the capsule are two separate colors.

The FDA has approved the use of sonic welding, gelatin banding, thermal sealing, and solvent bonding of the two halves of the capsule as an additional tamper-evident feature of the product. This method of enhancement to the packaging was recognized by the FDA as a way of improving the overall integrity of a product and viewed these features as an additional improvement that did not change or mitigate the tamper-evident regulations for the packaging of the product. The agency was very specific in stating that the use of these techniques did not constitute a substitute for tamper-evident features. Later the FDA went further with regulations requiring packagers using unsealed capsules to incorporate two tamper-evident features in the packaging of the product. The regulations permit the removal of one of these features if the sealed capsules are used.

Heat Sealing

Heat sealing is based on the thermoplastic properties of packaging materials (Table 3). This property permits the material to melt and re-harden back to its original state. While molten, the material can form an adhesive bond between two parts of a container and produce a complete hermetic seal. The molten thermoplastic acts just like glue in joining two parts together to complete a container.

Heat sealing is done using materials that are highly specialized in their construction. These materials not only provide the heat sealing material in one layer of their construction, they typically contain additional layers of plastic, metal, or woven material that deliver other properties necessary for the package to perform properly.

Heat seals are most often used with flexible packaging. The flexible packaging material is a laminate consisting of two or more layers of material bonded together with an adhesive or bonded together through direct extrusion of the layers of plastic and nonplastic materials.

An example of a typical foil and plastic multilayer film consists of (from inside to outside of the film) a heat-sealing layer, which most often is also the product-contact layer of the structure, an adhesive, a very thin aluminum foil, adhesive, and an outside layer of PET. Each of the materials in the structure contributes properties needed to successfully package the product. The inside contact layer and the heat-sealing layer is most often low-density PE. This material is approved for food and drug contact, is relatively inert, and melts and reforms, making the heat seal possible. The next layer may be an adhesive if a PE film is laminated to the foil layer, or the PE may be extruded directly on the surface of the foil. The aluminum foil material is an extremely thin and fragile material that provides a light and gas barrier needed for the package. Aluminum foil as thin as 0.5 mm (0.0005 in) is used for this layer. Thinner foils are available as are metallized plastic materials that resemble foil. Both the thinner rolled foil and metallized foil materials contain defects called pinholes that defeat the barrier properties. The foil does not have any structural strength and is used only as a barrier material. In some cases one side of the foil is coated to provide a background color for graphics. This is called a basecoat, and most often it is only a single color with no printing as part of its application to the foil. The next layer of the structure is an adhesive used to bond the foil to the outer layer of material, most often a PET film. The plastic PET film provides the laminated structure with physical strength and integrity and also functions to resist puncturing of the structure. On the inside of the PET film the remainder of the printing is done prior to lamination to the foil with an adhesive.

An extrusion-laminated film is another example of a heat-seal material, only this material is produced by co-extruding dissimilar plastics together in a single process that combines the melted streams of plastic into a single (unsupported) film.

Table 3 Heat-Sealing Methods

Name or type of sealing	Method of operation
Bar seal	1 or 2 heated and opposed bars clamp on the materials sealed, melting the sealant.
Band seal	Similar to a bar sealer. One or two heated bands with the addition of a cooling section clamp the materials, melting the sealant and then cooling in place.
Impulse seal	Similar to a bar or band sealer. A pulse of electric current through a heating element heats the material being sealed. The jaws are coated or covered with a material that with a high temperature release material.
Friction or spin-welded seal	2 components are rubbed together to produce the heat necessary to melt the plastic and weld them in place. Most often used with 2 round pieces being sealed together.
Ultrasonic seal	High-frequency sound waves rub the 2 mating surfaces together, melting the plastic and creating the seal.
Contact seal	2 surfaces are heated separately by a plate or other heating device. After the surface is molten, the heating device is removed and the 2 pieces are pressed together.
Induction seal	Alternating current is induced in a metallic material that heats because of the molecular movement created by the alternating field. The hot metal melts the plastic it is in contact with.
Solvent seal	A solvent is used to soften or melt the surface of a material and then the 2 pieces are pressed together.
Hot melt	Application of molten dots or a strip of molten plastic through a nozzle. The 2 surfaces are then pressed together to form a seal.
Dielectric	A high-frequency electric field melts the materials while they are held together under pressure.
Magnetic	A gasket or sealant containing a high percentage of iron is pressed between 2 surfaces then placed in a strong magnetic field.
Radiant	Infrared heating of the plastic without pressure melts the plastics and they fuse together.
Gas	A gas flame or very hot air is applied to the surfaces to be mated, melting the plastic material. The heat is removed and the two surfaces are pressed together.
Wire or knife	A hot wire or wires melt the plastic forming the seal and continue heating the material also effecting a cut or separation between pouches or bags.
Pneumatic	Air pressure is applied to a heated plastic film, forcing the 2 surfaces to mate.

The foil structure described above may be used as an insert inside a closure for tamper evidence and also to provide barrier properties to a closure. The laminated structure is punched into disks that are inserted directly into linerless closures or are attached to liners inside a closure with adhesive.

It normally is not necessary when heat sealing two materials to heat the material from both sides. When thick materials are used or if the packaging line is required to run at high speed, then the structure is heated from both sides. The variable factors in any heat seal are heat, pressure, and time.

Heat at the surface of the material to melt the thermoplastic material is an obvious consideration, but any description of the process also requires time to permit the heat to permeate the materials and melt the heat-sealing layer. The materials also must be held together under pressure during the heating and melting cycle to force the melted materials on both sides of the seal to intermingle and fuse together. This means that any heat seal has a dwell or cycle time required to produce the seal. This can be shortened with a number of techniques, but a specific length of time is always required to effect the process.

A key property of any heat-sealing material is its hot tack. This is the physical bonding strength of the material while hot or semi-molten. This property permits the seal to hold together while the material cools to room temperature and becomes hard. Materials with very low hot-tack strength would simply pull apart before the seal was set. If the content of the package was filled with a hot liquid, the vapor escaping from the surface of the liquid will produce pressure inside the container until it cools to room temperature. Without hot tack strength in the sealing material, the heat seal would immediately separate.

There are a large number of heat-sealing techniques that use different principles to produce heat and make materials seal. These are listed in Table 3. Some of the more novel are friction welding (spin welding) and the variety of induction and ultrasonic techniques. The induction and ultrasonic techniques are used extensively in pharmaceutical and medical-device packaging and in the construction of many medical devices. Friction or spin welding is a technique that is used more often to weld a closure on a container.

Induction sealing is used extensively to bond the plastic tamper-evident seal to the lip of a bottle. This technique relies on the foil in the tamper-evident structure to heat when subjected to an electric current by induction. The sealing material is held into place by the application torque to the closure. The induced current heats the aluminum and melts the plastic. The fact that pressure has already been applied and mechanically the sealing material cannot move allows the seal to form and eliminates hot tack dwell time necessary for the heat-seal layer to cool and forms a strong bond with the lip of the bottle is just the time it takes the bottle to run through that section of the packaging line. The next time you open a pharmaceutical product or any other product with a seal across the top of a bottle or jar, this is how it was done.

A more unique method of forming a melted heat seal is through the use of friction to melt the plastic. This is done with two round parts that fuse together but when manufactured are too big or too small to fit together. This is called an interference fit, and it is also the kind of mismatch you find in overcaps on cans and bottles, where a lid can snap or fit and grab the lip of a container. One of the two parts that are slightly mismatched in size is spun at high speed in a chuck. When it reaches a set speed, the inertia of the chuck is used to keep the part spinning while the two parts are mechanically forced together. The spinning action of one of the parts produces friction that causes the plastic on the contact surface of both to melt. As the plastic melts, it robs heat from the system and the free spinning chuck suddenly stops when the plastic fuses sufficiently. Plastic ends on composite cans and odd-shaped containers are produced with this technique. The Yoplait Yogurt container is probably the most common example of the use of friction welding to produce a uniquely shaped container.

A complete listing of heat-sealing techniques and a brief description of each is contained in Table 3. Some of these techniques are quite common and some are obscure, but all have been used to produce heat seals.

Common seal problems found in heat-sealed containers are failure because of wrinkles across and in the seal area and the forced expulsion of the heat-seal material when two components are pressed together with too much force.

The first problem, wrinkles, has plagued pouches and flexible packaging since their introduction. Strict process control and various methods of maintaining tension of the films forming a pouch are two methods of minimizing the problem. Channels, notches, guides, chevrons, and other modifications the sealing bars that force the material into a specific shape are other methods used to overcome poor heat seals. Testing heat seals for failures is usually destructive, and the most common method of dealing with wrinkled seals in medical applications is by visual inspection.

The second problem of material expulsion is almost always found when a foil or metal-supported films are heat sealed. The operator of the equipment employs too much pressure and heat to the seal. High or extreme heat causes the heat-seal layer on the foil to be liquid or almost a liquid. Pressure applied to the two sides of the sealing surfaces forces the melted plastic to squeeze out of the seal area defined by each side of the jaws of a sealer. At the edges of the jaws the molten material produces a very narrow bead of sealing material on the two edges of the seal. The narrow band of material is not very strong. These seals are very fragile and are the types that can lead to extensive field failures in distribution if not caught during manufacture.

Peelable Seals

One expectation of heat seals by consumers is that the seals be peelable or easy to open. As described above, the melting and fusing of the plastic materials do not lend itself to easy opening. In fact to control or try to make a seal peelable by manipulating the sealing parameters of heat, pressure, and dwell time is nearly impossible. The best method to produce a peelable seal requires that the materials be manipulated to be somewhat incompatible or that the failure location of the peeling material is different from the sealing interface.

The first method for producing peelable seals is to use slightly incompatible materials in the sealant layer and the container. During the melting and fusing process the amount of material that actually fuses together is limited by the composition of the materials. This creates a "partial seal" that is relatively low in bonding strength compared with the same seal produced by materials that completely fuse together. As a result the reduced strength of seal can be separated or peeled, provided the base structures are stronger than yield strength of the seal.

The second method used to produce a peelable seal is most often found with foil or metal structures that are part of the materials being fused. The adhesive used between the foil and the sealing layer is designed to separate at a relatively low strength level when force is applied. The material is sealed to the container normally and the seal is quite strong with regard to maintaining product integrity. When the consumer opens the container, the fused seal remains intact while the structure separates at the junction the foil and adhesive. Seals of this type sometimes produce webs or membranes that remain across the opening of the container after the foil portion of the structure has been peeled away. The layer of adhesive remains intact and the strength of the adhesive in the non-seal areas of the foil is not sufficient to cause the seal to rip or tear away. Thus the sealant layer remains intact as a web or membrane across the container.

SUMMARY

Container closures are not simple components to a package. They are highly engineered and very specialized products that perform a number of functions. The marketplace needs for child resistance, tamper evidence, and easy access present a difficult to balance set of problems. None of today's solutions is perfect and improvements to all types of closures will be a standard expectation for every type of closure.

REFERENCES

1. United States Department of Health and Human Services, Food and Drug Administration, Center for Drug Evaluation and Research (CDER), Center for Biologics Evaluation and Research (CBER), Guidance for Industry, Container Closure Systems for Packaging Human Drugs and Biologics, Chemistry, Manufacturing, and Controls Documentation, May 1999. Available at http://www.fda.gov/cder/guidance/1714fnl.pdf.
2. Michael L. Berins Editor - SPI Plastics Engineering Handbook of the Society of the Plastic Industry, Inc., Van Nostrand Reinhold – Chapman & Hall, 1991.
3. United States Code of Federal Regulations, Title 16 Commercial Practices, Chapter II Consumer Product Safety Commission, Statements of Policy and Interpretation, Part 1701.1 and 1701.3, Title 16, Volume II. January 1, 2004, 749–750. Available at http://www.access.gpo.gov/nara/cfr/waisidx_04/16cfr1701_04.html.

4. Barone S. Don't Gamble with your Packages: Make them Senior-Adult-Use -Effective. Solutions97 Packaging and Processing Technology Conference, the Packaging Machinery Manufacturers Institute, October 15, 1997.
5. United States Code of Federal Regulations, Title 21, Parts 210 and 211, Part 210-Current Good Manufacturing Practice in Manufacturing, Processing, Packing, and Holding of Drugs, Part 211-Current Good Manufacturing Practice for Finished Pharmaceuticals. Available at http://www.fda.gov/cder/dmpq/cgmpregs.htm.
6. United States Code of Federal Regulations, Title 21-Food and Drugs, Chapter 1-Food and Drug Administration, Department of Health and Human Services, Subchapter B, Food for Human Consumption, Part 113 Thermally Processed Low Acid Foods Packaged in Hermetically Sealed Containers. Available at http://ecfr.gpoaccess.gov/cgi/t/text/text-idx?c=ecfr&tpl=/ecfrbrowse/Title21/21cfr113_main_02.tpl.
7. Joseph F. Hanlon, Handbook of Package Engineering. 2nd ed. Technomic Publishing Inc., Lancaster Basel, 1992 ISBN 0-87762-924-2.
8. United States Pharmacopeia (USP) Pharmaceutical Container Closure and Liner Testing, USP 371 Elastomeric Closures for Injections, USP 661 Containers-Plastics, USP 671 Containers Performance, USP Volume 28 NF 23, 2005.
9. Bakker M. The Wiley Encyclopedia of Packaging Technology. New York, Chichester, Brisbane, Toronto, Singapore: John Wiley and Sons, Copyright 1986, ISBN 0-471-80940-3.
10. Jenkins WA, Kenton R. Osborn, Packaging Drugs and Pharmaceuticals, Technomic Publishing Inc., Lancaster Basel, 1993 ISBN 1-56676-014-3.
11. United States Code of Federal Regulations, Title 16 Commercial Practices, Chapter II Consumer Product Safety Commission, Part 1700 et al., Poison Prevention Packaging, cite: 16CFR1700, revised January 1, 2004.
12. United States Code of Federal Regulations, 16 CFR 1700 Supplementary Information, Relevant Statutory and Regulatory Provisions, The Poison Prevention Act of 1970. 15 U. S. C. 1471-1476.
13. United States Food and Drug Administration, Center for Drug Evaluation and Research, Questions and Answers on Current Good Manufacturing Practices, Good Guidance Practices, Level 2 Guidance, Control of Components and Drug Product Containers and Closures. Available at http://www.fda.gov/cder/guidance/cGMPs/component.htm.

11

Labels and Labeling

INTRODUCTION

Labeling encompasses a wide range of technologies and is one of the broad and varied parts of pharmaceutical packaging that is particularly difficult and challenging. It consists of two very separate elements that are both required to produce a label. The first element is the actual construction of a label, the materials used in its construction, the adhesives, the inks, and all the other process requirements including printing processes needed to describe the physical construction and manufacture of a label. The second is the graphic elements of the label, both text and pictures, that communicate with everyone using the product.

"Labeling" when used in a pharmaceutical sense consists of far more than the information found on the product, package, or accompanying label materials such as the package or product insert. Labeling describes in the most up to date terms all information regarding a drug's efficacy, quality, safety, and usefulness. Efficacy describes the medical conditions the drug is designed to treat, the correct dosing needed to properly administer the drug, and the therapeutic effects and indications expected from its use. Quality describes the physical form of the drug, its exact chemical composition, and the strength and physical form in which it is supplied. It also describes the correct storage and handling procedures required by the product. Safety describes possible side effects associated with the medication, contraindications, and rules for monitoring a patient while taking the drug, and any known consequences or complications concerning the drug's use. The audience for labeling is very broad. It is directed at physicians, medical practitioners, and patients, each of who has different needs, different capabilities for understanding the information, and different expectations of what the label will provide as a guide for the safe and effective use of the product. The information is primarily in text form with multiple graphic elements such as chemical structures or diagrams for administration and use and in some cases extensive tables detailing proven clinical results obtained in characterizing the product.

Various regulatory bodies around the world have established rules and regulations that describe the content and format of labeling required in their jurisdiction along with key supporting data that must be part of the development of labeling. These rules and regulations cover not only the label on the bottle but all text and graphic elements including advertising and claims made by the manufacturer for the product. Most of the regulations contain some type of pro forma headings that describe the topics that must be covered along with rules and regulations describing the exact information required for a specific drug class or type of drugs. All regulatory bodies monitor adverse events and consumer complaints on drugs in an effort to uncover any unknown danger or risk not found during clinical testing or after initial introduction of the product.

Text elements used in product labeling are designed to be clear and concise summaries of data and performance gathered from the time the drug begins the development and approval process until it completes its useful life and is removed from the market. The text along with the graphics is subject to constant review and updating as more is learned about a product after it enters the marketplace.

Graphics are the pictorial and other non-text elements of labeling. Graphics are in the same constant flux as text elements of labeling, creating multiple revisions of package information about the drug over the lifetime of a product. In addition to the labels attached to the primary packaging of a drug, labeling on secondary and tertiary packaging plays an almost equally important role. Last and sometimes the most difficult piece of pharmaceutical labeling to complete is the insert or outsert for the product. This piece of labeling from a regulatory, marketing, and graphics standpoint is the most extensive overview of the product. It has comprehensive text, multiple sections detailing different aspects of the product, and the chemical structure of the compound as part of the minimum required content. It also states what disease conditions the product is used to treat, how it is intended for use, and any contraindications known about the product. The insert or outsert, depending on the construction and placement of the information on or in a package, is usually the piece of labeling most reviewed and last approved by the FDA. Labeling and the development, application, and maintenance of the text and graphic information on any drug package is difficult, demanding, and absolutely necessary to provide the doctor, patient, and all other involved health care providers such as nurses, nurse practitioners, emergency medical technicians, and pharmacists information vital to the description, use, sale, and packaging needed for a drug.

HISTORY OF DRUG LABELS

Labels used for marking goods have an extremely long history. The use of marks to identify the source of an item can be traced back into antiquity before the time of Christ. During Roman times the herb Lycium was dispensed in small jars that bore the name of the drug and the name of the seller. The more modern expectation of labeling that conveys a large amount of descriptive information is relatively new. For most of history literacy was limited to a small group of people, and the number of items requiring tags or labels to identify or differentiate drug products was limited.

The first set of drug labels was attributed to a group of druggists in London in 1819. They adopted a set of regulations for marking poisons. One of their regulations stated, "That on every wrapper or vessel containing any drug or preparation likely to produce serious mischief, if improperly used, the name of the article be affixed in legible form, and as many persons can read print who cannot read writing, they would recommend that printed labels be used where possible, in preference to written ones." This is also when the skull and crossbones symbol for poison began to be commonly used.

Labels have evolved considerably from this point, and for pharmaceuticals they are a critical element of packaging. Regulatory bodies have also evolved from this loose confederation of London druggists to government agencies that review, modify, approve, monitor, and audit label manufacturing and label graphic communication.

LABELING REQUIREMENTS

Prescription Drug Labeling

There are few things in the world that are more scrutinized, revised, and monitored than prescription drug labeling. The regulations covering drug labeling are voluminous, detailed, and demanding. New and existing drug products are under review from the time of their initial clinical trials until they are removed or replaced in the marketplace. The FDA must approve each and every piece of labeling used to identify and describe a product (1). This includes the container label, the labeling on any secondary packaging like the carton, and most significantly the information contained on the product insert or outsert. The insert or outsert is designed to place a complete description and summary of all pertinent facts regarding the drug in the hands of a health care professional. The process of reviewing and revising the labeling content goes on throughout the life cycle of a drug. The drug manufacturer must monitor and maintain a file of all the adverse events reported about the drug and submit this information to the FDA on a regular basis as detailed in the FDA code. This data is used to monitor the drug throughout its life and to drive changes in labeling as more information becomes known about the drug over time.

All pharmaceutical and biologic drug manufacturers devote significant resources and intellectual capital to the development, control, and revision of product labeling. Subject matter experts from many corporate divisions: pharmacology, medicine, manufacturing, marketing, legal, and regulatory affairs participate in the creation and revisions of labels.

There are a number of major changes underway regarding electronic management and revision of labeling at the FDA. The widespread use of computers, the ease with which information can be shared, and the need to standardize electronic information have led the FDA, the European Community and other regulatory bodies to develop and set standards for electronic submission of information. These changes, referred to as structured product labeling (SPL), and the backbone that underlies its use, extensible markup language (XML), were required by the FDA in 2005 for submitting label information. These procedures and methods are replacing the interchange of information in PDF formats originally accepted by the agency for data transmission and review. These changes, which permit an interchange of information by the manufacturer and the FDA via electronic means, improve the ability of all parties to track and manage changes, manage revisions, and communicate changes to the field. PhRMA, the Pharmaceutical Research and Manufacturing Association, supports this change in process. This trade association represents the broad pharmaceutical and biological industry in the United States.

Another major change in the FDA's approach to prescription drug–labeling content took place in 2006 (2). After years of comments and complaints by physicians, pharmacists, health care providers, and consumer groups, the first major overhaul of prescription drug labeling in more than 25 years was enacted (3). The changes were directed at the patient insert only, while additional change to package labeling and the information provided to the patient is still under review. The new labeling is designed to provide health care professionals with better and more easily accessed information. The idea is to provide the same benefits that drug facts labeling provided to labeling of over-the-counter (OTC) nonprescription drug products to the physician, pharmacist, nurse, and health care professional on prescription products. This change emphasizes that patient safety is truly in the hands of professionals, and these changes to package inserts are designed to make the information they needed to improve patient safety easier to find and use. There is a large amount of debate going on regarding how much of the information will filter down to the patient and how soon additional changes will be required on information for the patient.

The changes to the patient insert are found in 21 CFR 201.56 and 21 CFR 201.57 and are covered in the overview of prescription labeling that follows (4,5).

This overview of prescription drug labeling and OTC labeling is illustrative of the regulations and requirements applied to label development, claims, advertising, and other supporting materials used with a drug product. It is not designed as a reference, and anyone requiring definitive information on this

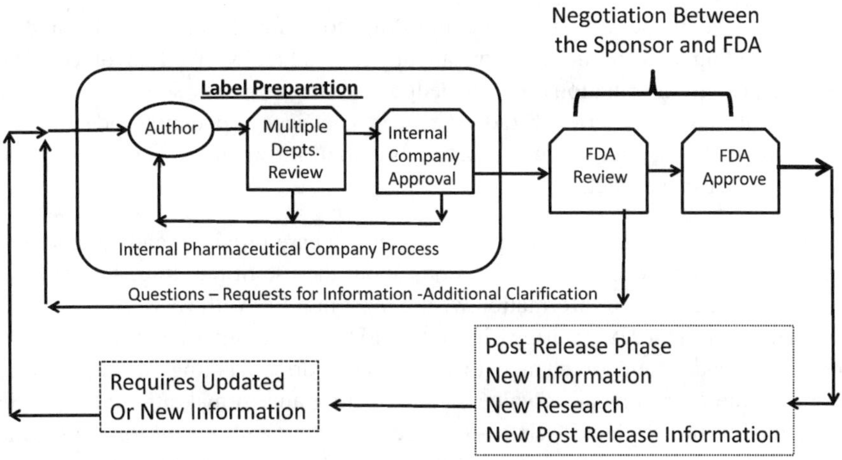

Figure 1 The drug labeling process.

complex subject must consult the Code of Federal Regulations (CFRs) or a professional trained in review and interpretation of the regulations.

Label Information

Labeling embodies significant intellectual property and significant liability in case of error. This is not an easy, brief, or inexpensive process for a pharmaceutical or biologics company to develop and maintain for each product it markets. Labeling goes through a number of iterative stages in its development and revision. The steps are similar throughout the world. A diagram of the process is found in Fig. 1.

The order of the processes includes the following:

1. Initial development: The discoverer, sponsor, or company authors, reviews, and approves the labeling internally in preparation for submission to the approving agency.
2. Negotiation and approval: The labeling is submitted for approval to the appropriate agency within the FDA. This phase includes review by the agency and subsequent negotiation of language, format, presentation of data, warnings, etc., that establish acceptable finished product labeling as viewed by both sides in its development. Part of this process normally requires back and forth responses to questions, requests for additional data, and clarification of positions by the parties.
3. Review, revision, and updating: A post-release monitoring cycle that continuously collects and evaluates information about the drug and triggers revisions to the labeling. The revisions may be required by the agency or

may be made voluntarily by the drug company to improve the understanding and communication about the product with all involved. The types of information or knowledge that triggers this sequence include post-market experience (adverse events, efficacy, potential side effects, interactions, etc.), new research, or other information germane to the subject or product.

This process provides the physician, health care provider, and patient with information they need to make decisions about the use of the drug. The process and the presentation of information we are most familiar with was developed and implemented in 1979. During 2006 the FDA, after significant review and research, made a major change to prescription drug labeling in an effort to improve the communication value of the labeling and to make the information contained in the product insert easier to read, access, and review for physicians and health care providers. This change was to help health care professionals make better decisions when prescribing a drug or multiple drugs for a patient.

Part 201 of the CFRs is broken into a number of subparts that address different requirements for labeling. The subparts are organized as follows:

Subpart A—General labeling provisions
Subpart B—Labeling requirements for prescription drugs and/or insulin
Subpart C—Labeling requirements for OTC drugs
Subpart D—Exemptions from adequate directions for use
Subpart E—Other exemptions
Subpart F—Labeling claims for drugs in drug efficacy studies
Subpart G—Specific labeling requirements for specific drug products

As referred to earlier in this chapter, the FDA issued a final rule in January 2006 that changed the format and content requirements for labeling as described in 21 CFR 201.56 and 201.57 (6,7,4). These changes applied to all human prescription drug products and biologic products. The requirements reflect an improved labeling format and the additional information required for the package insert. Table 1 compares the old and new labeling headings and information for the patient insert to help the reader better understand the changes that were made. Each of these sections has numerous subsections that describe issues and requirements for proper labeling of the specific prescription drug product or specific class of drug.

Some of the key changes to the labeling format include the addition of introductory prescribing information identified as "Highlights of Prescribing Information" ("Highlights), a "Table of Contents" (Contents) and "Full Prescribing Information" (FPI). The "Highlights" section is designed to excerpt from the FPI items health care practitioners most commonly referenced and considered most important. The "Contents" lists the sections and subsections of the FPI. The changes also reorder some of the sections found in the insert, make

Table 1 Prescription Drug–Labeling Sections—Comparison of Major Changes

Labeling format prior to 2006[a]	New labeling format 2006[b]
Description	Highlights of prescribing information
Clinical pharmacology	Product names, other required information
Indications and usage	Boxed warning
Contraindications	Recent major changes
Warnings	Indications and usage
Precautions	Dosage and administration
Adverse reactions	Dosage forms and strengths
Drug abuse and dependence	Contraindications
Overdosage	Warnings and precautions
Dosage and administration	Adverse reactions
How supplied	Drug interactions
	Use in specific populations
Optional sections:	
Animal pharmacology and/or animal toxicology clinical studies	FPI: contents
References	FPI
	Boxed warning
	1. Indications and usage
	2. Dosage and administration
	3. Dosage forms and strengths
	4. Contraindications
	5. Warnings and precautions
	6. Adverse reactions
	7. Drug interactions
	8. Use in specific populations
	9. Drug abuse and dependence
	10. Overdosage
	11. Description
	12. Clinical pharmacology
	13. Nonclinical toxicology
	14. Clinical studies
	15. References
	16. How supplied/storage and handling
	17. Patient counseling information

[a]As required by 21 CFR 201.56(e) and 201.80.
[b]As required by 21 CFR 201.56(d) and 201.57.
Abbreviation: FPI, full prescribing information.
Source: From Ref. 6.

them easier to find, and also set minimum graphic standards for the format on different parts of the labeling. Graphic standards refer to the size of type, the methods for listing items, and other elements needed to make the final presentation of labeling consistent from insert to insert.

A comparison of the locations between the old and new formats provides additional understanding of what was changed and how to reorder sections between the old and the new.

Location in old format	Location in FPI in new format
Boxed warning	Boxed warning
Description	Description
Clinical pharmacology	Clinical pharmacology
Indications and usage	Indications and usage
Contraindications	Contraindications
Warnings	Warnings and precautions
Precautions	
General	Warnings and precautions
Information for patients	Patient counseling information
Laboratory tests	Warnings and precautions
Drug interactions	Drug interactions
Drug/laboratory test interactions	Warnings and precautions
Carcinogenesis, mutagenesis, impairment of fertility	Nonclinical toxicology (carcinogenesis, mutagenesis, impairment of fertility)
Pregnancy	Use in specific populations (pregnancy)
Labor and delivery	Use in specific populations (labor and delivery)
Nursing mothers	Use in specific populations (nursing mothers)
Pediatric use	Use in specific populations (pediatric use)
Geriatric use	Use in specific populations (geriatric use)
Adverse reactions	Adverse reactions
Drug abuse and dependence	Drug abuse and dependence
Overdosage	Overdosage
Dosage and administration	Dosage and administration
How supplied	Dosage forms and strengths
	How supplied or storage and handling
Animal pharmacology and/or animal toxicology	Nonclinical toxicology (animal toxicology and/or pharmacology)
Clinical studies	Clinical studies
References	References

Source: From Ref. 6.

The FDA has stressed that most of the information found in the earlier form of an insert may be reorganized when labeling is revised. The FPI in the new format contains essentially the same information found in the old format, just reordered and reorganized as described in the comparison above. For example, information found in the new sections, such as "Drug Interactions," "Use in Specific Populations," and "Patient Counseling Information," are extractions of information formerly found in the "Precautions" section of the older labeling. Some sections such as "Clinical Studies" and "Nonclinical Toxicology" that previously were optional are now required. Two separate headings found in the older labeling, "Warnings" and separately "Precautions" have been consolidated as one section "Warnings and Precautions." This along with older information found in the "Precautions" section, such as Information for Patients, Drug Interactions, Pregnancy, Nursing Mothers, and Pediatric Use have been moved to the new "Patient Counseling Information," "Drug Interactions," and "Use in Specific Populations" sections as detailed in the above comparison of location.

Because there are so many similarities between the old and new labeling, the FDA expects that much of the information will be moved with little or no change into the new section. If the older labeling lacks information now required in the regulations, the sponsor of the product is required to develop the new sections and include them in the revised labeling. If the content of an old section is inadequate, it must be updated and revised. If older labeling totally lacks a section now required in the new regulations, it must be developed unless it is clearly inapplicable.

One of the most important changes and expectations in the new labeling is the change to the "Highlights" section. This section is to provide the information practitioners view most often and consider important in a prominent and easy-to-use presentation. The content summarized in the "Highlights" must be consistent with that found in other sections of the insert but not a verbatim repetition of the information. It is a concise summary of the information presented in a bulleted and tabulated way that provides the important information practitioners want and directs them to the sections of the FPI that provide the information in detail. Developing the "Highlights" section requires professional judgment about the data presented, its importance, and the clinical settings where the drug is used. Judgment is critical because the information presented will vary with different drugs and different classes of drugs. Safety information, for example, will vary on the basis of different dosing regimens for different indications or populations. The goal is always to present information in "Highlights" in direct and succinct language while cross-referencing and directing the practitioners to the location of more detailed information.

Other changes worth noting in this overview are the requirement to show the initial U.S. approval date for the product, boxed warnings, and a summary of recent major changes. When significant or substantive changes are made to the "Boxed Warning," "Indications and Usage," "Dosage and Administration," "Contraindications," or "Warnings and Precautions" sections of the product labeling, these items must appear in the "Highlights" under the heading "Recent Major Changes." Labeling changes to correct typographical or grammatical errors are not considered substantive or major label changes.

The FDA has set up a procedure for submission of labeling in the new format and commented on their understanding of the difficulty involved in making the changes. They recognize the considerable time and effort required to design, develop, and submit labeling for review and approval (Table 2). In this

Table 2 FDA Implementation Plan for New Prescription Drug Labeling

Applications (NDAs, BLAs, and efficacy supplements) required to conform to new labeling requirements	Time by which conforming labeling must be submitted to the agency for approval
Applications submitted on or after June 30, 2006	Time of submission
Applications pending on June 30, 2006, and applications approved any time from June 30, 2005, up to and including June 30, 2006	June 30, 2009
Applications approved any time from June 30, 2004, up to and including June 29, 2005	June 30, 2010
Applications approved any time from June 30, 2003, up to and including June 29, 2004	June 30, 2011
Applications approved any time from June 30, 2002, up to and including June 29, 2003	June 30, 2012
Applications approved any time from June 30, 2001, up to and including June 29, 2002	June 30, 2013
Applications approved prior to June 30, 2001	Voluntarily at any time

Source: From Ref. 6.

vein they have provided guidelines on how labeling may change. They have limited their focus to drugs approved in the last five years as the ones with the most interest and need by practitioners for improvement. Table 2 lists the FDA's timetable for implementation of new prescription drug labeling.

The FDA indicates that this final rule applies to all prescription drug NDA, BLA, or efficacy statements that have been approved in the five years prior to the effective date of the new rules and regulations.

When an efficacy supplement or statement is changed, the following criteria are applied to determine if the labeling must be submitted in the new format.

- A new indication or a significant modification of an existing indication, including removal of a major limitation of use.
- A new dosage regimen, including an increase or decrease in daily dosage or a change in frequency of administration.
- A comparative efficacy or comparative pharmacokinetics claim naming another drug.
- A change expected to significantly affect the size of the patient population to be given the drug, either broadening or narrowing the population (e.g., pediatrics, geriatrics).
- Clinical data to verify and describe the clinical benefit for a drug approved on the basis of a surrogate endpoint or on an effect on a clinical endpoint other than survival or irreversible morbidity (see 21 CFR 314.510 and 601.41).
- A labeling supplement with clinical data.

For the last bullet point the FDA defines clinical data in a draft guidance entitled "Submitting Separate Marketing Applications and Clinical Data for the Purposes of Assessing User Fees." This guidance when finalized would present the agencies' thinking on this topic (7,8).

Another key point to the changes in labeling must be emphasized. That key point is regarding the definitions of the graphical elements including the minimum font or type sizes that must be used throughout the new labeling (Table 3). Here again the agency is attempting to address complaints by seniors and practitioners that information in the past was physically hard to read. Table 3 provides a complete summary of the FDA requirements for graphic standards. Again this guidance when finalized would present the agencies' thinking on this topic.

This table also emphasizes a number of challenges involved with labeling pharmaceutical products. The amount of area or the size of the package when labeling a drug is normally very limited. To provide the patient and the practitioner all the information they need to use the drug in safely creative ways must be found to increase the total label area. The minimum font sizes only compound the space problem. Booklet labels, labels with multiple folds, and labels with accordion style folds all have been used successfully to increase total label area. These specialized labels combined with improved patient insert materials and other instructive materials prepared for the patient permit the prescriber to effectively deliver all the information needed by a patient and all the reference material they need if questions arise after they leave the doctor's office. They still have the option to contact the practitioner, but many times can find the information they need in simple straightforward language in the materials supplied to them with the drug.

Drug Facts Labeling—OTC Pharmaceutical Products

Communicating in today's busy world is difficult. Communicating well with the general public to help them make good decisions on how to take care of themselves and their families is critically important. Contrary to common thinking, OTC products contain powerful active pharmaceutical ingredients (APIs) that when used properly greatly improve people's lives; however, those same products used improperly or by the wrong group of people can seriously harm or even kill them. This concern for people's safety was behind the FDA's revision of OTC drug labeling in 2002. The label on a pharmaceutical product must communicate who should and who should not take the product, how to use the product, and most importantly what the product is supposed to do. OTC drug labels have always contained information needed by consumers regarding the product's safety, intended use, and efficacy. Unfortunately the information was not provided in a uniform way, and at times the labeling used ambiguous terms that made it generally hard to read and understand by many people needing clarity in the information.

Table 3 Type Size Requirements for Labeling and FDA-Approved Patient Labeling Included in the Packaged Product

	Type size requirements for labeling	FDA-approved patient labeling included with labeling	Type size requirements for FDA-approved patient labeling
New format (21 CFR 201.57)			
Trade labeling (i.e., labeling on or within the package from which the drug is to be dispensed)	Minimum 6-point type	FDA-approved patient labeling that is not for distribution to patients	Minimum 6-point type
		Any FDA-approved patient labeling (except a medication guide) that is for distribution to patients	Minimum 6-point type[a]
		Medication guide that is for distribution to patients	Minimum 10-point type
Other Labeling (e.g., labeling accompanying promotional materials)	Minimum 8-point type[a]	FDA-approved patient labeling that is not for distribution to patients	Minimum 8-point type
		Any FDA-approved patient labeling (except a medication guide) that is for distribution to patients	Minimum 8-point type
		Medication guide that is for distribution to patients	Minimum 10-point type
Old format (21 CFR 201.80)			
Trade labeling and other labeling	No minimum requirement	FDA-approved patient labeling that is not for distribution to patients	No minimum requirement
		Any FDA-approved patient labeling (except a medication guide) that is for distribution to patients	No minimum requirement[a]
		Medication guide that is for distribution to patients	Minimum 10-point type

[a]FDA does not require, but encourages a minimum type size of 10 points for this information.
Source: From Ref. 6.

In March 1999, the FDA finalized rules and regulations for OTC drug labeling designed to make it more uniform and easier to read and understand. This was not a small undertaking. More the 100,000 products were sold OTC at the time the rule was enacted. The FDA regulations provided three years of time for manufacturers to change the labeling on all OTC products and required that all products sold after May 16, 2002, carry the new form of labeling.

Most manufacturers began the change of labeling immediately by introducing new products with the new labels while work to change the older style of label was underway. The FDA permitted the sale of products with the older labeling after the May 16, 2002, date to deplete existing inventories, but all newly manufactured products after that date carried the new form of labels.

The FDA conducted extensive research with consumers about how they used existing OTC labels. They found that consumers felt the older labels were hard to read and far too technical. Terms like "precautions," "indications," and "contraindications" were too technical and confusing for many consumers. They also found older Americans who consume approximately 30% of the non-prescription drugs had a difficult time reading many of the labels. Previously the location of information about directions, warnings, and approved uses for a product appeared at different locations on a label depending on the manufacturer, brand, and product itself. Many times products manufactured by the same company had different styles and different formats for their OTC product labels. This made vital communication about ingredients, both active and inactive, difficult to determine. For people suffering from severe allergies, this type of information is critical.

The drug facts label introduced by the FDA was designed to easily and clearly communicate with people (Fig. 2). The label uses easy to read and simple to understand language in an easy to follow and uniform format to eliminate the problems with earlier labels. The language and format permit people to easily compare and select products. It makes dosing instructions easy to understand and follow. The information contained on a drug facts label is as follows:

- The active ingredient of a product and its dosage unit.
- The purpose of the medication.
- The uses (indications) for the drug.
- Dosage instructions that clearly communicate when, how, and how often to take the product.
- Specific warnings that state when the product should not be used under any circumstances. When the person taking the medicine should consult with a doctor or pharmacist and warnings that describe side effects of the product and substances or activities to avoid while taking the product.

The standardized formats make it easy for consumers to find the information in the same place on every product. It uses direct plain speaking terms to eliminate confusing or hard-to-understand terms. Previously used word headings

Drug Facts

Active Ingredient (in each tablet) **Purpose**

Chlorpheniramine maleate 2 mg . . . . Antihistamine

Uses temporarily relieves these symptoms due to hay fever or other upper respiratory allergies:

■ sneezing ■ runny nose ■ itchy, watery eyes ■ itchy throat

Warnings

Ask a doctor before use if you have

■ glaucoma ■ a breathing problem such as emphysema or chronic bronchitis
■ trouble urinating due to an enlarged prostate gland

Ask a doctor or pharmacist before use if you are taking tranquilizers or sedatives

When using this product

■ You may get drowsy ■ avoid alcoholic drinks
■ alcohol, sedatives, and tranquilizers may increase drowsiness
■ be careful when driving a motor vehicle or operating machinery
■ excitability may occur, especially in children

If pregnant or breast-feeding, ask a health professional before use.

Keep out of reach of children. In case of overdose, get medical help or contact a Poison Control Center right away.

Directions

adults and children 12 years and over	take 2 tablets every 4 to 6 hours; not more than 12 tablets in 24 hours
children 6 years to under 12 years	take 1 tablet every 4 to 6 hours; not more than 6 tablets in 24 hours
children under 6 years	ask a doctor

Other information store at 20–25° C (68–77° F) ■ protect from excessive moisture

Inactive ingredients D&C yellow no. 10, lactose, magnesium stearate, microcrystalline cellulose, pregelatinized starch

Figure 2 Example of a drug facts label.

like indications are replaced by the word "uses," and the words contraindications and precautions are eliminated entirely. The new label mandated the use of larger type sizes making it easier to read, and also introduced the standardized use of bullets, spacing between lines, and clearly marked sections to walk the consumer through all the critical information about the product. All of these changes were designed to improve readability and communication with the consumer. The label also makes it very clear that when questions arise, the consumer should contact a doctor, pharmacist, or health care professional for answers.

NDC NUMBER—THE NATIONAL DRUG CODE

The National Drug Code (NDC) is the method used by the FDA to identify drugs in the United States. The agency requests but does not require this number to appear on all drugs and all drug labelings. The requirement is the same for any prescription drug dispensed to a consumer. The number must appear as required by 21 CFR 207.35 (B) (9) of the code, and this part of the regulation is the key identifier for any registrant regarding the number's use. It should be reviewed whenever this question arises.

The number has a unique structure that is undergoing change to each of its three parts as more drug establishments (manufacturer, distributor, or repackager) are provided numbers.

The number began as a 10-digit number with the first four digits used to identify the drug establishment. The number assigned to each establishment is permanent and does not change. You will note some of the older established pharmaceutical manufacturers have very low establishment numbers listed on the labels. Wyeth and one of their acquisitions, Lederle Laboratories, both use single-digit (000X) registration numbers, as an example.

One confusing aspect behind the numbers used as NDC identifiers comes from the evolution or change in the number. Originally, alphanumeric numbers were assigned, but as the number of labelers (manufacturers, distributors, and repackagers) of drugs grew, the alphanumeric system was replaced with an all-numeric system. The original number assigned to manufacturers or labelers of drugs has grown from three digits to four digits, and now to five digits as more and more establishments register with the FDA.

When one of the original labelers (manufacturers or distributors) uses a number already assigned under the NDC or the National Health Related Items Code System, the original three-character labeler code is increased to four characters with the addition of a lead zero (0) to that portion of the number. The four-character labeler code is followed by a four-character product code and a two-character package code. The labeler is permitted to change alphanumeric characters in those codes to all-numeric characters by notifying the agency. Once the number is changed to all-numeric characters, it remains in this form. This change to all-numeric characters for NDC numbers is universal and the old alpha characters do not appear on drug labeling. Also, the original three-digit identifier for labelers has been superseded by a universal four-digit code and now as a five-digit code as the number of registered labelers of drugs has grown. The agency has provisions in the regulations to change this to a six-character number when all the five-digit numbers are exhausted.

The NDC number itself consists of three separate identifying parts and is limited to 10 numeric characters at present. The first section of the number is called the labeler code and identifies the manufacturer, distributor, or repackager of the product. As described earlier, this number remains constant for the labeler regardless of how many different drugs the entity produces or handles. This portion of the NDC number consists of four, five, or possibly in the future, six numeric characters with each unique number representing a specific drug manufacturer, distributor, or repackager.

The second and third sections of the number identify the drug product and the package type. With the original codes, all of the older establishments used an arrangement of 4-4-2 (e.g., 1234-1234-12) for their NDC numbers. This permitted the maximum flexibility for the number of different product formulations or types of formulations (solid and liquid, for example). With the advent of five-digit labeler designations, the labeler must decide and then commit to using either a three-digit product number and a two-digit package number (e.g., 123-12) or a four-digit product number and a one-digit package identifier (e.g., 1234-1). The labeler must use this format consistently for the drug throughout its use, and its complete NDC number is registered this way with the agency. In the first example an individual manufacturer could theoretically have 999 formulation variations in the drug and 99 packaging presentations of the drug, each one a unique product.

The middle section of the NDC number identifying the drug formulation is known as the product code. The product code numbers are designed to identify any change in the drug's strength or formulation. A change in the amount of active ingredient requires the assignment of a new number designation for the product and a change in the product code portion of the drug's NDC number. Changes in excipients used in the formulation of an existing drug would also trigger a change in the product number.

The final section of the number and the third section of the NDC number identifying a trade package size and type is known as the package code.

LABEL CONSTRUCTION

Types of Labels

Labels can be attached to almost anything to describe the package contents, use, manufacturer, owner, or its path and destination through the supply chain. Labeling includes all the methods used to attach information to the surface of an item, product, or package (10). It can include marking the item directly or it can mean attaching a label, a packaging element unique in its construction and marking, to the item. Labeling identifies, communicates, and markets products. Labeling is a unique product manufactured as a packaging component for almost every type and form of packaging.

A wide variety of materials can be used to produce labeling. Common labeling usually is made from foil, paper, fabric, laminates, or plastic. Multiple substrates may be combined with extrusion or adhesive lamination processes to produce label substrates with multiple properties (Figs. 3 and 4). Labels can also be printed directly on the container or the package. These distinctions and whether a label is coated with an adhesive or is nonadhesive begin to describe

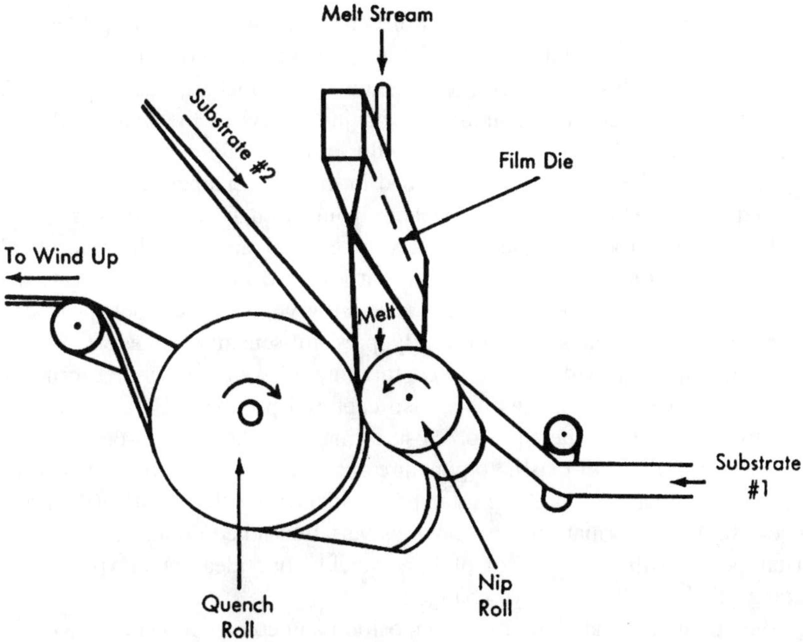

Figure 3 Extrusion lamination process.

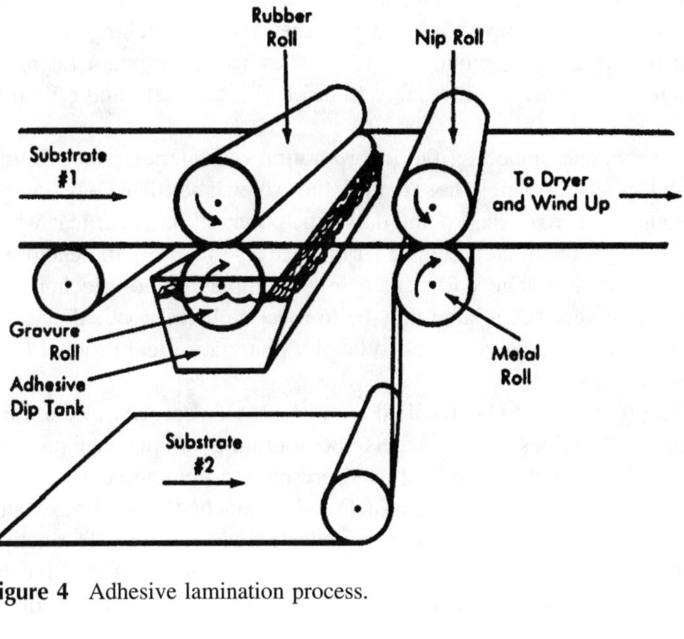

Figure 4 Adhesive lamination process.

the many different types and subclasses of labels that are applied to packaging. These distinctions along with other divisions like (surface) coated or uncoated, pressure sensitive, heat sensitive, conventional gummed, particle gummed, transfer, removable, and holographic present just a few of the many different varieties of materials and constructions used to produce a label.

Along with the many forms described in the last sentence, the market for labels and the predominant types of labels being used is undergoing a major shift. This is true for consumer products in general and for pharmaceutical products as well. Originally paper labels that required moisture to activate the adhesive were the most common type. These are giving way to labels produced on a variety of substrates with primarily pressure-sensitive adhesive. Plastic films and laminates in particular have captured a wide share of market formally held by paper labels. Laminates have also captured portions of the pouch and closure marketplaces in the form of closure liners. In these cases the superior strength and resistance of plastic to puncture or its visible destruction or obvious change characteristic to highlight tamper evidence is clearly identified via graphics and text information. This obvious variation noted on an item that has been tampered with is a symbol of how quickly our ideas and expectations regarding labeling function have changed.

The quality of a label, both its composition and construction along with its ability to be decorated with graphic elements, is another point of distinction for the various types of labels. Modern labels with holograms, metallized films, foils, and other decorative elements present a highly evolved combination of materials and manufacturing processes used in label manufacture (9). Pharmaceutical labels often are no different than consumer labels for high-value and high-quality products. They convey information, they are eye catching, and they may be constructed in a very unique way (11). They usually contain far more information than a consumer product label to aid and insure safe and effective use of the product (12).

For tamper evidence labeling, the incorporation of materials and printing techniques found in shrink bands, tax stamps (the paper label that goes up one side of a bottle closure across the top and down the other side), which tear when the closure is turned for opening, complete shrink sleeves that label and enclose a product, and labels that change color are a few examples of how technically complicated and how versatile a label may be to meet a pharmaceutical need or application. There are more clever ways to label a pharmaceutical product than almost anything else in packaging.

Labels recently have also been called upon to provide security in the form of anticounterfeiting devices. These labels incorporating unique designs and graphic elements are difficult to duplicate and present readily apparent symbols to the consumer that the article with them included or attached is genuine. Many labels contain both active and passive anticounterfeiting elements. The active elements are those visible and identifiable by the consumer. The passive or layered elements are additional items in the label's construction that require other techniques to observe and verify. This provides additional security in the supply chain to prevent or detect introduction of a counterfeit product. It also provides the manufacturer a way to be certain that the product in question is indeed the one that they had produced.

Today it is not unusual to see labels with multiple pages or leaves built into one small panel exposed to the consumer. The need to convey more and more drug and usage information in a limited space is a common labeling problem. Their adoption also acknowledges that secondary labeling such as cartons or product inserts may be discarded. This need led to a proliferation of new materials, printing methods (9), and adhesives that made booklets and expandable labels possible. Expandable labels, and there are many clever ways to make them, increase the amount of area for graphics and text. As legislation and regulations increase the amount of information that must be made available to the consumer, along with the need to keep or attach that information to the product for its useful life, the proliferation of labels that create more area for information will continue to grow in the general consumer marketplace and particularly in the pharmaceutical market. This chapter will highlight some of the technologies and types of labels now in use.

Plain Paper Labels

The plain paper label has been around from the beginning of labeling as we know it. The London druggists referred to this type of printed item for labels back in 1819.

The conventional plain paper unglued label is widely used in all types of formats and applications. Plain paper labels, both coated and uncoated, are still the predominant type of labeling used for food, cosmetic, industrial, and pharmaceutical products. They are supplied with adhesives that are activated by water, heat, or by removal from a backing material in the case of pressure-sensitive labels.

In pharmaceutical packaging the adhesive is of particular study and concern. Materials used in the adhesive can in some cases migrate through a plastic bottle wall and adulterate a product. Care in selection and rigorous testing to make sure components of the label adhesive do not move through the sidewall of a plastic bottle are a necessary part of qualification of labeling for any plastic primary packaging.

Paper labels with water-based adhesives are used on cans, bottles, and plastic containers. They are most prevalent in high-volume high-speed applications. So their use in pharmaceutical applications tends to be more limited. In high-speed applications these labels can be applied at rates of 100,000 or more per hour.

The paper used to produce this type of label employs a paper grade with one side coated to produce a smooth easily printed surface. These grades may be colored or possess other characteristics that produce a unique look or effect. An example of this would be a manufacturer wanting to create an old-fashioned label that had an antique feel to its appearance. An old fashioned label application for pharmaceuticals would be labeling on historic or OTC products that have been in existence for very long periods of time.

Gummed paper labels had been declining steadily in use until problems with tampering and possible terrorism gave them new life. These labels when used for address labeling on cases or cartons, or as distribution labels, are hard to remove without obvious tearing and damage. Gummed paper labels continue to be used on displays and other applications where automation is difficult or unnecessary.

There are two types of conventional gummed plain paper labels. The most common is the one we associate with any paper label. One side of the paper is decorated with all the color and printed elements we look for and the back or reverse side is coated with a film or layer of water-activated adhesive. Everyone has licked a stamp or a label to activate the adhesive before placing it on a package. There is one problem created with this type of adhesive on paper labels referred to as curling. The application of the adhesive and its drying would cause labels to "curl" or become rounded making them difficult to apply particularly at high speed.

This problem was overcome by using a particle gum application of adhesive on the backside of the label. Paper labels with a particle gum application of adhesive look and feel very different compared with labels with conventional gum adhesive. Particle gum adhesive is applied in the form of small minute granules of adhesive instead of a continuous film of adhesive. This minimizes potential curling problems of the label. Most of the newer applications of both conventional or particle gum adhesive labels incorporate a level of dry tack or "stickiness" as part of the adhesive's properties. The amount of dry tack or "stickiness" must be balanced against the way the labels are dispensed. The labels must still be easy to separate from each other and have the ability to cling to the application surface. This property speeds application and improves security when the gum or glue is completely activated.

Self-Adhesive Labels

During the last 10 to 15 years self-adhesive labels, primarily pressure-sensitive labels, have become much more common (Fig. 5). The labels' basic construction starts with any suitable substrate like paper, plastic, foil, laminate, or foil laminate material (13). The point of differentiation for self-adhesive labels comes from the type of adhesive used to attach the label to a package surface. Self-adhesive labels come in two types, pressure sensitive and heat sensitive. The terms refer to how the adhesive activates or what makes it stick and adhere to the surface being labeled. Again, as with paper labels the selection of adhesive, both heat activated and pressure sensitive, for plastic containers must be done with care. Pressure-sensitive and heat-activated low-molecular-weight adhesive components can migrate through plastic bottle sidewalls and adulterate the product.

Pressure-sensitive labels stick or adhere to the application surface using an adhesive already sticky or having "tack," a term used to describe the ability of

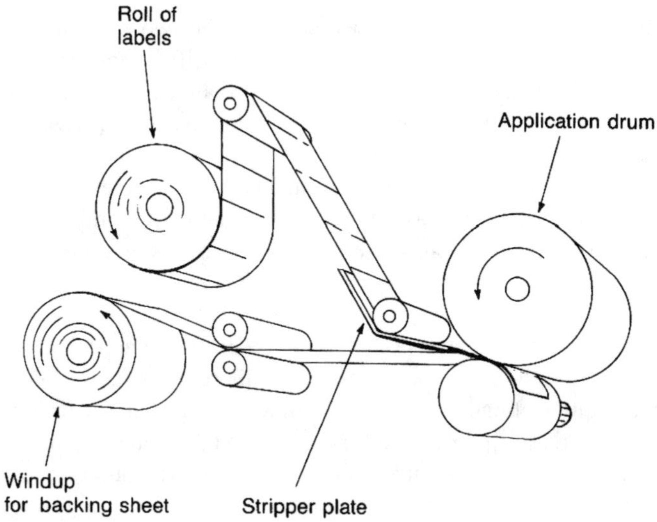

Figure 5 Schematic of a pressure-sensitive labeler.

the adhesive to grip an application surface on its own. These labels are mounted on a backing that does not adhere to the label adhesive and are peeled back from the mounting material or web just prior to label application (Fig. 5). Newer forms of the labels eliminate the web, and through the use of coatings on the front surface of the label use that surface as the release mechanism in a stack arrangement to permit separation of individual labels just prior to application. In a linerless pressure-sensitive label system the labels are separated from each other in the same way you separate each individual note in a stack of Post-It Notes®.

The level of performance that describes how aggressive the adhesive becomes is an additional way to describe or classify pressure-sensitive labels. For this classification the label adhesive is identified as permanent or removable.

The other type of self-adhesive label uses a completely different mechanism to activate the adhesive. Activation is the result of heat applied to the label. Again there are two different identifications for heat-activated adhesives. They are heat-activated instantaneous or heat-activated delayed action labels. Heat-activated labels are made from thermoplastic and thermosetting materials. Mold labeling of blow-molded plastic bottles is one of the more common applications for heat-activated labels. The label is placed in an open molding cavity, and the heat of the hot plastic activates the label adhesive.

Permanent adhesives on pressure-sensitive labels are necessary for round, irregular, or flexible surfaces. They are also used in applications where the label must be attached to the surface so tightly that the label itself is damaged, torn, or defaced if someone tries to remove it. For pharmaceutical products aggressive labels are used almost exclusively.

The two classifications of labels described speak to the two endpoints of distinction between label adhesives. Within each of these classifications there are a wide variety of label adhesives that meet unique or specialized requirements. Some of the variations are labels that can withstand an autoclave, or other wet or dry high-temperature environments. The amount of specialization is extensive, and specific adhesives for unusual end-use requirements are available.

All pressure-sensitive labels are available in sheet and roll form. Label manufacturers offer a wide variety of substrates to meet the needs of almost any application.

The substrate used for a label and the appearance of the label are undergoing significant change. As more and more prescription drugs are approved for OTC sales, brand and identification becomes much more important. Many manufacturers now design labeling to develop brand awareness when the drug is first introduced as a prescription product and carry it through to the consumer. Branding is credited with helping to maintain sales of a drug after patent expiration when generic drugs are introduced to compete with the former branded product.

Label and Package Printing

Printed packages are very different from printed labels in how the printed object is developed. Because product protection and promotion are two attributes of a product considered very important, the printing on the package receives considerable attention. Typically design shape or functional design attributes of a package are not based on the design and execution of the graphics and printing. The development of shape and function, particularly shapes that are irregular or possess compound curves, make an article difficult to directly print (Table 4). Most package converters for pharmaceuticals do not view printing as a primary function in their operations. This means that difficult or unusual printing techniques are seldom employed to directly label a container. Two-piece can manufacturers are probably the most proficient of the packaging converters in decorating their products but still see themselves as primarily can makers not printers. Packaging manufacturers are considered converters of multiple materials into packaging, and printing is not the most required or most important expertise in the complex package-manufacturing process. It is one of a number of key skills that must come together to produce a high-quality printed package. Different packaging materials require different printing techniques (Table 4). It is unusual to see multiple printing techniques at one packaging component manufacturer.

Pharmaceutical manufacturers who traditionally purchased preprinted labels and packaging and added variable information (lot, expiration dates, manufacture dates, etc.) on their manufacturing lines are beginning to become printers using electrostatic, laser, and ink-jet printing in a variety of ways. Electrostatic printing has moved quickly into blister applications. Traditionally

Table 4 Examples of Standard Package and Label Printing Processes

Package type	Printing process
Labels	Flexo, gravure, offset, letterpress, screen
Tags and product wraps	Flexo, gravure, offset, letterpress, screen
Corrugated	Flexo, letterpress, offset
Top liner	Flexo, gravure
Folding cartons	Offset, gravure, flexo
Flexible packaging	
Foil	Flexo, gravure, hot stamping
Plastic	Flexo, gravure
Paper bags	Flexo, gravure
Grocery bags (plastic or paper)	Flexo
Metal cans (2 piece)	Offset lithography, dry litho (offset letterpress)
Metal cans (3 piece)	Offset lithography, screen
Plastic bottles	Flexo
Caps and closures	Offset, flexo
Plastic cups, tubs, and assorted shapes	Flexo
Squeezable tubes	Flexo
Blister packs	Offset, litho, electrostatic

the information assigned to a product at the time of manufacture called variable print information was the only printing done on a pharmaceutical packaging line. Variable print information on a pharmaceutical package would include the lot number, expiration date, and other information assigned or required at the time of manufacture. This can vary dramatically depending on the countries outside the United States that import products manufactured here. Electrostatic printing eliminates preprinted foil by completely printing the outside of the lidding material used to seal tablets in blisters. Laser printing, another new technique, has been used in combination with preprinted substrates to produce bar codes along with required variable print information.

Labels are printed on a wide variety of substrates. Paper labels are most often printed on top-quality pigment-coated (impregnated) paper. Pressure-sensitive labels are printed on specially prepared multiple-layer paper and plastic substrates that consist of a facestock, adhesive, and a release liner. The release liner is a backing or web that holds the labels on a flat sheet or a roll and permits release of the actual label from the web. Facestock for labels is most often a grade of paper, but it may be polyvinyl chloride (PVC), polyethylene terephthalate (PET), polystyrene (PS), polyethylene (PE), polypropylene (PP), acetate, and other polymer films. Foil facestock may be foil laminated to paper or plastic, or it may be a plastic film that is metallized. Metallizing is a sputtering or deposition process that deposits a thin noncontinuous film of metal on one side of a clear or translucently colored plastic substrate to produce a foil-like appearance. Paper used for pressure-sensitive labels is typically a strong kraft-type paper with a clay coating that provides an excellent printing surface and a smooth high-quality look to the label.

Tamper-evident labels or security labels are manufactured with an adhesive that is stronger than the substrate used for printing. Any attempt to remove the label from the package results in paper or plastic tearing and general destruction or defacement of the label.

Printing creates the look, style, and feel that we associate with a product. Labels can be printed directly on the package, or they can be printed on a separate label component that is added or attached to the product during its manufacture. Cans and bottles have developed specialized techniques that permit printing directly on metal or glass. Plastic when used as the material for the container can be decorated or printed directly on the outside surface, or a separate label using plastic as the substrate may be applied to the container as the label. Color is very important, and a consumer shopping a specific product, including pharmaceutical products, by color first is quite common.

Every commercial printing process is used to produce labels. The most common printing processes employed are flexography, letterpress, gravure, lithography, silkscreen, and hot foil stamping (Fig. 6). All of these processes remain the same whether the label is produced on a roll or on a flat sheet. Different printing techniques are better adapted to roll or sheet operation, so typically sheet labels would be done using a process like lithography, while high-speed high-volume roll labels would be printed by flexography, letterpress, or roto-gravure methods. Other printing methods are used for labels, including ink-jet, xerography, laser, and thermal methods, but these are targeted at niche or specific manufacturing needs, not the broad production of labels. Many of these processes and the newer processes add variable print information on a label, while the fixed information is produced using high-speed printing techniques. Variable information on labels includes lot number, expiration date, date of manufacture, etc., and is added when the pharmaceutical product is produced. All of the printing processes used in label manufacture employ machinery that applies four to eight different colors in one pass through the machine and in many cases also has the capability to overcoat the printed surface with a varnish

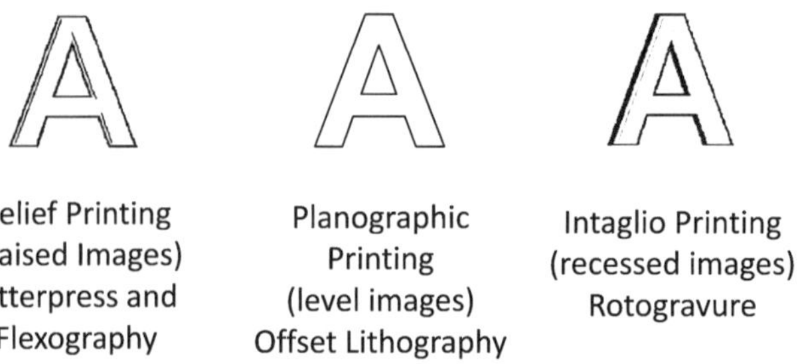

Figure 6 Characterizations of the printing process.

or wax. When this color capability is combined with a colored substrate, either plastic or paper, the effect is striking. Even more striking is the use of foil in the substrate along with multiple colors. The final clear coat or overcoat of the entire label surface increases the gloss of the label, the bright shiny surface appearance of the label.

Flexography

Flexography is probably the most widely used printing process for labels. The ability of this printing process to produce high-quality images on a wide variety of materials at economical conversion costs is the reason this process is the preferred method for package printing. This printing process produces approximately 65% of package label impressions, and the percentage share of market is growing. Offset lithography and gravure make up almost all other package printing including label printing. Flexography offers high output speeds and the ability to print on rough or varied substrates. It is fully adapted for rotary printing and die cutting in one operation. Paper, foil, or plastic substrate is converted to a finished multicolored label in one integrated operation. Flexography shares the flexible package printing market with gravure printing and dominates the corrugated printing of packages. Flexographic presses and the equipment needed to handle different substrates for labels are not expensive, and the low capital cost coupled with high unit output keep label prices low. Flexography also offers the fastest makeready options because one set of printing cylinders can be prepared for the press by the mounting of the plates and proofing, while a second set on the equipment is busy carrying out another function.

Flexographic printing has been gaining in finished print capability needed to produce top-quality labels. The process cannot match the fine images produced by letterpress; however, high-quality flexographic printing is becoming harder to distinguish in quality from gravure and offset printing unless a magnifying glass is used. The difference in printing method quality is lost on the shelf, and consumers cannot tell which printing process is used. Flexographic printing is considered in most printing circles as not as good as offset or letterpress but superior to gravure. It is interesting to note that this ranking has qualifications. When large solid areas are required in the label makeup, flexography is considered better than offset but not as good as gravure.

The flexographic process can print on smooth surfaces such as plastic films and cast-coated (clay coated) paper as well as rough surfaces such as kraft paper bags, corrugated, and liner board. The ability to print on rough surfaces comes from the rubber or photopolymer plate used to transfer the image. A foamed material is used behind the printing plate, and this helps the plate conform to the surface being printed. This does not mean that the flexographic process can overcome wide variations; the amount of variation in substrate thickness that can be tolerated by a flexographic press is strictly limited. A thick or heavy high spot in the substrate creates high pressure on the printing plate causing the ink to squeeze out. An overly thin spot creates poor or incomplete ink transfer.

Flexographic printing uses anilox rolls to meter and distribute ink in the printing plates. The anilox roll is made with a ceramic material or is a chrome-coated metal roll engraved with many small typically square cavities that pick up and transfer ink to the printing plate. The anilox roll can be thought of as a small bucket brigade carrying ink from the reservoir to the printing plate. The depth and shape of the engraving measure or meter the amount of ink delivered to the printing plate. The number of lines per inch in the engraving (fineness) determines the ability of the press to print fine patterns.

The ink for flexographic printing is a liquid. The ink contains pigments suspended in a polymer binder that is miscible in water or organic solvent. The water or organic solvent is used to adjust the viscosity of the material. The ink is very low in viscosity, very close to water, and similar to the ink used in a gravure process. The inks are referred to as fluid inks or liquid inks. The use of water as a solvent for the ink reduces air pollution but creates problems with wetting or properly laying down (producing a smooth thin film and not beading up) on many substrates. Ultraviolet (UV) inks, that is, inks that cure or harden when exposed to UV light, are used with hard to wet substrates and are replacing both solvent-based and water-based inks as a major way to reduce air pollution.

The use of liquid ink, and a relatively simple transfer mechanism compared with letterpress, and the soft printing plates are the main contributors to this printing technique's ability to handle rough and somewhat uneven substrates.

Advantages of flexographic printing

- Print color consistency
- Capable of using water base or solvent base inks.
- Prints well on rough surfaces
- Prints well on a wide variety of substrates including low-strength and lightweight paper substrates
- Low waste
- Fast changeover and setup times

Disadvantages of flexographic printing

- Lower speed than gravure or web offset printing
- Difficult to adjust color on the press
- Shallow relief plates can plug with dust and dirt
- Halftones and highlights tend to fill in or disappear
- Careful control of printing pressure is required

Letterpress Printing

Pharmaceutical manufacturers have traditionally used letterpress printing as their preferred method of producing labels. Letterpress is used for labels, blister cards (consumer product blisters), and decals. The reason behind pharmaceuticals' use of letterpress printing lies in its ability to produce very high-print resolution. Pharmaceutical labels, until relatively recently, have been produced with a large amount of very small type. Before the advent of booklet labels and other techniques used to increase the area available for printing, the only way to provide the patient and the doctor with important information was to squeeze it into the area available. The FDA has also recently added regulations, which specify the minimum font size permitted in various label applications. This meant very small type faces, sometimes as small as 4-point type were used to fit all the information onto a very small label. Letterpress printing was the only technique that could consistently reproduce fine lines and small type faces effectively. Labels for vials and ampules are excellent examples of very small packages with very small labels requiring excellent reproduction to remain legible.

Letterpress uses a rigid relief image on a metal or photopolymer plate to produce the image. It can be used as a rotary printing method or as a flatbed sheet printing process. The process requires a very smooth substrate surface to print on and it works very well for pharmaceutical labels printed on high quality clay coated paper. Both letterpress and flexography are relief printing processes. The major difference between the two print methods lies in the printing plates, the ink type, and the method of inking the plates on the press.

Letterpress uses a stiff plate, originally a metal plate, but this has been replaced more recently by plastic materials. The hardness of the printing plate called durometer produces a typical Shore A hardness on a letterpress plate of 90 versus a Shore A hardness of 30 to 60 for flexographic plates. Letterpress uses a high-viscosity paste ink that is very stiff. Because the process uses very stiff ink, the inking process on a press is a complicated series or array of rollers that work to both soften the ink by shear and distribute the ink uniformly to the printing plates.

Because letterpress can produce very sharp lines and letters, fonts were developed for the process featuring sharp serifs (Fig. 7). Serifs on letters are the small sharp lines at the bottom of a letter that makes it appear more finished or crisp, while fonts that are sans serif display a block letter with no added embellishment.

Letterpress operations run at 200 to 250 ft/min or 1.0 to 1.25 m/sec on narrow web label presses. This slow speed is one of the reasons it is being superseded by flexographic printing, which has much higher printing speeds.

Figure 7 Examples of letters with serifs and without serifs.

Because the printing plate used with letterpress operations is very hard and incompressible, the substrate must be relatively smooth and compressible. This needed characteristic of a substrate limits letterpress applications to paperboard and paper, primarily coated paper, coated paperboard, and plastic. Articles printed by letterpress have hard, sharp edges, and process color halftone screens are smooth and uniform.

Advantages of letterpress

- Excellent print resolution
- Excellent halftones for process color reproduction

Disadvantages of letterpress

- Cannot print on rough or uneven surfaces
- Relatively high-cost process due to limited speed

Lithography

Pharmaceutical packaging uses lithography or, more correctly, offset lithography in printing a variety of packages and labels. Offset lithography is used to print labels, folding cartons, wrappers, and metal cans. The process gets its name from the type of printing plates originally used. Bavarian limestone was the material chosen for making the original printing plates, and the name lithography in German literally means stone writing.

Lithography, more correctly called offset lithography, is a method of indirect printing. The plates used to produce the image are neither raised nor engraved. The plates separate the image areas from the nonimage areas chemically. The nonimage areas are wet by water, and the image areas are wet by ink on the printing press. The image is transferred from the ink- and water-wet plates to a rubber blanket, and this blanket with ink on its surface prints the substrate (Fig. 8). The inks used for this process only wet the print areas of the plate, while water only wets the nonimage portions of the plate. The inks are very stiff pastes that must be worked by a complicated series of rolls, which also meter it to the printing plate. The printing plate is first wet with a water solution, and then ink is transferred to the plate. The ink only adheres to the areas of the plate chemically treated to accept the ink. The water used to wet the plates is rarely plain water because that does not work well. Normally the water used on a lithographic press is mixed with phosphoric acid or gum Arabic to produce what is called a fountain solution or a dampening solution. Although direct lithography is possible, it is not used in packaging. In packaging, the terms "offset" and "offset lithography" refer to the use of a blanket or soft medium to pick up the image from the printing plate and transfer it to the package component being printed.

Sheet-fed offset lithography offers high-speed printing and is the predominant method for printing cartons for pharmaceutical packages. A typical

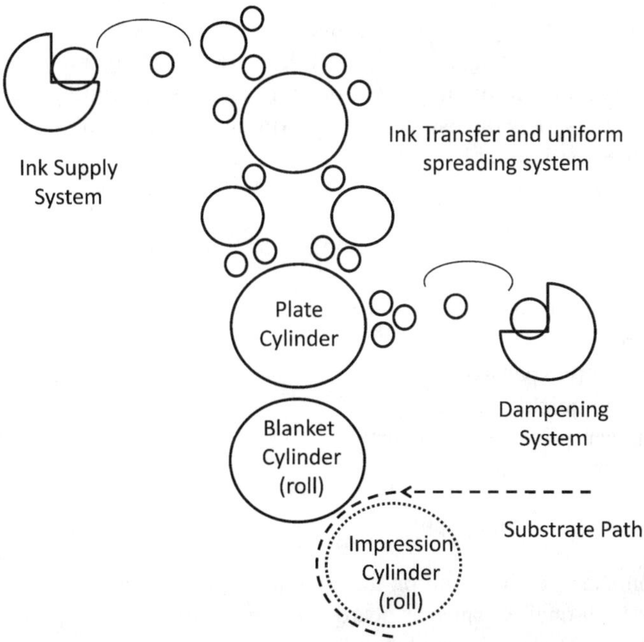

Figure 8 Roll diagram of an offset press.

carton press can produce 10,000 to 15,000 printed carton sheets per hour. This output rate correlates to a linear speed of 300 to 400 ft/min. Web- or roll-fed offset printing is approximately five times faster than sheet-fed printing, and unless the package has an extremely high-volume requirement, the speed is not an advantage. High-volume pharmaceutical packages are more likely to be produced on roll-fed flexographic presses with die cutters. Sheet-fed offset processes can print a wide variety of package sizes by varying the size of the sheet used in the process. Carton presses are commonly 60 in wide, and layouts that take advantage of the maximum print area are sought by the carton converters to maximize efficiency. This is a reason to place multiple labels or cartons on a sheet to create the maximum use of the print area. Offset provides a large printing area for label or carton production, and the ability to place an assortment of labels or cartons of different sizes on one sheet and print them simultaneously is more efficient than underutilizing the printing area available. The combination of many different labels or cartons on one printed sheet is called "gang" printing, and it permits economic production of small-volume labels and cartons with offset lithography. The practice is strictly regulated by the FDA. The FDA does permit limited use of the process in controlled circumstances, which are explained clearly in Current Good Manufacturing Practices (CGMP) regulations.

Offset lithography and flexographic printing offer many of the same advantages to the package converter: profitable printing for both short and long runs, high productivity, the ability to print on rough surfaces and metal surfaces, and easily made low-cost printing plates. Offset lithography offers multiple colors and coating capabilities, normally eight or more coating or printing stations. Offset lithography is the simplest method for printing a "perfected" sheet, which is a sheet printed on both sides simultaneously.

Advantages of offset lithography

- Economically practical for small to large print runs (less than 1000 to over 1,000,000)
- Good image resolution, including good halftone reproduction
- High-speed capability
- Printing plates are inexpensive
- Printing plates are quickly and easily made
- Offset blankets can print on rough surfaces
- Minimal setup time on press

Disadvantages of offset lithography

- Color variation sheet to sheet and on different areas of a sheet
- Requires stiff, high-quality paper or board
- Difficult printing of lightweight substrates
- Difficult printing of plastic films or plastic substrates
- More complex than other processes to operate

Rotogravure Printing

Rotogravure printing is considered a high-quality method for printing high-volume labels on plastic, foil, and paper substrates. It uses an engraved metal cylinder incorporated in a direct rotary printing process to print on smooth substrates. There are a number of variations to the gravure process, including the use of an offset blanket for transfer of the image. These variations make up a very small portion of gravure printing.

Rotogravure printing uses an engraved cylinder that is immersed in an ink bath to fill the cells created by the engraving, excess ink is removed with a doctor blade, and the image is transferred to the substrate (Fig. 9). Inks used in rotogravure are of very low viscosity, normally matching water, and consist of pigment, a binder made of a polymer, and a volatile organic solvent or water.

Gravure was until recently considered a time-consuming process because of the preparation of the gravure rolls and cylinders and a costly process if changes of the label were constantly taking place. The cylinders used to separate and produce the image took a long time to engrave and were very expensive. Electromechanical engraving and direct engraving with chemicals have replaced this slow process. Cells on the cylinder are engraved to different depths to control the amount of ink transferred to the substrate. A cylinder or roll is needed

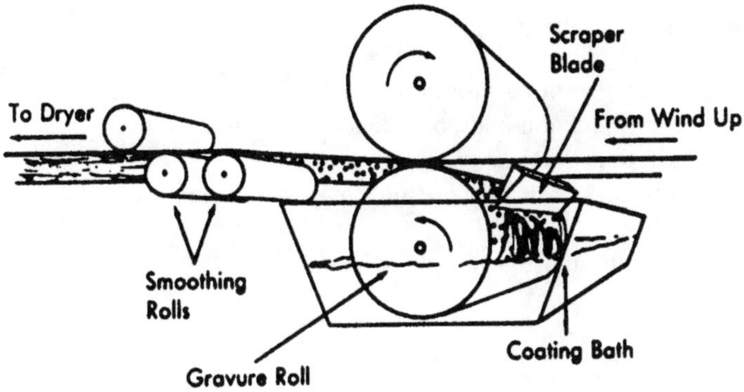

Figure 9 Rotogravure coating and printing process.

for each color to produce the image. Gravure is the only process that permits the production of halftones based on planned variations in ink film thickness. The gravure image is created on a steel cylinder that has a heavy layer of copper on its surface. The copper is engraved or etched to produce one color of the image, and then the cylinder is chromed to maintain the etched image and provide durability in the printing process. A gravure cylinder can produce millions of impressions.

A color separation is done to prepare the image and text for printing. A scanner, similar to a desktop unit, maps the image and records the color and its density. This is then broken down into primary colors, and a cylinder is prepared for each color. Color is placed on a web beginning with the lightest and progressing to the darkest.

Contact of the printing cylinder with the substrate is done with a backup or impression cylinder that presses the substrate against the engraved cylinder. The substrate in a gravure process must be compressible and smooth.

Gravure cylinders are produced by screening, both type and pictures, and the artifacts of the screen are seen as a characteristic of the gravure process. Screening introduces an effect called scallop into type and fine reverses, and it controls the reproduction of solid areas. If a very fine screen, 150 lines per in, is used, the scallop pattern is invisible to the eye and can only be seen using a magnifying glass. Gravure cannot produce fine features as well as flexographic or lithography printing.

Ideally, gravure cylinders are prepared for jobs that have large production volumes and have little or no change over long periods of time. It is normal to see a high number of cylinders in some type of storage arrangement at a gravure printer or converter because the cylinders are expensive and take a long time to prepare. This puts gravure at a severe disadvantage to other printing processes.

Few pharmaceutical packaging applications have the need for the high volume or high speed offered by gravure reproduction.

Advantages of rotogravure printing

- Highest speed of any printing process
- Excellent color reproduction and consistency
- Excellent for skin tones
- Durability of cylinders permits very long high-speed runs in millions of impressions
- Prints on low strength and lightweight substrates
- Fast press setup

Disadvantages of rotogravure printing

- High cost of print cylinders
- Time to prepare print cylinders
- Small type is ragged and poorly reproduced
- Resolution of fine details
- Changes and corrections to print require new cylinder production
- Rough or poor surface flatness paper, foil, or plastic substrate is hard to print

Screen Printing and Hot Foil Stamping

Screen printing is a technique that produces brilliant colors, high opacity, and black solids all applied by a single application of each color. For packaging, this process most often produces tags and labels where its heavy ink laydown produces standout logos in brilliant color. The process also adapts well when labeling an unusually shaped package. Examples would be cylindrical or conical packages, where screen printing is the best and possibly the only way to print directly on the package. This printing method is also used for printing instructions or logos on medical devices and equipment. This method of printing uses a stencil supported by a fabric or wire screen to separate image from nonimage area. The process is sometimes referred to as "silk screen" because the earliest support screens were made of silk. This process can print on almost any substrate.

Screen printing begins with the preparation of a stencil. The stencil uses synthetic fibers woven into a mesh that looks the same as a standard textile or piece of cloth. A key consideration in the choice of a mesh screen is the film thickness required to produce the image. Along with the screen choice, the viscosity of the ink or ink paste, the nature of the substrate, and the size of the particles contained in the ink or paste all must be part of the decision process. The common method of producing the image on a screen is through the use of a standard photographic process. The screen is filled with a photosensitive polymer and exposed the same way a camera exposes a picture. The difference comes in processing the screen as a photographic positive, that is, the areas where ink passes through are removed in the development process, a photo negative is the exact opposite. The amount of ink flowing through the screen is controlled by the thickness or diameter of the mesh and fibers used to make up the screen, the angle of the wiper or squeegee that pulls the ink across the screen, and the pressure exerted on the screen by the wiper or squeegee.

Screen inks are paste like in their consistency. They are formulated in many different ways and may be solvent based or water based if the image is baked or air-dried. Inks that are 100% liquid solids and cured by exposure to a high-energy source are called UV or EB (electron beam) inks. When printing posters or very large presentations, the ink may be oil based with the oil acting like a solvent or as reactant with the polymer in the binder. Solvent-based and water-based inks dry by evaporation but are accelerated by placing the printed product on a belt or web and then passing it through an oven to drive off the solvent to set or react the ink. UV and EB inks cure through exposure to UV light or an EB. An EB is a stream of α-particles. These inks are the fastest ways to cure or dry the very high film thickness or deposition of inks found on screen-printed articles. A high film thickness or application of ink on an article is an asset in instances when "feel" or "texture" is needed on an article. This characteristic may become important in the near future if Braille characters are needed to comply with labeling regulations. The European Union has requirements for Braille labeling.

Advantages of screen printing

- Large solid areas are uniform, opaque, and display high ink film thickness
- Produces vivid colors, metallic, and fluorescent effects
- Ability to label odd and unusually shaped objects
- Printing "plates" are inexpensive and easy to make
- All types of ink can be used
- Thick ink or coating applications to the substrate

Disadvantages of screen printing

- Slow-speed process
- Resolution of halftones and fine details is difficult

Foil Stamping

Foil stamping or hot foil stamping is a process that uses a thin metallized foil coating or an actual piece of foil laminated to a backing that permits its release when it is stamped or embossed on specific areas of the label to produce a visually appealing high-quality image. A platen press is used to place the metallic material on specific areas of the label. The platen is heated to melt the plastic adhesive backing and then pressed into the label surface to secure it to the substrate. The web of material containing the metallic material is normally as wide as the item being labeled to achieve accurate register. The portions of the material embossed or driven into the label pull away from the areas not subjected to heat. A separate press or a hot stamp foil unit is placed in line with an existing label press and timed to emboss the metallic web just as each color is printed in register. The used foil containing web is discarded even though it may contain a large metallic area that was unused. The foil-stamping material is costly to produce, and the process imposed discards off a large percentage of it after printing drives up the cost of hot stamped metallized labels. Because this process is expensive, it is used only occasionally and selectively in pharmaceutical packaging. It is very much a printing method that appeals to graphic designers.

Electrostatic and Laser Printing

Electrostatic printing is a process familiar to everyone as xerography (Fig. 10). It is the same process found in office copiers. It has been upgraded over time, and lasers are now used to place the image on the semiconductor drum. This process has been gaining prominence in pharmaceutical packaging with print on demand units used to print the foil or plastic lidding or backing found on tablet blisters. Initially the process was limited to one color, black, but has slowly evolved to offer a range of colors. The process provides manufacturing flexibility not found with other processes because it permits the manufacturer to purchase unprinted or blank lidding material for blisters that is primed or base-coated to accept and adhere to the ink (powder) and the substrate. This one material is adaptable to all products manufactured on a line equipped with this printing equipment.

Previously the foil lidding was preprinted by the packaging converter prior to shipment to the pharmaceutical manufacturer. This process was difficult to manage particularly for small-volume products because the minimum preprinted unit volume for most foil-lidding label material far exceeded the needed

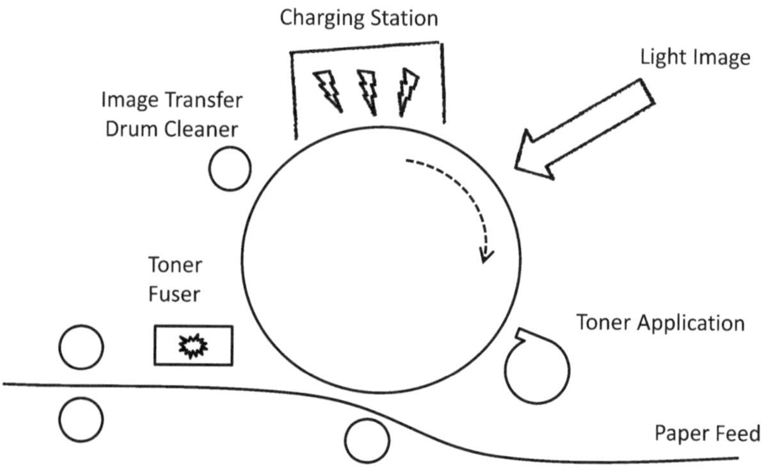

Figure 10 Diagram of an electrostatic printing unit.

quantity. This made label revisions costly and time consuming. Preprinted foil blister lidding and preprinted foil label material in general has typically had long lead times as part of their production. By placing a print-on-demand electrostatic or xerographic unit on the blister line, preprogrammed to produce the image necessary for the product, a small quantity of blisters can be produced and labeled at minimum cost.

Electrostatic or laser printing is a direct rotary printing process. An image, either scanned or generated by computer, is transferred to a semiconductor drum that is selectively charged and discharged to produce the image. The drum receives an application of a dry powdered toner (ink) that adheres to the charged areas of the drum and falls away from the non-charged areas. The toner is then applied (printed) to the substrate and fused (heated) into place. The toner, which is considered the ink in the printing process, is an extremely fine powder made up of pigment and resin (polymer) that softens and flows slightly when heated. The fusion (heating) process and the flow of the resin in the toner create adhesion to the substrate. Porous paper was the first and easiest substrate for the process, but foil and plastic materials can be used when base coated or modified on one surface with a material that will accept and adhere to the dry toner.

Because pharmaceutical packaging must deal with many small-volume products targeted to small patient populations, this print-on-demand printing method is extremely valuable in providing a manufacturer the ability to respond to small orders in a short period of time. It reduces supply chain complexity, inventories, and time to market. Each of these attributes is a highly prized capability in a world-class supply chain.

Advantages of electrostatic printing

- Direct digital reproduction of images
- Highly adaptable to all types of labels and the reproduction of bar codes
- Capable of handling any length of production run (blisters)
- Little training required for operation

Disadvantages of electrostatic printing

- Low speed compared with other printing processes
- Halftone images are difficult to produce
- Limited number of available colors

Ink-Jet Printing

Ink-jet printing has become common and universal with the use of computers. This type of printing has replaced dot matrix printing and is normally the low-cost alternative to electrostatic printing for computer applications. Pharmaceutical packaging and packaging in general have been using ink-jet printing for a long time to apply codes, lot numbers, expiration dates, and other variable information on a package or a label at the time of manufacturer. The area on the package or label that receives the variable information is left blank and uncoated or unvarnished to provide a porous substrate to accept the ink. Take a look at most expiration dates on packages, and you will note that they are applied using an ink-jet process, and you will note that the part of the label or package where they are applied is not shiny or has a different surface appearance compared with the rest of the label.

Ink-jet printing produces an image by atomizing and then depositing small droplets of ink on a substrate to form an image. The ink droplet is charged and then deflected in an electric field to produce the image (Fig. 11). The charging and deflection are controlled by the direct digital output of a computer. This makes it a simple and versatile method for converting digital information into printed images. Color printing, although available, is rarely used in packaging because the technique has traditionally been limited to placing variable information on the package. Most ink-jet printers used in pharmaceutical manufacturing operations are capable of producing two or three lines of information approximately 25 characters in length on each line.

Ink-jet printers are used on all types of packaging including metal, glass, plastic, paper, and composite rigid and flexible substrates. The ink used for most applications is solvent based or water based and is closer in composition to the ink found in a fountain pen than in other printing processes. The solution consists of a dye, not a true pigment, in a solvent. Pigment particles tend to clog the small ink-jet nozzles and are not used for that reason.

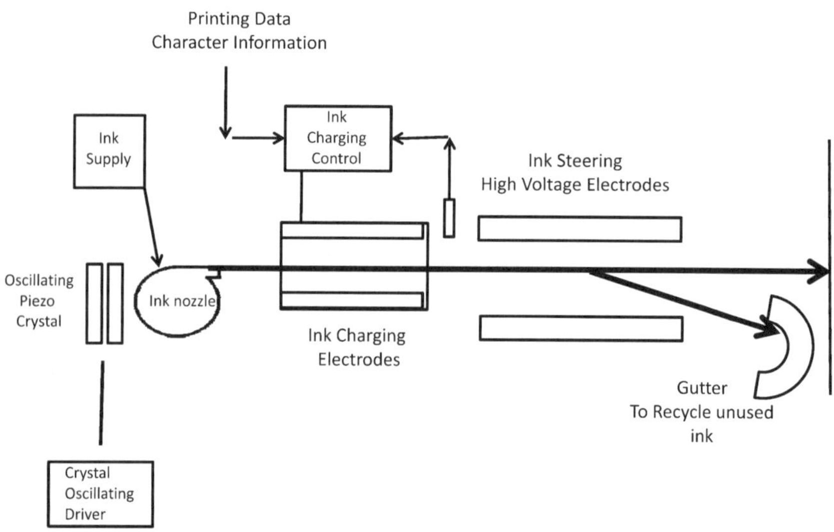

Figure 11 Diagram of an ink-jet printer.

Advantages of ink-jet printing

- Direct printing of digital information
- Continuously variable capability permits each image printed to be different from the last
- Fast and simple change of information being printed

Disadvantages of ink-jet printing

- Poor image quality at high speeds
- Environmental problems with solvent inks and cleanup materials

OVERVIEW OF BAR CODE ADMINISTRATION: GS1 DESIGNATIONS

GS1 is the international organization formed by the merger of the Uniform Code Council (UCC) U.S. and the Electronic Commerce Council of Canada with the European Article Number (EAN) International Organizations from Europe, Japan [Japan Article Number (JAN)], and other countries (14). The UCC in the United States is now known as GS1 U.S. The organization has headquarters in Brussels, Belgium, and Lawrenceville, New Jersey. It comprises local member organizations located in 100 different countries around the world. The GS1 set of standards is the most widely used set of supply chain standards in the world.

This new name GS1 replaced the former name, EAN/UCC, but its goals and functions remained the same. The organization is dedicated to the design, development, implementation, and maintenance of global standards that improve the functions of the global supply chain to improve efficiency. Its primary goal remains the development of the GS1 system, a complete series of standards for the improvement of supply chain management.

The organization has four products that touch on every area of the supply chain. The four products are

1. Bar codes
2. eCom
3. GDSN (global data synchronization network)
4. EPCglobal [radio frequency identification (RFID)]

Bar codes are used to automatically identify things (15). The unique identifier for any product is created by using GS1's company prefix or identification and the item reference number. This gives each product in the system a unique identifier that is useful and carries through to all part of commerce. The bar code that represents that number has many different forms, many of which are described later in this chapter. To maintain an orderly system, GS1 develops standards for bar codes. These standards cover all sectors of commerce and are the basis for how the system is used.

eCom is an extension of the bar code system. This system permits the interchange of all the electronic transactional information for an item. It extends and improves older electronic data exchange protocols that permit companies to transmit, share, and act on sales and inventory information based on the tracking of goods through the supply chain. At one time the use of electronic data transfer (EDT) was the province of large companies with the resources to set up the networks; the new standards have made electronic business messaging available to all.

GDSN is a service that maintains consistent information in the databases of all companies subscribing to a service. If important information such as size of a bundled unit, physical weight, or other property of a product changes, it is transmitted to all subscribers. Improvements and modifications to products happen constantly without changing the name or the number of the product; this makes sure that the minor changes, which may effect something in the supply chain, are known by all.

EPCglobal is the organization within GS1 responsible for developing the protocols for RFID chips, readers, and other equipments. This system is envisioned to function in much the same way the bar code systems function. It will eliminate the line-of-sight requirement needed by bar codes and provide additional information valuable to different parts of the supply chain. Most of the RFID labels and other equipments used in commerce subscribe to the EPCglobal standards. There are still problems and inconsistencies between these standards and those used by the Department of Defense and other governmental agencies, but it appears that harmonization will slowly take place between the different standards.

UNIVERSAL PRODUCT CODE NUMBERS

The Universal Product Code (UPC) administered by GS1 is the most widely used number and bar code found in commerce today (Fig. 12). Its counterpart is the EAN. This is the human-readable number that is seen primarily as a bar code configuration used at the point of sale for almost all products. The checkout scanner at any retail or commercial outlet is reading the UPC. This code and the idea behind it have the longest term of use of any of the codes used to manage products through the supply chain.

The UPC is numeric only. It was originally invented to speed up the checkout process at a supermarket and to track items to help manage inventories. Other companies quickly saw the merits of the system for managing their own

7 12345 12345 9

Figure 12 An example of a UPC number and bar code.

businesses because it simplified product management. The success of the system in all facets of the retail chain also demonstrated the power of the idea and is the reason you find bar code systems being used for non-retail inventory and supply chain management.

The UPC represents a UCC-assigned company prefix and a company-assigned product code. The UCC company prefix is the number assigned by GS1 (EAN/UCC) to a particular company. A company applies to GS1 (EAN/UCC) to enter the system and is assigned a number (the company prefix). The company pays an annual fee to the GS1 (EAN/UCC) for the right to use the number and the standards they have established for the proper use of the number. This number is only associated with the company and becomes its registered unique identifying number worldwide. The number can be six, seven, eight, or nine digits long. Originally all company prefixes were six digits, but more and more companies saw the merits of using bar codes for managing their products through the supply chain, and at the point of sale they began to adopt the UCC numbering conventions, and the UCC bar code standards used to change the number into a readable bar code.

The item reference number is the number assigned to the product or service by the company holding the GS1 (EAN/UCC) company prefix. This number identifies the individual product's unique identifier within the company and when combined with the company prefix becomes the unique identifier for the product or service throughout the supply chain to the consumer (Table 5). Originally the item reference number was six digits long and represented the number of products that could be designated in the system (99,999). The first

Table 5 GS1 (EAN/UCC) Number System Character Assignments

Prefix	Definition
0	Standard UPC number[a]
1	Reserved
2	Random weight items (fruits, vegetables, nuts, meats, etc.)
3	Pharmaceuticals
4	In-store marking for retailers[b]
5	Coupons[c]
6	Standard UPC numbers
7	Standard UPC numbers
8	Reserved
9	Reserved

[a]Must have a zero to do zero-suppressed numbers, a reduced number variation of the UPC number eliminating all zeros.
[b]Retailers can create a special code that can only be read in their stores and nowhere else.
[c]The coupon code alerts the scanner to look for the item in your checkout list and then applies the credit.
Abbreviations: UPC, Universal Product Code; EAN, European Article Number.

digit of the company prefix is special. It acts as an identifier and is called the number system character. Each of the numbers has a designation. Table 5 lists the assignments for the number system character. As company prefixes have grown in length, the item number length has varied as a function of the company prefix length.

The check digit is the number found at the end of the UPC. This number is calculated by the scanner each time an item is scanned and tells the scanner if it has scanned the item correctly or not. The number is calculated as follows:

1. Add together all the numbers in the odd positions of the number (digits 1,3, 5,7,9, and 11)
2. Multiply that number by 3
3. Add together all the digits in the even positions (digits 2,4,6,8, and 10)
4. Add this sum to the value from step 2
5. Using the number in step 4 determine the number, which when added to it will create a number divisible by 10. (E.g., the sum in step 4 is 106. By adding 4 to 106 you get 110, which is a number divisible by 10. Thus, for this example the check digit for the scanned number is 4.)

Each time the scanner scans an item, it makes this calculation. If the calculation does not compare the scanned check digit, the scanner knows that the scan was incorrect, and the item requires a rescan.

THE GLOBAL TRADE ITEM NUMBER

This is truly a global age with a global marketplace. "World class" in all commerce, not just pharmaceuticals, is measured by speed and efficiency, which are critical to a company's success and survival in the marketplace. With the increase in commerce and the need for increased information about products in the supply chain, the EAN/UCC has developed a unique number that permits users, when the number is used correctly, to identify their products as they move through the global supply chain from manufacturer to consumer. The global trade item number (GTIN) is considered the basis to identify trade items, both products and services, sold, delivered, warehoused, and billed through retail and commercial distribution systems (Table 6).

The GTIN number is an all-numeric identification system for monitoring and controlling the flow of products and information within the supply chain. This number when used on a product, case, or other levels of packaging like bundles or pallets is the basic identifier and the most important product and label identifier for e-commerce communications and transactions between the manufacturer, distributor, and retailer. The use of the number and the standards that surround its use permit those using the GS1 (EAN/UCC) system (Table 6) to have the confidence that their product will carry a unique identifier that permits fast and accurate movement of the product through the global supply chain. The GTIN number

Table 6 GTIN Data Structures

Code type	Digit position													
	1	2	3	4	5	6	7	8	9	10	11	12	13	14
UCC—12	0	0	X	X	X	X	X	X	X	X	X	X	X	C
EAN/UCC—13	0	X	X	X	X	X	X	X	X	X	X	X	X	C
EAN/UCC—8	0	0	0	0	0	0	X	X	X	X	X	X	X	C
EAN/UCC—14	X	X	X	X	X	X	X	X	X	X	X	X	X	C

0 = no digit.
X = digit.
C = check digit.

guarantees that the electronic communications and transactions conducted by automated means at different points in the supply chain will contain all the information needed at each step to correctly identify the company and the product.

The number permits the identification of the product at all levels of packaging.

The definition for the GTIN is a unique identification of trade items worldwide within the GS1 (EAN/UCC) system. A GTIN has a 14-digit data structure in its bar code, which may contain only 12 digits (the UPC number), 13 digits (EAN-13), or 8 digits (EAN-8). The GTIN is defined as a 14-digit number to include all the older numbers and their varying structures.

There are four data structures that are organized in different ways to accommodate inclusion in the GTIN (Table 6). All the numbers are right justified in terms of how the data is organized. The data structures are organized as follows:

1. UCC-12—the 12-digit structure
   This number represents a six-, seven-, eight-, or nine-digit UCC company prefix, a five-, four-, three-, or two-digit item reference number, and one digit representing the check digit.
2. EAN/UCC-13—the 13-digit structure
   This number represents the GS1 (EAN/UCC) company prefix and the item reference number and one digit representing the check digit.
3. EAN/UCC-14—the 14-digit structure
   This number consists of one digit representing the indicator digit assigned for the packaging level (product, bundle, case, pallet, etc.), 12 digits representing the GS1 (EAN/UCC) company prefix and item reference number, and one digit representing the check digit.
4. EAN/UCC-8—the 8-digit structure
   This number represents the seven-digit GS1 (EAN/UCC) company prefix and item reference number and one digit representing the check digit.

It is always a challenge to maintain records of an item as it moves through the supply chain and distribution channels. The GTIN was a response to this need, and is an attempt to make EDT of information needed to ship, track, bill, and inventory products in a uniform system. The GTIN makes available to small businesses the same capability to automatically track and bill products to customers in the same way large companies have been exchanging and managing their supply chains with the electronic interchange of data. The efficient transfer of data and products is becoming an area of intense focus. The advantage for a manufacturer and the supply chain of using the GS1 (EAN/UCC) GTIN is that it does not contain any new information. The GTIN is designed as a way of taking the existing company and item numbers, placing them under one all-encompassing standardized format, and then using the unique number that results as a pointer to multiple data bases within and outside the company. It is compatible with any system using the old UCC or EAN bar codes on the product or service without the development of another number. It gives companies the flexibility to continue using unique numbers for each packaging level (item, bundle, case, pallet, etc.) of their product or adopting the indicator digit to identify the packaging level.

Bar Codes

Bar codes are used to identify materials, packaging, and products in the many industries including pharmaceuticals. They have become the method of choice for controlling products through the supply chain to the customer. Very few packages used in the consumer marketplace come without a bar code somewhere on the label.

Bar codes are optically read symbols used for storing data. Data is recovered from the symbol by a variety of machine-reading techniques. Bar codes provide a fast accurate means for inputting data to a computer to record events or identify items. The events may be sales transactions, inventory recording, warehouse operations, dispatching, and recording the progress of an item in manufacturing or distribution operations. Bar code operation is extremely accurate with typical error rates far less than one per one million of scanned operations.

The FDA has published a rule: "Bar Code Label Requirements for Human Drug Products and Biological Products" to guide parties about how and where to use bar codes (16,17). The rule requires that all prescription drug products and OTC products used in a hospital or institutional setting and dispensed as a prescribed patient medication must display a linear bar code that contains the NDC number for the product (17,18). The reason behind this rule was based on the FDA's review of drug errors in hospitals and nursing homes that resulted in high numbers of adverse patient events and the number of deaths attributable to drug administration error. The FDA estimates that bar code labeling will result in 500,000 fewer adverse events or a reduction of 50% of the total adverse events now recorded. This FDA estimate is very conservative considering the data reported from hospitals using bar code technology to identify drugs. These institutions have reported dramatic reductions in medication error rates ranging between 71% and 86% according to figures reported by the FDA.

The use of bar codes in hospitals and other medical settings must be viewed as a total system. The system is designed to integrate the patient, his medical records, and his prescribed medication by computer (19). The hospital, for example, provides each patient a bar-coded wristband when admitted to the hospital. This bar-coded bracelet links the patient to his or her computerized medical records. The hospital has bar code scanners or readers available that are linked to the patient's medical records and to the drug identification data provided by the NDC number. The health care worker scans the patient's bracelet to access the medical records and then scans the drug being administered. The computer system compares the medical record with the drug being administered and if they match informs the worker that it is safe to administer the drug; if they do not match, it sends an error message to the worker, stopping the administration of the drug. The health care worker then investigates the cause of the error message. The problem may be one of the following:

- Wrong patient
- Wrong drug
- Wrong dose of the drug
- Wrong time to administer the drug
- The patient's chart had been updated and the prescribed medication was changed

The system would, for example, prevent a patient from receiving a drug dangerous to the condition being treated, or would prevent a person receiving multiple doses of the same drug from different health care workers at too short a time interval. This system, along with improving patient safety, also provides a transactional record of the drug, dose, and time of administration on the patient's medical record, making the record far more accurate and completely current in real time. It also means all records for the patient are in one place replacing multiple sheets of paper or multiple notations on a card. The virtual real-time record is always up to date and always available for review. Another benefit for an institution using a bar code system is the automatic generation of a transactional record for managing drug inventories in hospital or nursing home pharmacies.

The final rule on bar codes as written for blood product tracking is far more extensive. For blood products intended for transfusion, the rule requires codes in machine-readable format approved for use by the FDA. At a minimum, the machine-readable information must include the facility identifier, the lot number relating to the donor, the product code, and the donor's blood type (A, B, O, or AB) and Rh factor.

The 10-digit NDC number in bar code form with a check digit (total 11 digits) is too large for many small dose-packaging presentations such as parenteral vials used for drugs and vaccines. To address this problem,

RSS (reduced symbology set) codes were developed for small packages. The FDA extended the bar code rule to OTC products by making the requirement apply to drug products sold to hospitals or institutions in the same packaging used in the OTC consumer market. The key caveat here is that it only applies to OTC products that a physician would prescribe on the patient's record. Aspirin would require the code but toothpaste would not.

As part of this rule, the FDA also indicated that it would redefine the NDC number and make it more unique and useful for informational databases.

One should note that many drug products, particularly OTC products, use the UPC number to identify the product. This code was and continues to be used to permit assignment of different SKU (stock keeping unit) numbers to promotional variations of common medications. The unique code permits identification of products containing coupons, bonus sizes, and bundled products. A bundled product is usually described as including a product with a related product or promotional item. An example would be toothpaste with a toothbrush included as part of a promotion of the toothpaste product. These constantly changing versions of product are used to promote sales and would quickly exhaust the packaging component of the NDC number. With the new rule, the NDC number along with the UPC are required to appear as two separate codes on a package.

The NDC number can be contained in the 14-digit GTIN number to meet this regulatory requirement.

History of Bar Codes

Bar codes have been around for a long time. The first patent for a bar code was issued in 1930, and the first proposal to automate retail checkout came in 1932. The modern use of the code design, hardware for reading the codes, and the related infrastructure began in 1959 when Sylvania developed a railcar tracking system with complete installation and use on all railroads in the United States by 1961.

At this point, things began to move rapidly regarding the development of a fast accurate technology to identify goods. The first of the common bar code symbologies, called 2 of 5 symbology, was invented in 1968. The Plessy code, a modified pulse modulated bar code symbology, was developed in the United Kingdom in 1969, and the first laser scanner for bar codes was installed at General Motors in 1969. The first supermarket trial of bar codes was conducted in 1972. Interleaved 2-of-5 or I-2-of-5 symbology was invented in 1973 and is still in widespread use today. The UPC symbol was selected for supermarket use in 1973. From this point, codes began to expand and change rapidly. Code 39, the first alphanumeric code symbology made its appearance in 1975, and EAN symbology was invented in 1977 to extend bar code scanning to European supermarkets.

During this period of development, OCR (optical character recognition) and bar codes vied for the preeminent position. An in-store trial established bar codes as faster and more secure than OCR scanning, and after winning this head to head test, bar codes have supplanted other identification technologies and expanded to every facet of the supply and distribution chains.

During the late 1970s and 1980s, industries other than grocery and retail began to adopt bar codes as a more dependable method to identify and track items. These included the defense, banking, automotive, and health care industries, particularly blood banks within the health care group.

As sophistication and the need for increased information arrived, newer codes, new 1-D (one-dimensional or standard bar codes) and 2-D codes were invented. Code 128, which included the full American Standard Code for Information Interchange (ASCII) set of characters and some functional characters, was developed between 1980 and 1985. This is the most complete of the 1-D codes that appear as the vertical bars of a common bar code symbol. Code 128 still requires considerable space on a label for proper use, and as a result, smaller, more compact symbology was developed to increase the density of the amount of information conveyed. These codes, called 2-D codes, were invented between 1987 and 1994 to meet the need of placing more and more information on an article or a package.

GS1 STANDARDS ORGANIZATION

EAN International Article Numbering Association and UCC

The GS1 organization is the new name for the former UCC and EAN International Article Numbering Association, which controls the issuance of the numbers represented by the bar codes that tell or represent who the producing company is, while the company controls the second half of the number, which identifies the particular product and is called the SKU or product number found on the article or its packaging. The governing bodies specify and control the number of digits in the bar code format. EAN International and the UCC, a member organization of an expanded EAN International, are voluntary standards organizations charged by their respective boards with the comanagement of the GS1 (EAN/UCC) system of product identification and the global management process (GSMP) for products in the supply chain. The GS1 (EAN/UCC) system standardizes bar codes, electronic data interchange (EDI) transaction sets, XML schemas, and other supply chain solutions for efficient business transaction and tracking. By administering the assignment of company prefixes and coordinating the accompanying standards, EAN International and the UCC maintain the most robust item identification system in the world. The system is used for 23 different industries and in more than 140 countries. Additional information on these bodies is available at www.uc-council.org. These bodies assign numbers to manufacturers and create industry and worldwide standards for number use and specific standards for display of these numbers in bar code and human-readable formats for each level of product packaging (examples of levels of packaging are pallet loads, cases, and bundled items and individual items). For pharmaceutical products these two entities along with both the Health Industry Business Communication Council (HIBCC) and its affiliate International Organization, the European Health Industry Business Communications Council (EHIBCC), and the National Wholesale Druggists Association (NWDA) have published guidelines on the use of bar codes for transport labels. Their standards for generation of labels follow the guidelines of the first two organizations and are more concerned with placement of the information on a product case pack. The FDA becomes involved in the number and bar code identification administration when the NDC number, which contains the pharmaceutical manufacturer identification, is in the numbering systems of HIBCC, the NWDA, and UCC/EAN (16).

With automation and a need to have more information to effectively manage the supply chain, linear bar codes now used universally for product identification cannot encode and provide the amount of information necessary to satisfy all regulatory, quality, and supply chain needs. An example is encoded information identifying and tracking the lot and expiration date for each product from manufacture through distribution to the customer. Today this requires three separate codes and three separate scans. Many labels do not have the area or real estate necessary to print multiple bar codes. Many of the older 1-D linear codes require a large amount of physical space on a label. New codes, while carrying more information, still must maintain a minimum fixed size, required by the size and separation of the bars to be scanned or read.

Bar codes on packages follow a number of different conventions to encode the data. These various conventions somewhat track the evolution and increased sophistication of the codes and the scanning technology employed to read the code. Commonly used linear bar code conventions are called Interleaved 2-of-5 (also called 1-2/5), Code 39, and Code 128. Each system has its advantages and disadvantages. The oldest and least flexible is Interleaved 2-of-5. It is a numeric-only code and always contains an even number of digits (e.g., 367 becomes 0367). Code 39 (also called Interleaved 3-of-9) is a bar code symbology with a full alphanumeric character set. Code 128 is a bar code symbology capable of encoding the full ASCII character set; the extended ASCII character set, and four non-data function characters. It allows numeric data to be represented in a compact double-density mode, two data digits for every symbol character. The reason for describing the three most popular bar code symbologies is to provide a guide to understand how long and how high a bar code needs to be to encode the desired data. All of these codes; in addition to the information encoded, also contain stop and start characters, quiet zones, and one or two independent self-checking features based on parity bits and algorithms that determine if the code information as produced or as scanned (read) is correct.

All of the standards-setting agencies then place minimum sizes on the bar code based on the ability of various scanners to read the information. Typically, the key dimension, called the × dimension, is defined as the minimum width of the narrow bar and narrow space in a bar code symbol. The height of the bar code is a proportion based on the × dimension. For a 14-digit number such as the UCC/EAN number, the size of the bar code is dependent on the number of bars and spaces needed to produce a 14-character alphanumeric string. Both UCC/EAN and HIBCC place minimum sizes on the narrowest bar to achieve a code that can be scanned with a high degree of accuracy. The quality standard for the code is defined by ANSI. The smallest acceptable size for the × dimension is nominally 0.010 in. It can be reduced to 0.0075 in if very high-quality printing capability is employed. The minimum height of the bars in all cases is 0.2 in (5.1 mm). This size restriction means that for a 10-digit HIBCC number or a 10-digit NDC (UCC/EAN) number to appear on a label, the minimum size of the bar code on the label must be 1.1 in (28 mm) long. These minimum sizes represent real problems on small pharmaceutical labels or packages. On small packages where every dimension is less than 1.57 in (40 mm), it is permissible to encode a 12-digit UCC/EAN number in Code 128. There are restrictions on this format and the next format mentioned, but a useable number can be encoded. It is also permissible to encode only the 10-digit NDC number required by FDA on pharmaceuticals in RSS code.

All digits in a bar code must include the human-readable interpretation of the bar code regardless of which format is used. If the NDC number is used, the human-readable interpretation must be the full-expanded format for uniqueness and database compatibility. The NDC number consists of an FDA-assigned manufacturer's identifier followed by the manufacturer's assigned product code and package code. The manufacturer must ensure that the package code within the 10-digit NDC number is unique for the level of packaging and that the same NDC number is never used on two different packages. Normally the minimum size of the human-readable type is 6 point. On occasion, it can be reduced to 4.5 point, but this is difficult to read and must be avoided.

Additional difficulties in placement and the ability to read a bar code are introduced by round surfaces. When confronted with this problem, the bar code must be oriented for presentation to the scanner that allows the information to be read. The orientation on a round surface is usually vertical, and this creates problems for text and human-readable information that is normally aligned horizontally. Round surfaces coupled with glass, which introduces possible reflection or refraction of the scanning beam, can also produce false reads of a code by the automated scanner.

Adding an alpha character to a bar code will increase its size by as much as 40% in total length. Alpha characters (letters) should be avoided in codes when space limitations are part of the equation.

Two Dimensional Codes (2-D Data Matrix and other Matrix Codes)

The limitations of linear bar codes and the need for encoding increasing amounts of information into smaller and smaller spaces have lead to the development of 2-D code symbologies. Everyone who goes for shopping is familiar with 1-D or linear bar code symbologies. As the use of codes expanded to other industries and applications, the shortcomings of linear 1-D bar codes became evident. The most important limitation of a 1-D code is its restriction on data capacity to 12 to 15 bytes. In addition to this limitation, linear 1-D bar codes require space that may not be available on a label, and are intolerant of damage to the code. Linear 1-D bar codes are also difficult to read when printed on irregular product surfaces, or if insufficient contrast is present in the printing.

These shortcomings led to the development of 2-D codes. A 2-D code contains information on two axes, not one, as is the case with a linear bar code. All 2-D codes contain more information than 1-D codes, with some reaching 2000 to 3000 bytes of information depending on the size used. This larger data capacity is the primary reason for using 2-D codes.

A standard 1-D bar code is an item identifier (referred to at times in distribution operations as a license plate) that points to information contained in an external database. Examples would be price or inventory records for an item. When a linear bar code is scanned, the database is accessed and the transaction is completed using the external information.

When a 2-D symbology is used, the code can function as the database since the symbology is capable of holding 2 KB of data or more. For example, a label on an API could also contain the material safety data sheet for the material, eliminating the need for filing or repository systems at any location it is used, and eliminating the need to frantically search for the information in an emergency.

There are two primary types of 2-D codes, matrix and stacked bar. All of the symbologies to date use one or both of these technologies in the structure of the code. Two types of 2-D codes have garnered the widest acceptance in the industrial marketplace. These two code symbologies are referred to as Data Matrix and PDF417. There are at last count 24 other 2-D symbologies available but not widely used.

United Parcel Service (UPS) developed a third 2-D symbology called MaxiCode that appears widely on labels they generate and is used for sorting and tracking packages in their system. This code is limited to the tracking and sorting systems used in conjunction with UPS.

The advantages of using Data Matrix and PDF417 are the ability to encode large amounts of information and provide methods to maintain the security of the information even when the code is damaged (Fig. 13). Both symbologies have been placed in the public domain. Both symbologies can contain numerical data and pictures, and both use algorithms to allow data to be recovered or reconstructed if a portion of the code is damaged or unreadable. There are, however, other significant differences between the two codes.

PDF417 is in actuality a stacked linear bar code even thought it is considered a digital code (Fig. 14). It encodes its information in the width of the bars and spaces in the code. A bar or space can take on one of two values, thick or thin. PDF417 requires higher resolution printing than a Data Matrix code, and requires at least an 80% contrast ratio between the ink of the code and its background. This makes it difficult to use this code on colored base stock for

Selection of Codes for Product Identification & Security

Interleaved 2 of 5 - .013"
12345678

Interleaved 2 of 5 - .013"
12345678901234

Numeric Data Only - Even # of Digits

Code 128 Subset C - .013"
12345678

Code 128 Subset C - .013"
12345678901234

Numeric Data Only - Even # of Digits

Code 128 Subset B - .013"
AB123456

Code 128 Subset B - .013"
AB1234567890CD

AlphaNumeric - Variable

Data Matrix - ECC 200 @ .013" dot size — 12345678 — AlphaNumeric - Variable

Data Matrix - ECC 200 @ .013" dot size — 12345678901234 — AlphaNumeric - Variable

Data Matrix - ECC 200 @ .013" dot size — AB123456 — AlphaNumeric - Variable

Data Matrix - ECC 200 @ .013" dot size — AB1234567890CD — AlphaNumeric - Variable

Data Matrix - ECC 040 @ .013" dot size — AB123456 — AlphaNumeric - Variable

Data Matrix - ECC 040 @ .013" dot size — AB1234567890CD — AlphaNumeric - Variable

Data Matrix - ECC 200 @ .028" dot size — AB1234567890CD — AlphaNumeric - Variable

Figure 13 Code sizes and examples.

Figure 14 Example of a PDF417 code.

Figure 15 Example of a Data Matrix code.

labels. Some pharmaceutical companies have internal procedures to use colored base stock on labels to differentiate frozen, refrigerated, and room temperature storage requirements for cased product. These characteristics of the code cause scanners or readers to have limited tolerance for skew, printing on 3-D surfaces, or inaccurate bar width due to printer variance or stray pixels created by dirt or smudges when reading the code.

A Data Matrix code is a binary code, not a digital code like PDF417, and this difference is significant in understanding how the codes perform (Fig. 15). Data Matrix, being a binary code, has a number of advantages for industrial applications. The code requires only a 20% contrast ratio, which means it can be printed with a number of techniques such as chemical etching or laser etching on surfaces of materials. It can be printed with UV inks. A reader processing a Data Matrix code looks at each cell in the code to determine if it is light or dark. This is the binary structure of the code. Data Matrix codes are almost always read by a camera or CCD (charge-coupled device). Because Data Matrix is binary, it makes it easy to contain 16 bit characters such as Japanese Kanji. It also permits a Data Matrix code to contain executable instructions within its encoded information, allowing the product to specify and direct other related operations unique to this particular product or item. In contrast, PDF417 requires an external lookup table to decode segments into one of the 929 possible executable values.

A wide range of printers support printing of 2-D codes, and any label printer producing linear bar codes can also support 2-D code printing. Conventional offset printing, laser printing, and high-resolution ink-jet printing are all used to print 2-D symbologies on labels or packaging. The need or requirement to print a 2-D code directly on the surface of a product or part is much more difficult and problematic. Data Matrix is best suited for printing by laser etching, chemical etching, or embossing. This is important for identifying medical devices, or parts used in medical devices, such as electronic components or 3-D mechanical parts critical to proper operation of the device. Many parts look alike, but a code with a reader can eliminate any confusion surrounding what is right or wrong.

For pharmaceuticals, this ability to encode irregular items could be extended to etching a code on the surface of a tablet, vial, or other dose presentation of a pharmaceutical. Because the type or nature of the material when marked may result in a low-contrast ratio between the code and the material, it would require the use of a CCD scanner with high-speed image processing capability to read the etched symbol.

Equipment for reading codes, particularly 2-D codes can influence the code selected for a particular application. This distinction is disappearing as devices capable of reading both 1-D and 2-D symbols are available.

There are three types of readers used for 1-D and 2-D codes. These are area CCDs or cameras, laser scanners, and linear CCDs. Area CCDs or cameras acquire the entire code at once and are well suited for use on production lines or packaging lines where a fixed distance between the product and package can be mechanically built into the reading station. This insures that the image is always in proper focus, making reading and comparing images fast and easy.

Laser scanners or linear CCDs read codes row by row and are termed "linear" readers or scanners. These two types of scanners are familiar to anyone shopping in a grocery store where fixed scanners are used at the checkout station, and handheld moveable scanners are used to read items too bulky or heavy to put on the checkout conveyor.

Newer readers or scanners incorporate both technologies in their design. The biggest drawback to area scanners is the relatively limited focal length required to read a code correctly. This means that the scanner must be held closer to the object being read compared with a linear scanning device. Cost is also an issue for devices providing both types of capability when compared with a linear scanner.

The structure of 2-D codes presents severe problems to linear readers. Only Data Matrix and PDF417 provide sufficient markers at the corners or edges to allow the linear reader to orient itself and read the code. PDF417 gained a large amount of early acceptance because of the availability of handheld laser scanners used to read the code. Data Matrix gained acceptance because its use of fixed CCDs was ideally suited to production operations. Resistance came for Data Matrix because the CCDs were more expensive and less flexible than linear laser scanners.

While both symbologies can be read by all three types of readers, PDF417 is a much more difficult image to acquire if it is moving.

A number of pharmaceutical companies including Barr Laboratories, Abbott, and Merck are using 2-D symbology codes, usually the Data Matrix code on various pieces of labeling to manage version or revision control information. This is important when labeling, particularly inserts, have only minor variations from one revision to another or when similar inserts are used for different countries or regions and space is a problem. Data matrix 2-D codes also permit better lot control of components supplied by multiple vendors.

Table 7 Data Capacity of Different Sizes of Data Matrix Symbols ECC 200 Error Correction Capability

Square symbol size	Data capacity			
Row and column	Numeric	Alphanumeric	8 bit ASCII	ECC (%)
10 × 10	6	3	1	62.5
12 × 12	10	6	3	58.3
14 × 14	16	10	6	55.6
16 × 16	24	16	10	50.0
18 × 18	36	25	16	43.8
20 × 20	44	31	20	45.0
22 × 22	60	43	28	40.0
24 × 24	72	52	34	40.0
26 × 26	88	64	42	39.9
32 × 32	127	91	60	36.7
36 × 36	172	127	84	32.8
40 × 40	228	169	112	29.6
44 × 44	288	214	142	28.0
88 × 88	1152	862	574	28.0
144 × 144	3116	2335	1556	28.5

The actual size of the symbol is based on printer resolution capability to produce each individual square, or the fixed dot size used for each section of the code.

Both of these 2-D codes can be used on individual small packages where bar codes will not fit. The codes employ sophisticated algorithms for checking information and the ability to encode 10 to 100 times the amount of information contained in a bar code in a fraction of the space. In the case of Data Matrix and other 2-D symbologies, trade and governing associations have not adopted an accepted standard. Data Matrix is under copyright by International Data Matrix Incorporated. This group maintains the standards and has kept the standard consistent to allow the automotive and electronic industries to adopt and use the symbology on parts and computer chips. The standards are now being moved to GS1 (EAN/UCC) and ISO organizations (Table 7).

RSS Codes

The RSS code provides a method of bar coding identification information on very small products that do not have the label area necessary for a standard bar code. It has been used on parenteral vials and other small-dose pharmaceutical packaging as an identifier for the product. The FDA, recognizing the seriousness of drug-dosing errors, requires easily identifiable unit dose markings that permit the correct drug to be matched to the correct patient and for the correct strength of the drug to be identified. An RSS code designed for small packages can be read by conventional 1-D scanners found in many hospitals and health care establishments and provides a low-cost bridge to fuller implementation of bar codes in hospitals.

The RSS code is a linear code. One variant of the RSS code set is a stacked code that incorporates a 2-D format (PDF417) into the code, making it a hybrid or combination of 1-D and 2-D codes. The use of a 2-D portion of the code is important because the FDA has considered but has not required adding a lot number and expiration date to the bar codes.

The RSS symbology family has three main components and three sub-iterations. These three iterations are RSS truncated, RSS stacked, and RSS stacked omnidirectional. All codes can be combined with a composite component designated A, B, or C. The three main symbologies are as follows:

1. RSS-14
2. RSS limited
3. RSS expanded

RSS-14

RSS-14 is a linear symbology that facilitates omnidirectional scanning. It encodes 14 digits of numerical data used to identify GTIN for scanning in the supply chain. RSS-14 can be scanned and decoded in four segments and then reconstructed. This aids in facilitating omnidirectional scanning. Each segment consists of a data character and its adjacent finder pattern.

RSS-14 is dimensioned as 96× wide by 33× high (× = narrow bar width). RSS-14 is the full-height version of this part of the group of codes.

RSS-14 Stacked Omnidirectional

This form of RSS allows the full-height RSS-14 to be printed in two rows of two segments each, which reduces the overall length of the linear bar code so that it can fit better on certain packaging configurations. The separator pattern between the two rows is designed to eliminate cross-row scanning errors. RSS-14 stacked omnidirectional is dimensioned as 50× wide by 69× high.

RSS-14 Truncated

This form of RSS allows truncation (shortening the height of a bar code) of the height to 13 times the nominal printing density (× dimension multiplied by 13). The normal RSS-14 symbol has a height of 33× (× dimension multiplied by 33). Because of this truncation of the height, RSS-14 truncated does not retain the same "over-square" attribute that RSS-14 does and therefore is best scanned with a handheld scanner such as a wand, handheld laser, or CCD scanner.

RSS-14 Stacked

This RSS-14 stacked form of RSS allows the truncated RSS-14 to be printed in two rows of two segments each. It has the same application as RSS-14 truncated but is used where a narrower symbol is needed. The separator pattern between the two rows is designed to eliminate cross-row scanning errors. RSS-14 stacked is dimensioned as 50× wide by 13× high.

RSS Limited

RSS limited is a new linear symbology that encodes the same data as defined for all four types of RSS-14 previously described. The encodation process is different and limits the values assigned for indicator digits to 1 or 0. This form of RSS can be printed very small and is not generally intended for omnidirectional scanning. RSS limited contains two large data characters and a MOD 89 symbology check character. It does not have stacked or omnidirectional formats and is designed for small-item identification. RSS limited is dimensioned as 71× wide by 10× high.

RSS Expanded

RSS expanded is a variable length linear symbology that is encoded differently from RSS-14. This symbology allows up to 74 numeric or 41 alpha characters. This form of RSS can be printed at densities that can be omnidirectionally scanned. It is used to encode 14 digits of numerical data used to identify GTIN for scanning in the supply chain. It can also be used to encode all other types of GS1 (EAN/UCC) application identifiers, as used logically in GS1 (EAN/UCC) implementations of Code 128.

RSS expanded is designed for scanning in segments similar to RSS-14. It is variable in length with 4 to 22 segments available for use depending on the amount of data to be encoded. The symbology check character encodes a MOD 211 check value and the number of segments in the symbol.

RSS expanded encodes all GS1 (EAN/UCC) application identifier element strings. Special compaction methods have been designed to decrease the symbol size for common element strings.

Unlike UCC/EAN-128, the RSS expanded symbol is designed for reading by omnidirectional point of sale scanners. Code dimension is not specified because RSS expanded is a variable-length symbology.

RSS Expanded Stacked

This RSS expanded stacked form of RSS allows the RSS expanded to be printed in 2 to 11 rows of two segments each, which reduces the overall length of the linear bar code. So it can fit better on certain packaging configurations.

Composite Components of the Codes

There are three members of the composite component family: composite component A, B, and C. They are chiefly distinguished by their capacity to encode data. They are not designed as stand-alone symbols, but must be combined with GS1 (EAN/UCC) linear symbols that encode GS1 (EAN/UCC) system keys in the RSS codes.

Composite component is a 2-D bar code: It is not a true matrix code, but a stacked linear bar code. Composite component is bidirectionally decodable and therefore can be read with laser scanners and does not require the use of imaging scanner technology as true matrix codes do. They have a binary-based encoding scheme optimized for GS1 (EAN/UCC) application identifier strings. Special compaction methods have been designed to decrease the symbol size for commonly used element strings such as lot number and expiration date.

All composite symbols, except EAN/UPC composites, encode an image flag in both the linear component and the 2-D composite component to tell the scanner to continue scanning for the other component. Scanners that read EAN/UPC composite symbols should be set to look for both the linear and 2-D components.

Composite Code A (CC-A)

CC-A can encode up to 56 digits of alphanumeric data. CC-A can be combined with any of the GS1 (EAN/UCC) system symbols except for ITF-14. This code is based on Micro PDF417.

Composite Code B (CC-B)

CC-B can encode up to 338 digits of alphanumeric data. CC-B can be combined with any of the GS1 (EAN/UCC) system symbols except for ITF-14. This code is based on Micro PDF417. It is only used if the data string is too long to be encoded in CC-A.

Composite Component C (CC-C)

CC-C must be combined only with UCC/EAN 128. CC-C can encode up to 2361 digits of alphanumeric data. This code is based on PDF417, has multiple widths to match the UCC/EAN-128 width, and can have from 3 to 90 rows.

Bar codes have a number of unique attributes that make it possible to characterize the codes into simple categories that describe the symbologies. The following outline will be useful for understanding where and how the different codes, symbologies, and structures fit in the broad picture. By adding the characteristics of each code type and symbology, it also permits the reader to understand the code, which best fits the needs and the requirements of the overall system. Please remember that the older codes in this list are limited in the amount and density of data they employ for use. It is almost always better to err on the side of increased capability of the code and the system if a choice must be made.

CODE CATEGORY OVERVIEW

1. Narrow-width symbologies
2. Pulse-width modulated bar codes
3. Multi-width modular bar codes

Narrow-Width Bar Code Symbologies

1. 2 of 5 code
2. Interleaved 2 of 5 code
3. Code 39
4. Codabar

Characteristics of Narrow/Wide Bar Codes

Code 2 of 5 characteristics

1. Wide to narrow ratio of spaces and bars (2:1–3:1)
2. Five bars and five spaces per character
3. Continuous code
4. Self-checking characters
5. Numeric only
6. Variable length
7. Nonunique stop-start patterns

Interleaved 2 of 5 code characteristics

1. Wide to narrow ratio of spaces and bars (2:1–3:1)
2. Five bars and five spaces per character interleaved
3. Continuous code
4. Self-checking characters
5. Numeric only
6. Variable length
7. Nonunique stop-start patterns

Code 39 characteristics

1. Wide to narrow ratio of spaces and bars (2:1–3:1)
2. Five bars and four spaces per character
3. Inter-character gaps
4. Self-checking characters
5. Alphanumeric

6. Variable length
7. Unique stop and start code

Codabar characteristics

1. Wide to narrow ratio of spaces and bars (2:1–3:1)
2. Four bars and three spaces per character
3. Inter-character gap between characters
4. Self-checking characters
5. Numeric with two special characters and four alpha characters
6. Variable length
7. Alpha characters provide four unique stop and start codes

Pulse-Width Modulated Bar Code

1. Plessey

Plessey Code Characteristics

1. Fixed 3:1 ratio
2. Four bars and four spaces per character
3. Self-checking characters
4. Numeric plus six special characters
5. Variable length
6. Unique stop and start characters
7. Compulsory CRC check digit

Multi-Width Modular Codes

1. UPC/EAN
2. Code 128
3. PDF 417
4. RSS

UPC/EAN Codes

UPC/EAN code characteristics

1. Numeric
2. Continuous
3. Fixed lengths
4. Modulo 10 check digit

5. Two bars and two spaces per character
6. Seven modules per character
7. One, two, three, or four modules per element

Code 128 characteristics

1. Full ASCII character set
2. Continuous
3. Variable length
4. Modulo 103
5. Three bars and three spaces per character
6. 11 modules per character
7. Elements one, two, three, or four modules wide

REFERENCES

1. Department of Health and Human Services, U.S. Food and Drug Administration, Title 21 Chapter 1 Food and Drug Administration, Subchapter C, Drugs General, Part 200 Labeling Subpart A General Labeling Provisions, 21 CFR 200. Available at: http://www.accessdata.fda.gov/scripts/cdrh/cfdocs/cfcfr/CFRSearch.cfm?fr=201.2&SearchTerm=national%20drug%20code%20number.
2. U.S. Food and Drug Administration, Office of Public Affairs, FDA Consumer Magazine, July–August 2002, New OTC Drug Facts Label, June 21, 2002.
3. U.S. Food and Drug Administration. FDA News. FDA Announces New Prescription Drug Information, Format to Improve Patient Safety. January 18, 2006.
4. Department of Health and Human Services, U.S. Food and Drug Administration, Title 21 Chapter 1 Food and Drug Administration, Subchapter C, Drugs General, Part 201 Labeling, 21 CFR 201.
5. Department of Health and Human Services, U.S. Food and Drug Administration, Title 21 Chapter 1 Food and Drug Administration, Subchapter H Medical Devices, Part 801 Device Labeling, 21 CFR 801, April 1, 2006. Available at: http://www.accessdata.fda.gov/scripts/cdrh/cfdocs/cfcfr/CFRSearch.cfm?CFRPart=801.
6. FDA. Final Rule January 2006 21 CFR 201.56 and 201.57 (6, 7, 4) and "Labeling for Human Prescription Drug and Biological Products – Implementing the New Content and Format Requirements" FDA Guidance for Industry, Draft Guidance, January 2006, Labeling.
7. Department of Health and Human Services, U.S. Food and Drug Administration, Guidance for Industry, Labeling for Human Prescription Drug and Biological Products-Implementing New Content and Format Requirements, CDER, CBER January 2006, Labeling. Available at: http://www.fda.gov/OHRMS/DOCKETS/98fr/05d-0011-gdl0001.pdf.
8. Department of Health and Human Services, U.S. Food and Drug Administration, Guidance for Industry, Warnings and Precautions, Contraindications, and Boxed Label Warning Sections of Labeling for Human Prescription Drugs and Biological Products-Content and Format, CDER, CBER, January 2006.

9. Eldred NR. Package Printing. New York: Jelmar Publishing Co., Inc., 1993, ISBN 0-9616302-5-6.
10. Hanlon JF. Handbook of Package Engineering, 2nd ed. Lancaster Basel: Technomic Publishing Inc., 1992, ISBN 0-87762-924-2.
11. Kovel R, Kovel T. The Label Made Me Buy It. New York: Crown Publishers, Inc., 1998, ISBN 0-609-60168-7.
12. Jenkins WA, Osborn KR. Packaging Drugs and Pharmaceuticals. Lancaster Basel: Technomic Publishing Inc., 1993, ISBN 1-56676-014-3.
13. Bakker M, Eckroth D, eds. The Wiley Encyclopedia of Packaging Technology. New York, Chichester, Brisbane, Toronto, Singapore: John Wiley and Sons, Copyright 1986, ISBN 0-471-80940-3.
14. UCC Standards and Technology Development for Automatic Identification and Electronic Scanning. UCC Council Website: www.uc-council.org. This convention is the same for EAN and the standards are published as GS1 (EAN/UCC) Standards.
15. Lamoreaux RD. Barcodes and other Automatic Identification Systems. Leatherhead, Surrey, United Kingdom: Pira International, 1996, ISBN 1-85802-095-6.
16. Final Rule: Bar Code Labeling Requirements for Human Drug Products and Biological Products. February 2004. Available at: http://www.fda.gov/cber/rules/barcodelabel.htm.
17. Guidance for Industry: Bar Code Label Requirements: Questions and Answers. April 2006. Available at: http://www.fda.gov/cber/gdlns/barcode.htm.
18. FDA Issues Bar Code Regulation. February 25, 2004. Available at: http://www.fda.gov/oc/initiatives/barcode-sadr/fs-barcode.html.
19. Frequently asked Questions: Bar Code Label Requirements for Blood and Blood Components. April 2004. Available at: http://www.fda.gov/cber/faq/barcodefaq.htm.

12

Issues Facing Modern Drug Packaging

INTRODUCTION

There are many issues facing pharmaceutical packaging that play a role in keeping the products safe while being responsible to the patient and the community at large. In all the issues, packaging plays and will continue to play a significant role in their resolution. The problems range from patient compliance, use of unit dose packaging, anticounterfeiting measures, and broad environmental issues regarding material choice and sustainability. Counterfeiting has also become a major problem that continues to grow. In a broader context, both the environmental issues and the counterfeiting issues come from public perceptions and public misconceptions about materials used in packaging. Environmental issues stem from the idea that packaging is a "necessary evil" needed to deliver a product and that it uses more materials and is not efficiently designed. Anticounterfeiting measures, like tamper-evident indications for packaging, are much harder to quantify, but basically are expected to provide immediate evidence to the consumer that a product is the genuine article.

"Compliance or adherence," the new term for people sticking to a drug or therapeutic regimen, is the third requirement for packaging that needs better understanding and improvement because it is extremely important in the outcome of a course of treatment. It costs huge amounts of money to treat the results of chronic conditions that can be mitigated with a drug regimen. Hypertension and hyperlipidemia and high blood pressure and high cholesterol, respectively, are two examples of conditions that left untreated can lead to heart attack or stroke, and both can be treated with regular use of pharmaceutical products.

The public, and the health care community to varying degrees, view these issues as packaging problems. Each problem is something that can be improved with intelligently designed packaging. It may be possible to eliminate some of the problems if the patient community embraces the packaging and its resulting presentation of the product.

Packaging professionals are going to be called upon to lead the changes in packaging described in this chapter. They are going to oversee in many cases a complete change in the way pharmaceutical products and medical devices are presented to caregivers. Caregivers in this case are defined as not only health care professionals but also the untrained members of the public at large. In some cases, they are going to package products and develop the simple communication methods for products that can and will be used by the general public to treat emergency conditions. Defibrillators are a good example of a product that is expected to save a person's life while being administered by a member of the public at large.

Compliance or adherence, with the second term "adherence" gaining in general use, are now applied as terms describing patient maintenance of a drug dosage schedule for a chronic condition. This is probably the single biggest problem in pharmaceutical packaging. It can affect the outcome of clinical trials of new drugs, and it has a significant impact on the treatment of chronic disease in all segments of the population. Packaging, particularly interactive packaging, which combines graphics and other electronic features, is being explored as a way to remind and highlight to the patient is the time for them to take their medicine. This technology is also being combined with features that monitor when the drug is removed from the packaging, providing the clinical trial monitor, the pharmacist, and the physician a way to track the patient's dosage regimen.

Counterfeiting is an issue that is constantly in the news, and it will not go away anytime soon. The profits made by counterfeiting a legitimate drug are greater than those made when dealing with illegal drugs. This type of counterfeiting is far more damaging to people. There is a large and growing number of tragic stories of patients being severely injured or dying because the drug they thought they were taking was counterfeit. Even more tragic is the fact that many of the counterfeit drugs contain some active pharmaceutical ingredient (API) that makes their detection more difficult. Law enforcement, the Commerce Department, and the Food and Drug Administration (FDA) are working hard to develop methods to counteract counterfeiting, but with multiple outlets for drugs, particularly over the Internet, it is becoming a very difficult problem to combat (1).

Packaging is on the frontline as one of the key technologies needed in fighting this problem. Packaging that communicates directly to the consumer the "pedigree" or validity that it is the real thing and packaging that communicates in multiple ways, both overtly and covertly, that the product is genuine is seen as a must to stem or reduce the influx of bogus drugs entering the distribution system.

Environmental concerns and sustainability are issues that are broader than pharmaceutical packaging but still impact pharmaceutical packaging. The choice of materials is always difficult and a topic of much debate and concern in food

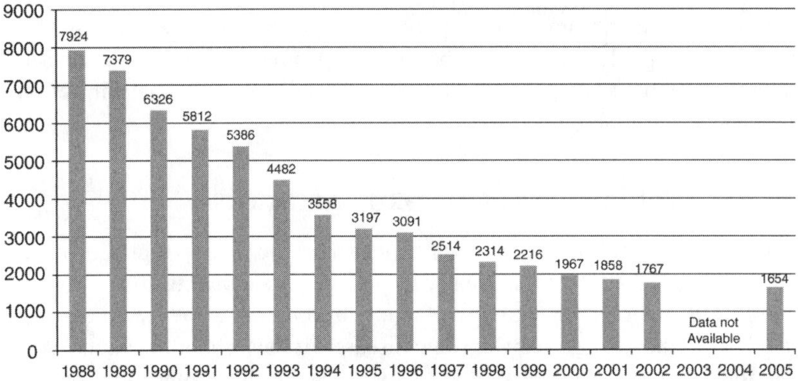

Figure 1 Number of U.S. landfills by year.

packaging. This concern about environmental responsibility is slowly creeping into decisions regarding pharmaceutical packaging.

Biopolymers are an exciting and new avenue of material application being opened by their increased availability for packaging end use. These polymers, which have both pluses and minuses, have captured the public's imagination, prompting questions about why they are not receiving more widespread use.

The overall perception that landfills are shutting down or the idea that we are running out of landfill space is another question that plagues all types of packaging (Fig. 1). Packaging, and particularly plastic packaging with a high volume-to-weight ratio, is viewed as a major contributor to the problem. As more and more pharmaceutical products move to over-the-counter status and as crossover products that combine the properties of foods and pharmaceuticals (nutraceuticals) begin to capture larger segments of the market, the same environmental questions that hang over food products will extend to pharmaceuticals or nutraceuticals.

Hospitals and health care facilities that rely on incineration for the disposal of infectious waste and other pharmaceutical waste are demanding materials that burn clean-meaning products that only produce carbon dioxide and water with no dioxins. This is another aspect of environmental packaging not generally known.

In Europe, the ability of the local governments to site landfills is almost nonexistent. These entities are going to demand more participation by all packaging manufacturers in providing solutions to any solid waste disposal issue. Composting, which is already a major method of disposal in Europe, will become more significant worldwide as a way to minimize the problem. The Duales system sometimes called the Green Dot system in Germany and the other systems that duplicate it all over Europe will shift the burden of eliminating packaging waste to companies and ultimately force solutions that are more efficient for all package disposal (2). This system has the capability to change the material mix, that is, the types of materials chosen for a package significantly, and to improve the environmental efficiency of any material chosen for use. This has always been a goal of packaging, but it will become far more important with the multiple incentives of marketplace costs and indirect taxes that impact and change what is acceptable and efficient for use.

COMPLIANCE OR ADHERENCE TO DRUG REGIMENS

The term "compliance" when used with pharmaceuticals has typically meant drug compliance. This term traditionally means the patient agrees to and follows the doctor's instructions for taking medicines for as long as the treatment regimen requires. In the case of many chronic diseases, it means that the patient must continue to take a pharmaceutical product for the rest of his or her life. The term can also refer to other doctor-prescribed requirements for the patient to follow in his or her treatment of a condition or disease. Examples of other compliance interventions include wearing a brace or other appliance to support a part of the body, caring for chronic wounds, or changing dressings after discharge from a health care facility. It also means following a regimen of required exercises as part of physical therapy or attending a therapy based on a series of meetings (e.g., counseling). The compliance or adherence requirement is essential for the patient to speed recovery or to maintain a healthy biochemical or physiological level of body function that could be life threatening if left untreated. The three most common conditions that require drug treatment are hypertension (high blood pressure), hyperlipidemia (high cholesterol), and diabetes (3).

The use of the term compliance, which was generally referred to as drug compliance, has been changing in recent years and is being replaced by the term "adherence." Previously the terms used were "compliance" or "noncompliance." Many people thought this was a deliberate attempt by patients to avoid following a doctor's instructions. In reality, it referred to a number of different reasons that caused patients to miss or not take their medications. A 1999 study published in the *Journal of the American Geriatrics Society* identified 26 factors that lead to noncompliance (3). A partial list of some of the common reasons to be out of compliance or adherence with a drug therapy includes the following:

- Forgetfulness
- Not getting a prescription filled
- Side effects, real or perceived
- No noticeable effect from the treatment, asymptomatic conditions
- Poor understanding of how to follow through with the treatment
- Poor or unclear instructions about how to take the medication (s)
- Complicated regimens for the drugs
- High cost of the drugs
- Physical difficulty in taking the drugs (e.g., hard to swallow, or hard to manipulate)
- Unpleasant taste or odor from the medication

- Taking an incorrect dose of medication (too much or too little)
- Forgetting to take one or more doses of the medication
- Stopping the medication too soon (common problem with antibiotics)
- Convenience factors

As you can see from the list, the reason patients do not always take their medication can be complicated and may involve many different factors as highlighted by the list. People rarely willfully avoid taking a medication, but many times think they know more than the doctor does or become doctors themselves. A good case in point is former President Clinton. He was prescribed statins to control his cholesterol, but after following the regimen for a period of time, he found his cholesterol levels were in a normal range and so discontinued taking the prescribed drug without telling his doctor; this led to coronary complications and eventually a coronary bypass operation (3).

It is estimated that adherence falls off rapidly after the initial treatment. For the statin drugs, the adherence rate is 97% at the beginning of treatment, but six months later the level of adherence is found to have fallen to 50% (3).

Two types of noncompliance are identified as the reasons behind adherence or compliance problem. These are unintentional compliance and intentional compliance. With unintentional compliance or unintentional adherence, the reason is as simple as "I forgot to take my medication." This forgetfulness can be driven by complicated routines, the burden of arranging and taking many medications on different schedules and plain old forgetfulness. These simple factors account for 25% of nonadherence (4).

Intentional nonadherence or noncompliance is different. Here the patient makes a conscious decision to stop taking the medication or doing the therapy prescribed by the doctor. Again, there are a number of reasons why patients make this decision. The most common factors cited are the fear of side effects (20%), drug costs (17%), lack of continued perceived need (14%), inconvenience (10%), and a combination of confusion, uncertainty, and others (14%) (4).

Packaging is one dimension of the product presentation that is expected to improve patient compliance and outcomes. Drugs with proven compliance-enhancing attributes would be strongly preferred by doctors and health care providers. Considering the fast and steep drop-off rate for most long-term adherence regimens, even small improvements will save money and help many patients avoid serious complications with their therapy. The problems with producing packaging to improve compliance are its ability to work with a multidrug regimen, its ability to provide the benefit, particularly with generic drugs, at a minimal cost penalty, and, finally, its ability to communicate in a way that is easy for patients to understand and use. These questions have not received proper investigation during the drug product development phase. Answers to these questions should be considered in the clinical trials of any new product. By testing and evaluating new packaging during clinical trials the key factors that enhance adherence can be identified and used, while the factors that confuse or discourage adherence can be modified or discarded.

Cost is always a very difficult factor to overcome when additional features are added to packaging. A typical bottle and closure used for drug packaging costs somewhere between 30 and 60 cents ($0.30–$0.60). This low cost translates into a much smaller cost per dose when you divide the package price by the number of tablets in the bottle. For comparison, blister packaging or any other form of packaging that organizes the medication into unit doses is going to be more expensive to manufacture because it uses more material and requires more handling for each individual dose.

In many cases, particularly with the elderly, it is a detriment for use to patients with arthritis or other problems because it requires more manipulation for opening. In the case of child-resistant blisters, the products are for the most part considered difficult to use by the elderly, even after significant improvements to the test protocol by the Consumer Product Safety Commission (CPSC) that focus on making packaging senior friendly. In order to produce products in individual dose containers, many of the pharmaceutical manufacturers would be required to invest in package manufacturing using new materials and methods not part of their manufacturing expertise. This investment and the development of in-house expertise needed to operate new packaging lines successfully are factors that will slow the development of adherence packaging.

Beyond the necessity to develop unit dose package manufacturing capacity, the ability to bundle multiple doses of products is a big problem. Drugs are developed one at a time. Many drugs are used in combination with others, but this type of bundling for multiple products comes later, after the drug is on the market. Some work is done regarding drug interactions to minimize dangerous drug interactions. In many cases, new drugs work in combination with others to treat multiple problems. This need to bundle and mix drugs is one of the most difficult hurdles to overcome in improving adherence. Dosing requirements for many drugs cause confusion in patients and most often do not fall into simple, easy to schedule and remember times. The confusion and the need to separate and take different products at different times lead to forgetfulness in patients, the number one reason they do not follow the doctor's instructions for taking their medications. Currently the only way this problem can be solved is by the patients separating and organizing multiple medications into some type of organizer for each day of the week (Fig. 2). They must count out the number of tablets or capsules from bottles. If the patients do not organize their multiple medications in a separate organizer, it is very difficult to ascertain if they have taken the required dose at the proper time without counting the number of tablets remaining in the bottle.

There are significant consequences for nonadherence or noncompliance. A list of the potential problems includes the following:

- Hospitalizations
- Surgical procedures required for treatment of the condition
- Potential transplant surgery (heart, liver, etc.)

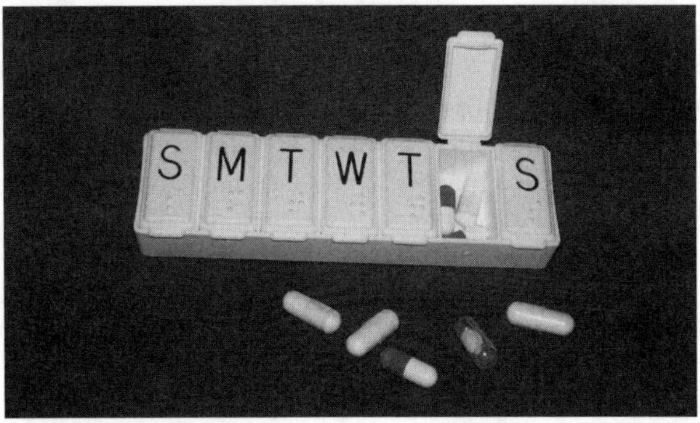

Figure 2 Picture of a weekly pillbox organizer.

- High rates of relapse or reoccurrence
- Reduced quality of life
- Emergency room visits
- Visits to doctors' offices
- Lost productivity

These problems are serious. All of them are much more costly than treating a condition at an early stage with a pharmaceutical product.

Packaging offers a method to correct or at least minimize these adherence problems. It must be emphasized that counseling the patient on a regular basis is another element equal in importance to packaging for improving adherence. The solutions, which may be advantageous to the patient, create a number of the problems in manufacturing and distribution that hinder implementation.

The biggest problem with developing and implementing a new package is proving the packaging makes a significant difference in patient outcome. Traditionally pharmaceutical development has focused on the safety, efficacy, and patient outcome based almost solely on treatment with a single drug (API molecule). The early developers (research and development laboratory personnel) are far removed from interaction with people farther down the distribution chain, like the marketer within their company, the patient, the provider, and the payer. Each of the downstream groups can and do have a more significant impact on adherence than does the developer. Each of these downstream group members in the health care delivery system has a significant stake in making sure that adherence is a prime focus for the product. The pharmaceutical manufacturer has incentive along these lines as well. The effectiveness of a new drug treatment is measured by the amount of costs it avoids in hospitalizations and other more expensive treatments. They want the patient to adhere to the regimen required by their drug for treatment of a chronic condition.

The companies that pay for and manage health care, the health care insurance companies and Health Maintenance Organizations (HMOs), are the groups most skeptical of new packaging. Adherence packaging typically adds cost, particularly to generic products, that they try to avoid. Payers want evidence-based proof that new packaging can and does make a significant difference to the patient. Data that support this position can be used to convince not only the payer but also the doctor and pharmacist along with the patient that one product is preferred over another because the benefits derived over time are greater.

It is rare in clinical trials or trials designed just to test packaging that this dimension of adherence is considered, even with significant benefits available to so many different stakeholders in the drug and health care delivery process. The idea of developing packaging to meet unmet clinical and economic needs is novel. It also requires that pharmaceutical companies and packaging suppliers to the pharmaceutical industry begin to think in a more holistic manner about how to deliver the complete spectrum of care needed for chronic disease.

Adherence forces the entire system to begin to consider what needs to be done over a long period of time that will keep the patient on his or her drug regimen. The site where the drugs are administered is a consideration for packaging that has not been fully explored. At home, care is very different from the care administered in a hospital, nursing home, or an assisted living facility. Current trends are to move patients to more and more at home care. A good example is self-injection of a number of biologic drugs by the patients themselves.

Using packaging to improve compliance is an idea that means one drug may be substituted for another on the basis of packaging to improve the patient's therapeutic outcome. This is especially true for the elderly. They almost always rely on multiple medications, not one product, and the idea of how to package and present multiple medications directly from the pharmacy has not been explored. As mentioned earlier, many seniors and patients with long-term drug-dosing needs rely on small pillboxes divided into days of the week to help them organize their medications. Unfortunately, the patient may have difficulty organizing six, seven, or more drugs into one of these containers. The containers do not have interactive features such as alarms, and the containers have no way of recording when a tablet or capsule was removed for consumption. From a doctor's and a pharmacist's point of view, a critical component lacking in packaging is a feedback mechanism that can help them monitor a patient's regimen and permit them to adjust that regimen on the basis of outcome. Many difficulties with medications for hypertension and other ailments are the result of a faulty memory on the part of the patient and the doctor responding to faulty information and adjusting the parameters of medication such as dose or frequency. This repeated unneeded adjustment of medication causes problems for the patient.

The current paradigm or algorithm for drug development starts with the drug researcher and moves to the developer. These two groups may spend years developing a drug before consulting or engaging packaging or marketing. They are the most removed from the patient and may not identify the patient's needs correctly. They may not understand the need to communicate with the doctor, the pharmacist, or the health care organization to prove that their product, packaged properly, produces a superior outcome to other competing products.

UNIT DOSE PACKAGING

Unit dose packaging is a complementary technology that may aid in improving adherence or compliance. A few basic studies are available that examine unit dose packaging as a way to improve adherence, and the data on the effectiveness of this type of packaging are very strong. Unfortunately, there are not enough of those, and the study participants are part of very controlled groups, both factors that put the findings in question and prevent more general extrapolation.

Many patients using multiple medications that require a multidose regimen or schedules for taking the products organize their medications in boxes labeled with each day of the week. These plastic boxes, purchased at the local pharmacy, permit the patient to efficiently organize his or her medications for the week. This simple container permits them to remove medication from a bottle or a blister and place it into a container for easy tracking and provides a very good memory aid that tells the patient that they have or have not taken their medication that day. A problem with these systems is that they do not provide a reminder about the time for the patient to take the medication and do not provide a method that permits the doctor or pharmacist to determine if all the medication was taken as prescribed.

When the responsibility falls on the patient to sort and organize medication, there are bound to be errors. Diabetics, for example, require four to five oral medications on average each day, and the elderly even more. With the elderly, confusion about how to separate and organize multiple medications can be a daunting task. These patients, because of natural loss of cognitive ability due to aging, are the most likely to make errors in dividing medications from bottles and placing them into a container that provides day-of-use certainty. Many others leave the medications in the bottles they receive from the pharmacy and rely on calendars and other forms of paper diaries to track their medication regimen.

In Europe and most of the rest of the world, blister packs are the accepted form of packaging for pharmaceuticals. This type of packaging for the most part promotes the proper dose regimen by bundling a dose of product into one blister cavity. This small improvement, which seems trivial, actually helps many patients always take the right amount of a drug. It eliminates one potential issue from patients' minds.

The biggest drawback of the use of unit dose packaging is that most drugs are studied and prescribed as a single entity. Drug interactions and potential problems are identified in the development process so there should be no issue with what can be taken with what. The problem comes from the need to bundle product for the user. As was described earlier, the weekly pillbox with separate compartments for each day of the week is an excellent way to organize medications for patients. This concept is difficult to translate into blister packaging.

If all of a patient's medications were bundled into one simple blister compartment, the patient would benefit in three ways. First, the potential for making an error when separating drugs would be reduced or eliminated. Second, there would be no chance that a patient could not track and maintain a daily schedule for drug use because each blister would provide a visual marker that all the drugs were taken. Finally, the presentation of multiple drugs in a blister package would eliminate the possibility of taking a double dose of product because they forgot or were unsure whether they actually took their medication. This last point is very important. Many patients just keep all the bottles for their medications in one location and then rely on their memory to take what is needed at the proper time directly from the bottle. This is not the most reliable way to track drug usage.

By placing the proper dose of each medication a patient must take in a single blister cavity or a pouch, questions of what to take and when to take are answered (Fig. 3). The patient opens and takes all the medications contained in

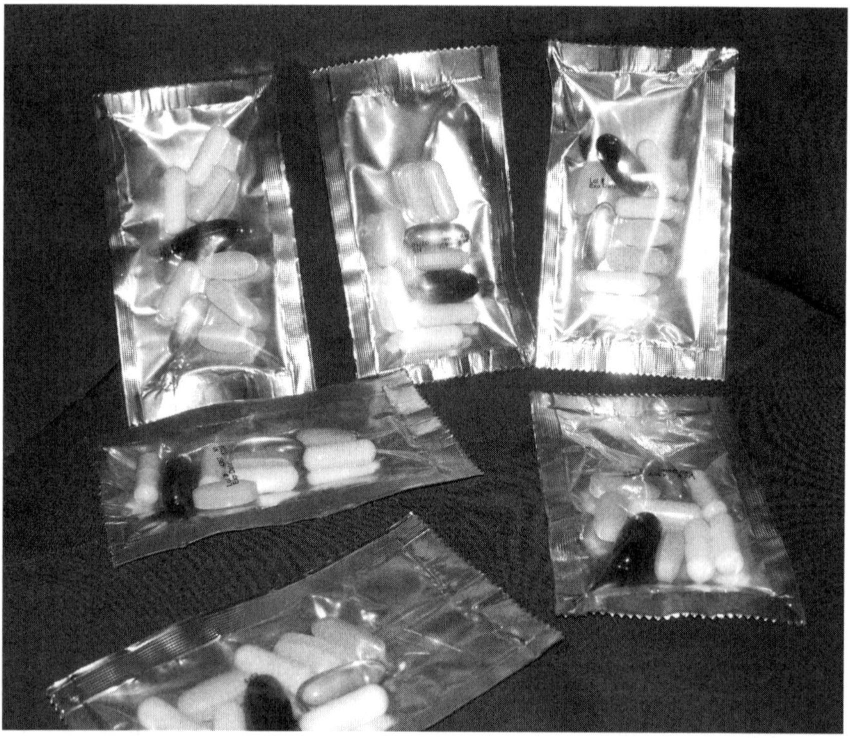

Figure 3 Picture of bundled unit doses in pouches. Each pouch corresponds to one dose of multiple medications. Example one pouch each morning, one pouch at noon, and one pouch each night. The pouches can contain the same or different medications required by the prescribed drugs.

the blister or pouch. This approach is flexible enough to cover the multiple times and the multiple drugs that must be taken during the course of a day.

Packaging, offering the benefits of the package just described, will be required to play a larger and larger role in improving patient outcomes. It will be viewed as an enabling technology for patients that have difficulty sorting and keeping track of medications. A method must be devised that would make it easy and convenient for local pharmacies to customize dose packs for patients. The customized packaging would eliminate many problems for the patient and would be beneficial to the health care insurance company paying for medications by maintaining treatment of the patient by the lowest cost means and avoiding more costly surgery or other interventions. The package would make a difference in patient hospitalizations and other dollar measures an insurance company would use to justify the increased cost.

In the future, bundled dose packs may become the standard for unit dose packaging. These packages would be the first step to improving outcomes and improving convenience for the patient; however, they fall short on one key element of adherence or compliance, feedback to the doctor or pharmacist.

One of the main benefits of unit dose packaging is the way it is designed to communicate with the patient. This must be extended beyond the patient back to the doctor and the pharmacist. From clinical trials through the ongoing treatment of a patient, most actionable feedback to the health care professional is based on and relies on the patient communicating with the doctor and professional on exactly what was taken in the course of treatment. This can create many problems. Memories are faulty, and many times the information relayed back up the health care chain is inaccurate. Besides memory, many patients do not want to tell the health care professional that they are no longer taking a medication or that they forgot to take their medication. This also contributes to the doctor's difficulty in managing a chronic condition because the results of testing and the patient's own physical well-being are not in line with the expected results or outcome produced by the drug.

Using multiple technologies, such as packaging of a complete unit dose of a drug and combining it with a bar code that could be scanned by a cell phone, is one idea on how to track patient adherence. By requiring the patient to scan and text message, each unit dose to a central source the feedback mechanism up the health care chain could be greatly enhanced. This use of electronic technology could go both ways, with a cell phone call, text message, or e-mail to the patient each time a dose of medication was required. Both these ideas would be onerous on a long-term basis for any patient and difficult to manage while traveling. They highlight a possibility that needs considerable improvement before any introduction.

A number of specific technologies have been developed for unit dose packaging. They range from programmable bottle caps and customized blister packs with built-in electronics to design layout of the tablets on the blister card and graphics (printing) to further refine the organization and design. All the electronic technologies share the same elements to make them work: an electronic chip that gathers and stores the information, a reader that converts the data into transmissible information and moves them to a personal computer, and a software that compiles and analyzes the data (5). The electronic chips track and store the time and event, information doctors need to follow a patient's adherence to using a medication. In difficult to follow or complex routines, only about one-sixth of the patients adhere to the dosing schedule for the drug. This is only for one drug, and when more than one drug is part of the regimen, the adherence rate continues to fall. By using technology to record the time when each tablet is dispensed from a blister card or a bottle, the pharmacist and the doctor gains a better understanding regarding adherence to the prescribed regimen by the patient and can more closely observe and adjust the medication to the individual's need. It also permits doctors and pharmacists to remove or discount unusual results that may arise when a patient tells the health care professional what they want to hear, not what they actually did in taking their medication.

There are a number of unit dose packaging studies that support and prove that systemic nonadherence or noncompliance to a prescribed drug regimen is not unusual. This increases with the number of oral medications the patient must take. A few of the studies now available are worth mentioning. These are relatively new and provide specific indications to how patient populations respond to unit dose packaging. One governmental study, called the "FAME" study, which stands for the Federal Study of Adherence to Medications in the Elderly, conducted at Walter Reed Army Hospital in Washington D.C. provides insight into the elderly and their needs to maintain uniform medication for chronic disease. This study highlighted two critical shortcomings regarding adherence; first, there is no effective way to improve adherence to a drug regimen during clinical studies that is in widespread use, and second, no meaningful studies are available that examine this problem among the elderly.

This federal study combined intensive patient education with customized blister packs that greatly improved convenience and organization of medications for the participants. The customized packaging combined all medications needed in a simple way to follow arrangement of doses for morning, afternoon, evening, or bedtime. This study took into account the need to manage multiple medications and provided blisters containing as many as nine medications in a single sealed blister unit. The study focused on men and women older than 65 years living independently. It excluded nursing home and other elderly patients who normally received help with using medications. Each of the patients was taking a minimum of four different drugs each day.

This study documented "substantially" improved adherence rates among participants, from 61% to 96% adherence. This study also documented significant improvements in the control of hypertension and cholesterol in the population involved in the study. It also highlighted the complementary role of medical counseling and packaging in significantly improving the measured patient outcomes (6).

This study also noted the labor-intensive shortcomings of the packaging used in the trial. This problem is one key barrier to the use of multiple medication dose packaging and the reason why this complementary solution to a difficult problem still needs to be solved. Until pharmacies can combine simplified product recognition systems with easy-to-use packaging, the patient will be left with personal sorting as the only solution that can significantly help them. It will remain an issue; it will become a larger issue as the baby boomers age, and thus highlights the need for more chronic care of blood pressure, high cholesterol, and other diseases of old age.

ANTICOUNTERFEITING PACKAGING

Counterfeiting of drugs has become one of the most significant problems facing the safe delivery of medications and health care worldwide. The explosion of drug manufacturers, particularly overseas drug manufacturers, the widespread use of the Internet, and the mind-boggling profits that can be made by counterfeiting drugs are all part of the problem. In many parts of the world, one has a better chance of getting a counterfeit product than the real thing. Counterfeiting is a growing problem within the United States and throughout the world (Fig. 4). It violates the rules of the World Trade Organization and most free-trade agreements between countries. Over the past 10 years, the counterfeiters have been gaining against law enforcement as detailed by the U.S. Department of Commerce. This assessment is based on information developed by the Department of Commerce that more and more counterfeit drugs are being consumed each year (7).

Packaging is one of the tools that need significant improvement to combat the counterfeiting problem. Originally, the counterfeit products were easy to identify because so many aspects of the packaging and graphics were obviously wrong. It was not unusual to see incorrect colors, misspelled words, and other easily identified problems with counterfeit products. This has changed. Now, many counterfeit products are impossible to detect on the basis of graphics and

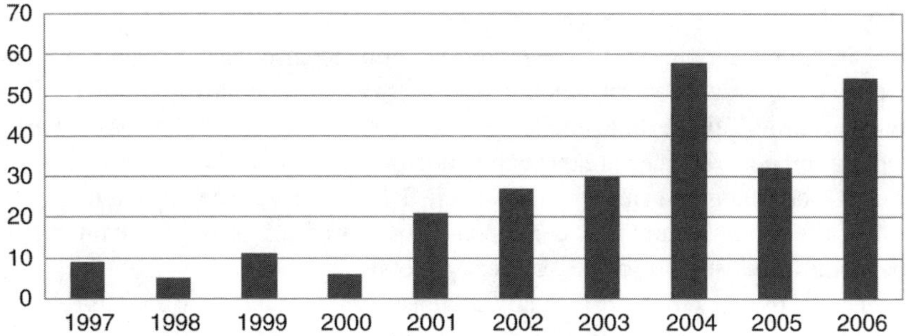

Figure 4 Graph FDA counterfeit drug cases by year.

labeling. That does not mean that they cannot be identified, just that it is more difficult to do. It also means that the manufacturers must add additional layers of security into the production of the product and packaging and must provide pharmacists and patients an easy method to determine if the product they receive is indeed the product they ordered.

This last point is one where significant work remains. The FDA requires a paper or an electronic "pedigree" that follows a drug and the details of each transaction, including the participants, and the specifics of what is being bought and sold to be recorded each time it moves in the supply chain. In addition to lot codes and other identifying information from the manufacturer, the products must be serialized, that is, a unique number is placed on each individual unit for distribution or sale in order to track each unit of a production lot from manufacturer to final end user. The ability to verify a drug lot number, serial code, and other identifying numbers against a secure database is going to require the development of encrypted product identification information on packages. It is also going to require the addition of other layers of security, both overt and covert, to make it extremely difficult to substitute a bogus product into the product supply and distribution stream.

The Internet and the explosion of Internet pharmacies online make policing almost impossible because this supply chain falls outside the licensed and regulated wholesaler and repackager in the standard supply chain. The Internet makes buying and selling of pharmaceutical products a global business, no longer limited to a country, and puts many counterfeiters beyond the reach of local law enforcement in most countries. It also provides anonymity to the perpetrator that permits him or her to operate with little fear of identification or exposure.

The World Health Organization (WHO) defines a counterfeit drug as ..."one that has been deliberately and fraudulently mislabeled with respect to identity and/or source." In the United States, this is considered a substandard drug by regulation and is actively pursued by the Department of Commerce, Federal Bureau of Investigation (FBI), and other investigative agencies along with the FDA. In China, a substandard drug would be called "fake," in India, it would be called "spurious," and in all cases, regardless of what the counterfeit product is called, it does not work or cause harm to the patient taking it.

The FDA and the Department of Commerce also classify a drug that actually contains the correct amount of API from an unapproved source as a counterfeit drug. Counterfeit products that fit these descriptions have never been tested for bioequivalence, are probably not produced in a facility that meets Current Good Manufacturing Practices (CGMP), have never been tested to prove that it works correctly, and may contain any number of unsafe contaminants such as pesticides, industrial solvents, or heavy metals.

The FDA has recorded a steady increase in the number of counterfeiting cases. In the past 10 years, they have seen the number of cases increase more than five times.

They have also recorded that the quality of the counterfeit product and its packaging has improved significantly. The counterfeits' exterior appearance, and the normal distinguishing marks that a consumer might observe, equal that of the real product obtained from the established supply chain. The FDA has been actively involved in an initiative to stop the introduction of counterfeit drugs into the supply chain with the establishment of their Counterfeit Drug Task Force. This group issued the first comprehensive action plan in 2004 and has continued to issue updates and reports on the current state of the problem.

Currently, there are a number of ways that products can be provided to a patient (consumer) (Fig. 5). The first way is for the manufacturer of the product to supply the finished product to the consumer in a ready-to-use dose-specific (unit dose) package. This is not always the best or least expensive way to supply a product. It requires that one size or a few sizes of product presentation fit all users. It also requires that the drug have only one use or only one set of warnings or instructions.

Another way drugs are distributed is through an FDA-registered and FDA-approved wholesaler. The manufacturer supplies the drug wholesaler with the product in bulk. An example would be the supply of tablets in bulk. The wholesaler takes the bulk product and either in their own operations or in the operations of an FDA-approved repackager breaks down the bulk shipment of

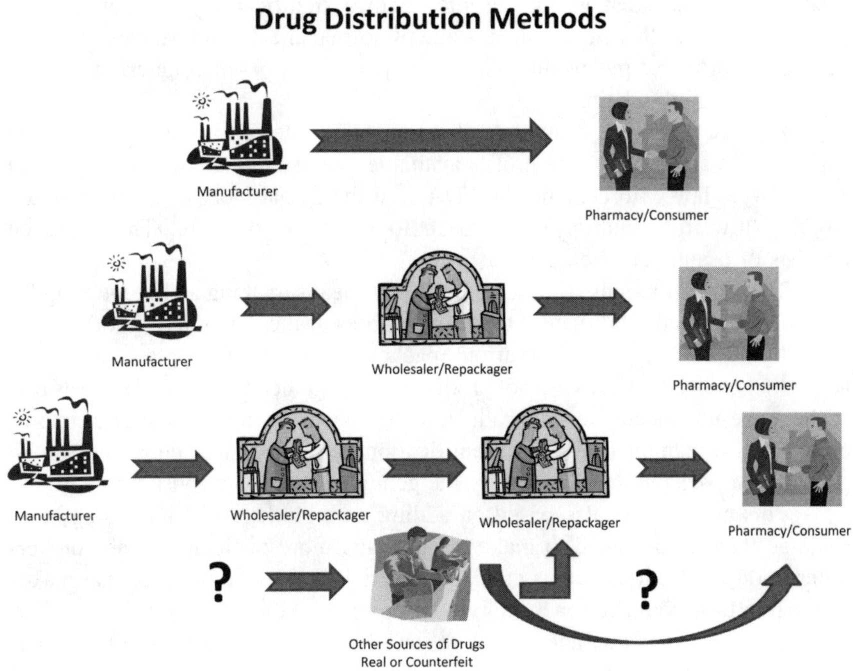

Figure 5 Schematic of various distribution channels and the opportunity for fraud.

the product into multiple doses for the patient. This is exactly what happens on a smaller scale at your local pharmacy where a product is in bulk; for example, a 500- or 1000-tablet count bottle is parceled out to customers in small bottles as specified on the prescription obtained from the doctor. This model, like the first, is very secure, with very limited number of locations or people actually handling or moving the drug.

The opportunity for fraud in the system comes when one drug wholesaler sells to another wholesaler bulk or finished product. This movement of product among wholesalers may happen multiple times depending on inventories and business needs at various companies in the distribution system. As the drugs pass from wholesaler to wholesaler, the possibility of other suppliers entering into a transaction with one of the wholesalers in the system grows. Legitimate end users, hospitals and nursing homes, use this system for reselling excess inventory or unneeded product back into the supply chain. It makes them efficient users of inventory and avoids drugs going out of date because there was no need in their facility for a particular product. It also improves their operational costs because it permits them to avoid losing the cost of unneeded inventory.

Unfortunately, a counterfeiter looking to introduce a bogus product into the system can enter this practice of reselling to wholesalers. The buyer inspects the product, and if the quality of the packaging and all the easily identified physical attributes of the product seem to be in order, the transaction takes place. That is not to say that when presented with a significantly lower price for a product, or some other inducement, some of normal checks and balances may not be employed. It also means that this is a way for improper or counterfeit product to enter the supply chain.

Because there are a number of ways for counterfeit product to enter the supply chain and because the profits available from selling a counterfeit product are so large, law enforcement, the FDA, and the Department of Commerce are working toward a multilayer approach to solve the problem. The approach consists of a number of elements.

The first is to secure the packaging. This means tracking and accounting for all packaging used legitimately to produce drugs at the manufacturer or repackager. It means audits of the various package component suppliers to ensure additional empty packages are not produced that can then be sold and used as part of the make-up of a counterfeit product. It also means securing any special inks or other hard-to-manufacture overt identification items like holograms that can be placed on a package to help identify the genuine article to a consumer.

This approach is also based on adding multiple layers of technology to a package that are designed to make the package more difficult for the counterfeiter to duplicate. It also adds an external serial code to the package that makes it possible to account for each individual unit of the dispensed product.

The FDA and other agencies view the addition of a unique serial number to each package and encoding that number in either a radio frequency identification (RFID) chip and/or in a bar code, or both, as the strongest level of protection that can be added to an individual package. This information combines with the "pedigree" that follows the package from manufacturer/repackager to consumer and permits the tracing of the product through the supply chain. The pedigree contains not only detailed transaction information but also other information and seeks to identify the individual buyer and seller by name in the transaction along with the product. This further secures the transaction because each buyer and seller, both corporate and individual, is required to be registered with the FDA.

Beyond the bar codes and RFID chips, pharmaceutical manufacturers are adding other overt measures to combat counterfeiting, like holograms, color-changing ink, and watermarks to the package. These measures are only partially effective. They fall short when the consumer is unaware of their use and does not know what to look for or what constitutes the correct overt symbol or marking. In the past few years, counterfeiters have been able to duplicate many of these technologies and incorporate them into packaging for pharmaceuticals. In addition to these overt measures designed to discourage counterfeits, covert technology can be implemented by the manufacturer and used by law enforcement, correctly equipped pharmacies, and others in the supply chain to test packages. The covert technologies include inks or dyes that fluoresce or absorb UV light and respond to specific wavelengths of light, invisible bar codes, and some watermarks. Other covert technologies, usually called forensic technologies include chemical markers, taggants, and other unique chemical-identifying features of a product. Many of these measures taken together can discourage counterfeiters but in many cases have only made them more adept at duplicating the genuine article.

The responsibility for implementation of almost all of these technologies falls to the packaging group or department in most companies. The use of many different available technologies makes the problem appear to be one that can be solved at the package level. This is not the case, and again packaging is expected to include other technologies that can combat the problem. The most notable of these technologies is the development of encodeable RFID labels that can be written with secure information at each step in the supply chain.

Securing the supply chain is seen as the primary way to fight counterfeiting. It probably will be the only way to eliminate the introduction of product from false sources as material moves along the chain. Securing the supply chain is seen as supplying a unique pedigree to the product at each transaction point. This paper or electronic trail can be verified to validate that only registered and approved wholesalers, repackagers, and retailers have access to product. It is designed to track the quantity of product shipped for distribution to each location, further securing the supply chain. An electronic pedigree contains at a minimum the following elements:

1. Detailed product information
2. A unique pedigree serial number

3. Transaction details
4. Recipient details
5. Lot, quantity, and expiration information on the specific product
6. Certification signature (electronic from the seller)
7. Authentication signature (electronic from the buyer and the final dispenser, e.g., pharmacy)

Detailed Product Information

The detailed product information includes the trade name and the chemical name of the product. The manufacturer of the product, the specific strength of the product (e.g., 65 mg), and the total amount of the product (e.g., 90 tablets), the NDC number, and other information about the manufacturer are all details of the record. This saleable unit of product would also receive a unique pedigree serial number.

Transaction Details

At each point in the supply chain, the buyer and seller are recorded with complete information about each transaction element included in the electronic pedigree record. An example would be a listing that says who is selling (the manufacturer) and who is buying (the wholesaler) and all the certification information necessary to ensure that the transaction was correct. This would include the license number of the manufacturer, the license number of the wholesaler, the invoice number, the ship date, the received date, and the names of both buyers and sellers that have checked and verified the transaction to be legitimate and the quantity of the shipment accurate.

The wholesaler and the pharmacy would duplicate this same information if that were the next step in the transaction. The wholesaler would add the lot number, the expiration date, and the quantity of drug shipped to the individual pharmacy. This permits the same lot of product to be shipped to multiple pharmacies and still be tracked. The serialized package information sent to that particular pharmacy would also be recorded. Again, a signature of an individual authorized to ship or receive the product would be included as part of the record, further certifying that the transaction was legitimate.

The FDA requires this type of record keeping in either paper or electronic form to be applied to all drug shipments. Individual states, notably California and Florida, have also adopted this policy as a way to combat bogus drugs from reaching consumers.

It is easy to see how Internet pharmacies located outside the end users' country could substitute counterfeit product into a transaction. It also highlights how price can and is used as an inducement to "go outside the system." Unfortunately, mail-order pharmacies offering low prices have created a political problem. Legitimate pharmacies deliver the correct drugs at a low cost, something very important to seniors or anyone on a fixed income. The agencies responsible for maintaining a safe supply of the drugs have opposed this form of supply because of the possibility for fraud. This has lead to an entrenched battle over the supply of drugs from outside the country.

It should be noted that organized crime has begun to move into drug counterfeiting. There are a number of reasons behind this shift in focus. Surprisingly, the penalties for counterfeiting drugs are far less than those imposed for dealing in illegal narcotics. The profits gained by counterfeiting drugs are significant. The Department of Commerce estimates that for every $1000 invested in illegal narcotics a return of $3000 is obtained. That same $1000 invested in branded counterfeit drugs returns $30,000, or a 10-fold increase over what is thought to be an extremely profitable trade in illegal drugs (7).

ENVIRONMENTAL ISSUES

The public and political focus on the environment, brought about by the ongoing debate on global warming, has created concern about packaging and any "synthetic" material. Packaging for pharmaceutical and medical device products are lumped in the general category of packaging by the public and receive the same scrutiny as all other forms of consumer packaging in terms of the environment. From time to time, concern about infectious waste is also raised, but this is a much smaller issue than the idea of packaging waste and the environment. The ideas of sustainable packaging and sustainable development are topics that address these public concerns. These ideas will not go away anytime soon. Renewable resources, which tend to mean things grown versus things mined or produced from oil, are very real and very strong demands by the public. These ideas coincide with the desire to use naturally occurring and renewable materials like corn or wheat to produce plastics. The idea that we can grow plastics instead of refining them from oil is certainly something that will continue to be a highly desirable goal by the public at large (8).

The public consciousness about environmental issues has been cyclic over the past 20 years. It has waxed and waned depending on environmental issues in the news and political issues about public environmental policy. One major trend has remained constant during this time. That trend is the expectation by the public that things will get cleaner and better and be more sensitive to the environment. Improved air and water standards coupled with an awareness for better maintenance and conservation of all natural resources are behind the ideas of sustainability and are driving the change in packaging.

Modern drug packaging is under the same scrutiny that all packaging and particularly consumer packaging receive. The ideas and the drive behind sustainable packaging along with sustainable solutions to manufactured products add a dimension to packaging that does not always follow a logical or scientific path. Public perceptions of science and evidence produced by research are cast aside by claims and rumors on the Internet that have no basis. This public perception bias can force changes in packaging even when they do not make scientific or business sense.

This conscious concern by the public for environmental sensitivity extends to all the materials used for packaging. It makes people examine and determine in their own minds if a product is "overpackaged," even if they do not understand the requirements needed for protection, distribution, and compliance. Sustainable packaging, packaging that is neutral to the environment, and recycled packaging and packaging materials are packaging design questions that must be addressed if a pharmaceutical package is going to be accepted as good packaging. Materials that have questions marks about possible environmental or physiological problems will always face public questioning and scrutiny, even after scientific evaluation and testing confirms they are safe to use. The awareness of these issues and an understanding of how to incorporate them into packaging design is an unwritten expectation all pharmaceutical companies have of their packaging departments.

On top of the environmental concerns, other issues face packaging. Questions about specific materials affecting our health are constantly making headlines. The public is concerned about heavy metals, phthalates, and materials that mimic human hormones. This too constitutes an area where the packaging engineer must consider carefully the design options available when a new product is under development. The choice made, albeit scientifically sound, may be viewed as environmentally insensitive or even dangerous. Avoiding these questions or being prepared for them is an important change in how pharmaceutical packaging is done.

Packaging and the Environment

Packaging is considered a big problem when it comes to solid waste, and although pharmaceutical packaging is exempt from most laws, it is still lumped together with all packaging in the public's perception of waste (Fig. 6). As more and more products fall into the gray areas between food and pharmaceutical products, and as more and more products are administered in the home, questions about pharmaceutical packaging will arise. The overview of the problem that follows quantifies the issue and provides a dimension of the concern the public carries about plastic packaging (Fig. 7).

Pharmaceutical packaging waste must not be confused with infectious waste from health care facilities. These wastes fall into a different category and will be discussed later in this chapter. Pharmaceutical wastes are the bottles, cans, blisters, and other packaging material used to provide over-the-counter products, infant formulas, medical nutritional products, and other drug- or wellness-related items that are sold as consumer products. To understand this issue, an overview of waste disposal in the United States is the best way to quantify the problem. Issues related to European solid waste disposal are also discussed because the majority of pharmaceutical packaging is international in nature.

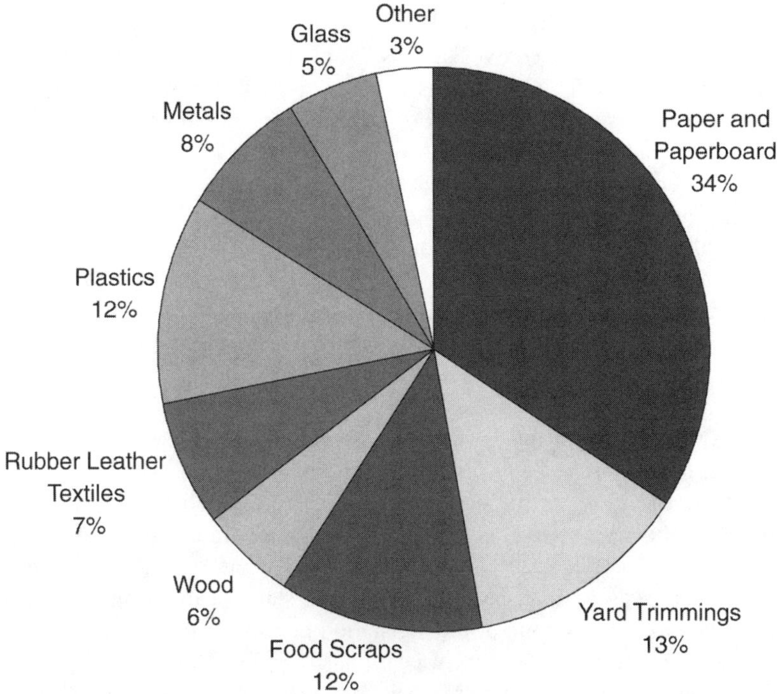

Figure 6 Total MSW Generation—246 million tons. *Abbreviation*: MSW, municipal solid waste.

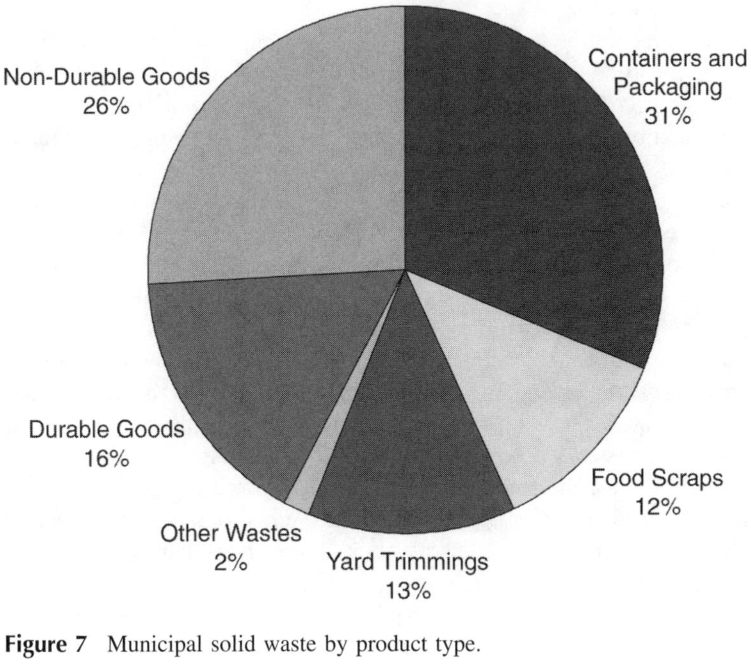

Figure 7 Municipal solid waste by product type.

Solid waste is a big issue in the United States. The number of landfills in the United States has decreased over the past 20 or so years from approximately 8000 to about 1650 (9). This statistic is misleading in that newer landfills are significantly bigger than their predecessors, meaning the overall volume capacity of all open landfills has not changed.

The concern about what to do with our trash began in the 1980s because news reports indicated we were running out of landfill room, and this report made the public aware of municipal solid waste (MSW) as an environmental issue. Pharmaceutical packaging, particularly packaging used in the home for medical nutritional products, along with packaging used for high volume over-the-counter medicines and medical products contributes to and is considered part of the packaging problems surrounding MSW.

The concern about MSW comes at the state and local level. This is where the term "NIMBY" (not in my back yard) originated. The idea that we were running out of landfill space gained a lot of publicity. The idea that new landfills had to be sited and built in the local community created a backlash that was summed up in the term NIMBY. The U.S. packaging industry came under legislative attack as part of this concern about MSW. What made the issue very difficult for the packaging industry was that the legislation came from the local and state level with almost none of the legislation proposed at the federal level. Since the problem of landfill space was primarily a local or city issue, and particularly a big city issue, it is natural to understand why this became the focus of local and state regulation. The northeastern United States, the area of greatest population density in the country viewed this as an acute problem in the late 1980s and early 1990s.

The environmental protection agency (EPA) began to provide information on MSW about this time (Figs. 6 and 7), first through contract with private firms and later through a yearly publication detailing the types and amounts of solid waste, the generation of trash per capita, and trends in recycling and source reduction (Fig. 8). The data developed showed that packaging accounted for approximately one-third of solid waste (10). Since the initial reports in the early 1990s, this figure has remained relatively constant.

The concern about landfill space coincided with the demise of incineration. During this period, the idea of disposing of waste by burning and then recapturing the heat for the generation of electricity was examined extensively (Fig. 9). A number of cities, most notably, Columbus, Ohio, built a trash to steam incineration facility and operated it successfully for a number of years. This effort to eliminate waste by incineration began to fail when the public became alarmed about the emissions produced by these plants and there was difficulty in finding site locations for these facilities. The facilities were costly to build and operate and had to be constantly upgraded to reduce emissions. These plants had long lead times in construction, and the payback on the investment was also long term and based on the price of electricity in the region. The plants did demonstrate that the volume of solid waste could be dramatically reduced, and the

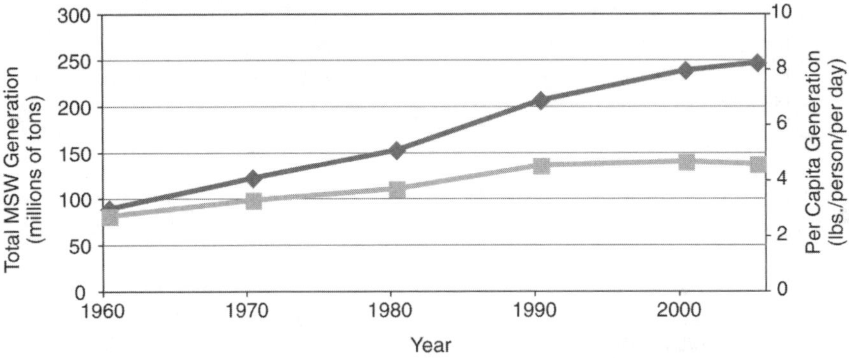

Figure 8 Municipal solid waste generation trends (lb/person/day) and total MSW generation. *Abbreviation*: MSW, municipal solid waste.

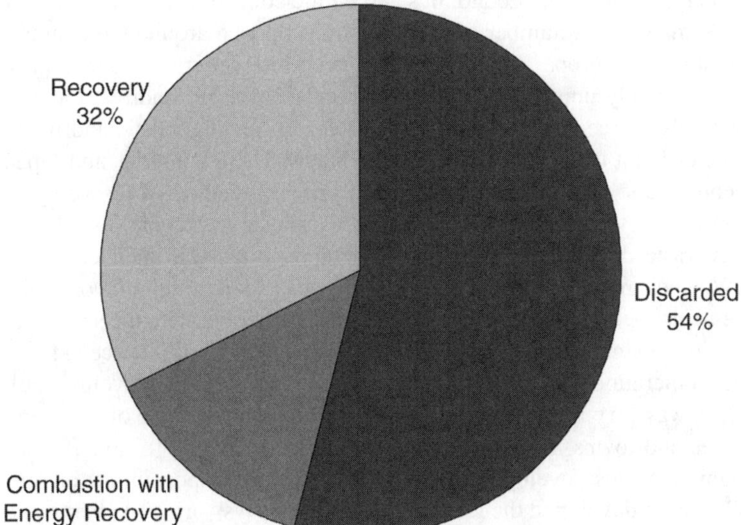

Figure 9 Management of municipal solid waste in the United States.

energy contained in plastics, paper, paperboard, yard waste, rubber, wood, and textiles were enough to make the incineration practical (Fig. 10). Recovery of glass and metals from the process was another income stream (Fig. 9). After incineration, the amount of ash and residual materials that actually had to go to a landfill was reduced by 80% to 90% by volume. This reduction, offered at a time of diminishing landfill space, was extremely attractive to many communities short on open land. For big cities and areas in the northeast, this was the attraction that fueled studies and proposals for the facilities, not to mention

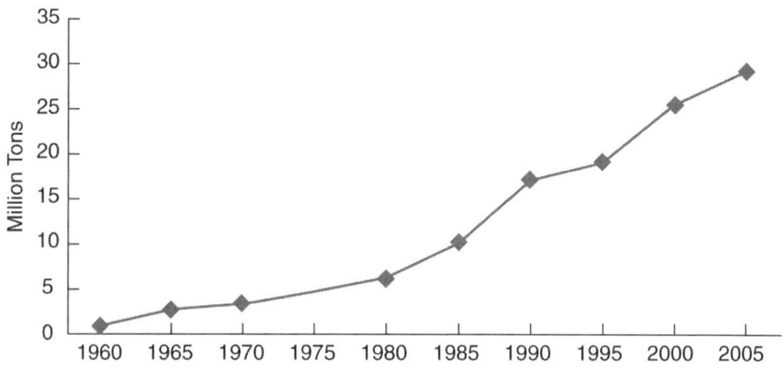

Figure 10 Plastics in U.S. municipal solid waste growth.

reducing the transportation cost and fuel burned to truck trash long distances from large metropolitan areas. Economics and public concern about the safety of the facilities remain, but a number of these facilities operate around the country.

Incineration in Europe is a different story. Most countries do not have abundant and relatively uninhabited areas for site placement of landfills. Landfill space is extremely scarce, and the political obstacles to creating a new landfill are possibly more difficult than those in the United States. These countries and Japan relied and continue to rely on incineration as a primary method of disposing of waste. Europeans viewed packaging, and primarily plastic packaging, as a major culprit in the waste disposal problem, and they targeted these in their efforts to reduce MSW. Their remedy was the same as what had found public favor in the United States. Europe embraced recycling somewhat earlier but with the same zeal as that found in the United States because of problems with landfill space (11).

When incineration was no longer a possibility, the idea of recycling took hold. This idea was very popular with the American public. The use of recycling by major cities and towns increased dramatically during the 1980s and 1990s.

The United States developed the idea of "producer responsibility or producer pays" when it developed the idea of the "polluter pays" principle that is the basis for superfund legislation in the United States. Superfund sites are those requiring significant cleanup and the federal laws in the United States require producers to fund the cleanup. This idea was originally only for hazardous wastes, but it evolved, as will be evident in the next section of this chapter on Europe's Green Dot program.

United States Recycling Programs

Recycling of packaging materials has been going on for a long time in the United States (Fig. 11). Metal, particularly steel, was always easy to separate by magnetism, and paper, particularly newsprint, was an easy and bulky item to separate. Progressive municipalities, looking for revenue sources to offset solid waste

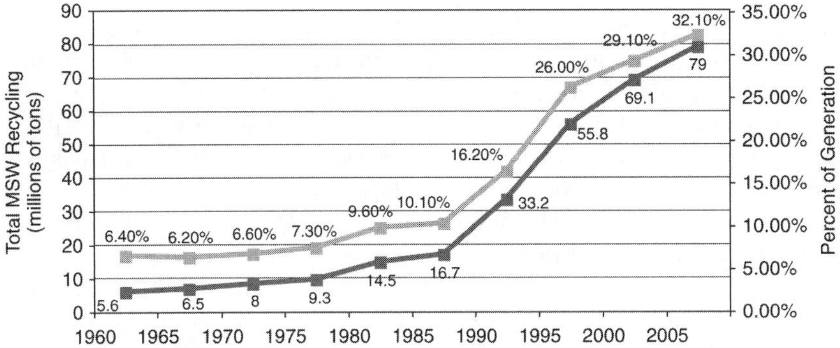

Figure 11 Municipal solid waste recycling rates.

Table 1 Packaging Components of the Municipal Solid-Waste Stream

Containers and Packaging	Generation of weight (million tons)	Recovery of weight (million tons)	Percent of generation of each product
Steel	237	150	63.3%
Aluminum	1.90	0.69	36.3%
Total metals	4.27	2.19	51.3%
Glass	10.9	2.76	25.3%
Paper and paperboard	39.0	22.9	58.8%
Plastics	13.7	1.28	9.4%
Wood	8.56	1.31	15.3%
Other materials	0.24	Neg.	Neg.
Total containers and packaging	76.7	30.5	39.8%

Source: From Ref. 1.

collection costs, began collecting newspapers separately from other trash and began to use some rudimentary separation techniques to recover steel and aluminum from the waste collected (Table 1). They then sold this waste to scrap dealers at a profit. Early on, this was very successful because there were a limited number of municipalities supplying the separated materials and the value of the recycled material was high.

During the 1980s, targeted programs attempting to recycle PET soft drink bottles and high density polyethylene (HDPE) milk bottles began. Both these materials had a high value, and soft drink bottles were always a target for public criticism because they were exclusively a one-way container. Most municipalities used a drop-off program for the bottles that worked well for the conscientious citizens who wanted to do something about waste. Because the idea was new and the packaging was not designed for easy recycling, a number of problems developed in the programs. The biggest problem was the fact that labels could not be easily removed from milk bottles, making their recycling difficult or impossible. Some municipalities asked residents to remove the labels while others went to extreme measures to try to get the labels off the HDPE bottles. Grand Rapids, Michigan, in one of the early programs, had workers cutting the label sections from the milk bottles in their recycling facility with a knife. Needless to say, the amount of material collected in these early programs was lower than the amount of material that was being produced.

PET soft drink bottles had much more success in recycling. Initially this was made possible through bottle-deposit laws in 9 states—10, if one wants to count California's refund system. PET is a material with a higher value because it has a number of superior performance properties when compared with HDPE that make it more valuable to recycle. The threat of additional deposit legislation in the states prompted the beverage industry recycling program to begin. This program was not as successful as the program instituted in the recycling states, but did contribute to the overall amount of PET being recovered from waste for recycling. This effort by both the states and industry pushed the recycling rate for PET bottles to approximately 20% in the United States. The deposit states provided the majority of this gain in total recycling with approximately 90% redemption and recycling rates. Although no new states have passed deposit laws since the 1990s, the recycling rate for PET and other plastics continues to rise.

Collection Methods for Recycling

With so much interest in recycling first spawned, the crisis in landfill space local municipalities began to change the way trash was collected and disposed. The most obvious method is to take the whole garbage stream and process and separate the recyclables from the trash before disposal (Fig. 11). In a system like this, the recovery rates are limited to the efficiency of the separation process. Surprisingly, the U.S. Bureau of Mines did a large and extensive review of separation processes on the basis of their experience with separating minerals. Their study determined that the cost of recycling waste in this manner was too high on the basis of average efficiency that could be expected from the process. The relatively low efficiency of the process was primarily due to the inability to identify and separate packaging into the component plastic material. PET and PVC bottles look much the same to people, and trying to make a distinction between two plastic materials at any speed is simply not possible.

This problem pushed the separation problem back to the individual households and consumers. The consumer was now asked to keep their recyclable materials separate from the rest of the trash. Communities that instituted recycling programs faced a number of behavioral problems and challenges. Many sociology studies have examined people's willingness or unwillingness to participate in recycling programs. Most of the studies pinpoint convenience and motivation as the two key factors in making a system work. People with an environmental concern display high motivation in handling recyclable materials. They will take the time to determine the type of material being recycled, will clean it and remove any food residues, and will separate and store it for long periods of time if a common collection point is not open on a regular basis. They are also the people who will drive to collection centers to dispose of their trash.

Obviously, there are people at the other extreme who will not participate in any program. These people even when pushed by legislation and mandatory compliance rules still resist complying with the process.

The good news is that most people fall between these two extremes of behavior. This makes the need to combine motivation and convenience into something that is "user friendly" to the majority of people a must. Communities have met this need in a number of different ways across the United States.

Today, most recyclable collection is done on the same day the other trash is collected. Consumers are requested to place their commingled recyclable materials in a container separate from the other trash. The blue box or blue bin has been the most successful method of differentiating the trash separated by the consumer.

In apartment complexes and multifamily dwellings, the system is a little different but equally as effective. The collection system consists of a number of recyclable containers positioned next to the central trash collection containers. Residents put their recyclables in the proper containers at the same time they dispose off their trash.

Many communities have instituted education programs for their citizens to highlight the need for participation. This not only highlights the economic benefits in the form of lower taxes for garbage removal services but also points out to all of us the savings in energy and water use and other resource conservation savings that recycling provides. Ongoing education in the form of advertising and direct mailings to the home are a must for any community wanting to maintain and grow total recycling rates. New residents need to learn how the community recycles and older residents must learn the value of their participation. Education of schoolchildren has also been very effective in influencing families to participate in recycling while it builds recycling habits into the children.

Communities have also recognized that requiring citizens to remove labels or crush cans reduces the number of people willing to participate actively in the program. Removal of food wastes from plastics in commingled recycling materials is a problem, but one that has been slowly overcome by evolving sortation and cleaning techniques. The idea of commingled recycling that permits the bundling of all recyclable materials directly from curbside gets the highest number of participants because it is viewed as a simple and easy way to help the environment and the community (Table 1). Most communities highlight their efforts by providing each household with a bin, with blue being the most universally adopted color for a recycling container. The bin permits families to accumulate and store recyclables with the normal trash until each normal pick-up. The MSW collector keeps the separated recyclables from commingling with the normal trash either by using a compartmentalized truck or by having two different trucks work each route, one for normal trash and one just for recyclables.

Mandatory recycling programs tend to have higher participation rates than voluntary programs with the public. It is interesting to note that a well-run voluntary program can equal or surpass the amount of material recycled by emphasizing the positive aspects of the process and working diligently on continued education of the community. Many people participate in mandatory programs by recycling only a small amount of the possible packaging they consume, and many others may not participate at all if there is no enforcement. Success in a program whether it is voluntary or mandatory is influenced more by how the program is run, and there is evidence to suggest that mandatory programs with no enforcement achieve a lower participation rate from the public than similar voluntary programs.

Recycling programs must be evaluated on the basis of cost/benefit they bring to a community. The more convenient a program is, the more costly it tends to be. The plastics container industry has added symbols to plastic packages to make it easier and more convenient for consumers to identify and separate plastics for recycling (Fig. 12). As communities increase advertising

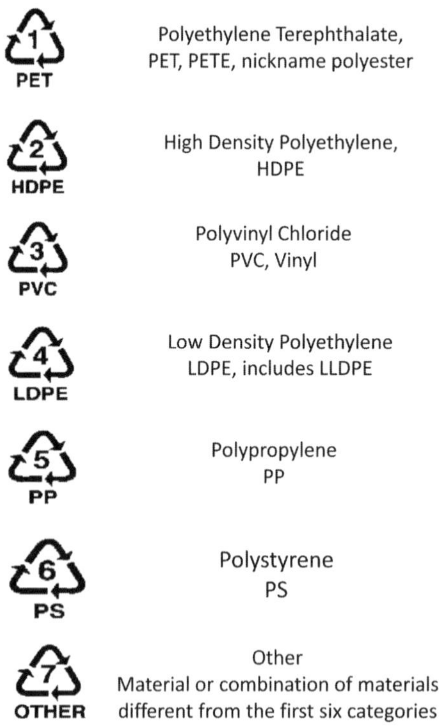

Figure 12 Plastic recycling symbols and codes.

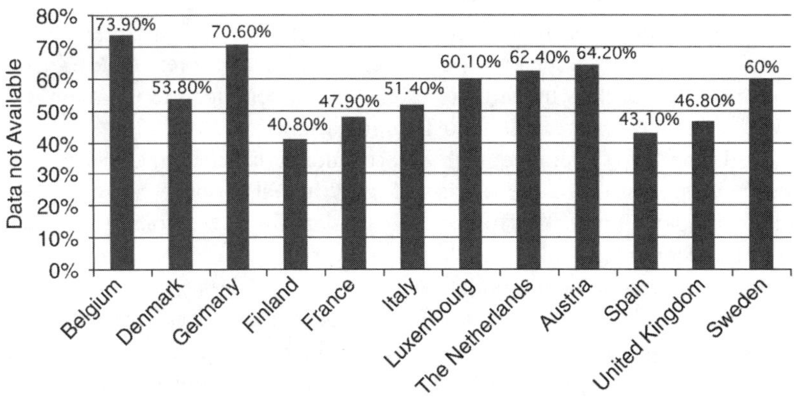

Figure 13 Recycling rates by country in the EU 2003. *Abbreviation*: EU, European Union.

and direct communication, which add costs to a program paid for by their constituents, they must evaluate if the additional amount of material collected provides sufficient justification to continue the communication or education programs.

It is interesting to note that Europe has achieved a much higher recycling rate than that in the United States using a mandatory program that shifts the costs of the program from direct taxation to a packaging "producer pays" model (Fig. 13). This may be more of a cultural factor in countries like Japan and Germany where respect for authority and a willingness to conform to the expectations of the community at large has a stronger influence.

European Recycling Programs

Germany led the way in establishing the model for recycling in Europe. They instituted a simple idea that has gained favor in parts of the United States and across the world. That idea is "producer responsibility." It is simple and means that the person who creates the waste is responsible for cleaning it up at the end of its life cycle. It also included the thought that it is the responsibility of the producer to ensure that at least some portion or minimum fraction of the waste created is recycled and to pick up the cost of the recycling. The producer's ability in product or package design to significantly effect recycling was the one driving point behind this idea. Design decisions can make recycling easy or hard, depending on the choices made.

These ideas led to the establishment of the Duales system Deutschland GmbH (Dual system of Germany). The more common name for this system is the "Green Dot" system. The government is responsible for developing and issuing legislation that requires producers to take responsibility for used packaging and to take the steps necessary to see that it is collected and recycled. The Green Dot system is now active in 25 different countries, and more than

460 billion packaging items have been labeled with it (12). Industry response to the system led to the development of a collective action to address the problem. This collective approach is much more effective and beneficial to the members than trying to collect and recycle their individual products (Fig. 13).

The idea and the system met with a large amount of criticism and comment in the early years, but it became successful over time. The success in Germany led to the European Union (EU) adopting the plan along with other countries interested in becoming members of the EU. The plan has expanded to include consumer electronics and automobiles.

The plan is not without its critics. Manufactures that want to participate in the program must pay a fee to the Duales system in order to use the Green Dot. Fees are based on the type and weight of the packaging being recycled. Generally, the heavier or the more difficult it is to recycle the packaging, the higher is the fee. Fees vary on the basis of the material in question. Plastic packaging has the highest fees, with more "natural" materials like glass having a much lower fee. License fees based on materials vary widely, with plastics carrying a fee of €3.0/kg or more, while glass carries a fee of €0.15kg.

The Duales system uses the consumer as its first method of sortation. There are other methods that do not require the consumer to separate packaging waste, but the majority of systems employed in Germany require the household to separate their waste into yellow bags or separate bins for collection (13). Glass, paper, and cardboard (corrugate) are not put into the yellow bags; they are taken to collection bins placed in neighborhoods. Glass bins further separate the waste by having divided bins or separate bins for clear, brown, and green glass.

The alternative to this system is one where the household brings all its Green Dot packaging waste to a central collection station. Duales employees then separate the waste into the various categories for recycling.

The Duales system collects all types of packaging, including composite materials. The list of materials includes glass, paperboard, cardboard (corrugate), lightweight polystyrene foam, composite mixed material flexible packaging, aluminum, and tinplate. The categories and materials chosen were based on their ability to be recycled and their environmental impacts. The Duales system establishes criteria for the various materials that the manufactures are required to adhere to in order to obtain the Green Dot license. Life cycle analysis of the materials as well as other studies of the materials and independent testing were and are used to determine the criteria for a package to meet the regulations. All criteria are peer reviewed, and all peer-reviewed criticisms of the government criteria as well as the Duales responses to that criticism are available to the public.

The new packaging laws and the Green Dot system have been successful in reducing the amount of materials going to landfills. According to Duales System Deutschland, all recycling goals have been met since the start of the program. The weight of packaging consumed and not recycled has steadily declined. One thing this legislation has done in Germany is force manufactures to reduce the weight of their packaging and the amount of packaging they use. It is not unusual to see a bare toothpaste tube or bottle of eyedrops on the shelves in Germany and other countries that have adopted the Green Dot.

Plastic Packaging and the Environment

Environmental concerns, negative publicity on the use of plastics, and the idea that plastic is not natural and that using plastic is wasteful and pollutes the environment are a few of the opinions people have against plastics.

Many consumers in general view plastic packaging as bad. In many ways, this packaging is superior to glass or metal in performance and protection of the product, but many consumers do not share this view. Many of the concerns about plastics in the environment stem from misconceptions and rumors that persist in the blogosphere and the Internet. Ideas that PET when placed in the refrigerator or in a microwave oven creates dangerous chemicals in food and pharmaceutical products harmful to humans is one example. The idea that burning plastic automatically creates dioxin is another. The American Plastics Council and many major chemical and plastic manufacturers have worked to educate consumers and the public at large about the benefits of plastic packaging and the benefits that plastic brings to our lives in many other products. This has softened some of the criticism, but they remain.

All packaging has an environmental impact. Each material, be it glass, metal, paper, or plastic has impact on the environment from the time it is made until it is disposed.

Recycling Rates for Plastic Packaging

Plastic, a major packaging material, is a smaller percentage by weight of total solid-waste generation, but because of its low density, the volume of waste is much higher than its weight fraction in MSW. This factor is the main reason that environmental groups and legislative bodies tend to focus on plastic packaging as a major contributor to the MSW disposal problem. The fact that the use of plastic packaging was and continues to grow at a rate significantly greater than other packaging materials also added to the concern over the use of plastics in packaging.

The issue of landfill space and site location of landfills is not unique to the northeastern United States. In many countries, there are inadequate land resources available for locating and operating solid-waste disposal sites. In Europe, landfill space is very scarce, and, politically, it was impossible to develop increased landfill disposal sites for solid waste. Countries in Europe and elsewhere that cannot bury their solid waste rely heavily on incineration. During the 1980s and 1990s, major concerns about possible hazardous emissions and the amount of carbon dioxide created by incineration forced a very hard look at what was needed to reduce MSW and particularly plastic packaging waste.

These requirements led to a focus on both source reduction and recycling in the United States and Europe. In the United States, curbside programs that focus on cans, bottles, and other rigid or semirigid containers are used in most communities. The focus on bottles in these programs brings in most plastic bottles, and the bulk of the plastic materials are recycled. In addition to these programs, supermarkets and other stores that use plastic for shopping bags also provide drop-off points for the used bags. In the deposit states, PET bottles are collected along with cans and glass. It is interesting to note that only 3 states now have deposits on plastic water bottles compared with the 11 that use bottle deposits to promote recycling (14). The states that include water bottles are California, Hawaii, and Maine.

U.S. Municipal Solid Waste: An Overview

The United States generated 245.7 million tons of MSW in 2005 (9). This was a decrease of over 1.6 million tons of waste from that in 2004 (9). When composting is excluded from the figures, the amount of MSW that was recycled in the United States was 58.4 million tons, an increase of 1.2 million tons from that of the previous year (9). These data highlight how effective the programs for dealing with MSW have been in the United States and that the plastics contained in the MSW is one of the most significant components of the recycling stream. More significant in the figures is the trend over the past few years that the amount of MSW generated per capita has flattened and has decreased. (Fig. 8)

Containers and packaging make up 31.2% of the MSW stream or 76.7 million tons of material. This is driven by paper, metal, glass, and plastic, with the first three materials the primary drivers by weight (Fig. 7).

Infectious Waste

Medical waste is very different from the packaging waste discussed earlier in the chapter. Medical waste refers to all the waste generated by hospitals, clinics, physician's offices, dental practices, blood banks, medical research facilities, veterinary hospitals/clinics, and any other facility that treats patients outside the home (15).

The waste generated by these sources is regulated by the Medical Waste Tracking Act of 1988 administered by the U.S. EPA. The act defines medical waste as "any solid waste that is generated in the diagnosis, treatment, or immunization of human beings or animals, in research pertaining thereto, or in the production or testing of biologicals." A list of some of these items is as follows:

- Any material contaminated with blood (e.g., bandages)
- Culture dishes for growing pathogens
- Surgical waste, including gloves, sponges, and other contaminated consumable from an operation

- Disposable surgical instruments
- Needles, both for injection of drugs and those used for blood collection or transfusion
- Lancets
- Body tissue and organs
- Samples or stocks of disease cultures.

All these wastes are generally classified in one of four ways:

- Infectious
- Radioactive
- Hazardous
- Other general waste

Only a small fraction, estimated by EPA at between 10% and 15%, of medical waste falls into the first three categories, the remainder of the waste is similar to waste generated in homes and offices.

Beyond EPA regulation, most states have their own set of regulations for health care facilities that further classify and regulate how these materials are handled and disposed.

Approximately 90% of potentially infectious medical waste is incinerated (16). The incineration is subject to strict EPA guidelines for emissions, particularly emissions of mercury or other heavy metals.

For packaging this means, the materials chosen for use should avoid creating as by-products any of the regulated materials as defined by the EPA. By selecting polymers consisting of only carbon, oxygen, and hydrogen, the disposal of the materials by this method result in by-products of carbon dioxide and water. This may be a competitive advantage for companies when competing with products produced with materials that create other regulated by-products of incineration.

BIODEGRADABLE PLASTICS

Biodegradable plastics and packaging is a subject that intrigues the imagination and offers a quick, easy fix to solid waste. The idea of using a biodegradable material for packaging really became a focus of attention when the "solid-waste crisis" of diminishing landfill space was announced in the 1980s. The idea behind biodegradable materials was the thought that plastics could be made with materials or structures that after a "reasonable period of time" would begin to break down when exposed to sunlight or when exposed to microorganisms and water. A number of materials made it to the market with these designations, but after a short period of time two facts emerged that put this problem into perspective. Many of the materials that were touted as biodegradable were not, and the ones that really were biodegradable were very expensive. Another annoying fact also emerged as people studied these materials, that is, they did not possess the performance attributes needed for packaging.

A major problem behind the use of biodegradable polymers was that the majority of people involved with the polymer synthesis and development did not understand how a modern sanitary landfill is designed to work. Thus many of the necessary conditions that they highlighted about how these polymers would degrade and about the biodegradable aspects of the polymer were not present in modern landfills and did not permit the degradation process to take place. After all these failings and problems were exposed, biodegradable materials quickly exited the stage as a possible solution to the solid-waste crisis.

An explanation and background is needed to place biodegradable polymers into perspective. A biodegradable polymer is one that with the addition of water and microbes will break down over time (17). Generally, they are made from naturally occurring materials like corn, soybean oil, starch from wheat, and other organically grown materials. There is some controversy surrounding the manufacture of a biodegradable polymer when one considers the fossil fuel energy needed to grow and fertilize the crop and the energy needed to convert the organic material into a useful material. However, once the material is converted, it offers a number of advantages for reducing the solid waste going to landfill. The idea that the material comes from a crop fits well with the idea of sustainability for the environment.

Biodegradable materials now are beginning to find their way back into favor because of the knowledge gained about disposing of solid waste and because another idea has become quite popular in the United States and in Europe. That idea is composting.

Composting is a way of recovering a lot of solid waste that was previously dumped in a landfill. It makes use of the energy and food value present in yard waste, food waste, and other organic materials that are harvested, cut, or picked in our gardens, lawns, or our homes. In a composting operation, the whole idea is to present a way to enhance the growth and activity of microorganisms. These organisms and water convert the waste materials into useful fertilizer and humus.

Many of the biodegradable materials require much more aggressive conditions than those found in a backyard composting pit. The International Standards Organization (ISO) has published a standard, EN 13432:2000, which details how a material must behave if it is considered biodegradable or compostable. The standard includes a test to measure actual metabolic conversion of the material into carbon dioxide by microbes. The test standard is EN 14046 (also published as EN 14855 biodegradability under composting conditions) requires 90% conversion of the material in less than six months. The standard also requires that the material lose its physical identity and contain no heavy metals or materials harmful to the environment. At the end of the process, the material is passed through a 2-mm sieve. Any residue greater than 2 mm is considered to have failed, and the total amount of material greater than 2 mm cannot exceed 10% of the total. Other conditions that must be maintained in the control compost are pH, salinity, and amount of nitrogen (N), potassium (K), phosphorus (P), and magnesium (Mg).

By substituting biodegradable packaging materials for traditional packaging material to this mix is a significant improvement in the ultimate disposal of waste because so little needs to be buried.

In Europe, composting is much more prevalent than in the United States. France and Germany both have separate collection systems designed to pick up and process what some call "wet organics," which is a common part of the waste stream. Biodegradable materials are placed in this waste stream and avoid the high taxes and fees that are part of the Green Dot program for disposing of plastic waste. Europe will see continued growth in composting of organic wet food wastes in the coming years as they limit the amount of biodegradable waste permitted in normal landfills. The fact that these plastics are not subject to the Green Dot fees highlights an interesting possibility. The total cost of the material through the complete production, use, and disposal system may actually be less than that through the standard packaging. It would permit the use of a higher-priced packaging material when the total system economics and costs are considered.

Biodegradable Materials

There are a number of materials that are biodegradable, and a short review of them is in order. These materials while being presented in a packaging context also offer some interesting possibilities in terms of resorbable materials, that is, materials that can be placed in the body and then absorbed over time. Examples of places to use materials like this would be in promoting bone growth at the point of a tooth extraction or promoting bone growth over a large area.

Biodegradable polymers fall into a number of categories, polyesters, starch-based materials, lactic acid polymers, polyvinyl alcohol materials, and protein-based polymers (18). Examples of protein-based polymers would be polysaccharides and wood-derived plastics. The polyvinyl alcohol materials are produced in commercial quantities, and their properties, primarily their water solubility, are covered in chapter 7 on materials.

Starch-Based Plastics

Starch is a linear polymer made up of repeating (polysaccharide) glucose units using glucosidic linkages in the 1and 4 carbon positions on the molecule. These polymers typically have 500 to 2000 glucose units in their composition. Amylose and amylopectan are the two main molecules of a starch. The α-linkage in the amylase starch allows the polymer to be flexible and digestible. Starch-based polymers are derived from corn, wheat, and potatoes.

A polymer can be produced that mimics a standard thermoplastic material by selecting and mixing a number of starch feedstocks, water, and a plasticizer. The starch content of the polymer must exceed 60% before breakdown can occur. As the starch content is increased, the polymer becomes more and more biodegradable, but begins to lose properties. Blending starch-based polymers with aliphatic polyesters or polyvinyl alcohol, which is also biodegradable, produces polymers with the necessary properties for some applications.

Biodegradation of starch-based polymers occurs when enzymatic attack from microbes breaks down the glucosidic linkages between the sugar groups. This produces a breakdown of the polymer into monosaccharides, disaccharides, and oligosaccharides that are digested in metabolic pathways. If less than 60% starch is used, the polymer breaks down into fragments, and the entire polymer does not degrade beyond this point. Microbes can also break down the typical plasticizer, glycerol, which is used to produce the starch polymer. These polymers require a plasticizer to reduce brittleness and improve flexibility.

There are four types of starch-based polymers:

- Thermoplastic starch products
- Starch synthetic aliphatic polyester blends
- Starch polybutylene succinate (PBS) or polybutylene succinate adipate (PBSA) polyester blends
- Starch PVOH blends

These materials are almost all starch or they are blended as mentioned earlier with other biodegradable materials listed above. The materials are water soluble along with being biodegradable. These materials are targeted as a replacement for polystyrene foam, in cushions, and as the bulk fill or cushioning "peanuts" one finds in a corrugated box to protect a fragile item in shipment.

Lactic Acid Polymers

Lactic acid polymers received a lot of attention when Wal-Mart highlighted their use as part of their sustainable packaging initiative. These polymers, which are clear and look like PET, are used to replace polystyrene and other materials in clamshells and films. The polymer is used in clamshells for fresh produce at Wal-Mart. Wal-Mart also switched from using PET to a sandwich of paperboard and PLA for packaging their high-end electronic products. Bottles are now being produced from this polymer for water. There is a high degree of resistance to introducing this into the recycle stream because it is mistaken for PET and, when recycled, causes severe problems in the PET-recycling process.

Lactic acid polymers have been around for a long time, but they have always had a problem of high cost of manufacture. Wallace Carothers at DuPont® produced low-molecular-weight versions of the material in 1932, and

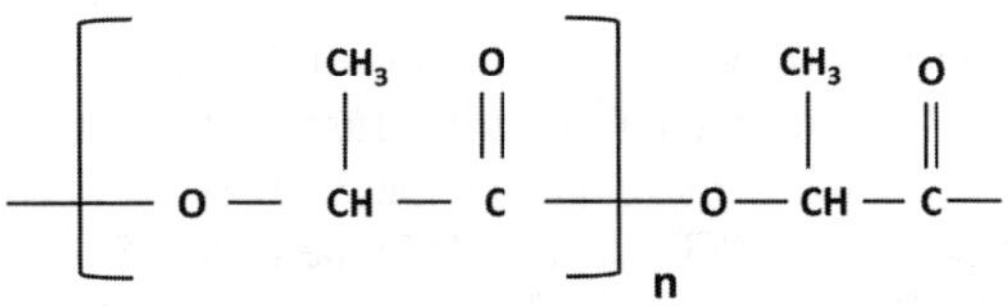

Figure 14 Polylactic acid.

DuPont patented his process in 1954. More recently, Cargill® has made an investment and commitment to produce polylactic acid and lactic acid polymers (Fig. 14).

The process starts with dextrose, which is fermented to make lactic acid. A condensation then turns two of the lactic acid molecules into a cyclic lactide molecule. The lactide is first purified by vacuum distillation and then subjected to a solvent-free melt polymerization that causes the lactide rings to open and attach to each other end to end, forming the polymer.

This polymer, which is compostable, requires more heat than is found in one's normal compost pile to degrade. Temperatures in the 140°F range for a number of days are used in commercial composting operations to degrade PLA back to lactic acid. It has a high demand for oxygen in the composting process, and this may be problematic for some operations. It also slows other biologic reactions when it becomes lactic acid and reduces the pH of the composting material.

Cargill along with NatureWorks®, located in Blair, Nebraska, is the leading producer and promoter of PLA polymers and has commercialized the material on a limited scale. The problem for expanding this market is the performance properties of the polymer (19). Temperatures above 114°F cause it to soften and melt. This limits its use to all but the most benign conditions for food and beverages.

Polyesters

Although celluloid and celluloid-based plastics were the earliest of the biodegradable plastics, the first biodegradable plastic with no celluloid component was polyhydroxybutyrate/valerate (PHBV). This plastic has properties that mimic polypropylene, and it is produced by a bacteria. Imperial Chemical Industries (ICI) of Great Britain, which called this product Bipol, originally developed the plastic. It suffered from the same problems as most bacterially produced products, high costs and marginal performance properties, when compared with synthetic materials produced from hydrocarbons. The product was dropped during the 1990s, but has recently been revived by a company called Metabolix®, and PHBV has been produced on a commercial scale in a fermentation plant. One attraction of this technology is that a surplus of fermentation capacity that was originally designed to make animal feed (e.g., lysine) and food additives like monosodium glutamate (MSG) is available in the United States.

Technically biodegradable aliphatic copolyesters made from bacterial fermentation are called polyhydroxyalkanoates or PHAs. The polymer is produced in the bodies of bacteria fed with glucose. According to the literature, hundreds of these polyesters have been identified and some have been produced with genetically altered bacteria.

A number of synthetic biodegradable polyesters are also beginning to appear on world markets. The synthetic polyesters are produced in modified PET polymerization facilities from standard petrochemical monomers. These materials differ from normal polyesters by exhibiting the ability to break down into carbon dioxide and water quickly when exposed to water and microbes in commercial composting facilities (20). The products meet U.S., European, and Japanese standards for composting materials suitable for introduction into MSW composting facilities, with a breakdown rate of approximately 12 weeks in the commercial conditions.

The materials produced in synthetic synthesis fall into two broad categories in terms of physical properties. One group is a set of amorphous polyesters that display good flexibility and clarity somewhat comparable to low density polyethylene (LDPE). A second group of more highly specialized polymers are semicrystalline and display properties more like that of PET, PP, or PS. All these materials suffer from a severe cost disadvantage when compared with PET, PP, PS, or LDPE. This fact probably more than any other prevents broader market penetration and use.

The promise of polyesters as biodegradable materials is large; the problem will be to evaluate their total costs in a way that puts them on par in a system (cradle to grave) comparison with standard polymers. Unfortunately, most companies only look at the cost to manufacture a product and do not have ways to adequately capture and assess the true system cost of a polymer when the environmental dimension is added. Possibly the Duales system, which charges manufactures directly for the waste they produce, will bring about a slow change in the way this is assessed. Up until now, the source reduction, the reduction in weight of packaging material used, or the elimination of other superfluous packaging has mitigated the need for this type of thinking. As these obvious improvements are exhausted, the broader systems' approach view may gain acceptance.

Other Biodegradable Polymers

There are a number of other potential biodegradable polymer options that may have application to packaging. These include wood-based polymers that are not cellulosic and protein-based plastics. One very large question remains regarding plastics produced by bacteria in fermentation reactions. What are the implications of genetically modified bacteria used to produce plastics or other materials when introduced into the food and pharmaceutical packaging chain? This question has been answered to some extent by the region, and the answers show surprising divergence between the regions.

Europe has banned genetically modified crops, and European consumers have shown a strong aversion to anything with genetic modifications. In the case of grain crops, this has caused problems for countries wishing to export to the EU, and it remains a question to be answered for "nonnatural" materials. Part of the questions will revolve around the composition and performance of the materials produced; however, if the materials produced by genetically altered bacteria do not have a naturally occurring analogue, they may face a difficult road to acceptance.

The United States on the other hand has not had the same aversion to genetically modified foodstuffs. Genetically modified grain that is resistant to pests, drought, and chemicals has been used in animal feed, and although not listed or marked directly on products, has not created the same aversions and fear as seen in Europe. It follows that materials produced by genetic modifications to bacteria will be accepted in the same way.

Naturally Occurring Biodegradable Polymers

There are four types of polymers that occur naturally. These polymers fall into four broad groups:

- Polysaccharides
  - Starch
  - Cellulose
- Proteins
  - Gelatin
  - Casein
  - Silk
  - Wool
- Polyesters
  - Polyhydroxyalkanoates
- Others
  - Lignin
  - Shellac
  - Natural rubber

As you can see from this list, there are a number of ways to produce biopolymers that are "natural" in terms of polymers produced from a number of common sources. One problem with these materials is the way in which they are produced. None of these materials are produced by a biologic entity that creates a large amount of material quickly. Rubber plantations are very large because the amount of material they produce on a per tree basis is relatively small. The same is true for the number of sheep used to produce wool. Modern synthetic fibers are produced at a much faster rate and on a much larger scale than wool. Again, the efficiency of the animal process is low, considering the amount of time and the amount of feed required to grow a wool coat on an animal.

Production of these natural polymers with modern production techniques poses its own problems. Using genetically modified bacteria or other genetically modified materials carries many uncertainties, not the least of which is how the public would react to materials coming from these sources.

The biopolymers that are on the market, polylactic acid, polysaccharides, and others are the only materials that anyone will work with in the foreseeable future. These materials have performance shortcomings and cost disadvantages. Improvements in both these areas will be the only change that will spur their use. It will be awhile before many of these materials make it into accepted and widespread commercial packaging and pharmaceutical packaging use.

OTHER PHARMACEUTICAL PACKAGING ISSUES

This chapter has tried to highlight some of the issues facing packaging that are apparent to all those involved in the industry. Many others deserve listing because they also could become issues in the near future.

The cost of health care will remain an issue. This fact will focus a spotlight on every aspect of pharmaceutical packaging, manufacturing, and supply chain. It may mean that changes in testing requirements, validations, and other material qualifications will be streamlined. The cost to qualify or update existing packaging to take advantage of new materials and new manufacturing technologies is another issue that has no simple or easy solution. Once the packaging for a product is established during development and stability testing, there is little incentive and, rarely, an economic incentive to reconfigure to a newer lower-cost material. This is the exact opposite of food and beverage packaging. It is better understood in the context of relatively small lot quantities or volume of any single pharmaceutical product compared with that of a food or a beverage product and the amount of resources needed to complete testing and submit an abbreviated new drug application (ANDA) for a product. Savings that take years to produce are not as attractive to pursue compared with developing the next new product. The same resources are needed for both, and the reward of a new product always outweighs the cost savings available with a packaging modification.

Reimbursement for pharmaceuticals will also remain a key question. As highlighted in this chapter, improved compliance may mean it is easier for the drug to be listed on formularies or pharmacy qualification lists of the government and health care providers (21).

Another issue is going to be directed to patient shipment of products. Regardless of whether the shipper is a qualified wholesaler, repackager, or pharmacy, the demand for convenience, particularly having a product delivered directly to the home, is going to grow. This demand means that more packaging will be required to ship product to the patient because bundled product on a pallet, or multiple individual cases from a qualified distribution point, will always use less secondary or tertiary packaging than packaging needed to individually wrap and protect a product for small parcel shipment.

People on fixed incomes and the elderly perceive pharmaceuticals as a problem because these are one of the last things they still must purchase with their own money. Even for those in health plans with prescription drug benefits, the costs are almost always the largest out-of-pocket expense item they have for their health care. This means they, and all the processes and materials, will continue to be a cost containment possibility. The packaging for pharmaceutical products is required to help alleviate this issue.

Equipment and processes for pharmaceuticals have not undergone change in a long time. Bottles, blisters, pouches, and other packages that contain the products are things that have not seen significant equipment fabrication improvement in a long time. The basic designs of feeders, fillers, and package manufacturing equipment for pharmaceuticals is another area that can be improved by someone willing to invest in the design and development of equipment that can produce packages that are new, not variations of the same old theme.

Finally, materials must be mentioned as an issue. The materials used in all parts of pharmaceutical packaging have not changed in a long time. Minor modifications have occurred, but no new packaging resins or coating technologies for plastics have been accepted by the pharmaceutical industry. PET was the last material to displace another and achieve a large role in pharmaceutical packaging. Ultra-high barrier materials, materials that provide oxygen scavenging or other capabilities to improve efficacy and improve long-term shelf life, are still needed. Materials that combine low-cost and extreme barrier performance can and must provide improved protection for products that are extremely moisture sensitive. Materials that can provide protection for temperature-sensitive drug products and vaccines and materials that are better barriers for gases are an issue because the current materials must be stretched and modified considerably to produce the barrier required by many new biopharmaceutical products.

As one can see, there are plenty of issues facing pharmaceutical packaging. They range from materials to delivery methods. They require packaging improvements and the development of new packaging. They are problems that will take years to solve. Hopefully, innovations in materials, processes, and delivery methods will emerge quickly and be rapidly introduced to the marketplace.

REFERENCES

1. Bernstein, Elisa BG. FDA's counterfeit drug initiative. Healthcare Compliance Packaging Council Symposium on Compliance; Baltimore, MD; May 15, 2007.
2. United States Environmental Protection Agency. EPA Paper on Germany's Green Dot System, 2007. Available at: www.epa.gov/oppt/epp/pubs/envlab/greendot.pdf.
3. Commonwealth of Massachusetts, Group Insurance Commission. Keeping Up with Medication Dosage and Frequency is Vital to Your Health. For your Benefit Newsletter; 2005. Available at: http://www.mass.gov/gic/healthartdrugcompliance.htm.
4. BCG/Focus. The Hidden Epidemic: Finding a Cure for Unfilled Prescriptions and Missed Doses. The Boston Consulting Group, 2003, Harris Interactive www.bcg.com. Available at: http://bcgtest.bcg.com/impact_expertise/publications/files/Hidden_Epidemic_Finding_Cure_Unfulfilled_Rx_Missed_Doses_Dec2003.pdf.
5. Mohan AM. Electronic compliance packaging enhances drug efficacy. Packaging Digest August, 2005. Available at: http://www.packagingdigest.com/articles/200508/26.php.
6. Lee J, Grace K, Taylor A. Effect of a pharmacy care program on medication adherence and persistence, blood pressure, and low density lipoprotein cholesterol: a randomized controlled trial. JAMA 2006; 296(21): 2563–2571. Available at: http://jama.ama-assn.org/cgi/reprint/296/21/2563.
7. Gren J. International trade issues related to counterfeit medicines and the need for Global Cooperation. U.S. Department of Commerce, Healthcare Compliance Packaging Council Symposium on Compliance; Baltimore, MD; May 15, 2007.
8. Royte E. Corn plastic to the rescue. Smithsonian Magazine, August 2006. Available at: http://smithsonianmagazine.com/issues/2006/august/pla.htm.
9. United States Environmental Protection Agency. Municipal Solid Waste in the United States 2005. Facts and Figures. Office of Solid Waste. EPA530-R-06-011, October 2006.
10. U.S. Environmental Protection Agency. Municipal and Industrial Solid Waste Division. Municipal solid waste in the United States 2005. Facts and Figures. Office of Solid Waste. Executive Summary. October 18, 2006. Available at: www.epa.gov/osw.
11. Econominst. The print edition. The truth about recycling. Available at: http://www.economist.com/science/tq/displaystory.cfm?story_id=9249262. Accessed July 2007; published June 7, 2007.
12. Steven Boussemaere. Appliance of the Packaging and Packaging Waste Directive in Europe through the Green Dot system—presentation of the situation in Belgium. Belgrade, Serbia; Fost Plus, April 2007.
13. How to Germany. All about recycling. Expatriates Newsletter and the American Women's Club of Cologne 2007. Available at: http://howtogermany.com/pages/recycling.html 2007.
14. American Plastics Council. The Association of Postconsumer Plastic Recyclers, 2005. National Post-Consumer Plastics Bottle Recycling Report. Available at: www.americanchemistry.com.
15. Community College of Rhode Island. Information on regulated medical waste disposal. Available at: http://www.ccri.edu/Safety/regulated_medical_waste_disposal_policy.shtml, accessed July 2007.

16. Princeton University. Health and Safety, Biological and Medical Waste Disposal. Available at: http://web.princeton.edu/sites/ehs/biosafety/biowaste.htm. Accessed July 2007.
17. Nolan—ITU PTY Ltd. In association with ExcelPlas Australia. Biodegradable plastics—developments and environmental impacts October 2002 Available at: http://www.environment.gov.au/settlements/publications/waste/degradables/biodegradable/chapter2.html.
18. EPIC, Environment and Plastics Industry Council. Biodegradable polymers—a reviewNovember 2000.
19. Balcom M, Welt B, Berger K. Notes from the packaging laboratory polylactic acid: an exciting new packaging material, December 2002, Available at: http://edis.ifas.ufl.edu/BODY_AE210.
20. Leaversuch R, ed. Biodegradable polyesters: packaging goes green. Plastics Technology Online, September 2002. Available at: http://www.ptonline.com/articles/200209fa3.html.
21. Mohan AM. Electronic compliance packaging enhances drug efficacy. Packaging Digest 2005. Available at: http://www.packagingdigest.com/articles/200508/26.php.

Glossary of Terms

ABS: Acrylonitrile-butadiene-styrene polymer.

absorption: The assimilation of one substance by another, where the substance being absorbed diffuses into the absorbing material.

acid: A chemical compound containing a hydrogen atom that dissociates from a molecule to a hydronium ion in water.

acid etching: Glassware is immersed in an acid bath to create a smooth, frosted effect on the glass surface.

ACL: Applied color lettering. Colored lettering or design of ceramic nature permanently fused onto glass bottle surface.

ACL lug: A small protrusion or small depressed cavity in a base of a bottle to act as a guide in positioning the bottle in the decorating machine for application of ACL.

ACLAR®: The Honeywell trade name and registered trademark for polymerized chlorotrifluoroethylene.

adsorption: The attachment of one substance onto the surface of another by means of a strong interaction between the two substances or materials. This differs from absorption in that the substances are joined only at the surface.

aerosol: A dispersion of very small (submicron)-sized liquid droplets or solid particles into a gas. The term is used in packaging as a label for all liquid or semisolid solutions or suspensions dispensed under pressure.

amorphous: Not crystalline when used with plastics, it means molecules arranged in a random order, without structure.

ampule: Glass tubing sealed at both ends, containing a drug intended for injection.

analgesic: A substance that relieves pain.

ANDA: Abbreviated New Drug Application.

anesthetic: A drug that stops or suppresses sensations such as pain by affecting either the central nervous system (general anesthetic) or the peripheral nerve structures (local anesthetic).

angina: Any inflammatory disease of the throat or fauces. Often used as synonymous with angina pectoris, a heart disease characterized by intermittent chest pain coupled with feelings of suffocation.

angstrom: A unit of length equal to one hundred millionth of a centimeter.

annealing: A controlled and programmed temperature method of gradually cooling glass containers in oven or lehrs to relieve structural stresses.

annealing: A controlled temperature method of gradually cooling glass containers in ovens or lehrs to relieve structural stresses and to make the glass less brittle.

ANSI: American National Standards Institute.

antibiotics: Substances that inhibit the growth of or destroy microorganisms.

antihistamines: Substances that neutralize or inhibit the effects of histamine released by the body during allergic reactions or in response to a disease.

anti-inflammatories: Substances that neutralize or inhibit the inflammation of tissue.

antimicrobial: Refers to substances that destroy microorganisms.

antioxidant: A chemical substance that can be added to a plastic resin to minimize or prevent the effects of oxygen attack on the plastic (e.g., yellowing or degradation).

antioxidants: Compounds that react with oxygen. When blended into an oxidizable substance, antioxidants will react preferentially with oxygen and prevent oxidation of the substance.

antipruitic: A substance that relieves itching.

antiseptic: A substance that inhibits the growth of microorganisms.

antistatic agent: A chemical substance that can be applied to the surface of a plastic bottle or incorporated in the plastic from which the bottle is made. Its function is to render the surface of the plastic article less susceptible to an accumulation of electrostatic charges, which attract and hold fine dust on the surface of the bottle.

API: Active pharmaceutical ingredient—the active ingredient in a pharmaceutical product.

aseptic: Free from disease-producing microorganisms. In biologic or medical applications, it refers to an operation performed in a presterilized environment that is controlled to prevent contamination through the introduction of microorganisms. Sterile area that is controlled to remain sterile during operation.

aseptic filling: The process of combining sterilized pharmaceuticals with sterile packaged in a sterile environment.

assay: The determination of the concentration of the active ingredient in a pharmaceutical.

ASTM: American Society for Testing and Materials.

astringent: A substance that contracts tissues or canals, reducing the discharge of fluids.

atmosphere: Commonly used as a unit of pressure, referring to the air pressure at sea level. One atmosphere is equal to 14.7 lbs/ft^2.

autoclave: A vessel capable of containing high-pressure steam that is commonly used to sterilize pharmaceuticals, medical instruments, and medical devices.

B_2O_3: Boric acid.

bactericide: A substance that kills bacteria.

bacteriostat: A substance that is added to a drug formulation to control the growth of bacteria.

bacteriostatic: Inhibiting or retarding the multiplication of bacteria.

barrier: Used in packaging to denote resistance to the permeation of gasses.

base: A chemical compound that when dissolved in water dissociates to form a hydroxyl ion and raises pH above 7. Usually a metal oxide or hydroxide.

base box (tinplate): Defined as 112 sheets of tinplate 14 $\times$ 20 in in size or 31,360 in^2 of area on each side of the sheets.

basis weight: The weight in pounds of a ream (500) sheets of paper cut to a given standard size for that grade, e.g., 500 sheets 25 × 38 lb. Coated book paper weight 80 lbs.

BATF: Bureau of Alcohol, Tobacco, and Firearms.

beach puncture: The corner of a cube, mounted on a pendulum, is swung through the material. The upward swing compared with the downward swing gives the energy absorbed. ASTM D781.

bioavailability: The availability of an administered drug to the circulatory system.

BLA: Biologics license application (FDA–CBER).

bioburden: Population of viable microorganisms on a product. In the context of sterilization, it should be determined immediately before sterilization. Bioburden may be described for an item in its entirety, including external surfaces, or for fluid contact surfaces only.

bioprocess components: Single-use disposable parts, modules, or sections of a system, which may include but not be limited to filters, membrane chromatography units, tubing, connectors, fittings, flexible bags or rigid containers, and probes.

bioprocess system: An assemblage or combination of bioprocess components intended to facilitate the processing of pharmaceutical or biological solutions in a single-use disposable format. The system may be preassembled before a bioburden reduction/microbial control step or assembled by a user by connecting presterilized subassemblies through the use of aseptic connections.

blank: The mold parts used in all glass container machines for preliminary formation of glass in preparation for completion of the glass containers in the finish mold where the bottles are blown. The blank forms the parison; hence the parison itself is at times referred to as the blank.

blister package: A package consisting of a cavity thermoformed from a thermoplastic material and a flat lid stock designed to seal each of the cavities to the edges of the trimmed card.

blocking: The lowest temperature at which a material adheres to itself under pressure. ASTM D918.

blow molding: Blow molding regardless of the specific process refers to the production (extrusion of plastic) of a parison (preform), enclosing the hot parison or preform in a chilled or cooled female mold, and inflation with air of the parison in the female mold to expand the hot semi-molten plastic to conform to the shape of the mold producing a finished part or container.

blushing: A haze seen in coatings or plastics from the absorption of water.

boil-in: A test for boil-in-bag structures. One-inch strips are sealed across, and a weight puts peel stress on the bond in wet steam at 250°C for 30 minutes.

bottom plate (glass): The part of the mold equipment that forms the bottom of the bottle.

brick pack (Brik Pak®): A composite container formed from a laminate of paper, plastic, and foil, which produces a rectangular box-shaped container.

brightness: Reflectance of a surface compared with magnesium oxide and reported as a percentage. Thus 90 brightness is 90% of the reflectance of magnesium oxide. ASTM D985.

BTU: British thermal unit.

buccal cavity: The cavity formed by the cheek.

buffer or buffered: A buffer is a substance that when dissolved in water acts to resist changed in pH that would otherwise be caused by environmental factors (e.g., CO_2 in air or alkaline salts in glass containers).

buttress thread: A design of thread profile (cross-section), which takes the form of a right triangle or slight modification of that form. It is usually positioned so that the right angle is at the bottom of the thread cross-section and the adjacent to the neck of the bottle finish. The horizontal leg of the right triangle is the bearing surface for a matching cap thread.

CAD/CAM: Computer-assisted design/computer-assisted manufacturing (or in printing computer-assisted makeup).

CaO: Calcium oxide.

capacity: The amount of space inside a container provided for a given amount of product.

caplet: A tablet-shaped capsule.

capsule: A transparent or colored gelatin material, hard- or soft-shelled that contains a drug preparation (API + excipients).

carboxyl group: A group with the chemical structure —COOH.

carboy: A large, usually plastic, container used to transport acids and other chemicals.

catalyst: A chemical compound that accelerates the rate of a chemical reaction without being consumed in the process.

caustic: A strong base example sodium hydroxide.

CBER: FDA's Center for Biologic Evaluation and Research.

CDA: Clinical data architecture, FDA's system for data exchange.

CDER: FDA's Center for Drug Evaluation and Research.

CDRH: FDA's Center for Devices and Radiological Health, which regulates firms that manufacture, repackage, relabel, or import medical devices sold in the United States.

CFR: Code of Federal Regulations.

CGMP: Current Good Manufacturing Practice (as written and used outside the FDA).

chalking: For printing, a term that refers to the improper drying of ink. Pigment dusts off because the binder (vehicle) has been absorbed by the substrate too rapidly.

chipboard: A low-grade board made from waste paper.

clarity: Ratio of light transmitted through the specimen to light transmitted through air, times 100. ASTM D1746.

class 100, class 1000, class 10,000: Refers to work areas with the quality of air is specified by the FDA. The specification states for example no more than 100 particles 0.5 μm in size are larger may be present in one cubic foot of air in a class 100–designated work area.

class I recall: A recall of a drug or medical device in which there is a possibility that death could result if the product was not recalled.

CMC: Chemicals, Manufacturing, and Controls section of a new drug application to the FDA.

coefficient of thermal expansion: A dimensionless number that expresses the degree to which a material will expand when subjected to a known and specified increase in temperature.

coextrude: Simultaneous extrusion of more than one plastic material in a multiple layer or coextruded structure.

colitis: Inflammation of the colon.

color: Value, hue, and chroma are compared with a standard in north light. Gloss transparency and opacity are also noted. Tristimulus values are measured by instruments with filters that pass on X (*red*), Y (*green*), and Z (*blue*) elements. Metamerism is the match of color under one light and a mismatch of the same color when viewed under another light. Refer to the Munsell color solid.

comonomer: One component of a copolymer, chemically distinct from the other comonomers.

compliance: Adherence by a patient to the drug regimen prescribed by the doctor.

controlled substance: A drug that is subject to the Controlled Substances Act of 1970, which sets limitations on the prescription and requirements for storage and record keeping depending on the potential for physical or psychological dependence.

copolymer: A polymer made from at least two different comonomers.

corona treatment: Subjecting a polymer surface to an electrical discharge to alter its surface characteristics. Usually used to prepare the plastic surface for direct printing.

cosmetic: Formulate products used to decorate, adorn, or beautify but which have no therapeutic effect or purpose.

count: As in 50-count bottle, the number of items (e.g., tablets, capsules) in the container. Also used to specify that the container can hold this number of doses.

CPSC: Consumer Product Safety Commission.

cream: A medicated preparation based on an emulsion of oil in water.

creep: Also called cold flow or deformation under load. Plastic is subjected to steady tension or compression at a constant temperature. Movement is plotted against time, and the slope of the curve is reported at the beginning, in the middle, and near failure.

cross-link: To join separate polymer chains by means of short molecular fragments that chemically bond to each other (e.g., crosslinked or "cured" thermoset polymer materials).

CT: Continuous thread.

CT finish: Continuous-thread finish. An interrupted protruding helix on the neck of a container to accommodate a screw-type closure.

cul-de-sac: A sac-like cavity such as that formed between the eye and the eyelid.

DAM: Digital asset management.

deadfold: When a film or any web does not spring back after being folded, it is said to have "good deadfold." Aluminum foils have good deadfold characteristics compared with plastics with poor deadfold characteristics. The plastics have a tendency to return to their original shape prior to folding.

deliquescence: Refers to a substance that readily absorbs moisture. Becoming damp or liquid by absorption of water from the atmosphere and then dissolving in the water taken up. This property is found in salts (e.g., $CaCl_2$).

demulcent: A substance formulated to soothe the part of the body to which it is applied.

densitometer: For printing, a reflection densitometer is used to measure and control the density of color inks on a substrate.

density: Weight per unit of volume of a substance.

depyrogenation: The elimination of pyrogens by heat or chemical processes.

dermatological: Relating to the skin or diseases of the skin.

diacid: An organic acid containing two carboxyl groups.

dialysis: The process of separating mixed substances in solution by means of a membrane that is permeable only to some of the components and not to others.

die: Any tool or arrangement of tools designed to cut, shape, or otherwise form materials to a desired configuration.

dielectric heating: Creating heat in nonconductive substances such as plastics by subjecting them to a high-frequency electrical field, which causes rapid molecular vibration.

diester: An organic molecule containing two ester groups.

dioxin: A complex organic compound that can form during incineration of organic materials that contain chlorine atoms. Toxic to animals and a cause of chloracne in humans.

dispersion coating: A coating process in which the coating material is suspended in the coating bath, as contrasted to solvent coating in which the coating agent is dissolved in the bath. Nonaqueous dispersions of PVC are a common major ingredient of metal can coatings.

distillation: The process in which a liquid is purified by transforming it into a vapor, separating the vapor from the impure liquid and then condensing and collecting it.

diuretic: A drug formulation that increases urinary discharge.

DMF: Drug Master File, a blinded repository for proprietary information that permits the FDA to review the safety and adequacy of a component.

dosage form: The form of a drug preparation that determines how the drug is administered (e.g., tablet, oral liquid, suppository, parenteral liquid).

dosimetric release: The determination that a product is sterile based on physical irradiation process data rather than sterility testing.

durometer: Commonly referred to as hardness or softness measure of an elastomer or rubber. For a durometer test, a special point that extends below a pressure foot is pressed against rubber or other resilient material being tested, which is at least one-fourth inch thick. Penetration of the spring-loaded indenter is read on a dial with a scale from 0 to 100. ASTM D1706.

EAA: Ethylene acrylic acid.

ECM: Enterprise content management.

effervescent: Refers to substances that produce a gas, usually CO_2, upon mixing with water.

efficacy: The term used to describe the effectiveness of a drug in treating the disease or condition it is designed to treat.

efflorescent: Refers to substances that lose water on exposure to air.

EFPIA: European Federation of Pharmaceutical Industries and Associations.

EHIBCC: Health Industry Business Communication Council (HIBCC) and its affiliate International Organization, the European Health Industry Business Communications Council.

elastomer: A polymer with the elastic characteristics similar to rubber: the ability to be stretched to at least twice its original length without sustaining permanent deformation.

elixirs: Syrups containing 20% to 25% alcohol.

elongation: The maximum stretch in a film that will still return to its original size as a percentage of the original length.

EMEA: The European Agency for the Evaluation of Medicinal Products.

emollients: Substances that soften and relax the tissues when applied locally.

emulsion: A liquid consisting of a discontinuous, immiscible liquid phase dispersed in a continuous liquid phase.

enteric: Refers to coatings that delay dissolution until a solid dosage form reaches the intestine.

EPDM: Elastomers based on terpolymers of ethylene, propylene, and a diene monomer such as 1,4-hexadiene.

epidermis: The first inner layer of the skin.

ester: The reaction product of an acid and an alcohol that contains a —COO—group.

EVOH: Ethylene vinyl alcohol, a copolymer of ethylene and vinyl alcohol.

extraction: This test method is used to test plastics and coatings for leachable or soluble elements that can be removed from a standard area. It is done with both water and solvent. For example, 100 in^2 of plastic is exposed to 200 mL of n-heptane at processing temperature. If more than 5 mg is extracted, it is reported in milligrams per square inch. ASTM F34.

FD&C: Food, Drug, and Cosmetic Act.

FDA: United States Food and Drug Administration.

FFDCA: Federal Food, Drug, and Cosmetic Act.

false-positive sterility test: Test result that is incorrectly interpreted to indicate that the test article contained viable organisms and that the article was not sterile. Commonly, such a result will derive from faulty sample handling or test execution—or from an interaction between a sample and the microbial growth medium that causes the medium to become turbid.

filtration: The process by which solid particles are removed from a liquid by passing the liquid through a porous medium whose pores are so small that the solid particles will not pass through them.

finish: As applied to bottles and closures, describes the thread design: the size, pitch, profile length, and thickness.

flash: Extra plastic attached to a molding along the parting line of the die. It must be removed for a part to be considered finished or in final form.

flint: A term used to describe a glass color that is perfectly clear and transparent.

fluidized bed: A group of solid particles in a container that are agitated by the upward-flowing stream of gas. The particles appear as a cloud in the bottom of the confining space.

fluorocarbon: An organic compound containing fluorine.

flux: A substance or mixture used to promote the fusion of metals or minerals. Can be called a fluxing agent.

free radical: A highly reactive species formed by the rupture of a chemical bond.

fusion seal: A seal created by melting the adjacent surfaces that are to be sealed together.

g/cc: Grams per cubic centimeter.

gang printing: The practice of printing groups of several different labels consecutive on a sing roll or sheet of label stock. This is strictly regulated by the FDA in label and carton production.

gas transmission: A measure of the amount of a gas passing through a membrane. It is expressed as volume in milliliters per 100 in^2 per 24 hour at 1 atm pressure and 75°F, passing through a film 1 mil (0.001 in) thick. ASTM E96.

gastrointestinal: The system of body organs that included the stomach and small intestine.

gel: A colloidal semisolid consisting of a networked structure of suspended, fine, solid particles surrounded by a liquid. Differs from a colloidal solution, which has no solid particles to confer some rigidity to the structure.

gelatin: A water soluble substance extracted from animal tissue and bones and used in the manufacturing of capsules.

generic: Used in the pharmaceutical business to describe any drug that is labeled for sale with its technical name rather than a trade name, and usually manufactured by companies that were not the original developer of the product.

glass: The USP on the basis of performance in chemical durability tests specifies four types of glass. Type 1, 2, and 3 are intended for packaging parenteral preparations and type NP for nonparenteral products.

glassine: A highly compacted grease-proof paper produced by calendaring at very high pressure in the presence of steam.

GMP: See cGMP.

GRAS: Generally recognized as safe. The FDA list of ingredients acceptable for use.

haze: A term used to characterize the milky appearance of a polymer film that is caused when light is scattered by surface imperfections or film inhomogeneities.

HDPE: High-density polyethylene.

head space: The space between the level of the contents of a container and the closure of the container.

hermetic seal: Any seal or any container so sealed that is impervious to all gasses under normal conditions of handling and storage.

HF: Hydrofluoric acid.

HIBCC: Health Industry Business Communication Council and its affiliate international organization, the European Health Industry Business Communications Council (EHIBCC).

HIPS: High-impact polystyrene.

HL7: Health level seven.

hologram: The image formed by a lensless photographic process (holography) that uses laser light to produce three-dimensional images.

hormone: A substance formed in and secreted by the endocrine glands. May be made synthetically.

hydrocarbon: An organic compound containing only carbon and hydrogen.

hydrolysis: Reaction of a compound with water, resulting in destruction of the compound and the formation of at least two new ones.

hypertonic: Having a greater osmotic pressure than blood plasma, lacrimal fluid, or interstitial fluid. It can be applied more specifically to a fluid in which cells shrink.

hypoglycemia: An abnormally small concentration of glucose in the circulating blood.

in vitro: Refers to chemical or physical tests of drugs using laboratory procedures and apparatus (in glass).

in vivo: Refers to tests of drugs in laboratory tissue, animals, and humans (in life).

IND: Investigational New Drug. First review of drug information by the FDA prior to initiating clinical trials.

induction heating: Heating a metal object by application of an external magnetic field to generate heat-producing eddy currents in the object.

infusion: Introduction of a fluid other than blood into a vein (e.g., a saline solution drip).

inhalant: A substance that can be vaporized by mechanical means or by heat and carried into the respiratory tract by inhalation.

innerseal: A membrane sealed across the top of a container to act as a barrier and as a tamper-evident device.

intermediate: Used in manufacturing chemistry to denote a compound use in the manufacture of another. A precursor to another compound.

intracisternal: Introduction of a cannula into the cisterna cerebellomedullaris for aspiration of cerebrospinal fluid or injection of air into the ventricles of the brain.

intravenous: Administration of a drug by injection directly into a vein.

ion: A charged atom or group of atoms formed by the dissociation of a molecule, often in an aqueous medium.

ionomer: A copolymer-containing acid groups, some of which are neutralized with metal ions such as sodium or zinc.

IQ: Installation qualification. This is a review of the equipment that establishes that the equipment meets its design specifications, and was installed in accordance with the design specifications. A term used in validation.

IR: Infrared.

irradiation: Application of ionizing radiation capable of destroying microbial bio-burden—a process for sterilization.

irrigation: Washing out a cavity or wound with fluid.

isotonic: Solutions that have the same osmotic pressure. A more specific definition is a solution in which cells neither swell nor shrink.

isotonicity: The situation obtained when the colligative or osmotic properties of a pharmaceutical are matched with those of a biological site of administration, frequently mucous membranes.

IV: Intravenous.

kaolin: A family of clays containing combinations of hydrated alumina and silica.

keratolytic: A medication used to treat conditions that lead to horny skin growths.

kpsi: 1000 psi.

labile: Free to move. In chemistry, used to characterize a molecule or group that moves within the molecular surroundings to link up with other molecules or with other species like itself.

lactose: A crystalline sweet, water-soluble disaccharide found in milk.

laminate: A multilayer web with layers tightly bonded to one another with adhesives or by fusion bonding.

latex: The milky juice of exudation of plants obtained by tapping the trunk (e.g., the fluid from a rubber plant).

LDPE: Low-density polyethylene.

light-resistant container: A container that protects the contents from the effects of light.

lipophilic: Having a strong affinity for oily or fatty substances.

LLDPE: Linear low-density polyethylene.

lot or lot number: A lot refers to all the products made during a single run or manufacturing sequence on a piece of equipment or a complete production line. A run may last for a given quantity, for hours, or days, it normally denotes all products made in one sequence of starting and stopping the equipment when all raw materials are consumed or a given quantity is produced. A lot number is the assigned designation of that specific manufacturing sequence.

lyophilization: Freeze-drying. The removal of water or solvent from a substance by applying a vacuum to the substance after it has been frozen.

m^2: Square meter.

magma: Highly thickened suspensions for oral administration.

mandrel: A metal rod or bar used as a core around which metal, glass, etc., is cast molded or shaped.

melt index: The amount of plastic in grams that can be forced through a 0.0825-in orifice in 10 minutes when heated and under pressure of 2160 g. Viscosity measures of a polymer melt. The higher the melt index, the lower the viscosity.

melt viscosity: The viscosity of a molten polymer, i.e., its resistance to flow.

metallized: Possessing a submicron-thickness coating of a metal that has been applied by a vacuum process.

mg: Milligram.

microbial control: Assembly of products in a controlled clean environment, followed by exposure to gamma radiation. This process reduces bioburden load but does not support a "sterile" label claim.

microencapsulated: The encasement of small particles, either solid or liquid, within a shell that prevents their escape until the shell is ruptured by an external force or dissolved by a solvent.

micron: One ten-thousandth of a centimeter.

microorganisms, microbes: Living microscopic entities including bacteria and molds.

migration: The movement of substances through the wall of a container or package. An undesirable attribute of a container that permits the contents to become adulterated, or to lose potency by the migration of components, possibly from the plastic or label adhesive into the product, or the diffusion of the highly volatile components out of the package through the walls of the container.

mil: One thousandth of an inch.

MnO_2: Manganese dioxide.

Modulus: A general expression used to describe tensile, elastic, or plastic modulus. Modulus is the slope of the line generated when stress is plotted against strain. A high modulus material is a stiff material.

moisture permeability: The transmission of moisture through a membrane, film, or package. For a package, the rate of this process is expressed as milligrams per day per liter of container volume. See **gas transmission**, also referred to as MVTR or WVTR.

mucous membrane: The lining of the mouth, throat, and the nasal and bronchial passages.

Mullen or Mullen test: Pressure in pounds per square inch required to force a rubber diaphragm through a round hole against a specimen of paper firmly clamped around the edges of the hole. Results are reported as Mullen units. ASTM D774. A machine used for testing the bursting strength of corrugate.

Multiple-dose container: A multiple-unit container for parenteral or ophthalmic formulations.

Multiple-unit container: One that permits the withdrawal of part of the contents while containing and protecting the un-withdrawn balance.

NDA: New Drug Application (FDA). Submission of all information necessary for review by the agency prior to approval of a new drug.

NDC: National Drug Code.

NF: National Formulary.

NWDA: The National Wholesale Druggists Association.

OFAS: Office of Food Additive Safety.

offset: The process of using an intermediate cylinder to transfer an image from the image center to the substrate.

ointment: A medicated preparation with the oleaginous base. More generally, a semi-solid preparation intended for topical administration.

oleaginous base: A base with the nature or quality of an oil.

ONPLDS: Office of Nutritional Products, Labeling, and Dietary Supplements.

ophthalmic: Related to the eye.

OPP: Oriented polypropylene.

optical isomers: Isomers that have the same empirical and structural formula but which differ in the location of substituent groups around a central carbon atom called a chiral carbon. In their pure form they rotate the plane of incident polarized light passed through them.

OQ: Operational qualification. This confirms that all equipment adjustable and variable parameters are operational within design limits. It also confirms that all alarms, alerts, and other functional features of the system operate within their functional specifications. A term used in validation.

organosol: Colloidal dispersion of an organic solid in an organic solvent also referred to as nonaqueous dispersions or NADs.

orient, oriented: The alignment of polymer molecules in a film or container sidewall (e.g., PET bottles are biaxially oriented).

OTC: Over the counter.

otic: Related to the ear.

outsert: Supplementary labeling information in the form of a printed leaflet adhered to the outside of a package with adhesive or held in place by an overwrap.

overseal: In packaging, any device applied over the primary closure to hold it firmly in place and/or supplement its gas barrier. It can also refer to a tamper-evident feature of a package.

oxidation: Reaction with oxygen, more generally removal of electrons from an atom or molecule.

parenteral: Introduction of substances into an organism by subcutaneous, intramuscular, intravenous, or intramedullary injection. Introduction by some other means than through the gastrointestinal tract.

parison: A partially formed bottle of glass or plastic that is subsequently formed by blowing after confinement in a mold into the finished shape.

partial pressure: When a mixture of gasses is contained in a vessel or package, the pressure exerted on the vessel walls by just one of the gaseous species is known as the partial pressure of that species. The sum of all the partial pressures is equal to the total pressure exerted by the gas mixture.

PbO_2: Lead dioxide.

PBOM: Packaging bill of material. The document that defines all the components of a package and its identification codes for manufacturing.

PE: Polyethylene.

peptide: Compound made up of two or more amino acids joined through an amide linkage.

perfluorinated: An organic compound that has all the hydrogen atoms replaced by fluorine atoms.

peridural: Upon the outside of the dura mater, a tough fibrous membrane forming the outer envelope of the brain.

permeation: The movement of a labile substance through a solid by diffusion. In packaging, used to describe the movement of gasses through the package walls.

PET: Polyethylene terephthalate.

petrolatum: A gelatinous or oily translucent substance obtained from petroleum.

pH: A measure of the hydrogen ion concentration in and the acidity of an aqueous solution.

pharmaceutical: A manufactured, processed, or compounded form of a drug.

PIM: Product information management.

plasticizer: A substance mixed into a plastic to decrease its stiffness and increase its softness.

plastisol: A plastic used in the form of an emulsion. See also **organosol**.

polyglycols: Polymers made from glycols, which are organic molecules containing at least two hydroxyl (OH) groups.

polymer: A high molecular weight molecule formed by reacting small molecules (monomers) together to form a long chain consisting of many monomer units.

polyolefin: Any polymer whose monomer units are unsaturated hydrocarbons (olefins) containing only carbon and hydrogen. Polyethylene and polypropylene are the most common polyolefins used in packaging.

PP: Polypropylene.

ppb: Parts per billion.

PQ: Performance qualification. This testing confirms that the equipment is functioning within its operator requirement specifications and that the operation within those limits is repeatable. A term used in validation.

primary package: The package that contains the product being packaged.

process colors: For printing, the subtractive primaries: yellow, magenta, and cyan plus black in four color process printing.

prophylaxis: Prevention of disease or its spread by the administration of drugs and/or procedures.

protein: Complex organic compounds containing amino acids essential to tissue growth and repair.

protocol: A set of procedures. Test or validation protocols are the set of instructions that govern how a test is run and how the data is to be reported.

PS: Polystyrene.

PTFE: Polytetrafluoroethylene.

pulpboard: See **chipboard**.

PVC: Polyvinyl chloride.

PVDC: Polyvinylidene chloride.

pyrogen: Agent that causes a rise in body temperature, especially if injected. The most important pyrogen in sterile drug manufacture is endotoxins, a residue from gram-negative bacteria.

racemization: The conversion of a dextro (d) or a levo (l) optical isomer into an equilibrium mixture of the two.

rectal mucosa: The outer layer of the lining of the rectum.

resin: The term for a polymer in the form of small pellets that is packaged in a bag or in bulk and shipped to a processor. Sometimes a direct reference to the polymer itself.

respiratory system: The body system that includes the mouth, nose, throat, bronchial passages, and lungs.

reverse osmosis: The process in which the solute in a solution is removed by forcing the solvent, against the normal osmotic pressure to flow through a membrane that is not permeable to the solute. Used to remove salts from seawater.

RH: Relative humidity.

roll stock: A web wound up on a roll and fed into a process such as printing, can making, labeling, or wrapping. May also be referred to as coil stock.

SAL: Sterility assurance level. In practice, the sterile state cannot be proven. Instead, sterility is expressed as the probability of a single viable microorganism occurring on an item after sterilization. Normally the term is expressed as 10–n and not as an absolute.

saturated: When used to describe a type of chemical bond or molecule, the bonding is saturated if no double or triple bonds exist, i.e., each atom is joined within the molecule to other atoms only by single bonds.

scabacide: A substance that destroys the organism causing scabies.

secondary package: The package that contains the primary package. It is not in direct contact with the product. Usually a box or carton.

semipermeable-membrane: A membrane that permits the passage of one or more components of a solution but does not allow the passage of other components. Such membranes are usually permeable only by the solvent.

shelf life: The time required for the potency of a drug to drop to 90% of its labeled potency.

silicone: Polymers composed of molecular chains containing alternating silicon and oxygen atoms. These can be liquids or solids depending on the molecular weight and the groups attached to the chain.

single-dose container: A single-unit container for primarily used for parenterals but also for other dosage forms that contains only one dose of product.

single-unit container: A container that holds the quantity of drug intended for administration as a single dose promptly after the container is opened.

SiO_2: Silicon dioxide (silica).

SIP: Sample item portion. Defined portion of a product assembly or component.

SO_2: Sulfur dioxide.

sol: A colloidal solution or liquid phase of a colloidal solution.

solvent coating: Coating a web (e.g., paper, plastic) with materials by passing the web through a solution of the coating materials and then evaporating the solvent from the layer of liquid that remains on the web. Can also be used to refer to a can coating or other coating that uses an organic solvent as the carrier and evaporative portion of the formulation.

sorb, sorbtion: In packaging, the removal of pharmaceutical components by a package or adsorption onto the package surface or absorption onto the package wall of the API or other therapeutic agent.

SPL: Structured product labeling.

stearate: A waxy salt of stearic acid derived from animal fat.

sterile, sterility: The absence of microorganisms.

sterility testing: Tests performed to determine whether viable microorganisms are present. Commonly, the test involves immersing a component or system or flushing a fluid pathway with sterile microbial growth medium, incubation of the medium under conditions favorable for microbial growth, and observation of turbidity or other indication of microbial growth after a suitable incubation period.

sterilization: A validated process used to render a product free from viable microorganisms. It is generally accepted that a terminally sterilized unit purporting to be sterile attain a sterility assurance level of $\leq$10–6, i.e., probability of less than or equal to one chance in 1 million that a viable microorganism is present in the sterilized unit. Lower SALs may be validated as sterile in some cases.

steroid: Fat-soluble organic compounds such as sterols, bile acids, and sex hormones.

stratum corneum: The outermost layer of skin

strength: For pharmaceutical use the amount of active ingredient in a drug preparation.

stress crack: The solvent-induced generation of cracks in a plastic object. Typically occurs in areas of unrelieved stress.

strip package: A package made by enclosing an object to be packaged such as a tablet between two webs and then sealing the webs together so that the seal completely surrounds the object being packaged.

subcutaneous: Beneath the skin.

sublingual: Under the tongue.

substrate: Refers to the primary structural material or the surface of the primary material that is applied to other materials designed to alter the characteristics or properties of the original material.

suppository: The dosage form designed for insertion into the rectum.

surfactant: Any substance, normally a soap or detergent, that forms a compatibilizing boundary layer between two liquids or a liquid and solid. This layer leads to the staple dispersion one phase in another.

suspension: Solid particles dispersed in a liquid. If the suspension is stable, it will resist the normal gravitational separation into two phases.

systemic: Administration of a drug so that it gains access to the circulatory system. Can also refer to the introduction of a drug to all parts of the body to treat only one location.

tablet triturate: A tablet that dissolves readily in the mouth.

tack, tacky: The tendency for a material to stick to itself and other objects. Stickiness, sticky.

talc: A naturally occurring white mineral containing hydrated magnesium oxide and silica.

tear: Force required to continue a split in paper or film. Usually performed two ways, in the machine direction and in the cross direction. ASTM D689.

tensile modulus: See modulus.

tensile strength: The resistance of a specimen to breaking when stressed longitudinally.

tertiary package: The package that holds the secondary package. Usually the corrugated case.

therapeutic: Relating to the treatment of disease.

thermoform: To form a plastic sheet into a shape using pressure to force heat-softened plastic into a mold. A mechanical plug assist may be added to improve the process.

thermoplastic: Describes any substance that becomes more pliable as it is heated. In packaging and molding, it refers to a material that can be formed when hot but become rigid after cooling.

thermoset: Plastics that become rigid when heated or subjected to energy that initiates a chemical reaction at reactive sites linking all the individual polymer strands together permanently.

tight container: A term defined by the USP that describes a container that protects its contents from contamination by extraneous liquids, solids, or gases from physical loss of the drug and from efflorescence, deliquescence, or evaporating under ordinary or customary conditions of handling, shipment, storage, and distribution.

tincture: A solution of a drug in alcohol.

TiO_2: Titanium dioxide.

topical: Administration of a drug to the skin surface or the lining of body cavities. Its effectiveness is limited to the localized areas to which it is applied.

torque: Screw caps (closures) require a rotating force to affect a seal on a container. The removal torque is measured in inch-pounds, five minutes after application. For metal caps on glass containers, this should be at least one-fourth the cap size (e.g., 6 in-lb for a 24-mm closure).

toxin: A noxious or poisonous substance formed during the growth of certain microorganisms.

toxoid: A toxin that has been treated, e.g., with formaldehyde to destroy its toxic property but retain its antigenicity, i.e., its capability of stimulating the production of antitoxin antibodies and thus engendering immunity.

transdermal: Administration of drugs through the skin.

type I glass: Glass composed largely of silica and boric oxide that is very low in water-extractable impurities.

type II glass: Glass containing larger amounts of water-soluble sodium and calcium oxides than type I glass. Soda-lime glass with no boron-containing materials present.

type III glass: Glass containing even greater quantities of water-extractable oxides than type II glass. A different and lower grade of soda-lime glass.

unit-dose container: A single-unit container for products for administration by other than parenteral means as a singe dose direct from the container.

unsaturated: In chemistry, molecules that contain more than one bond between two atoms. In polymers, it is usually referring to double and triple bonds between carbon and another atom.

unsupported: An unsupported film is one with low mechanical strength that is not bonded to another film for strength. Foil materials that are not laminated to a plastic film for increased strength and resistance to tearing are referred to as unsupported films.

urethra: The tube that drains the bladder to the outside of the body.

USP: United States Pharmacopoeia.

USP-NF: United States Pharmacopoeia National Formulary.

UV: Ultraviolet.

validation: Testing and establishing documented evidence that provides a high degree of assurance that a specific process, component, or piece of equipment will consistently produce a product meeting its predetermined specifications and quality attributes.

vasoconstrictors: Drugs that reduce the flow of body fluids by constricting the ducts, tubes, and canals through which these fluids flow. Often referred to drugs that constrict the circulatory system.

vasodilators: Drugs that increase the flow of body fluids by relaxing the muscles surrounding the ducts, tubes, and canals through which those fluids flow.

vicat softening: The temperature at which a rod of 1 mm^2 cross-section can be forced 1 mm into plastic. ASTM D1525.

vinyl: A common name for PVC or polyvinyl chloride. It is also used to denote a specific monomer linkage in a polymer.

wavelength: The distance between identical points in a wave pattern. A measure of the energy content of light, the shorter the wavelength the higher the energy level.

web: Used to denote the unwound portion of paper or plastic film in a packaging process (e.g., a printed web would mean the roll and the unwound portion of printed film). For paper, the roll and unwound portion of the material in the printing or packaging equipment.

well-closed container: A USP term. A container that protects its contents from extraneous solids and physical loss under the ordinary and customary conditions of handling, shipment, and distribution.

XML: eXtensible Markup Language.

Bibliography

Bakker, Marilyn, Eckroth, David. The Wiley Encyclopedia of Packaging Technology. New York; Chichester; Brisbane; Toronto; Singapore: John Wiley and Sons, 1986. ISBN 0-471-80940-3.

Bartlett, Jimmy D, Jaanus, Siret D. Clinical Ocular Pharmacology. 3rd ed. Boston: Butterworth-Heinemann, 1996. ISBN 0-7506-9448-3.

Berins, Michael L (editor). SPI Plastics Engineering Handbook of the Society of the Plastic Industry, Inc. 5th ed. Van Nostrand Reinhold – Chapman & Hall, 1991.

Chambers, James V, Nelson, Philip E (editors). Principles of Aseptic Processing and Packaging. 2nd ed. Washington DC: The Food Processors Institute, 1993. ISBN 0-937774-03-0.

Crank, J. Mathematics of Diffusion. London: Clarendon Press, 1975.

Dominghaus, Hans. Plastics for Engineers: Materials, Properties, Applications. Munich; Vienna; New York; Barcelona: Carl Hanser Verlag, Hanser Publishers, 1993.

Doyle, Mona (editor). Packaging Strategy – Winning the Consumer. Lancaster; Basel: Technomic Publishing, 1996. ISBN 1-56676-298-7.

Eldred, Nelson R. Package Printing. Jelmar Publishing, 1993. ISBN 0-9616302-5-6.

Gruenwald, G. Plastics – How Structure Determines Properties. Munich; Vienna; New York; Barcelona: Hanser Publishers, 1993. ISBN 3-446-16520-7.

Hanlon, Joseph F. Handbook of Package Engineering. 2nd ed. Lancaster; Basel: Technomic Publishing, 1992. ISBN 0-87762-924-2.

Hernandez, Ruben J, Selke, Susan EM, Culter, John D. Plastics Packaging – Properties, Processing, Applications, and Regulations. Cincinnati; Munich: Hanser Gardner Publications, 2000. ISBN 1-56990-303-4 (Hanser Gardner), ISBN 3-446-21404-6 (Hanser).

Jenkins, Wilmer A, Osborn, Kenton R. Packaging Drugs and Pharmaceuticals. Lancaster; Basel: Technomic Publishing, 1993. ISBN 1-56676-014-3.

Kovel, Ralph, Kovel, Terry. The Label Made Me Buy It. New York: Crown Publishers, 1998. ISBN 0-609-60168-7.

Lamoreaux, Robert D. Barcodes and Other Automatic Identification Systems. Leatherhead: Pira International, 1996. ISBN 1-85802-095-6.

The National Formulary, NF 23 (23rd edition). Prepared by the Council of Experts and Published by the Board of Trustees, 2005, Official from January 1, 2005. Rockville: United States Pharmacopeial Convention.

Oberg, Eric, Jones, Franklin D, Horton, Holbrook L, Ryffel, Henry H (editor), Geronimo, James H (assistant editor). Machinery's Handbook. 22nd ed. New York: Industrial Press, 1984, ISBN 0-8311-1155-0.

Pötsch, Gerd, Michaeli, Walter. Injection Molding, An Introduction. Munich; Vienna; New York: Hanser Publishers; Cincinnati: Hanser/Gardner Publications, 1995. ISBN 3-446-17196-7.

Progelhof, Richard C, Throne, James L. Polymer Engineering Principles, Properties, Processes, Tests for Design. Munich; Vienna; New York: SPE Books, Hanser Publishers; Cincinnati: Hanser/Gardner Publications, 1993. ISBN 3-446-17337-4.

Rauwendaal, Chris. Polymer Extrusion. Munich; Vienna; New York: SPE Books, Hanser Publishers; Cincinnati: Hanser/Gardner Publications, 1990. ISBN 3-446-16080-9.

Rees, Herbert. Mold Engineering. Munich; Vienna; New York: SPE Books, Hanser Publishers; Cincinnati: Hanser/Gardner Publications, 1995. ISBN 3-446-17729-9.

Rees, Herbert. Understanding Injection Molding Technology. Munich; Vienna; New York: Hanser Publishers; Cincinnati: Hanser/Gardner Publications, 1994. ISBN 1-56990-130-9.

Rosato, Donald V, Rosato, Dominick V (editors). Blow Molding Handbook, Technology, Performance, Markets, Economics, The Complete Blow Molding Operation. Munich; Vienna; New York: Hanser Publishers, 1989. ISBN 3-446-15071-4.

Small, Bonnie B, Thomas, DW, et al. Statistical Quality Control Handbook. Charlotte: ATT Technologies, Delmar Printing Co., 1985. Book Code 700-444 (ATT Commercial Sales).

Stedman's Medical Dictionary. 27th ed. Philadelphia; Baltimore; New York; London; Buenos Aires; Hong Kong; Sydney; Tokyo: Lippincott Williams & Wilkins (a Wolters Kluwer Company), 2000. ISBN 0-683-40007-X.

Stevenson, James F. Innovation in Polymer Processing, Molding. Munich; Vienna; New York: SPE Books, Hanser Publishers; Cincinnati: Hanser/Gardner Publications, 1996. ISBN 3-446-17433-8.

Steward, Bill. Packaging as an Effective Marketing Tool. Leatherhead: Pira International, 1995, 2nd printing 1996. ISBN 1-85802-099-9.

Throne, James L. Thermoforming. Munich; Vienna; New York: SPE Books, Hanser Publishers; Cincinnati: Hanser/Gardner Publications, 1987. ISBN 3-446-14699-7.

The United States Pharmacopeia, USP 28 (28th revision). Prepared by the Council of Experts and Published by the Board of Trustees, 2005, Official from January 1, 2005. Rockville: United States Pharmacopeial Convention.

Vieth, WR. Diffusion In and Through Polymers, Principles and Applications. Munich; Vienna; New York; Barcelona: Carl Hanser Verlag, Hanser Publishers, 1991.

Index